Finite Element Analysis of Structures using MATLAB

Finite Element Analysis of Structures using MATLAB

M.V. RAMA RAO

Professor, Department of Civil Engineering,
Vasavi College of Engineering, Hyderabad-500006 INDIA

ANDREW POWNUK

Associate Professor, Department of Mathematical Sciences
University of Texas at El Paso, El Paso, Texas 79968-0514, USA.

 **BS Publications**

An imprint of **BSP Books Pvt. Ltd.**
4-4-309/316, Giriraj Lane,
Sultan Bazar, Hyderabad - 500 095.

Finite Element Analysis of Structures using MATLAB

by **M. V. Rama Rao, Andrew Pownuk**

© 2024, *by Publisher,* All rights reserved.

Disclaimer: The authors and the publishers have taken due care to provide the authentic, reliable and up to date information related to the subject. However, neither the authors nor the publisher shall be responsible for any liability for any damage caused as a result of use of this book. The respective user must check the accuracy from other sources too.

Published by:

BSP BS Publications

An Imprint of BSP Books Pvt. Ltd.
4-4-309/316, Giriraj Lane,
Sultan Bazar, Hyderabad - 500 095.
Phone: 040 - 23445688
e-mail: info@bspbooks.net
www.bspbooks.net

ISBN: 978-93-95038-72-0 (Hardback)

Preface

This book is primarily intended for the use of undergraduate and graduate students studying finite element analysis in their coursework. The objective is to explain the related concepts in a clear and succinct manner with relevant examples and figures.

The book presents working MATLAB code for using most of the elements discussed in the textbook. Example problems have been solved using the MATLAB code provided. The results obtained are validated using ABAQUS/CAE Learning Edition.

The textbook is divided into twelve chapters and two appendices. Appendix-A is devoted to the MATLAB code developed for the solution of problems given in the textbook. Appendix-B gives the step-by-step procedure for solving the same problems using ABAQUS.

The first chapter gives a brief introduction of the history of Finite Element Analysis (FEA). The second chapter introduces the concept of elastic continuum and related concepts. The third chapter gives a brief description of variational methods.

The fourth chapter explains the concepts of nodes and elements. The categorization of elements into one-dimensional, two-dimensional, and three-dimensional elements is introduced here. Interpolation functions and natural coordinates are explained along with Lagrangian and serendipity elements. The fifth chapter deals with formalization of FEA and associated concepts.

The sixth and seventh chapters describe in detail various aspects of one-dimensional and two-dimensional elements. The eighth chapter deals with numerical integration using Gaussian quadrature. The ninth and tenth chapters deal with axisymmetric and three-dimensional elements respectively. Three-dimensional elements are discussed in the tenth chapter.

Application of FEA to the analysis of thin and thick plates is discussed in the eleventh chapter. A few problems of vibration related to structural dynamics are explained in the twelfth chapter. The appendix A, divided into seventeen parts, presents the MATLAB code prepared for the solution of numerical examples presented in the textbook. Appendix B, divided into six parts, presents the step-by-step procedure along with the relevant screenshots, to solve the same problems.

Foreword

With immense pride and delight, I write this foreword for the textbook on Finite Element Analysis of structures using MATLAB, authored by my esteemed former student, Dr M. V. Rama Rao, along with Dr Andrew Pownuk.

Dr Rama Rao was among my earliest students at the Indian Institute of Technology Kanpur in the 1980s. Throughout his time as my student, I had the pleasure of teaching him two courses and guiding his Master's thesis. He was an exceptionally motivated student, one of those rare individuals who truly inspire their teachers. Over the years, I have witnessed with great pride his career growth and development. His unwavering commitment and passion for his subject are truly infectious. This book stands as a testament to his dedication and hard work.

To my former student, I extend my heartfelt congratulations for this extraordinary endeavour. I am confident that this textbook will leave a lasting impact, just like a dedicated educator, guiding and inspiring countless individuals along their academic and professional journeys.

The field of Finite Element Analysis has undergone significant transformations over the years, solidifying its position as a cornerstone of modern engineering and computational mechanics. The ability to analyse complex structures with precision and accuracy is a vital skill, empowering engineers to innovate and create robust designs that shape the world around us.

In this comprehensive textbook, the authors expertly combine the power of MATLAB and ABAQUS to provide readers with a well-rounded understanding of Finite Element Analysis. By integrating these two widely used software platforms, the book equips readers with a versatile toolkit, effectively serving them across a broad spectrum of engineering applications. The approach of this textbook strikes a perfect balance between theoretical foundations and practical implementation, making it invaluable for students, researchers, and industry professionals alike.

Congratulations once again to the authors for this remarkable contribution. Dr Rama Rao's dedication to his field and his collaboration with Dr Pownuk have produced a truly outstanding resource that will undoubtedly enrich the knowledge and skills of its readers.

Prof. Sudhir K. Jain
Vice Chancellor
Banaras Hindu University
Varanasi - 221005, U.P., India
(Former Director of IIT Gandhinagar)

ॐ असतो मा सद्गमय ।
तमसो मा ज्योतिर्गमय ।
मृत्योर्मा अमृतं गमय ।
ॐ शान्तिः शान्तिः शान्तिः ॥

Om Asato Maa Sad-Gamaya |
Tamaso Maa Jyotir-Gamaya |
Mrtyor-Maa Amrtam Gamaya |
Om Shaantih Shaantih Shaantih ||

Lead us from the unreal to the real
Lead us from darkness to light
Lead us from death to immortality
Aum peace, peace, peace!
From: Shanti Mantra, RIG VEDA

A Note To The Students

Students are required to note the following points:

1. It is essential to gain an insight into the behaviour of the structure before attempting to use a commercial software/MATLAB code provided in this textbook.

2. They are first advised to solve small scale problems that are easy to draw on a piece of paper. A good starting point is to understand the example problems presented in this textbook.

3. After a thorough understanding of the required concepts, students shall prepare the input data for running the MATLAB code and use it for solution.

4. It is essential to get familiarity of the MATLAB platform and then learn the fundamental concepts and important commands of MATLAB.

5. Students are also advised to get licence for the learning edition of ABAQUS and learn the basics thoroughly. Students may refer to the ABAQUS solution manuals are available online for this purpose.

6. Students are strongly encouraged to solve the same problem on ABAQUS and also solve using MATLAB code provided. This will help them to validate their results and gain confidence.

7. This textbook gives steps to solve several numerical examples using abaqus and as well as MATLAB for the benefit of the students.

8. Students should also learn how to understand and interpret the results of analysis using the approaches outlined above to gain an in-depth understanding of concepts.

Acknowledgements

Writing a textbook involves so much of sacrifice, especially on the part of the family members. The authors would like to thank their families for their help support and patience.

The first author fondly remembers and cherishes the influential role played by his late parents Mr. Narasimha Murthy Mallela and Mrs. Vijayalakshmi Mallela without whose love and affection, nothing would have been possible in this world. The first author owes a debt of gratitude to his wife Prameela, daughter Deepika, son Harsha for their patience, support, and encouragement throughout the course of this work. The first author also expresses his deep sense of gratitude to the God Almighty for making this work possible.

The second author would like to acknowledge the help and support of his wife during the preparation of this manuscript.

The first author would like to gratefully acknowledge the inspiration received from his Guru who taught him at IIT Kanpur: Dr. Sudhir Jain, currently the Vice Chancellor of Banaras Hindu University, Varanasi, India. The first author also thanks Prof. Jain for kindly agreeing to write the Foreword.

M.V. Rama Rao

Andrew Pownuk

List of Symbols

Description	Meaning
A	Area of cross section
$[B]$	Strain-displacement matrix
$[c]$	Damping matrix
C	Modulus of rigidity
C_x, C_y, C_z	Direction cosines
C^n	n^{th} order continuity
$[C]$	Constitutive matrix
$\{d\}$	Vector of nodal displacements
D_r	Bending stiffness of plate
$[D]$	Inverse of constitutive matrix
E	Young's modulus
$\{F\}$	Force vector
h	Thickness of plate
I	Area moment of inertia
$I(\ )$	Potential energy functional
$[J]$	Jacobian matrix
$\left[k^{(e)}\right]$	Element stiffness matrix
$[K]$	Structure stiffness matrix
L	Length
l, m, n	Direction cosines along x, y, z axes
$\bar{m}$	Linear mass
M_{xx}, M_{yy}, M_{zz}	Moments on plate
$[M]$	Structure mass matrix
$\left[m^{(e)}\right]$	Element mass matrix

N_i	Interpolation function at i^{th} node
$[N]$	Matrix of interpolation functions
$[N_S]$	Matrix of interpolation functions for surface tractions
$\{Q^{(e)}\}$	Element force vector
$\{Q\}$	Structure force vector
$[T]$	Rotation transformation matrix
$\{u\}$	Vector of displacements
$\{v\}$	Displacement vector
$\{\dot{v}\}$	Velocity vector
$\{\ddot{v}\}$	Acceleration vector
u, v, w	Displacements along x, y, z directions
S	Surface area
t	Thickness of plate
U	Total strain energy
U_0	Specific strain energy
W_e	Work done by external forces
W_i	Weights at gauss sampling points
X_b	Body force along x direction
X_S	Surface traction along x direction
$\{X\}$	Vector of body forces
Y_b	Body force along x direction
Y_S	Surface traction along y direction
Z_b	Body force along x direction
Z_S	Surface traction along z direction
χ	Curvature of plate
$\{\delta d\}$	Vector of virtual nodal displacements

$\delta u, \delta v, \delta w$	Virtual displacements along x, y, z directions
δS	Infinitesimal surface area
$\delta \Pi$	First variation of potential energy functional
ε	Strain
$\varepsilon_r, \varepsilon_\theta, \varepsilon_z$	Radial, circumferential, and axial strains
$\phi(x)$	Exact solution
$\tilde{\phi}$	Approximate solution
$\delta \tilde{\phi}$	Small variation in $\tilde{\phi}$
γ	Shear strain
η	Natural coordinate along y direction
μ	Poisson's ratio
Π	Potential energy
θ	Rotation
ρ	Mass density
σ	Stress
$\sigma_r, \sigma_\theta, \sigma_z$	Radial, circumferential, and axial stresses
τ	Shear stress
ω	Natural circular frequency
ξ	Natural coordinate along x direction
ζ	Natural coordinate along z direction

Contents

CHAPTER 1

Introduction

CHAPTER 2

Elastic Continuum

CHAPTER 3

Variational Methods

CHAPTER 4

Discretization of Domain

CHAPTER **5**

Formalisation of Finite Element Method

CHAPTER **6**

One Dimensional Elements

CHAPTER 7

Two Dimensional Elements

CHAPTER 8

Numerical Integration

CHAPTER 9

Axi-Symmetric Elements

CHAPTER 10

Three Dimensional Elements

CHAPTER 11

Analysis of Plates

CHAPTER 12

Free Vibration Analysis of Structures

1 Introduction

Often, the behavior of continuous physical systems can be better understood by constructing a mathematical model. The mathematical model is constructed by selecting various parameters that govern the behavior of the structural system. The mathematical model of a physical system is best understood in terms of the governing ordinary differential equations (ODE's) or partial differential equations (PDE's). These governing equations of the mathematical model need to be solved in order to predict the behavior of the physical system. The physical domain of this analytical solution to the mathematical model has an infinite number of points. The analytical solutions of governing ordinary differential equations (ODEs as well as PDEs) are given by a mathematical expression that yields the values of the desired unknown quantities at infinite number of locations in the body. However, these analytical solutions are not usually obtainable because of the complicated geometries, loadings, and material properties.

Often, it is very difficult to obtain a closed form solution to the governing equations. It is generally not possible to obtain analytical mathematical solutions for problems involving complicated geometries, loadings, and material properties. Hence the use of numerical methods, such as finite element method, needs to be relied upon in order to obtain acceptable solutions. In such cases, the use of finite element method is mandated and is invaluable. While the analytical approach requires the solution of differential equations, the finite element formulation of the problem results in a system of simultaneous algebraic equations for solution. These equations are solved to obtain approximate values of the unknowns at discrete number of points in the continuum.

The name *finite element* was coined by Clough(1960). Finite Element Method is a numerical technique used to solve a system of a system of governing equations over the domain of a continuous physical system. Areas of work where finite element method can be applied include structural analysis, heat transfer, fluid flow, mass transport, and electromagnetic potential.

Finite element method consists of discretizing the physical domain into simple geometric shapes called finite elements. The process of dividing a continuous physical body into an equivalent system of finite elements is called discretization. These elements are connected to one another at a finite number of points named as nodes. Thus, discretization involves the process of splitting up a continuous domain into an assemblage of finite number of finite elements. The process of dividing a body into an equivalent system of smaller bodies viz. finite elements is called discretization. These finite elements are interconnected at points common to two or more elements by nodes/ boundary lines / surfaces. While analytical solution is sought for the entire physical body as a continuum, in finite element method equations are formulated for each finite element separately and then are combined to obtain solution for the entire body. Thus, the finite element solution for a structure refers to the displacements at each node and stresses within each of the finite elements that make up the entire structure. In non-structural problems, the nodal unknowns may, for instance, be temperatures or fluid pressures due to thermal or fluid fluxes.

1.1   HISTORY OF FINITE ELEMENT METHOD

The basic ideas of the finite element method owe their origin to the seminal contributions made by Turner, Clough, Martin, and Topp (1956) and Argyris and Kelsey (1960). Topp's paper brings out the initial application of simple finite elements for pin-jointed bar and triangular plate with in-plane loads. It represents one of the key contributions in the development of finite element method.

The advent of the digital computer made finite element method practically viable as it provided a rapid means of performing many calculations involved in its implementation. With the development of high-speed digital computers, the application of finite element method also evolved at a rapid and impressive pace.

The pioneering contribution made by Przemieniecki (1970) presents finite element method for the solution of stress analysis problems. A broader interpretation of finite element method its applicability to any general field problem was provided by Zienkiewicz and Cheung (1967). This eventually led to the realization that finite element equations can also be derived using least squares approach or a weighted residual method such as Galerkin method. This created a widespread interest among applied mathematicians in applying the finite element method for the solution of linear and nonlinear differential equations. Over a period of time, there have been many research publications related to finite element method.

In the year 1941, Hrennikoff presented a solution of elasticity problem using one-dimensional elements. A similar problem was handled by Mc Henry (1943) in the year 1943. In the same year, Courant introduced shape functions over triangular subregions to model the whole region. Levy developed the flexibility and stiffness methods for structure problems in the years 1947 and 1953 respectively. Argyris and Kelsey (1960) developed matrix structural analysis methods using energy principles in the year 1954. In 1956, Turner, Clough, Martin and Topp derived stiffness matrices for truss, beam and 2D plane stress elements and direct stiffness method. It was in the year 1960 that Clough coined the name *"Finite Element"*. In the same year, Turner et al. took up work on large deflection and thermal analysis. Melosh developed stiffness matrix for plate bending element in the year 1961. Also, Martin developed the tetrahedral stiffness matrix for 3D problems in 1961. During the next year 1962, Gallagher worked on problems involving material nonlinearity. Grafton and Strome (1963) developed element stiffness matrix for a curved-shell bending element in 1963. In the same year, Melosh applied variational formulation to solve non-structural problems. Three dimensional elements of axisymmetric solids were developed by Clough et. al in the year 1965. Zienkiewicz published his first book on finite element method in 1967. He worked on viscoelasticity problems in the succeeding year. In 1969, Szabo and Lee adapted weighted residual methods to structural analysis. A book on nonlinear continua was published by Oden in 1972. Nonlinear dynamic behaviour of large-displacement problems was investigated by Belytschko in 1976. Since then, several new developments have taken place in the field of finite element analysis. These include:

(a) development of several new elements

(b) convergence studies

(c) developments of supercomputers

(d) availability of powerful microcomputers

(e) development of user-friendly general-purpose finite element software packages.

Non-deterministic Finite element models represent the latest developments in Finite element methods. The non-deterministic approach to finite element analysis considers uncertainty in the structural parameters that define the structural system. The structural behaviour of such an uncertain system is sought in terms of non-deterministic structural response. The following are various developments in the area:

(a) Stochastic finite element methods model uncertainty of structural parameters as stochastic variables. The uncertainty is defined using probability distribution functions.

(b) Interval finite element methods model uncertainty of structural parameters by closed intervals. They combine the concepts of interval algebra and classical finite element analysis. The structural response is sought in terms of interval variables.

(c) Fuzzy randomness is incorporated into structural behaviour to construct a fuzzy random model of finite element analysis. Fuzzy randomness is also modelled using probability boxes (p-boxes).

The development of high storage, and faster computers have removed all limitations imposed on earlier implementation of finite element models. The coming years will be sure to witness much rapid progress in the area of finite element method.

Elastic Continuum

2.1 EQUILIBRIUM CONDITIONS FOR THREE-DIMENSIONAL CONTINUUM

Consider a deformable body as shown in Figure 2.1(a) being subjected to a system of forces that keep it in equilibrium. Forces that act on the body can be grouped into two categories:

(a) **Body forces:** Forces that occur at every point of the body are called body forces.

(b) **Surface tractions:** Forces that are applied at specific locations on the surface (the boundary of the three dimensional solid) are known as surface tractions.

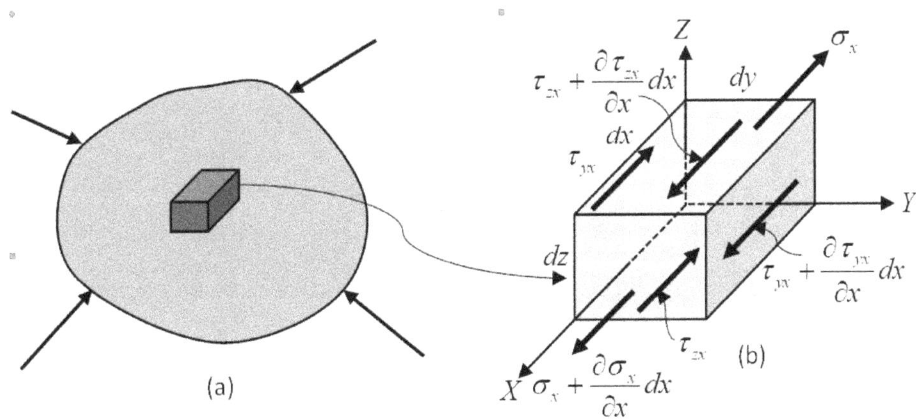

Figure 2.1 Three-dimensional body in a state of equilibrium.

Under the combined action of the body forces and the surface tractions that keep the body under equilibrium, the body undergoes deformation that causes strains and stresses in the body. To understand the state of stress at any given point in the body, it is customary to pick up a cuboid of infinitesimal dimensions $dx, dy \ and \ dz$ as shown in Figure 2.1(b). The volume of this cuboid is given by $dV = dxdydz$. This infinitesimal cuboid is a free body taken out of the system for the analysis of stresses and strains. As the body is in equilibrium, it implies that the cuboid is also in equilibrium.

First, the cuboid is analysed for equilibrium and the stresses and strains are calculated. Then the sides of the cuboid are let to tend to zero, so that the cuboidshrinks to zero volume and thus becomes a point.

Figure 2.1(b) shows the forces acting along the X direction only for the sake of clarity. Considering the equilibrium of the cuboid in the X direction and considering the body forces per unit volume along X-direction to be X_b ,

$$\sum X = 0 \Rightarrow$$

$$\left(\sigma_x + \frac{\partial \sigma_x}{\partial x}dx - \sigma_x\right)dydz + \left(\tau_{yx} + \frac{\partial \tau_{yx}}{\partial y}dy - \tau_{yx}\right)dxdz + \left(\tau_{zx} + \frac{\partial \tau_{zx}}{\partial z}dz - \tau_{zx}\right)dydz + X_b dxdydz = 0 \cdots (2.1a)$$

Similar equations of equilibrium can be written for Y and Z directions also. $\sum X = 0$, $\sum Y = 0$ and $\sum Z = 0$ lead to Eq. 2.1(a), Eq. 2.1(b), and Eq. 2.1(c). Thus, the equilibrium equations for a three- dimensional deformable body as given by

$$\frac{\partial \sigma_x}{\partial x} + \frac{\partial \tau_{yx}}{\partial y} + \frac{\partial \tau_{zx}}{\partial z} + X_b = 0 \qquad \qquad \dots (2.1b)$$

$$\frac{\partial \tau_{xy}}{\partial x} + \frac{\partial \sigma_y}{\partial y} + \frac{\partial \tau_{zy}}{\partial z} + Y_b = 0 \qquad \qquad \dots (2.1c)$$

$$\frac{\partial \tau_{xz}}{\partial x} + \frac{\partial \tau_{yz}}{\partial y} + \frac{\partial \sigma_z}{\partial z} + Z_b = 0 \qquad \qquad \dots (2.1d)$$

Here Y_b and Z_b are the body forces along Y and Z directions respectively.

The remaining equations of equilibrium, i.e. $\sum M_X = 0$, $\sum M_Y = 0$ and $\sum M_Z = 0$ lead to

$$\tau_{yx} = \tau_{xy} \qquad \tau_{zy} = \tau_{yz} \qquad \tau_{zx} = \tau_{xz} \qquad \qquad \dots (2.2)$$

Substituting Eq. 2.2 in Eq. 2.1 we obtain six components of stress as $\lfloor \sigma \rfloor = \lfloor \sigma_x \quad \sigma_y \quad \sigma_z \quad \tau_{xy} \quad \tau_{yz} \quad \tau_{xz} \rfloor^T$ related as shown below:

$$\frac{\partial \sigma_x}{\partial x} + \frac{\partial \tau_{xy}}{\partial y} + \frac{\partial \tau_{xz}}{\partial z} + X_b = 0$$

$$\frac{\partial \tau_{xy}}{\partial x} + \frac{\partial \sigma_{y}}{\partial y} + \frac{\partial \tau_{yz}}{\partial z} + Y_b = 0$$

$$\frac{\partial \tau_{xz}}{\partial x} + \frac{\partial \tau_{yz}}{\partial y} + \frac{\partial \sigma_{z}}{\partial z} + Z_b = 0 \qquad \qquad .....(2.3)$$

The surface tractions X_S, Y_S and Z_S act on an infinitesimal surface area ΔS as shown in Figure 2.2. The normal to ΔS has the direction cosines l, m and n about the X, Y and Z axes. The relationship between the stress components and the surface tractions can be given as

$$\sigma_x l + \tau_{xy} m + \tau_{xz} n = X_S$$
$$\tau_{xy} l + \sigma_y m + \tau_{yz} n = Y_S \qquad \qquad .....(2.4)$$
$$\tau_{xz} l + \tau_{yz} m + \sigma_z n = Z_S$$

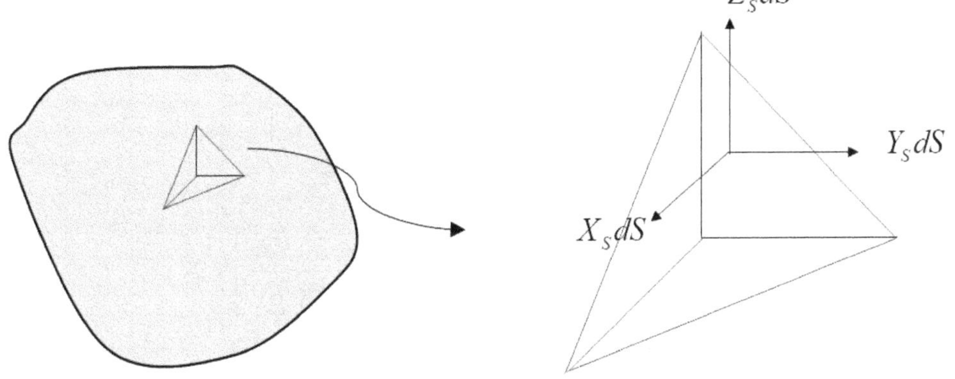

Figure 2.2 Components of surface traction on an element on the boundary.

2.2   EQUILIBRIUM CONDITIONS FOR TWO-DIMENSIONAL CONTINUUM

The equilibrium conditions for two-dimensional continuum can be obtained from Eq. 2.3 and Eq. 2.4 as follows:

$$\frac{\partial \sigma_x}{\partial x} + \frac{\partial \tau_{xy}}{\partial y} + X_b = 0$$

$$\frac{\partial \tau_{xy}}{\partial x} + \frac{\partial \sigma_y}{\partial y} + Y_b = 0 \qquad \qquad ......(2.5)$$

$$\sigma_x l + \tau_{xy} m = X_S$$
$$\tau_{xy} l + \sigma_y m = Y_S$$

.....(2.6)

2.3   STRAIN-DISPLACEMENT RELATIONSHIP

Whenever a body is subjected to a system of forces that keep it in a state of equilibrium, the body undergoes deformations. These deformations cause strains. In case of a deformable body undergoing strains u, v *and* w along the x, y *and* z directions respectively.

The following set of equations relate strains with the corresponding displacements.

$$\varepsilon_x = \frac{\partial u}{\partial x} + \frac{1}{2}\left[\left(\frac{\partial u}{\partial x}\right)^2 + \left(\frac{\partial v}{\partial x}\right)^2 + \left(\frac{\partial w}{\partial x}\right)^2\right]$$

$$\varepsilon_y = \frac{\partial v}{\partial y} + \frac{1}{2}\left[\left(\frac{\partial u}{\partial y}\right)^2 + \left(\frac{\partial v}{\partial y}\right)^2 + \left(\frac{\partial w}{\partial y}\right)^2\right]$$

$$\varepsilon_z = \frac{\partial w}{\partial z} + \frac{1}{2}\left[\left(\frac{\partial u}{\partial z}\right)^2 + \left(\frac{\partial v}{\partial z}\right)^2 + \left(\frac{\partial w}{\partial z}\right)^2\right]$$

$$\gamma_{xy} = \frac{\partial u}{\partial y} + \frac{\partial v}{\partial x} + \left[\frac{\partial u}{\partial x}\frac{\partial u}{\partial y} + \frac{\partial v}{\partial x}\frac{\partial v}{\partial y} + \frac{\partial w}{\partial x}\frac{\partial w}{\partial y}\right]$$

$$\gamma_{yz} = \frac{\partial w}{\partial y} + \frac{\partial v}{\partial z} + \left[\frac{\partial u}{\partial y}\frac{\partial u}{\partial z} + \frac{\partial v}{\partial y}\frac{\partial v}{\partial z} + \frac{\partial w}{\partial y}\frac{\partial w}{\partial z}\right]$$

$$\gamma_{xz} = \frac{\partial w}{\partial z} + \frac{\partial u}{\partial x} + \left[\frac{\partial u}{\partial z}\frac{\partial u}{\partial x} + \frac{\partial v}{\partial z}\frac{\partial v}{\partial x} + \frac{\partial w}{\partial z}\frac{\partial w}{\partial x}\right]$$

.....(2.7)

The strain components defined by $\{\varepsilon\} = \lfloor \varepsilon_x \quad \varepsilon_y \quad \varepsilon_z \quad \gamma_{xy} \quad \gamma_{yz} \quad \gamma_{xz} \rfloor^T$ is the vector of strains for the deformable body. These strains defined above are expressed in terms of partial derivatives of displacements as well as their powers and products. Thus, these strain components define the state of strain in a deformable body subjected to large deformations involving nonlinear strains.

However, for problems of linear elasticity with small deformations, squares and products of partial derivatives are much smaller in comparison to the partial derivatives and hence can be neglected, thus greatly simplifying the relationship as follows:

$$\varepsilon_x = \frac{\partial u}{\partial x} \qquad \varepsilon_y = \frac{\partial v}{\partial y} \qquad \varepsilon_z = \frac{\partial w}{\partial z}$$

$$\gamma_{xy} = \frac{\partial u}{\partial y} + \frac{\partial v}{\partial x} \qquad \gamma_{yz} = \frac{\partial v}{\partial z} + \frac{\partial w}{\partial y} \qquad \gamma_{xz} = \frac{\partial w}{\partial x} + \frac{\partial u}{\partial z}$$

$$.....(2.8)$$

These relations can also be expressed in the matrix form as follows:

$$\begin{Bmatrix} \varepsilon_x \\ \varepsilon_y \\ \varepsilon_z \\ \gamma_{xy} \\ \gamma_{yz} \\ \gamma_{xz} \end{Bmatrix} = \begin{bmatrix} \dfrac{\partial}{\partial x} & 0 & 0 \\[6pt] 0 & \dfrac{\partial}{\partial y} & 0 \\[6pt] 0 & 0 & \dfrac{\partial}{\partial z} \\[6pt] \dfrac{\partial}{\partial y} & \dfrac{\partial}{\partial x} & 0 \\[6pt] 0 & \dfrac{\partial}{\partial z} & \dfrac{\partial}{\partial y} \\[6pt] \dfrac{\partial}{\partial z} & 0 & \dfrac{\partial}{\partial x} \end{bmatrix} \begin{Bmatrix} u \\ v \\ w \end{Bmatrix} \qquad .....(2.9)$$

Eq. 2.9 can be expressed more concisely as follows:

$$\{\varepsilon\} = [B]\{u\} \qquad\qquad .....(2.10)$$

Here $\{\varepsilon\}$ is the vector of strains and $\{u\}$ is the vector of displacements. The connecting matrix of partial derivative operators represented by the $[B]$ matrix can be appropriately named as the *strain-displacement matrix* as it relates the strains at any point in the elastic continuum to the corresponding displacements.

In Eq. 2.8, the three displacement components u, v *and* w are subjected to the kinematic boundary conditions of the deformable body and are continuous and continuously differentiable functions of x, y *and* z. Thus, the six strain components can be uniquely evaluated from the values of the above three displacement components. However, in case the six strain components are specified, the three displacement components cannot be uniquely determined from these strain components. It is also observed from the Eq. 2.9 that the strain-displacement matrix is rectangular and thus not invertible. Thus in Eq.2.10, $\{u\}$ cannot be uniquely expressed in terms of $\{\varepsilon\}$.

Hence, in order to uniquely determine the displacement components, additional conditions are imposed on strains to compute the displacement components uniquely, ensuring continuity throughout the body during its deformation. These additional conditions imposed on strains are known as *strain compatibility conditions*. These conditions can be obtained by eliminating *u,v and w* by differentiating Eq. 2.8 and are listed below:

$$\frac{\partial^2 \varepsilon_x}{\partial y^2} + \frac{\partial^2 \varepsilon_y}{\partial x^2} = \frac{\partial^2 \gamma_{xy}}{\partial x \partial y}$$

$$\frac{\partial^2 \varepsilon_y}{\partial z^2} + \frac{\partial^2 \varepsilon_z}{\partial y^2} = \frac{\partial^2 \gamma_{yz}}{\partial y \partial z}$$

$$\frac{\partial^2 \varepsilon_z}{\partial x^2} + \frac{\partial^2 \varepsilon_x}{\partial z^2} = \frac{\partial^2 \gamma_{xz}}{\partial x \partial z}$$

$$2\frac{\partial^2 \varepsilon_x}{\partial y \partial z} = \frac{\partial}{\partial x}\left(-\frac{\partial \gamma_{yz}}{\partial x} + \frac{\partial \gamma_{xz}}{\partial y} + \frac{\partial \gamma_{xy}}{\partial z}\right)$$

$$2\frac{\partial^2 \varepsilon_y}{\partial x \partial z} = \frac{\partial}{\partial y}\left(\frac{\partial \gamma_{yz}}{\partial x} - \frac{\partial \gamma_{xz}}{\partial y} + \frac{\partial \gamma_{xy}}{\partial z}\right)$$

$$2\frac{\partial^2 \varepsilon_z}{\partial x \partial y} = \frac{\partial}{\partial z}\left(\frac{\partial \gamma_{yz}}{\partial x} + \frac{\partial \gamma_{xz}}{\partial y} - \frac{\partial \gamma_{xy}}{\partial z}\right) \qquad \text{.....(2.11)}$$

2.4 LINEAR CONSTITUTIVE RELATIONS

These relations connect the six components of stress given by $\{\sigma\}$ in terms of the six components of strain given by $\{\varepsilon\}$. For a linearly elastic material that obeys Hooke's law, these relations can be given by

$$\{\sigma\} = [C]\{\varepsilon\} \qquad \text{.....(2.12)}$$

where $[C]$ is the linear constitutive matrix.

The components of stress and strain in Eq. 2.12 can be represented as follows:

$$\begin{Bmatrix} \sigma_x \\ \sigma_y \\ \sigma_z \\ \gamma_{xy} \\ \gamma_{yz} \\ \gamma_{xz} \end{Bmatrix} = \begin{bmatrix} C_{11} & C_{12} & C_{13} & C_{14} & C_{15} & C_{16} \\ C_{21} & C_{22} & C_{23} & C_{24} & C_{25} & C_{26} \\ C_{31} & C_{32} & C_{33} & C_{34} & C_{35} & C_{36} \\ C_{41} & C_{42} & C_{43} & C_{44} & C_{45} & C_{46} \\ C_{51} & C_{52} & C_{53} & C_{54} & C_{55} & C_{56} \\ C_{61} & C_{62} & C_{63} & C_{64} & C_{65} & C_{66} \end{bmatrix} \begin{Bmatrix} \varepsilon_x \\ \varepsilon_y \\ \varepsilon_z \\ \gamma_{xy} \\ \gamma_{yz} \\ \gamma_{xz} \end{Bmatrix}$$

.....(2.13)

The matrix equation given by Eq. 2.12 can be inverted as shown in Eq. 2.14 below:

$$\begin{Bmatrix} \varepsilon_x \\ \varepsilon_y \\ \varepsilon_z \\ \gamma_{xy} \\ \gamma_{yz} \\ \gamma_{xz} \end{Bmatrix} = \begin{bmatrix} D_{11} & D_{12} & D_{13} & D_{14} & D_{15} & D_{16} \\ D_{21} & D_{22} & D_{23} & D_{24} & D_{25} & D_{26} \\ D_{31} & D_{32} & D_{33} & D_{34} & D_{35} & D_{36} \\ D_{41} & D_{42} & D_{43} & D_{44} & D_{45} & D_{46} \\ D_{51} & D_{52} & D_{53} & D_{54} & D_{55} & D_{56} \\ D_{61} & D_{62} & D_{63} & D_{64} & D_{65} & D_{66} \end{bmatrix} \begin{Bmatrix} \sigma_x \\ \sigma_y \\ \sigma_z \\ \gamma_{xy} \\ \gamma_{yz} \\ \gamma_{xz} \end{Bmatrix}$$

.....(2.14)

The above equation can be stated concisely as follows:

$$\{\varepsilon\} = [D]\{\sigma\}$$

.....(2.15)

The matrices $[C]$ and $[D]$ contain 36 elements each in case of a general non-linear, anisotropic material.

However, when the material is linearly elastic, both the above matrices become symmetric and contain 21 independent elastic constants (6 diagonal + 15 off-diagonal elements).

Whenever a material has three mutually perpendicular planes of symmetry, the material is said to be orthotropic. For an orthotropic material, only 9 constants are required to completely define $[C]$ and $[D]$ matrices as shown below:

$$\begin{Bmatrix} \sigma_x \\ \sigma_y \\ \sigma_z \\ \gamma_{xy} \\ \gamma_{yz} \\ \gamma_{xz} \end{Bmatrix} = \begin{bmatrix} C_{11} & C_{12} & C_{13} & 0 & 0 & 0 \\ & C_{22} & C_{23} & 0 & 0 & 0 \\ & & C_{33} & 0 & 0 & 0 \\ & & & C_{44} & 0 & 0 \\ & sym & & & C_{55} & 0 \\ & & & & & C_{66} \end{bmatrix} \begin{Bmatrix} \varepsilon_x \\ \varepsilon_y \\ \varepsilon_z \\ \gamma_{xy} \\ \gamma_{yz} \\ \gamma_{xz} \end{Bmatrix}$$

.....(2.16)

From Mechanics of materials, we know the following relations between six components of stress and strain.

$$\varepsilon_x = \frac{\sigma_x}{E} - \mu\frac{\sigma_y}{E} - \mu\frac{\sigma_z}{E} \qquad \varepsilon_y = -\mu\frac{\sigma_x}{E} + \frac{\sigma_y}{E} - \mu\frac{\sigma_z}{E} \qquad \varepsilon_z = -\mu\frac{\sigma_x}{E} - \mu\frac{\sigma_y}{E} + \frac{\sigma_z}{E}$$

$$\gamma_{xy} = \frac{\tau_{xy}}{C} = \frac{(1+\mu)}{E}\tau_{xy} \qquad \gamma_{yz} = \frac{\tau_{yz}}{C} = \frac{(1+\mu)}{E}\tau_{yz} \qquad \gamma_{xz} = \frac{\tau_{xz}}{C} = \frac{(1+\mu)}{E}\tau_{xz} \qquad \dots\dots(2.17)$$

The stress-strain relationship given by Eq. 2.17 can be written in matrix form as follows:

$$\begin{Bmatrix} \varepsilon_x \\ \varepsilon_y \\ \varepsilon_z \\ \gamma_{xy} \\ \gamma_{yz} \\ \gamma_{xz} \end{Bmatrix} = \begin{bmatrix} 1 & -\mu & -\mu & 0 & 0 & 0 \\ & 1 & -\mu & 0 & 0 & 0 \\ & & 1 & 0 & 0 & 0 \\ & & & 2(1+\mu) & 0 & 0 \\ & sym & & & 2(1+\mu) & 0 \\ & & & & & 2(1+\mu) \end{bmatrix} \begin{Bmatrix} \sigma_x \\ \sigma_y \\ \sigma_z \\ \tau_{xy} \\ \tau_{yz} \\ \tau_{xz} \end{Bmatrix} \qquad \dots\dots(2.18)$$

Inverting the above relation yields the following equation:

$$\begin{Bmatrix} \sigma_x \\ \sigma_y \\ \sigma_z \\ \tau_{xy} \\ \tau_{yz} \\ \tau_{xz} \end{Bmatrix} = \frac{E}{(1+\mu)(1-2\mu)} \begin{bmatrix} 1-\mu & \mu & \mu & 0 & 0 & 0 \\ & 1-\mu & \mu & 0 & 0 & 0 \\ & & 1-\mu & 0 & 0 & 0 \\ & & & \frac{(1-2\mu)}{2} & 0 & 0 \\ & sym & & & \frac{(1-2\mu)}{2} & 0 \\ & & & & & \frac{(1-2\mu)}{2} \end{bmatrix} \begin{Bmatrix} \varepsilon_x \\ \varepsilon_y \\ \varepsilon_z \\ \gamma_{xy} \\ \gamma_{yz} \\ \gamma_{xz} \end{Bmatrix} \dots\dots(2.19)$$

2.5 TWO-DIMENSIONAL STRESS DISTRIBUTION -PLANE STRESS CONDITION

The three-dimensional state of stress at a point in a deformable point can be greatly simplified in case of two-dimensional objects. In case of a thin lamina or plate subjected to forces in its own plane, the state of stress is termed as plane stress condition. *Plane stress condition* is characterised by a very small dimension along one of the coordinate directions. Thus, two dimensions alone are significant, and thickness (the third dimension) is quite small. *Thus, in the case of plane stress, as the name implies, stresses exist only in the plane of the lamina and all out-of-plane stresses are zero.*

Consider a thin lamina in the *XY* plane as shown in the Figure 2.3. The thickness of the lamina is along the z-direction and is very small compared to the dimensions of the lamina in the *XY* plane.

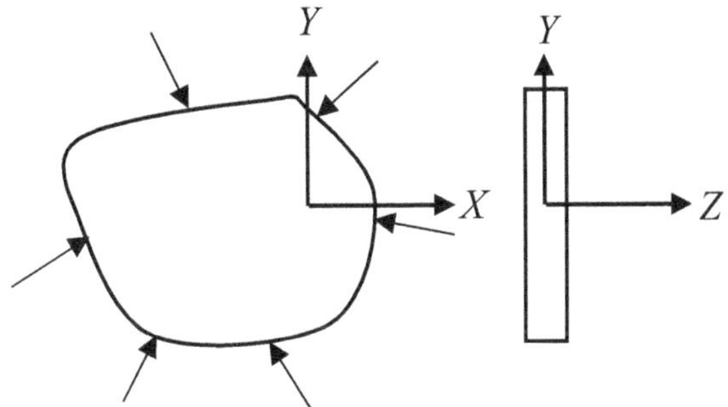

Fig. 2.3 Plane Stress.

The in-plane stresses σ_x, σ_y and τ_{xy} are functions of *X* and *Y* alone. Thus, for plane stress, the situation can be represented shown below:

$$\sigma_z = \tau_{xz} = \tau_{yz} = 0 \qquad\qquad .....(2.20a)$$

$$\sigma_x,\ \sigma_y,\ \tau_{xy},\ \varepsilon_x,\ \varepsilon_y,\ \varepsilon_z \text{ exist.} \qquad\qquad .....(2.20b)$$

The stress-strain relationship in case of plane stress can be obtained as follows by substituting Eq. 2.20a in Eq. 2.12 as shown 2.21:

$$\begin{Bmatrix} \varepsilon_x \\ \varepsilon_y \\ \varepsilon_z \\ \gamma_{xy} \\ \gamma_{yz} \\ \gamma_{xz} \end{Bmatrix} = \frac{1}{E} \begin{bmatrix} 1 & -\mu & -\mu & 0 & 0 & 0 \\ & 1 & -\mu & 0 & 0 & 0 \\ & & 1 & 0 & 0 & 0 \\ & & & 2(1+\mu) & 0 & 0 \\ & sym & & & 2(1+\mu) & 0 \\ & & & & & 2(1+\mu) \end{bmatrix} \begin{Bmatrix} \sigma_x \\ \sigma_y \\ \sigma_z = 0 \\ \tau_{xy} \\ \tau_{yz} = 0 \\ \tau_{xz} = 0 \end{Bmatrix} \qquad .....(2.21)$$

Eliminating rows and columns 3, 5 and 6 from Eq. 2.21 leads to the following reduced system of equations:

$$\begin{Bmatrix} \varepsilon_x \\ \varepsilon_y \\ \gamma_{xy} \end{Bmatrix} = \frac{1}{E} \begin{bmatrix} 1 & -\mu & 0 \\ -\mu & 1 & 0 \\ 0 & 0 & 2(1+\mu) \end{bmatrix} \begin{Bmatrix} \sigma_x \\ \sigma_y \\ \tau_{xy} \end{Bmatrix} \qquad\qquad .....(2.22)$$

14 | **Finite Element Analysis of Structures using MATLAB**

Inverting the above equation leads to

$$
\begin{Bmatrix} \sigma_x \\ \sigma_y \\ \tau_{xy} \end{Bmatrix} = \frac{E}{\left(1-\mu^2\right)} \begin{bmatrix} 1 & \mu & 0 \\ \mu & 1 & 0 \\ 0 & 0 & \dfrac{(1-\mu)}{2} \end{bmatrix} \begin{Bmatrix} \varepsilon_x \\ \varepsilon_y \\ \gamma_{xy} \end{Bmatrix}
\qquad \ldots\ldots(2.23)
$$

In addition, the third row in Eq. 2.21 can be expanded as

$$
\varepsilon_z = \frac{1}{E}\left(-\mu\sigma_x - \mu\sigma_y\right) = \frac{-\mu}{E}\left(\sigma_x + \sigma_y\right)
\qquad \ldots\ldots(2.24)
$$

But from Eq. 2.23,

$$
\sigma_x = \frac{E}{\left(1-\mu^2\right)}\left(\varepsilon_x + \mu\varepsilon_y\right) \text{ and}
$$

$$
\sigma_y = \frac{E}{\left(1-\mu^2\right)}\left(\mu\varepsilon_x + \varepsilon_y\right)
\qquad \ldots\ldots(2.25)
$$

Substituting Eq. 2.25 in Eq.2.24 yields

$$
\varepsilon_Z = \frac{-\mu}{\left(1-\mu\right)}\left(\varepsilon_x + \varepsilon_y\right)
\qquad \ldots\ldots(2.26)
$$

2.6 TWO-DIMENSIONAL STRESS DISTRIBUTION PLANE STRAIN CONDITION

Consider a three-dimensional body with one dimension predominantly long in comparison to the other two dimensions. Some examples are a dam, a wall-footing, or an embankment whose cross sections are shown in Figure 2.4:

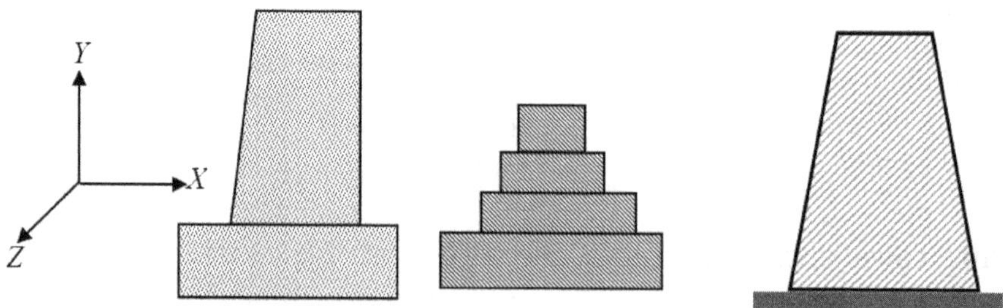

Figure 2.4 Examples of Plane Strain.

In all the above cases, the cross-sectional area is in the XY plane and the length dimension is along the Z axis. Thus, all the strains along the Z-direction are zero. Thus, only in-plane strains exist, and all out-of-plane strains are zero. This is characterised as *plane strain condition.*

The strains and stresses in the *plane strain condition* are as follows:

Out-of- plane strains $\quad \varepsilon_z = \gamma_{xz} = \gamma_{yz} = 0$

in-plane strains $\qquad \varepsilon_x, \varepsilon_y, \gamma_{xy} \quad$ exist

in- plane stresses $\qquad \sigma_x, \sigma_y, \tau_{xy}$ exist

out-of- plane stresses $\quad \sigma_z$ exists

$$\qquad\qquad\qquad\qquad\qquad\qquad .....(2.27)$$

The stress-strain relationship in case of plane stress can be obtained as follows by substituting Eq. 2.27 in Eq. 2.19 as shown below:

$$\begin{Bmatrix} \sigma_x \\ \sigma_y \\ \sigma_z \\ \tau_{xy} \\ \tau_{yz} \\ \tau_{xz} \end{Bmatrix} = \frac{E}{(1+\mu)(1-2\mu)} \begin{bmatrix} 1-\mu & \mu & \mu & 0 & 0 & 0 \\ & 1-\mu & \mu & 0 & 0 & 0 \\ & & 1-\mu & 0 & 0 & 0 \\ & & & \frac{(1-2\mu)}{2} & 0 & 0 \\ & sym & & & \frac{(1-2\mu)}{2} & 0 \\ & & & & & \frac{(1-2\mu)}{2} \end{bmatrix} \begin{Bmatrix} \varepsilon_x \\ \varepsilon_y \\ \varepsilon_z = 0 \\ \gamma_{xy} \\ \gamma_{yz} = 0 \\ \gamma_{xz} = 0 \end{Bmatrix} .....(2.28)$$

The reduced system of equations can be written as

$$\begin{Bmatrix} \sigma_x \\ \sigma_y \\ \tau_{xy} \end{Bmatrix} = \frac{E}{(1+\mu)(1-2\mu)} \begin{bmatrix} (1-\mu) & \mu & 0 \\ \mu & 1-\mu & 0 \\ 0 & 0 & \frac{(1-2\mu)}{2} \end{bmatrix} \begin{Bmatrix} \varepsilon_x \\ \varepsilon_y \\ \gamma_{xy} \end{Bmatrix} \qquad .....(2.29)$$

Inverting the above relationship leads to

$$\begin{Bmatrix} \varepsilon_x \\ \varepsilon_x \\ \gamma_{xy} \end{Bmatrix} = \frac{(1+\mu)}{E} \begin{bmatrix} (1-\mu) & -\mu & 0 \\ -\mu & 1-\mu & 0 \\ 0 & 0 & 2 \end{bmatrix} \begin{Bmatrix} \sigma_x \\ \sigma_y \\ \tau_{xy} \end{Bmatrix} \qquad .....(2.30)$$

Also, from the fourth row of matrix equation in Eq. 2.28

$$\sigma_z = \frac{E\mu}{(1+\mu)(1-2\mu)}\left(\varepsilon_x + \varepsilon_y\right) \qquad .....(2.31)$$

From Eq. 2.29,

$$\left(\sigma_x + \sigma_y\right) = \frac{E}{(1+\mu)(1-2\mu)}\left(\varepsilon_x + \varepsilon_y\right) \qquad .....(2.32)$$

From Eq. 2.32 and Eq. 2.31 we get

$$\sigma_z = \mu\left(\sigma_x + \sigma_y\right) \qquad .....(2.33)$$

3 Variational Methods

Differential equations are used to describe various phenomena appearing in the physical world. These differential equations are formulated based on basic physical principles, namely conservation laws in many mechanical engineering problems. Alternately, it is possible to use the concept of minimization of potential energy. This approach is called "variational formulation". Finite element formulation can be derived by this variational formulation if there exists a variational principle corresponding to the problem of interest.

3.1 POTENTIAL ENERGY FUNCTIONAL

It is a very important to describe the behaviour of an elastic body subject to a system of forces. When a body is subjected to a system of forces that keep it in equilibrium, the body deforms. As a result, the body stores strain energy because of the deformations. In addition, when the forces attached to the boundary of the body move through deformations work is done by these external forces. As a result, potential energy Π is stored in the body that can be described as follows:

$$\Pi = U + \left(-W_e\right) \qquad \qquad \dots\dots(3.1)$$

where U is the strain energy stored in the body and $-W_e$ is the work done by the external forces.

Principle of minimum potential energy states that the potential energy is minimum with respect to the state variables or function variables at the equilibrium state. This principle can be easily applied to deformation of elastic bodies by identifying the strain energy and potential energy due to external forces which are assumed to be fixed during the deformation.

The total potential energy can be defined as a function of several functions, viz. as a functional as shown:

$$\Pi = \Pi\left(u_1, u_2, u_3, \dots u_n\right) \qquad \qquad \dots\dots(3.2)$$

where $u_i, i = 1., n$ being the state variables.

Now the objective is to minimize Π with reference to the state variables. **_Calculus of variation_** deals with such problems to minimize a functional. Principles of calculus of variations is described in the next section.

3.2   CALCULUS OF VARIATIONS

Consider the problem of minimizing the functional $I\big(\phi(x)\big)$ described as follows:

Minimize $I\big(\phi(x)\big)$

where
$$I\big(\phi(x)\big)=\int_{x_1}^{x_2} f\big(x,\phi,\phi_x,\phi_{xx}\big)dx$$
.....(3.3)

where
$$\phi_x=\frac{d\phi}{dx}\ ;\ \phi_{xx}=\frac{d^2\phi}{dx^2}$$
.....(3.4)

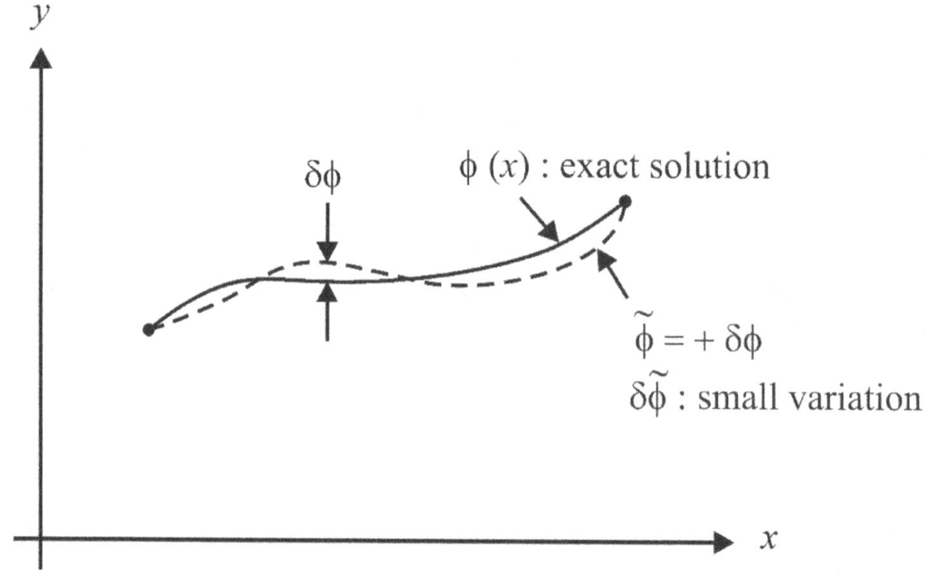

Figure 3.1 Approximating $\phi(x)$.

Consider an approximate solution $\tilde{\phi}$ which has a small variation $\delta\tilde{\phi}$ over an exact solution $\phi(x)$ as seen in Figure 3.1, that is

$$\phi(x)=\hat{\phi}(x)+\delta\phi$$
.....(3.5)

The approximate solution is substituted to the functional expression yielding,

$$I\left(\tilde{\phi}(x)\right) = \int_{x_1}^{x_2} f\left(x, \tilde{\phi}, \tilde{\phi}_x, \tilde{\phi}_{xx}\right) dx = I\left(\phi(x)\right) + \delta I \qquad \qquad .....(3.6)$$

In order that $\phi(x)$ is a solution, $\delta I = 0$ for any $\delta\phi$.

$$\delta I = \int_{x_1}^{x_2} \left(\frac{\partial F}{\partial \phi} \delta\phi + \frac{\partial F}{\partial \phi_x} \delta\phi_x + \frac{\partial F}{\partial \phi_{xx}} \delta\phi_{xx} \right) dx \qquad \qquad .....(3.7)$$

Integration by parts yields

$$\delta I = \int_{x_1}^{x_2} \left(\frac{\partial F}{\partial \phi} \delta\phi - \frac{d}{dx}\left(\frac{\partial F}{\partial \phi_x}\right)\delta\phi - \frac{d}{dx}\left(\frac{\partial F}{\partial \phi_{xx}}\right)\delta\phi_x \right) dx + \frac{\partial F}{\partial \phi_x} \delta\phi \Big|_{x_1}^{x_2} + \frac{\partial F}{\partial \phi_{xx}} \delta\phi_x \Big|_{x_1}^{x_2} = 0 \quad .....(3.8)$$

Performing integration by parts one more time, we get

$$\delta I = \int_{x_1}^{x_2} \left(\frac{\partial F}{\partial \phi} - \frac{d}{dx}\left(\frac{\partial F}{\partial \phi_x}\right) + \frac{d^2}{dx^2}\left(\frac{\partial F}{\partial \phi_{xx}}\right) \right)\delta\phi\, dx + \left(\frac{\partial F}{\partial \phi_x} - \frac{d}{dx}\left(\frac{\partial F}{\partial \phi_x}\right) \right)\delta\phi \Big|_{x_1}^{x_2} + \frac{\partial F}{\partial \phi_{xx}} \delta\phi_x \Big|_{x_1}^{x_2} = 0$$

$$.....(3.9)$$

As $\delta\phi$ is arbitrary, it implies that

$$\int_{x_1}^{x_2} \left(\frac{\partial F}{\partial \phi} - \frac{d}{dx}\left(\frac{\partial F}{\partial \phi_x}\right) + \frac{d^2}{dx^2}\left(\frac{\partial F}{\partial \phi_{xx}}\right) \right) dx = 0, \ x_1 \leq x \leq x_2 \qquad .....(3.10)$$

over the domain. The boundary conditions are given by

$$\left(\frac{\partial F}{\partial \phi_x} - \frac{d}{dx}\left(\frac{\partial F}{\partial \phi_x}\right) \right) \Big|_{x_1}^{x_2} = 0 \quad \text{or} \quad \delta\phi = 0$$

$$.....(3.11)$$

$$\frac{\partial F}{\partial \phi_{xx}} = 0 \qquad \qquad \text{or} \qquad \delta\phi_x = 0$$

3.2.1   Essential and Natural Boundary Conditions

Boundary conditions that are specified over the primary unknowns and their derivatives are known as *essential boundary conditions*.

E.g. $\delta\phi = 0$ and $\delta\phi_x = 0$ $\qquad \qquad .....(3.12)$

On the other hand, boundary conditions that are specified over the secondary unknowns are known as *natural boundary conditions*.

$$\left. \left(\frac{\partial F}{\partial \phi_x} - \frac{d}{dx}\left(\frac{\partial F}{\partial \phi_x} \right) \right) \right|_{x_1}^{x_2} = 0 \ \text{ and} \frac{\partial F}{\partial \phi_{xx}} = 0 \qquad .....(3.13)$$

3.3 RAYLEIGH-RITZ METHOD

Rayleigh-Ritz Method is one of the historically famous approximate methods to minimize a functional $\Pi(\phi(x))$. The trial functions used are

$$\tilde{\phi}(x) = \sum_{i=1}^{n} c_i \phi_i(x) \qquad .....(3.14)$$

$$\Pi(\tilde{\phi}) = \Pi(c_1, c_2, c_3, .., c_n)$$

$$\frac{\partial \Pi}{\partial c_i} = 0, i = 1, .., n$$

This method is very simple and easy to understand. However, it is not easy to find a family of trial functions for the entire domain satisfying the essential boundary conditions when geometry is complicated.

In the next section, we shall explore the Galerkin's method.

3.4 GALERKIN'S METHOD

Consider the solution of a differential equation defined by Eq.3.15 over an elastic continuum domain Ω.

$$L(u) = P \qquad .....(3.15)$$

Here L is the differential operator. Further, u is a continuous variable defined over the domain Ω and needs to satisfy specified values over the boundary of the domain.

We now seek the solution of Eq. 3.15 over Ω by considering an approximation function $\bar{u}$ defined as

$$\bar{u} = \sum_{i=1}^{N} u_i N_i \qquad .....(3.16)$$

The error $\varepsilon(x)$ at a point x is defined as

$$\varepsilon(x) = L(\bar{u}) - P \qquad .....(3.17)$$

The central idea is now to minimize the error. This is done by multiplying the error with a weighting function $w(x)$ and put it under the integral sign and equate it to zero as follows:

$$\oint_V w_i(x)\varepsilon(x)\,dV = \oint_V w_i(x)\big(L(\overline{u})-P\big)\,dV = 0, i = 1...N \qquad .....(3.18)$$

In Galerkin's method, same basis functions N_i chosen to approximate $\overline{u}$ in Eq.3.16 is used to define the weighting function ϕ also, as shown:

$$\phi = \sum_{i=1}^{N}\phi_i N_i \qquad .....(3.19)$$

It is to be noted that the choice of ϕ_i is arbitrary except at the nodes where ϕ needs to take specified values. As both $\overline{u}$ and ϕ are constructed in a similar manner, implementation of Galerkin's method is quite simple and straightforward. This is illustrated by an example problem as follows:

Consider the infinitesimal bar-segment as shown in Figure 3.2. The bar segment is a free body taken out of a bar held in equilibrium under the action of axial forces. The segment has a length of dx and is subjected to a surface traction of $q(x)$ per unit lateral surface area. Areas of cross-section at the left and right ends of the bar are A and $A+dA$ respectively. Axial stress at the left and right ends of the bar are σ and $\sigma+d\sigma$ respectively.

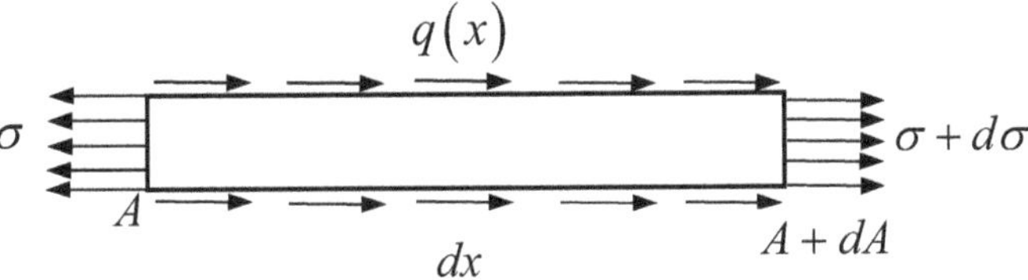

Figure 3.2 Bar segment.

Considering equilibrium along the longitudinal axis of the element, we have

$$-\sigma A + (\sigma+d\sigma)(A+dA) + q(x)\,dx = 0 \qquad .....(3.20)$$

$$\sigma\frac{dA}{dx} + A\frac{d\sigma}{dx} + q(x) = 0 \Rightarrow \frac{d}{dx}(\sigma A) + q(x) = 0 \qquad .....(3.21)$$

But $\sigma = E\dfrac{du}{dx}$

$$\frac{d}{dx}\left(AE\frac{du}{dx}\right)+q(x)=0 \qquad\qquad .....(3.22)$$

is the governing differential equation for the deformation of the bar. Here, AE is the axial rigidity and $u(x)$ is the axial deformation of the bar. The next step is to define test function ϕ to approximate $u(x)$ over the domain $\Omega(0\leq x\leq L)$.

For the error to be zero over the domain Ω,

$$\int_{\Omega}\phi\left[\frac{d}{dx}\left(AE\frac{du}{dx}\right)+q(x)\right]dx=0 \qquad\qquad .....(3.23)$$

In Eq. 3.23, it is observed that the test function ϕ is a zeroth order derivative whereas the expression in the brackets contains a second order derivative. We use integration by parts to change both to first order derivatives. This is done as follows:

$$\phi\left(AE\frac{du}{dx}\Big|_{0}^{L}\right)-\int_{0}^{L}AE\frac{d\phi}{dx}\frac{du}{dx}dx+\int_{0}^{L}\phi q(x)dx=0 \qquad\qquad .....(3.24)$$

The above equation is called a *weak form* of Eq. 3.23 because, Eq.3.24 is a first order differential equation obtained from Eq.3.23 which is a second order differential equation.

In Eq. 3.24, inside the parenthesis of the first term we have $AE\frac{du}{dx}\Big|_{x=L}$ and $AE\frac{du}{dx}\Big|_{x=0}$ which represent axial forces at the ends of the bar and are equal for the equilibrium of the bar. Thus, $AE\frac{du}{dx}\Big|_{0}^{L}$ from the first term of Eq. 3.24 gets evaluated to zero.

Thus Eq. 3.24 gets modified as

$$\int_{0}^{L}AE\frac{d\phi}{dx}\frac{du}{dx}dx-\int_{0}^{L}\phi q(x)dx=0 \qquad\qquad .....(3.25)$$

Taking

$$\phi=\phi_{1}\left(1-\frac{x}{L}\right)+\phi_{2}\left(\frac{x}{L}\right)=\lfloor N_{1}(x)\quad N_{2}(x)\rfloor\begin{Bmatrix}\phi_{1}\\\phi_{2}\end{Bmatrix}=\lfloor N(x)\rfloor\{\hat{\phi}\} \qquad\qquad .....(3.26)$$

and

$$u = u_1\left(1-\frac{x}{L}\right) + u_2\left(\frac{x}{L}\right) = \lfloor N_1(x) \quad N_2(x) \rfloor \begin{Bmatrix} u_1 \\ u_2 \end{Bmatrix} = \lfloor N(x) \rfloor \{\hat{u}\} \qquad \ldots\ldots(3.27)$$

where

$$\lfloor N(x) \rfloor = \left\lfloor 1-\frac{x}{L} \quad \frac{x}{L} \right\rfloor \Rightarrow \lfloor N'(x) \rfloor = \left\lfloor -\frac{1}{L} \quad \frac{1}{L} \right\rfloor \qquad \ldots\ldots(3.28)$$

$$\frac{d\phi}{dx} = \{\hat{\phi}\}^T \lfloor N'(x) \rfloor^T$$

$$\frac{du}{dx} = \lfloor N'(x) \rfloor \{\hat{u}\} \qquad\qquad\qquad\qquad\qquad\qquad \ldots\ldots(3.29)$$

Eq. 3.25 can now be written as

$$\int_0^L AE\{\hat{\phi}\}^T \lfloor N'(x) \rfloor^T \lfloor N'(x) \rfloor \{\hat{u}\}dx - \int_0^L \{\hat{\phi}\}^T \lfloor N(x) \rfloor^T q(x)dx = 0$$

$$\{\hat{\phi}\}^T \left(\int_0^L AE\lfloor N'(x) \rfloor^T \lfloor N'(x) \rfloor \{\hat{u}\}dx - \int_0^L \lfloor N(x) \rfloor^T q(x)dx \right) = 0 \qquad \ldots\ldots(3.30)$$

As the choice of $\{\hat{\phi}\}$ is arbitrary, Eq. 3.30 can be stated as

$$\int_0^L AE\lfloor N'(x) \rfloor^T \lfloor N'(x) \rfloor \{\hat{u}\}dx = \int_0^L \lfloor N(x) \rfloor^T q(x)dx \qquad \ldots\ldots(3.31)$$

$$\int_0^L AE \begin{Bmatrix} -\dfrac{1}{L} \\ \dfrac{1}{L} \end{Bmatrix} \left\lfloor -\frac{1}{L} \quad \frac{1}{L} \right\rfloor \{\hat{u}\}dx = \int_0^L q(x) \begin{Bmatrix} 1-\dfrac{x}{L} \\ \dfrac{x}{L} \end{Bmatrix} dx \qquad \ldots\ldots(3.32)$$

$$\frac{AE}{L} \begin{bmatrix} 1 & -1 \\ -1 & 1 \end{bmatrix} \begin{Bmatrix} \hat{u}_1 \\ \hat{u}_2 \end{Bmatrix} = \begin{Bmatrix} Q_1^{(e)} \\ Q_2^{(e)} \end{Bmatrix} \qquad \ldots\ldots(3.33)$$

Eq. 3.33 can be stated as

$$\left[K^{(e)} \right] \{\hat{u}\} = \{Q^{(e)}\} \qquad \ldots\ldots(3.34)$$

where $\left[K^{(e)} \right]$ is the element stiffness matrix and $\{Q^{(e)}\}$ is the vector of element forces.

Eq. 3.34 represents the stiffness equation for the element.

4 Discretization of Domain

The process of a dividing a domain into a finite number of elements associated with nodes is called discretization. Each element so divided from the domain is representative of the actual physical behaviour of the system. The number and type of elements into which the domain is divided is a matter of engineering judgement. Usually, the solution obtained with the least possible number of elements with desired behaviour uses less computational effort and produces the desired result. Figure 4.1 illustrates a domain discretized into triangular elements, each with three nodes.

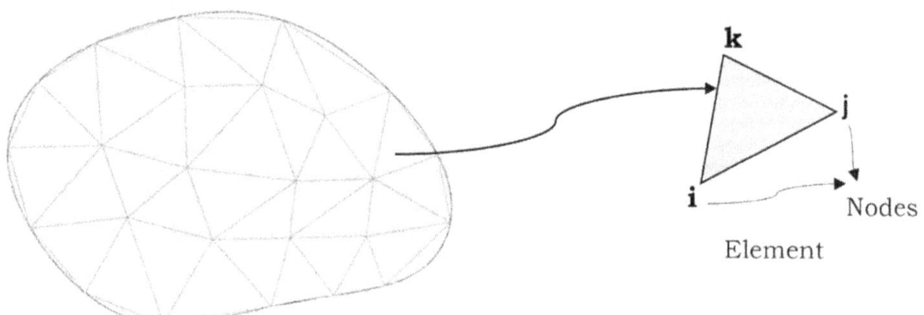

Figure 4.1 A discretised domain.

4.1   TYPES OF ELEMENTS

Elements are categorised based on their geometry. We have the following classification of elements:

a)   One dimensional elements

b)   Two-dimensional elements

c)   Three-dimensional elements

4.1.1   One Dimensional Elements

These elements have the length dimension predominant in comparison with the cross-sectional dimensions. The element may be along the x-axis or may be inclined. The primary line elements are bar, truss, and beam elements. They do have a cross-sectional area but are usually represented by line segments. In general, the cross-sectional area within the element can vary, but throughout this text prismatic elements alone are used.

The simplest linear element has two nodes, one at each end, although higher-order elements having three nodes or more (called quadratic, cubic, etc., elements) also exist. Section 4.2.4 includes discussion of higher-order line elements.

Examples: Bar element, plane truss element, beam element, plane frame element, space truss element, space frame element. These are the simplest of elements to consider and will be discussed in Chapter 6. Many of the basic concepts of the one-dimensional elements are discussed in Chapter 6.

One dimensional linear element

4.1.2   Two-Dimensional Elements

As indicated by its name, two dimensions of the element are predominant in comparison with its thickness in the third dimension. These elements are further categorised based on their geometry as follows:

(a) Triangular elements

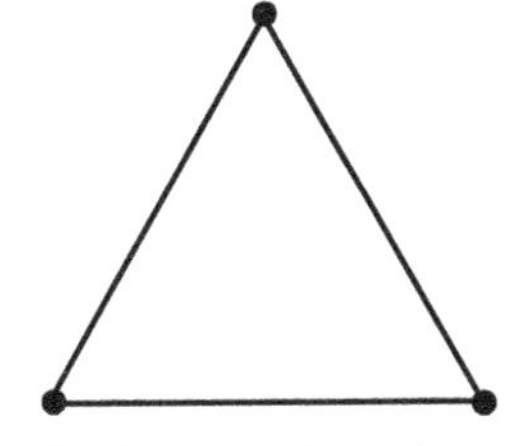

3-node triangular element

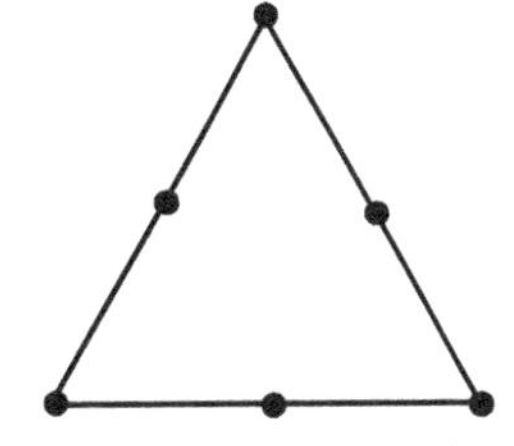

6-node triangular element

(b) Rectangular elements

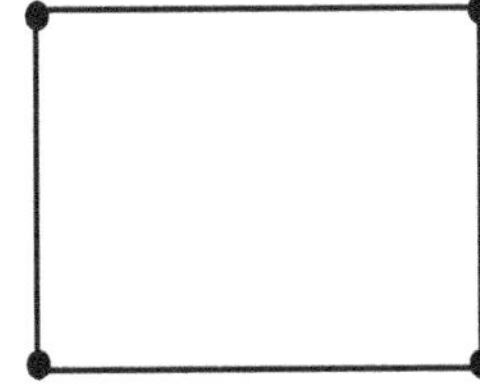

4-node rectangular element

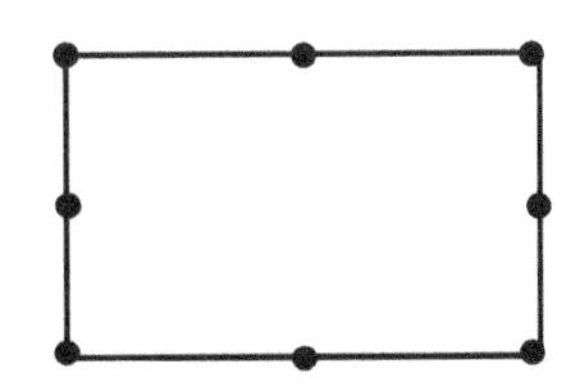

8-node rectangular element

(c) Quadrilateral elements

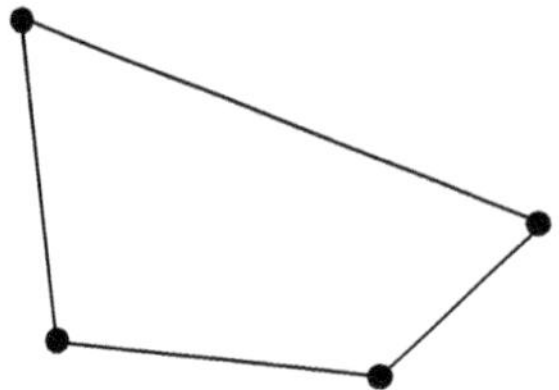

4-node quadrilateral element

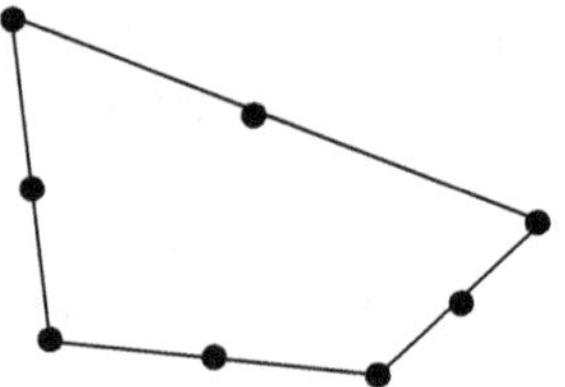

8-node quadrilateral element

These elements are discussed at length in Chapter 7.

4.1.3   Axisymmetric Elements

These are planar elements used in axisymmetric problems. These elements take advantage of the axi-symmetry of the problem. The following are the important categories of axisymmetric elements:

(a) Axisymmetric triangular elements

(b) Axisymmetric rectangular elements

These elements are discussed in detail in Chapter 9.

4.1.4   Three-Dimensional Elements

Most common three-dimensional elements are brick elements and tetrahedral elements. They are used when it is required to perform three-dimensional stress analysis. The basic three-dimensional elements have corner nodes only and straight sides, whereas higher-order elements with mid-side nodes (and possible midface nodes) have curved surfaces for their sides. These elements are taken up for study in Chapter 10.

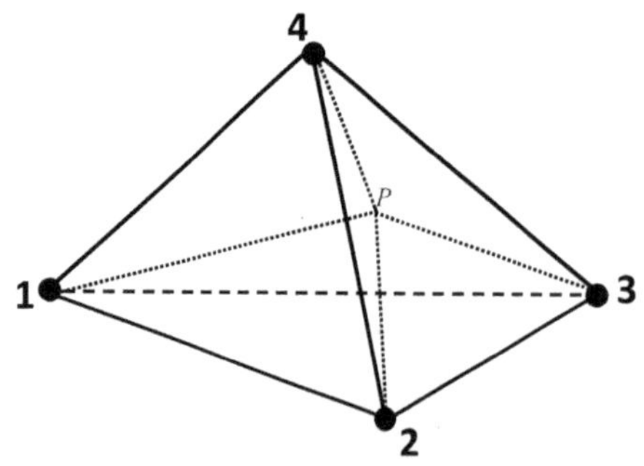

4-node tetrahedral element

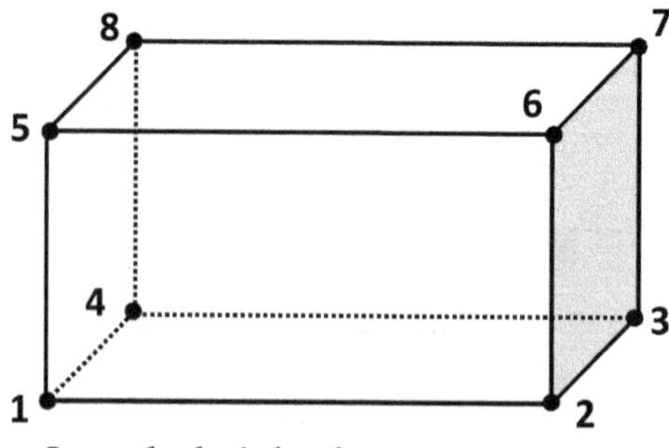

8-node brick element

4.1.5 Higher Order Elements

Higher order elements are used when it is required to achieve higher accuracy even at a high computational cost. *The following are the examples of higher order triangular elements:*

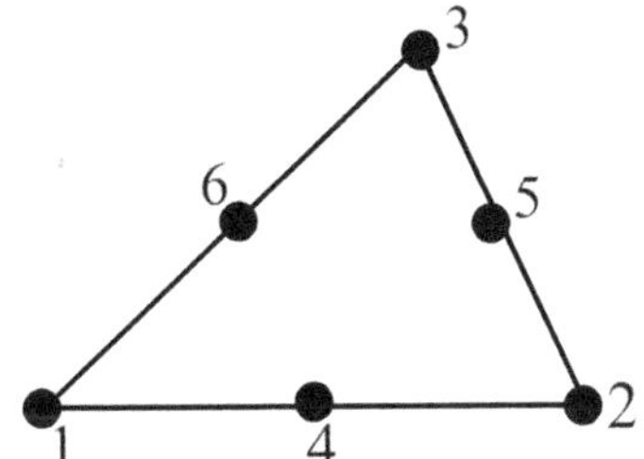 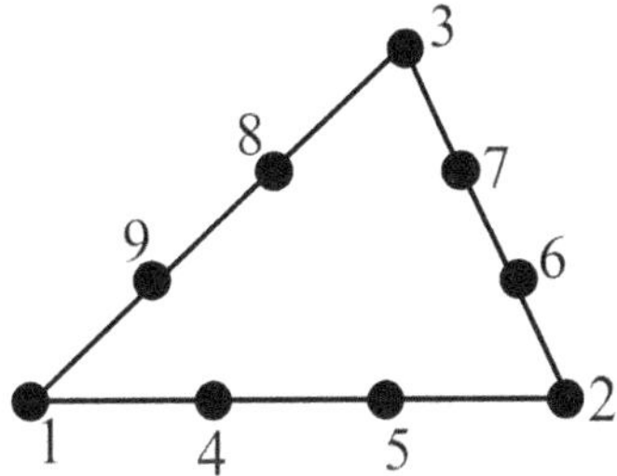

The following are the examples of higher order rectangular elements

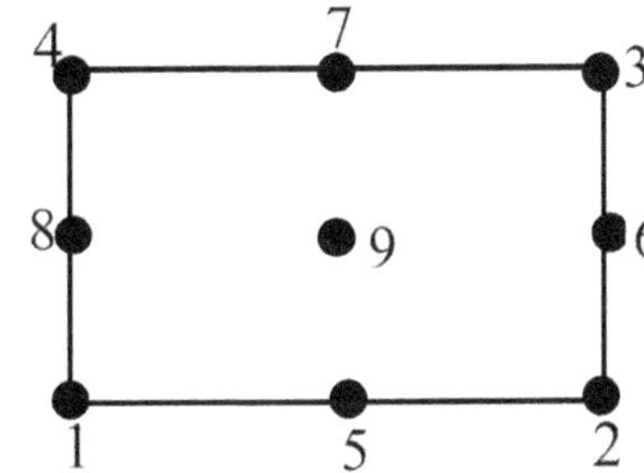 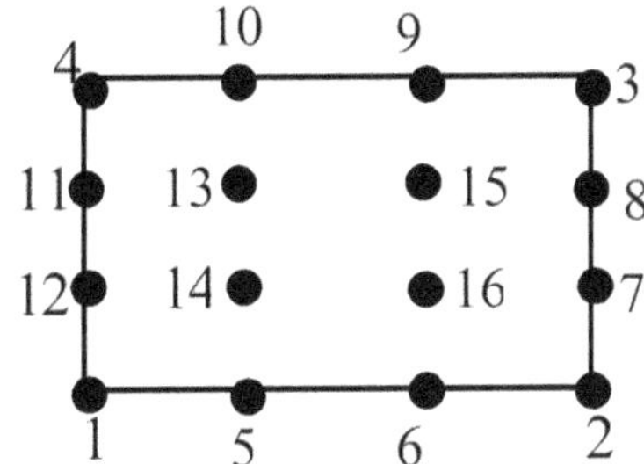

4.2 INTERPOLATION FUNCTIONS

In the process of discretisation, the domain is divided into a finite number of elements that are connected at nodes. In a continuum, solution can be obtained at each point of the domain, the number of such points at which solution can be sought is infinite.

However, in case of the discretized domain consisting of finite elements, solution can be obtained only at a finite number of nodal points. However, there is no information regarding the behaviour of the points within the element.

For example, in case of a plate is subjected to transverse loading, the solution domain consists of infinite number of points while obtaining the solution using a partial differential equation. However, when the plate is divided into a finite number of, say, triangular or rectangular elements, displacements are available only at the nodes of the elements and the behaviour of points inside the element is unknown.

The behaviour of points inside the element is modelled using interpolation functions. *Interpolation functions* seek to express the displacements at any point within the elements in terms of its nodal displacements. Thus, in a way, the interpolation functions serve the purpose of defining the **shape** of displacement field inside of the element. Thus, interpolation functions are also called *shape Functions*.

Several functions including trigonometrical functions have been used as interpolation functions in the past. However, polynomial interpolation functions became popular because of their simplicity, ease of implementation and ability to model complicated displacements fields.

4.2.1   Polynomial Interpolation Functions

A polynomial in x of degree n is defined as follows:

$$u = a_0 + a_1 x + a_2 x^2 + a_3 x^3 + \ldots\ldots + a_n x^n \qquad \ldots\ldots(4.1)$$

where $a_0, a_1, a_2, \ldots a_n$ are constants, some of which can be zero. The total number of terms in an nth degree polynomial is $(n+1)$.

Selection of an appropriate polynomial to define a given displacement field is based on various considerations discussed in the next section.

4.2.2   Criteria for Selection of Interpolation Functions

The following are the criteria for selection of interpolation functions.

(a) *Continuity*: The interpolation function shall be continuous within the boundaries of the element. This requirement is satisfied by choosing polynomial interpolation functions. The interpolation function and its derivatives shall be continuous across the inter-element boundaries. In general, if the function and its derivatives up to the nth derivative are continuous across the boundary of an element, the interpolation function is said to possess C^n continuity. Thus, for an element having C^0 continuity, only the interpolation function will be continuous across the inter-element boundary. Similarly, an element with C^2 continuity, the function and its first and second derivatives will be continuous across the inter-element boundary. For example, the interpolation function for a bar element has C^0 continuity while interpolation functions for beam and plate elements have C^1 and C^2 continuity respectively.

(b) *Rigid body modes*: The interpolation function must be capable of modelling rigid body modes in a displacement field. This is because when a structure is subjected to forces that make it undergo rigid body motion, the nodal displacements shall be the same. It is possible only when the interpolating polynomial contains a non-zero constant.

(c) *Representing constant strain state*: The interpolation function shall be capable of representing constant strain state within the element. As the level of discretization becomes finer, each element eventually shrinks to an infinitesimal size. Thus, the interpolation function shall be capable of providing a constant solution as each element shrinks to a point. This is satisfied only if the interpolating function contains non-zero coefficients for linear terms in x, y and z.

(d) *Convergence*: The interpolation function shall allow the displacement to converge to a solution as the domain is divided into more and more elements, each of smaller size. The resulting displacement shall approach the displacement obtained from the continuum model.

To illustrate this, consider a chain with several links, and suspended from two supports as shown in the Figure 4.2. The chain deflects at the junction of the links, but each individual link only rotates but does not deform as it is rigid. *Thus, the chain is stiffer than a cable of a similar geometry hung from the same supports.*

Figure 4.2 A suspended Chain.

However, if the size of each link is reduced and the number of links is increased, the chain will become more flexible and will have more deflection. As the number of links is increased, the chain becomes more flexible, and its stiffness reduces leading to a larger value of deflection. *Thus, we can say that the deflection of the chain approaches that of a cable from the below as it becomes more and more flexible.*

Following the same argument, one can say that the displacements obtained from the finite element model of a structure are lesser than the displacements obtained from the continuum model and *approach it from below* as the level of discretization becomes finer.

The stiffness of the cable is a lower bound to the stiffness of the chain while the deflection of the cable is the upper bound for the deflection of the chain.

This allows us to experiment with various levels of discretization and choose the appropriate number of elements yielding desired level of accuracy with the least number of elements and computational effort.

(e) *Compatibility*: The displacements shall be compatible between adjacent elements such that the elements do not separate or overlap during deformation. Elements which do not satisfy this requirement are known as *incompatible elements*. Compatibility in case of beam, plate and shell elements ensures that there is no sudden change of slope across the inter-element boundaries during deformation.

(f) *Geometric invariance*: The chosen polynomial interpolation functions shall be *geometrically invariant i.e.*, they shall not have any preferred direction. This means that the displacements shall not change with the choice of coordinate system. This is usually ensured by considering complete polynomials that include all possible terms. In case, when all terms cannot be included, geometric invariance can be achieved by having balanced terms with reference to all coordinate directions. This is usually achieved by using terms from the Pascal's triangle, explained in the next section.

4.2.3   Pascal's Triangle

For the convergence of solution in finite element analysis, the assumed displacement field should be isotropic. This is achieved if the components of the displacement field are complete polynomials. This is achieved by using Pascal's triangle. It is a technique of including the appropriate number of terms in the displacement model. It also gives an idea about the number of nodes required for the completion of the selected polynomial. Figure 4.3 shows the Pascal's triangle for two dimensions in Cartesian coordinates.

$$1 \qquad \text{Constant term(1)}$$
$$x \qquad y \qquad \text{Linear terms(2)}$$
$$x^2 \qquad xy \qquad y^2 \qquad \text{Quadratic terms(3)}$$
$$x^3 \qquad x^2 y \qquad xy^2 \qquad y^3 \qquad \text{Cubic terms(4)}$$
$$x^4 \qquad x^3 y \qquad x^2 y^2 \qquad xy^3 \qquad y^4 \qquad \text{Quartic terms(5)}$$
$$x^5 \qquad x^4 y \qquad x^3 y^2 \qquad x^2 y^3 \qquad xy^4 \qquad y^5 \qquad \text{Quintic terms(6)}$$

Figure 4.3 Pascal's Triangle in Cartesian Coordinates.

4.3   NATURAL COORDINATES

In Finite element analysis, we need to compute the stiffness matrix and force vector of each element by numerical integration. When a structure is discretized into an assemblage of elements of different types (one dimensional/two dimensional/three dimensional) and shapes (linear, triangular, rectangular, tetrahedral), numerical integration using Gaussian Quadrature is usually adopted to compute the stiffness matrices and force vectors of various elements. For example, by an appropriate process of transformation, variables with reference to a given coordinate system (cartesian, cylindrical polar or spherical polar coordinate system) are converted into equivalent non-dimensional variables.

Gaussian Quadrature uses integration of non-dimensional variables between limits -1 to +1 and performs integration in a very fast and efficient manner. Thus, it is desirable to have a **natural** coordinate system for the element in terms of dimensionless variables to facilitate numerical integration.

Figure 4.4 shows Pascal's triangle for natural coordinate system. Natural coordinates for one dimensional, two dimensional and three-dimensional elements are formulated and discussed in the succeeding sections.

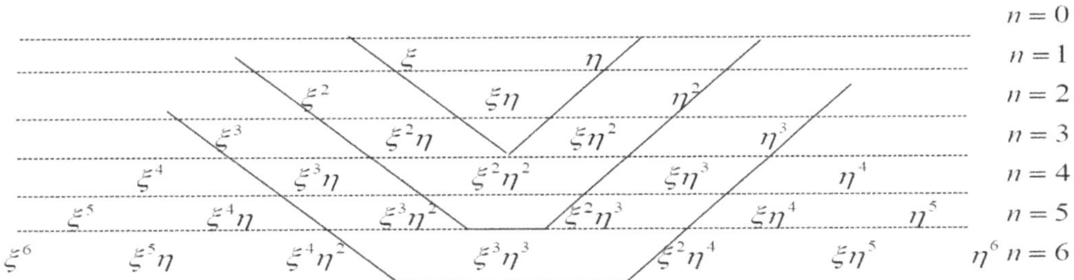

Figure 4.4 Pascal's Triangle in natural coordinates.

4.3.1   Natural Coordinates for a One-Dimensional Element

Consider a one-dimensional element with two nodes at either end. The length of the element is L . The element is oriented along the x direction. The first and second nodes of the element correspond to $x=0$ and $x=L$ respectively, as shown in the Figure 4.5 below:

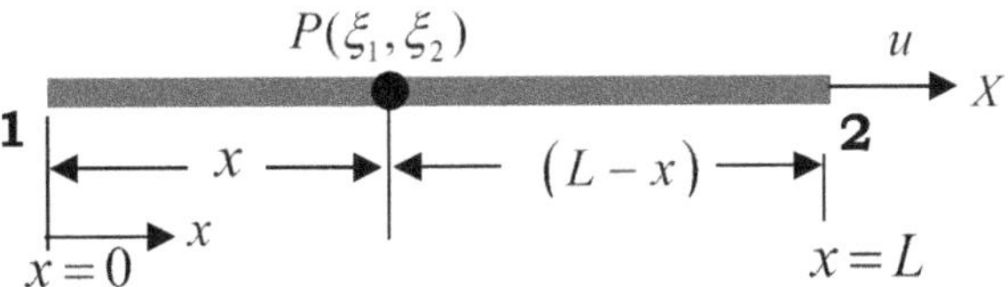

Figure 4.5 Bar Element.

Natural coordinates ξ_1 *and* ξ_2 for an arbitrary point P are defined as follows:

$$\xi_1 = \frac{(L-x)}{L} = 1 - \frac{x}{L}$$

$$\xi_2 = \frac{x}{L}$$

$$.....(4.2)$$

It is readily observed that $\xi_1 + \xi_2 = 1$ or $\sum_{i=1}^{N} \xi_i = 1$ where N is the total number of natural coordinates for the element.

Eq. 4.2 can be more conveniently expressed in a matrix equation shown below:

$$\begin{bmatrix} 1 & 1 \\ 0 & L \end{bmatrix} \begin{Bmatrix} \xi_1 \\ \xi_2 \end{Bmatrix} = \begin{Bmatrix} 1 \\ x \end{Bmatrix} \qquad .....(4.3)$$

The inverse of this relation between natural and Cartesian coordinates can be stated as

$$\begin{Bmatrix} \xi_1 \\ \xi_2 \end{Bmatrix} = \begin{bmatrix} 1 & 1 \\ 0 & L \end{bmatrix}^{-1} \begin{Bmatrix} 1 \\ x \end{Bmatrix} = \begin{Bmatrix} \dfrac{(L-x)}{L} \\ \dfrac{x}{L} \end{Bmatrix} \qquad .....(4.4)$$

Eq. 4.4 expresses natural coordinates ξ_1 and ξ_2 in terms of the Cartesian coordinate x. It is also important to express the partial derivatives with reference to cartesian coordinates with the corresponding partial derivatives with reference to the natural coordinates. Thus

$$\frac{\partial}{\partial x} = \frac{\partial}{\partial \xi_1} \frac{\partial \xi_1}{\partial x} + \frac{\partial}{\partial \xi_2} \frac{\partial \xi_2}{\partial x} = \frac{1}{L} \left(\frac{\partial}{\partial \xi_2} - \frac{\partial}{\partial \xi_1} \right) \qquad .....(4.5)$$

4.3.2   Area Coordinates for a Triangular Element

Figure 4.6 shows a triangular element with nodes 1, 2 and 3 in cartesian coordinate system. The nodes have the coordinates $(x_1, y_1), (x_2, y_2)$ and (x_3, y_3) respectively. Now consider a point $P(x, y)$ inside the element. Imagine that this point P is attached to the nodes 1, 2 and 3 by means of rubber bands that can stretch or shrink to adjust themselves to the distance between P and the respective nodes. Thus, the area of the triangle is trifurcated into three smaller triangular areas A_1, A_2 and A_3 respectively that are defined as follows:

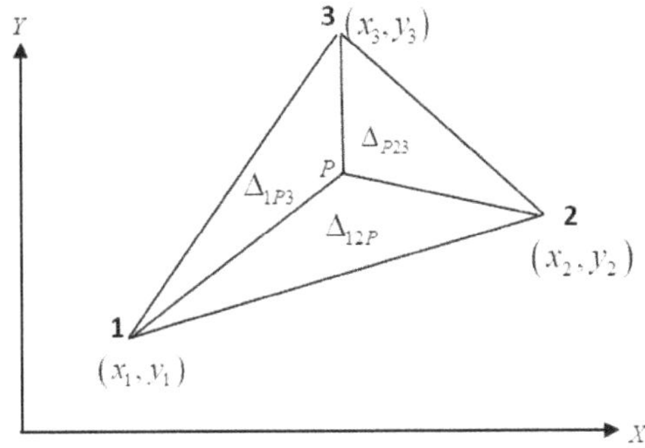

Figure 4.6 A 3-node triangular element.

$$A_1 = Area(\Delta_{P23}) \, , \, A_2 = Area(\Delta_{1P3}), \, A_3 = Area(\Delta_{12P}) \, .$$

Now, natural coordinates are defined as follows:

$$\xi_1 = \frac{A_1}{A} \; ; \; \xi_2 = \frac{A_2}{A} \; ; \; \xi_3 = \frac{A_3}{A} \qquad\qquad .....(4.6a)$$

with the relationship

$$A_1 + A_2 + A_3 = A \; \Rightarrow \xi_1 + \xi_2 + \xi_3 = 1 \Rightarrow \sum_{i=1}^{N} \xi_i = 1 \qquad .....(4.6b)$$

As observed from Eq. 4.6, these natural coordinates are defined as ratios of areas. Thus, they are also referred to as **Area Coordinates**. The area coordinates for the nodes 1, 2 and 3 are (1,0,0), (0,1,0) and (0,0,1) respectively.

Eq. 4.6(b) shows that the sum of the natural coordinates is unity where N is the number of nodes of the element.

The area of the triangle A and its component areas A_1, A_2 and A_3 are expressed in terms of the nodal coordinates by the following determinants:

$$A = \begin{vmatrix} 1 & 1 & 1 \\ x_1 & x_2 & x_3 \\ y_1 & y_2 & y_3 \end{vmatrix} ; A_1 = \begin{vmatrix} 1 & 1 & 1 \\ x & x_2 & x_3 \\ y & y_2 & y_3 \end{vmatrix} ; A_2 = \begin{vmatrix} 1 & 1 & 1 \\ x_1 & x & x_3 \\ y_1 & y & y_3 \end{vmatrix} ; A_3 = \begin{vmatrix} 1 & 1 & 1 \\ x_1 & x_2 & x \\ y_1 & y_2 & y \end{vmatrix} ; \qquad .....(4.7)$$

The relationship between cartesian and natural coordinates can be stated as follows:

$$\begin{Bmatrix} 1 \\ x \\ y \end{Bmatrix} = \begin{bmatrix} 1 & 1 & 1 \\ x_1 & x_2 & x_3 \\ y_1 & y_2 & y_3 \end{bmatrix} \begin{Bmatrix} \xi_1 \\ \xi_2 \\ \xi_3 \end{Bmatrix} \qquad\qquad .....(4.8)$$

The inverse relationship of the natural and cartesian coordinates is given as follows:

$$\begin{Bmatrix} \xi_1 \\ \xi_2 \\ \xi_3 \end{Bmatrix} = \begin{bmatrix} 1 & 1 & 1 \\ x_1 & x_2 & x_3 \\ y_1 & y_2 & y_3 \end{bmatrix}^{-1} \begin{Bmatrix} 1 \\ x \\ y \end{Bmatrix} = \frac{1}{2A} \begin{bmatrix} c_1 & b_1 & a_1 \\ c_2 & b_2 & a_2 \\ c_3 & b_3 & a_3 \end{bmatrix} \begin{Bmatrix} 1 \\ x \\ y \end{Bmatrix} \qquad ....(4.9)$$

where $c_i = x_j y_k - x_k y_j$; $b_i = y_j - y_k$ and $a_i = x_k - x_j$ are cyclic expressions with $i, j, k = 1, 2, 3$ taken cyclically.

Eq. 4.9 can be expressed as

$$\xi_i = \frac{1}{2A}\left(c_i + b_i x + a_i y\right), i = 1, 2, 3 \qquad (4.10)$$

The partial derivatives with reference to cartesian coordinates can be expressed in terms of the corresponding partial derivatives with reference to natural coordinates as follows:

$$\frac{\partial}{\partial x} = \frac{\partial}{\partial \xi_1}\frac{\partial \xi_1}{\partial x} + \frac{\partial}{\partial \xi_2}\frac{\partial \xi_2}{\partial x} + \frac{\partial}{\partial \xi_3}\frac{\partial \xi_3}{\partial x} = \sum_{i=1}^{3} \frac{b_i}{2A}\frac{\partial}{\partial \xi_i}$$

$$\frac{\partial}{\partial y} = \frac{\partial}{\partial \xi_1}\frac{\partial \xi_1}{\partial y} + \frac{\partial}{\partial \xi_2}\frac{\partial \xi_2}{\partial y} + \frac{\partial}{\partial \xi_3}\frac{\partial \xi_3}{\partial y} = \sum_{i=1}^{3} \frac{a_i}{2A}\frac{\partial}{\partial \xi_i}$$

$$.....(4.11)$$

4.3.3   Volume Coordinates for a 4-node Tetrahedral Element

Consider a 4-node tetrahedral element as shown in the Figure 4.7:

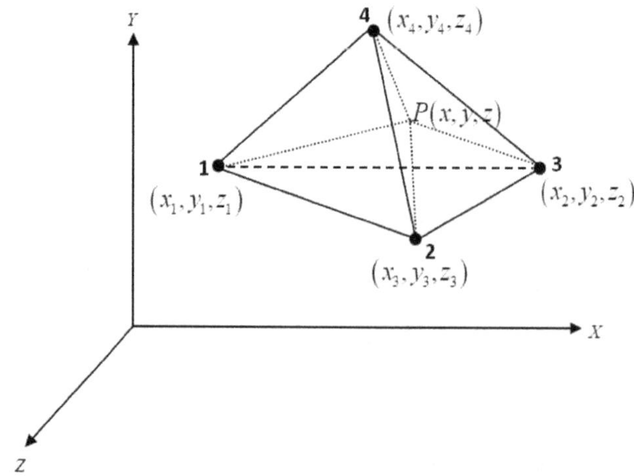

Figure 4.7 Four node tetrahedral element

The element has 4 nodes with coordinates $(x_1, y_1, z_1), (x_2, y_2, z_2), (x_3, y_3, z_3)$ and (x_4, y_4, z_4) respectively. The nodes are numbered in such a way that the first three nodes are numbered in a counterclockwise sense when viewed from the last node.

The volume of the element is given by

$$V = \frac{1}{6} \begin{vmatrix} 1 & 1 & 1 & 1 \\ x_1 & x_2 & x_3 & x_4 \\ y_1 & y_2 & y_3 & y_4 \\ z_1 & z_2 & z_3 & z_4 \end{vmatrix} \qquad \qquad .....(4.12)$$

Consider a point $P(x, y, z)$ inside the element which is assumed to be connected to all four nodes of the element by extensible rubber bands. The four sub-volumes formed within the element as a result, are shown below:

$$V_1 = vol(P234) = \frac{1}{6} \begin{vmatrix} 1 & 1 & 1 & 1 \\ x & x_2 & x_3 & x_4 \\ y & y_2 & y_3 & y_4 \\ z & z_2 & z_3 & z_4 \end{vmatrix} ; \quad V_2 = vol(1P34) = \frac{1}{6} \begin{vmatrix} 1 & 1 & 1 & 1 \\ x_1 & x & x_3 & x_4 \\ y_1 & y & y_3 & y_4 \\ z_1 & z & z_3 & z_4 \end{vmatrix} ;$$

$$V_3 = vol(12P4) = \frac{1}{6} \begin{vmatrix} 1 & 1 & 1 & 1 \\ x_1 & x_2 & x & x_4 \\ y_1 & y_2 & y & y_4 \\ z_1 & z_2 & z & z_4 \end{vmatrix} ; \quad V_4 = vol(123P) = \frac{1}{6} \begin{vmatrix} 1 & 1 & 1 & 1 \\ x_1 & x_2 & x_3 & x \\ y_1 & y_2 & y_3 & y \\ z_1 & z_2 & z_3 & z \end{vmatrix} \quad .....(4.13)$$

The natural coordinates ξ_1, ξ_2, ξ_3 and ξ_4 are defined as

$$\xi_1 = \frac{V_1}{V} \quad \xi_2 = \frac{V_2}{V} \quad \xi_3 = \frac{V_3}{V} \quad and \quad \xi_4 = \frac{V_4}{V} \qquad .....(4.14)$$

$$\xi_1 + \xi_2 + \xi_3 + \xi_4 = 1$$

These natural coordinates are also called *volume coordinates* because they are obtained by taking ratios of volumes. The natural coordinates of nodes 1,2,3 and 4 are (1,0,0,0), (0,1,0,0), (0,0,1,0) and (0,0,0,1) respectively. The coordinates of point P can be expressed as

$$x = \xi_1 x_1 + \xi_2 x_2 + \xi_3 x_3 + \xi_4 x_4$$
$$y = \xi_1 y_1 + \xi_2 y_2 + \xi_3 y_3 + \xi_4 y_4 \qquad \qquad .....(4.15)$$
$$z = \xi_1 z_1 + \xi_2 z_2 + \xi_3 z_3 + \xi_4 z_4$$

The relationship between cartesian coordinates and natural coordinates can be expressed as

$$\begin{Bmatrix} 1 \\ x \\ y \\ z \end{Bmatrix} = \begin{bmatrix} 1 & 1 & 1 & 1 \\ x_1 & x_2 & x_3 & x_4 \\ y_1 & y_2 & y_3 & y_4 \\ z_1 & z_2 & z_3 & z_4 \end{bmatrix} \begin{Bmatrix} \xi_1 \\ \xi_2 \\ \xi_3 \\ \xi_4 \end{Bmatrix} \qquad \qquad .....(4.16)$$

The inverse relationship can be stated as follows:

$$\begin{Bmatrix} \xi_1 \\ \xi_2 \\ \xi_3 \\ \xi_4 \end{Bmatrix} = \frac{1}{6V} \begin{bmatrix} a_1 & a_2 & a_3 & a_4 \\ b_1 & b_2 & b_3 & b_4 \\ c_1 & c_2 & c_3 & c_4 \\ d_1 & d_2 & d_3 & d_4 \end{bmatrix} \begin{Bmatrix} 1 \\ x \\ y \\ z \end{Bmatrix} \qquad \qquad .....(4.17)$$

where

$$a_1 = \begin{vmatrix} x_2 & y_2 & z_2 \\ x_3 & y_3 & z_3 \\ x_4 & y_4 & z_4 \end{vmatrix}; \; a_2 = -\begin{vmatrix} x_1 & y_1 & z_1 \\ x_3 & y_3 & z_3 \\ x_4 & y_4 & z_4 \end{vmatrix}; \; a_3 = \begin{vmatrix} x_1 & y_1 & z_1 \\ x_2 & y_2 & z_2 \\ x_4 & y_4 & z_4 \end{vmatrix}; a_4 = -\begin{vmatrix} x_1 & y_1 & z_1 \\ x_2 & y_2 & z_2 \\ x_3 & y_3 & z_3 \end{vmatrix}$$

$$b_1 = -\begin{vmatrix} 1 & y_2 & z_2 \\ 1 & y_3 & z_3 \\ 1 & y_4 & z_4 \end{vmatrix}; b_2 = \begin{vmatrix} 1 & y_1 & z_1 \\ 1 & y_3 & z_3 \\ 1 & y_4 & z_4 \end{vmatrix}; b_3 = -\begin{vmatrix} 1 & y_1 & z_1 \\ 1 & y_2 & z_2 \\ 1 & y_4 & z_4 \end{vmatrix}; b_4 = \begin{vmatrix} 1 & y_1 & z_1 \\ 1 & y_2 & z_2 \\ 1 & y_3 & z_3 \end{vmatrix}$$

$$c_1 = \begin{vmatrix} 1 & x_2 & z_2 \\ 1 & x_3 & z_3 \\ 1 & x_4 & z_4 \end{vmatrix}; c_2 = -\begin{vmatrix} 1 & x_1 & z_1 \\ 1 & x_3 & z_3 \\ 1 & x_4 & z_4 \end{vmatrix}; c_3 = \begin{vmatrix} 1 & x_1 & z_1 \\ 1 & x_2 & z_2 \\ 1 & x_4 & z_4 \end{vmatrix}; c_4 = -\begin{vmatrix} 1 & x_1 & z_1 \\ 1 & x_2 & z_2 \\ 1 & x_3 & z_3 \end{vmatrix}$$

$$d_1 = -\begin{vmatrix} 1 & x_2 & y_2 \\ 1 & x_3 & y_3 \\ 1 & x_4 & y_4 \end{vmatrix}; d_2 = \begin{vmatrix} 1 & x_1 & y_1 \\ 1 & x_3 & y_3 \\ 1 & x_4 & y_4 \end{vmatrix}; d_3 = -\begin{vmatrix} 1 & x_1 & y_1 \\ 1 & x_2 & y_2 \\ 1 & x_4 & y_4 \end{vmatrix}; d_4 = \begin{vmatrix} 1 & x_1 & y_1 \\ 1 & x_2 & y_2 \\ 1 & x_3 & y_3 \end{vmatrix} \quad .....(4.18)$$

Thus, the natural coordinates can be expressed in terms of cartesian coordinates as

$$\xi_1 = \frac{1}{6V}\left(a_1 + b_1 x + c_1 y + d_1 z\right)$$

$$\xi_2 = \frac{1}{6V}\left(a_2 + b_2 x + c_2 y + d_2 z\right)$$

$$\xi_3 = \frac{1}{6V}\left(a_3 + b_3 x + c_3 y + d_3 z\right) \qquad\qquad (4.19)$$

$$\xi_4 = \frac{1}{6V}\left(a_4 + b_4 x + c_4 y + d_4 z\right)$$

4.4 CONSTRUCTION OF INTERPOLATION FUNCTIONS

In this section, construction of interpolation functions for elements is demonstrated by considering three elements *viz.* bar element, triangular element and the tetrahedron element.

The bar element is a one-dimensional element. The triangular element is two dimensional. The tetrahedral element is a three-dimensional element.

4.4.1 Interpolation Functions for Bar Element

Consider the bar element as shown in the Figure 4.8. The element has two nodes. At each node, there is a displacement u along the positive X direction. The solution domain of the element is given as $(0 \le x \le L)$

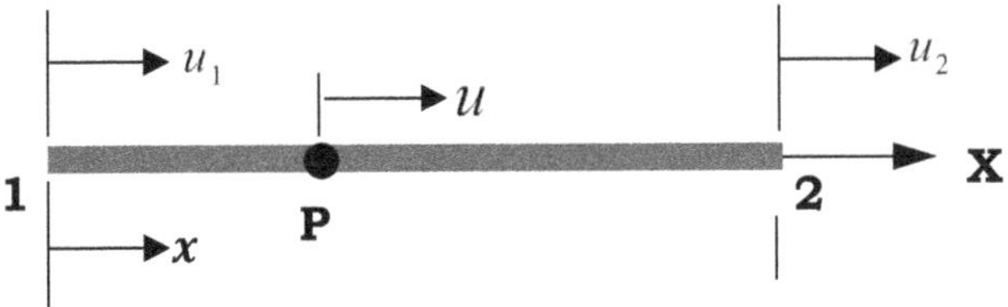

Figure 4.8 Bar Element.

The polynomial interpolation function $u(x)$ to be chosen for the bar element is defined as follows:

$$u(x) = a_0 + a_1 x = \lfloor 1 \quad x \rfloor \left\{ \begin{array}{c} a_0 \\ a_1 \end{array} \right\} \qquad \dots (4.20)$$

Here is $u(x)$ a linear function of x. Noting that at the first node corresponding to $x=0$, $u=u_1$ and at second node corresponding to $x=L$, $u=u_2$

$$\left\{\begin{matrix} u_1 \\ u_2 \end{matrix}\right\} = \begin{bmatrix} 1 & 0 \\ 1 & L \end{bmatrix}\left\{\begin{matrix} a_0 \\ a_1 \end{matrix}\right\} \Rightarrow \left\{\begin{matrix} a_0 \\ a_1 \end{matrix}\right\} = \begin{bmatrix} 1 & 0 \\ -\dfrac{1}{L} & \dfrac{1}{L} \end{bmatrix}\left\{\begin{matrix} u_1 \\ u_2 \end{matrix}\right\}$$ .....(4.21)

Substituting Eq. 4.21 in Eq. 4.20, we obtain

$$u(x) = \left(\lfloor 1 \quad x \rfloor \begin{bmatrix} 1 & 0 \\ -\dfrac{1}{L} & \dfrac{1}{L} \end{bmatrix}\right)\left\{\begin{matrix} u_1 \\ u_2 \end{matrix}\right\} = \begin{bmatrix} N_1(x) & N_2(x) \end{bmatrix}\left\{\begin{matrix} u_1 \\ u_2 \end{matrix}\right\} = \lfloor N(x)\rfloor\{d\}$$ .....(4.22)

where

$$N_1(x) = 1 - \frac{x}{L} \quad \text{and}$$ .....(4.23)

$$N_2(x) = \frac{x}{L}$$

It is observed from Eq. 4.23 that these interpolation functions are linear in x and are the same as the natural coordinates of any point P in the element as defined by Eq. 4.2.

These linear interpolation functions are plotted as shown in Figure 4.9. It is further observed that at any point P along the element,

$$N_1(x) + N_2(x) = 1 \Rightarrow \sum_{i=1}^{2} N_i(x) = 1$$ .....(4.24)

The plots of these interpolation functions are as shown in Figure 4.9:

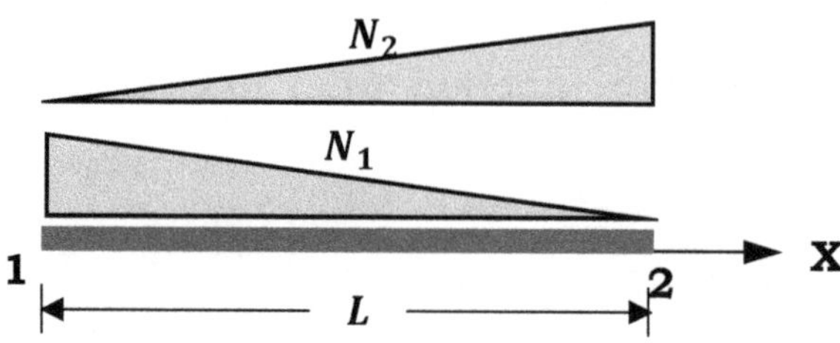

Figure 4.9 Interpolation functions for bar element.

The displacement at any point P on the element located at distance of x is described using interpolation functions N_1 and N_2 as

$$u = u_1 N_1 + u_2 N_2 = \lfloor N_1 \quad N_2 \rfloor \begin{Bmatrix} u_1 \\ u_2 \end{Bmatrix} = \lfloor N \rfloor \{d\} \qquad \qquad .....(4.25)$$

The shape of displacement function u defined by Eq. 4.24 is depicted graphically in the Figure 4.10 as shown below:

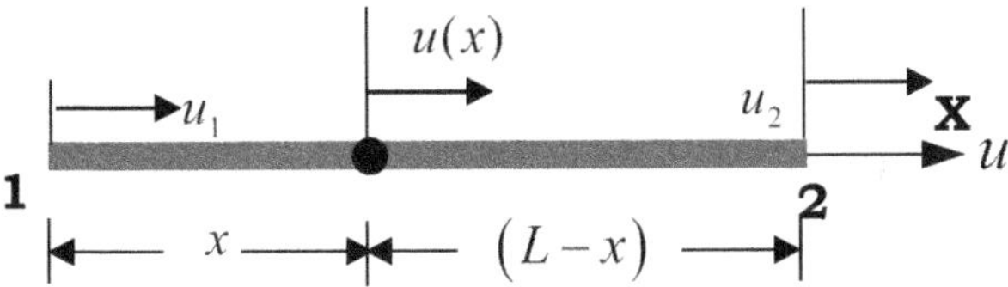

Figure 4.10 Variation of displacement u along the element.

4.4.2   Natural Coordinates for Beam Element

Consider the beam element as shown in the Figure 4.11. The element has two nodes. At each node, there is a displacement V along the positive y-direction, and an in-plane rotation θ (anticlockwise) about the z-axis. The solution domain of the element is given as $(0 \le x \le L)$

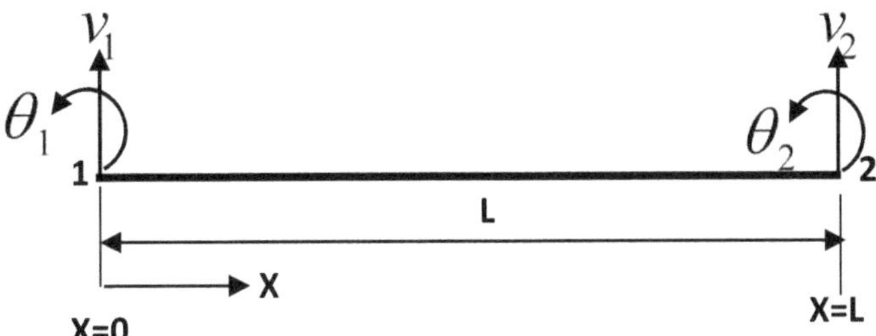

Figure 4.11 Beam Element.

The polynomial interpolation function to be chosen for the beam element must satisfy the following criteria over its domain.

Table 4.1 Requirements of unknown parameters over solution domain of element.

Parameter	Requirement
Vertical displacement v	Continuity
In-plane rotation θ	Continuity
Curvature $\dfrac{d^2v}{dx^2}$ (as bending moment is $EI\dfrac{d^2v}{dx^2}$)	Existence
$\dfrac{d^3v}{dx^3}$ (as shear force is $EI\dfrac{d^3v}{dx^3}$)	Existence

In view of the above, it is most appropriate to choose a cubic polynomial in x as interpolation function in order to satisfy the requirements outlined in Table 4.1.

$$v(x) = a_0 + a_1x + a_2x^2 + a_3x^3 = \lfloor 1 \quad x \quad x^2 \quad x^3 \rfloor \begin{Bmatrix} a_0 \\ a_1 \\ a_2 \\ a_3 \end{Bmatrix} = \lfloor \phi(x) \rfloor \{\alpha\} \qquad .....(4.26)$$

where $\lfloor \phi(x) \rfloor$ is the vector containing various powers of x and $\{\alpha\}$ is the vector of undetermined coefficients.

Differentiation of Eq. 4.26 with reference to x would yield

$$\theta(x) = \frac{dv}{dx} = a_1 + 2a_2x + 3a_3x^2 = \lfloor 0 \quad 1 \quad 2x \quad 3x^2 \rfloor \begin{Bmatrix} a_0 \\ a_1 \\ a_2 \\ a_3 \end{Bmatrix} \qquad .....(4.27)$$

Using Eq. 4.26 and Eq. 4.27 at the nodes 1 and 2, the following relation is obtained:

$$\{d\} = \begin{Bmatrix} v_1 \\ \theta_1 \\ v_2 \\ \theta_2 \end{Bmatrix} = \begin{bmatrix} 1 & 0 & 0 & 0 \\ 0 & 1 & 0 & 0 \\ 1 & L & L^2 & L^3 \\ 0 & 1 & 2L & 3L^2 \end{bmatrix} \{\alpha\} = [A]\{\alpha\} \qquad .....(4.28)$$

$$\{\alpha\} = [A]^{-1}\{d\} \qquad .....(4.29)$$

Substituting in Eq. 4.26, we get

$$v(x) = \lfloor \phi(x) \rfloor [A]^{-1} \{d\} = [N(x)]\{d\} \qquad \qquad .....(4.30)$$

where $\lfloor N(x) \rfloor$ is the shape function matrix and is evaluated as follows

$$[N(x)] = \lfloor 1 \quad x \quad x^2 \quad x^3 \rfloor \begin{bmatrix} 1 & 0 & 0 & 0 \\ 0 & 1 & 0 & 0 \\ 1 & L & L^2 & L^3 \\ 0 & 1 & 2L & 3L^2 \end{bmatrix}^{-1}$$

$$[N(x)] = \lfloor 1 \quad x \quad x^2 \quad x^3 \rfloor \begin{bmatrix} 1 & 0 & 0 & 0 \\ 0 & 1 & 0 & 0 \\ -\dfrac{3}{L^2} & -\dfrac{2}{L} & \dfrac{3}{L^2} & -\dfrac{1}{L} \\ \dfrac{2}{L^3} & \dfrac{1}{L^2} & -\dfrac{2}{L^3} & \dfrac{1}{L^2} \end{bmatrix} = \begin{Bmatrix} 1 - \dfrac{3x^2}{L^2} + \dfrac{2x^3}{L^3} \\[2mm] x - \dfrac{2x^2}{L} + \dfrac{x^3}{L^2} \\[2mm] \dfrac{3x^2}{L^2} - \dfrac{2x^3}{L^3} \\[2mm] -\dfrac{x^2}{L} + \dfrac{x^3}{L^2} \end{Bmatrix} \qquad .....(4.31)$$

The interpolation functions obtained in Eq. 4.31 are known as **_Herimite Cubic polynomial_** interpolation functions. Taking $\eta = \dfrac{x}{L}$, the interpolation functions can be written as $\lfloor N(x) \rfloor = \lfloor N_1(x) \quad N_2(x) \quad N_3(x) \quad N_4(x) \rfloor$,

where

$$N_1(x) = 1 - \frac{3x^2}{L^2} + \frac{2x^3}{L^3} = 1 - 3\eta^2 + 2\eta^3$$

$$N_2(x) = x - \frac{2x^2}{L} + \frac{x^3}{L^2} = L\left(\eta - 2\eta^2 + \eta^3\right)$$

$$N_3(x) = \frac{3x^2}{L^2} - \frac{2x^3}{L^3} = 3\eta^2 - 2\eta^3 \qquad \qquad .....(4.32)$$

$$N_4(x) = -\frac{x^2}{L} + \frac{x^3}{L^2} = L(-\eta^2 + \eta^3)$$

These interpolation functions are plotted as shown in Figure 4.12:

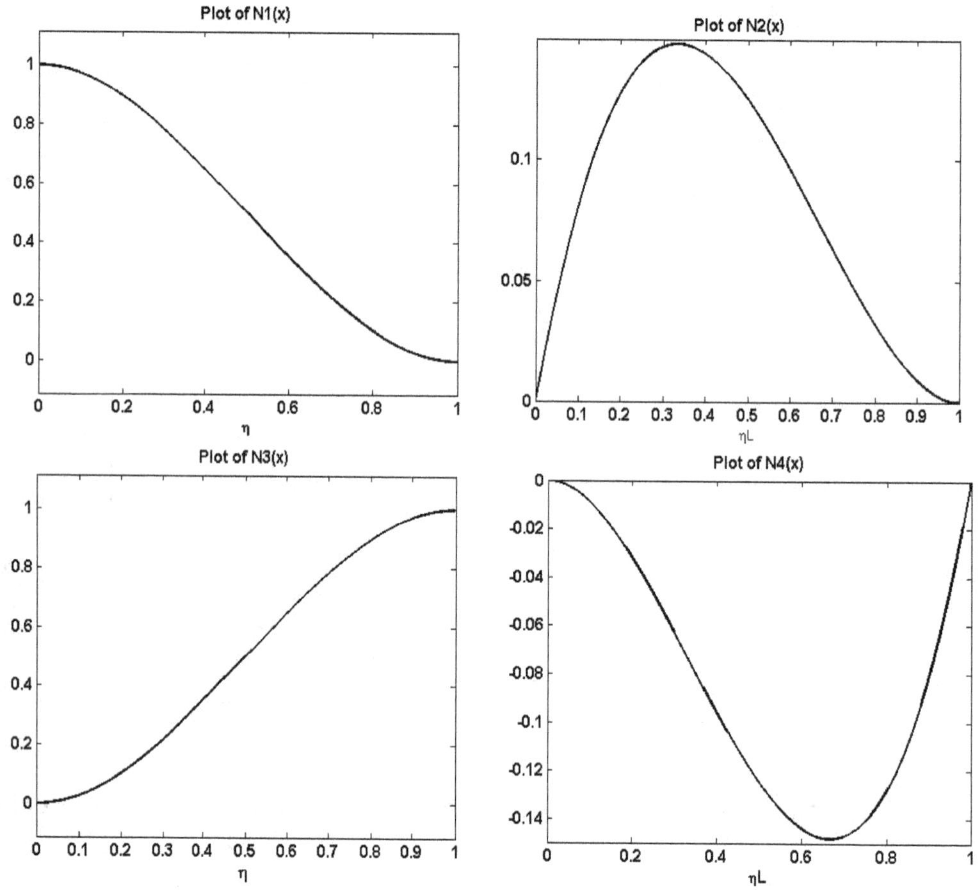

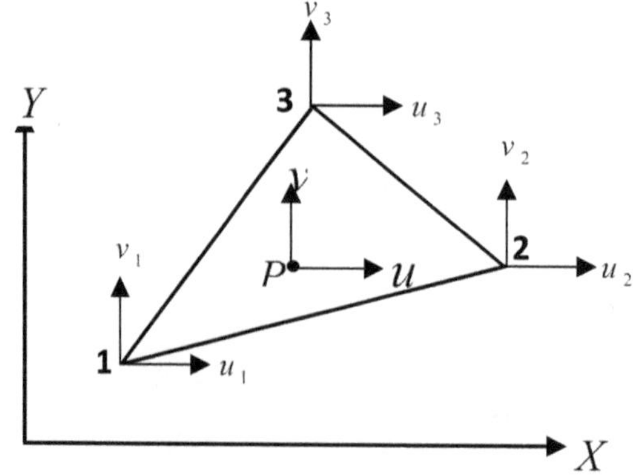

Figure 4.12 Plots of Hermite Cubic Polynomial Interpolation Functions.

4.4.3   Interpolation Functions for a 3-node Triangular Element

Figure 4.13 Three node triangular element.

Consider the 3-node triangular element shown in Figure 4.13, taken from section 4.3.2. The displacements along the X and Y directions at a point P inside the element are u and v respectively. The nodal displacements at nodes 1, 2 and 3 of the element along the X and Y directions are (u_1, v_1), (u_2, v_2) and (u_3, v_3) respectively. The natural coordinates derived in Eq. 4.8, are adopted as interpolation functions. Thus, the interpolation functions are expressed as $N_i \equiv \xi_i$. The displacements at the point P are expressed in terms of the nodal displacements by using interpolation functions as follows:

$$u = u_1 N_1 + u_2 N_2 + u_3 N_3$$
$$v = v_1 N_1 + v_2 N_2 + v_3 N_3$$

.....(4.33)

The sketches of shape functions N_1, N_2 and N_3 are given in Figure 4.14.

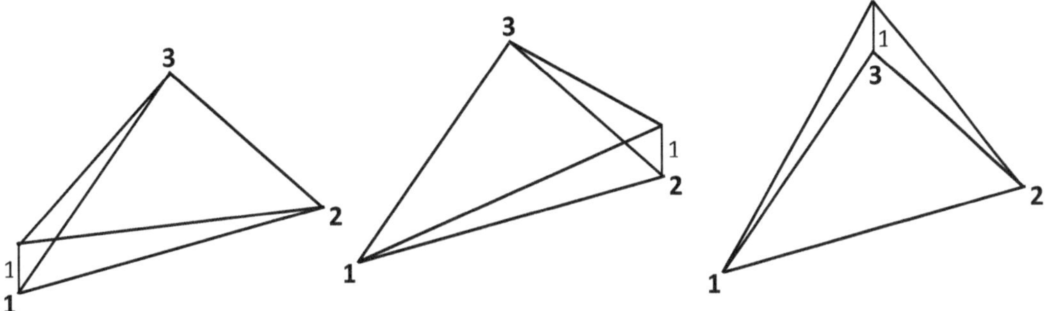

Figure 4.14 Interpolation functions for triangular element.

Eq. 4.33 can be expressed in matrix form as follows:

$$\begin{Bmatrix} u \\ v \end{Bmatrix} = \begin{bmatrix} N_1 & 0 & N_2 & 0 & N_3 & 0 \\ 0 & N_1 & 0 & N_2 & 0 & N_3 \end{bmatrix} \begin{Bmatrix} u_1 \\ v_1 \\ u_2 \\ u_2 \\ u_3 \\ u_3 \end{Bmatrix} \Rightarrow \{u\} = [N]\{d\}$$

.....(4.34)

Interpolation functions can also be obtained for the 3-node triangular element in terms of the natural coordinates by a process of transformation from a reference element to the parent element. The relationship between the set of points in the reference element and the corresponding set of points of the parent element is *bijective*. This means, for every point (x, y) within the domain of the parent element, there exists a unique point (ξ,η) in the reference element and vice versa, as shown in Figure 4.15. Each portion of the boundary of the parent element defined by the nodes attached to it correspond to the portion of the boundary of the reference element defined by the corresponding nodes attached to it. We define x in terms of the reference coordinates ξ and η as follows:

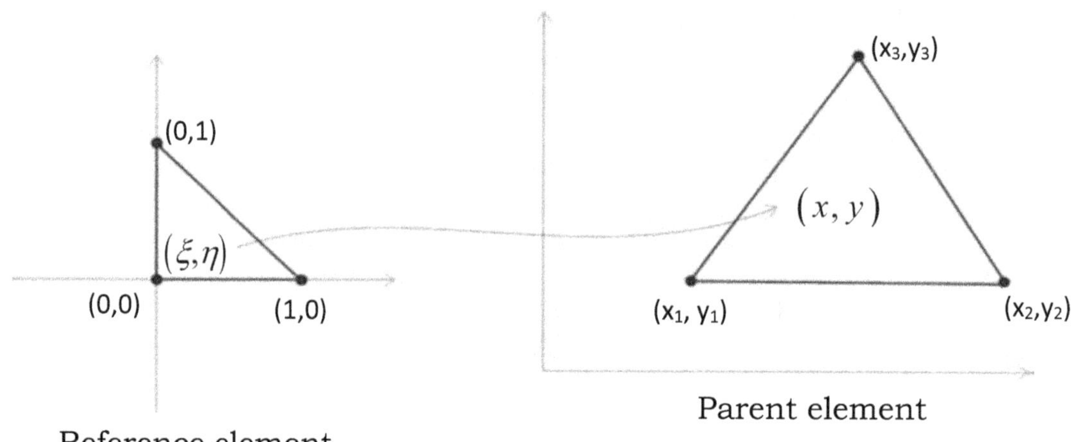

Figure 4.15 Geometrical transformation for a 3 noded triangular element.

$$x = \alpha_1 + \alpha_2\xi + \alpha_3\eta = \begin{bmatrix} 1 & \xi & \eta \end{bmatrix}\begin{Bmatrix} \alpha_1 \\ \alpha_2 \\ \alpha_3 \end{Bmatrix} \qquad \ldots\ldots(4.35)$$

Substituting the coordinates of the reference element, we obtain

$$\begin{Bmatrix} x_1 \\ x_2 \\ x_3 \end{Bmatrix} = \begin{bmatrix} 1 & 0 & 0 \\ 1 & 1 & 0 \\ 1 & 1 & 1 \end{bmatrix}\begin{Bmatrix} \alpha_1 \\ \alpha_2 \\ \alpha_3 \end{Bmatrix} \Rightarrow \begin{Bmatrix} \alpha_1 \\ \alpha_2 \\ \alpha_3 \end{Bmatrix} = \begin{bmatrix} 1 & 0 & 0 \\ -1 & 1 & 0 \\ -1 & 0 & 1 \end{bmatrix}\begin{Bmatrix} x_1 \\ x_2 \\ x_3 \end{Bmatrix} \qquad \ldots\ldots(4.36)$$

Substituting in Eq. 4.35, we obtain

$$x = \begin{bmatrix} 1 & \xi & \eta \end{bmatrix}\begin{bmatrix} 1 & 0 & 0 \\ -1 & 1 & 0 \\ -1 & 0 & 1 \end{bmatrix}\begin{Bmatrix} x_1 \\ x_2 \\ x_3 \end{Bmatrix} \qquad \ldots\ldots(4.37)$$

$$x = \tau_1(\xi,\eta)x_1 + \tau_2(\xi,\eta)x_2 + \tau_3(\xi,\eta)x_3$$
$$y = \tau_1(\xi,\eta)y_1 + \tau_2(\xi,\eta)y_2 + \tau_3(\xi,\eta)y_3$$

$$.....(4.38)$$

where

$$\tau_1(\xi,\eta) = 1 - \xi - \eta$$
$$\tau_2(\xi,\eta) = \xi$$
$$\tau_3(\xi,\eta) = \eta$$

$$.....(4.39)$$

Partial derivatives of x and y are obtained from Eq. 4.38 as follows:

$$\frac{\partial x}{\partial \xi} = \sum_{i=1}^{3} \frac{\partial \tau_i}{\partial \xi} x_i \qquad \frac{\partial x}{\partial \eta} = \sum_{i=1}^{3} \frac{\partial \tau_i}{\partial \eta} x_i$$
$$\frac{\partial y}{\partial \xi} = \sum_{i=1}^{3} \frac{\partial \tau_i}{\partial \xi} y_i \qquad \frac{\partial y}{\partial \eta} = \sum_{i=1}^{3} \frac{\partial \tau_i}{\partial \eta} y_i$$

$$.....(4.40)$$

The matrix containing the above partial derivatives, known as the Jacobian is constructed as follows:

$$[J] = \begin{bmatrix} \dfrac{\partial x}{\partial \xi} & \dfrac{\partial y}{\partial \xi} \\[2mm] \dfrac{\partial x}{\partial \eta} & \dfrac{\partial y}{\partial \eta} \end{bmatrix}$$

$$.....(4.41)$$

The Jacobian can be simplified as follows:

$$[J] = \begin{bmatrix} -1 & 1 & 0 \\ -1 & 0 & 1 \end{bmatrix} \begin{bmatrix} x_1 & y_1 \\ x_2 & y_2 \\ x_3 & y_3 \end{bmatrix}$$

$$.....(4.42)$$

A relationship like Eq. 4.38 can be obtained in the case of nodal displacements also as follows:

$$u = \tau_1(\xi,\eta)u_1 + \tau_2(\xi,\eta)u_2 + \tau_3(\xi,\eta)u_3$$
$$v = \tau_1(\xi,\eta)v_1 + \tau_2(\xi,\eta)v_2 + \tau_3(\xi,\eta)v_3$$

$$.....(4.43)$$

4.4.4 Interpolation Functions for a 4-node Rectangular Element

Consider a rectangular element with four nodes as shown in the Figure 4.16. The element has a width of $2a$ and a depth of $2b$ .The x and y coordinates of nodes can be expressed in natural coordinates using the following transformation:

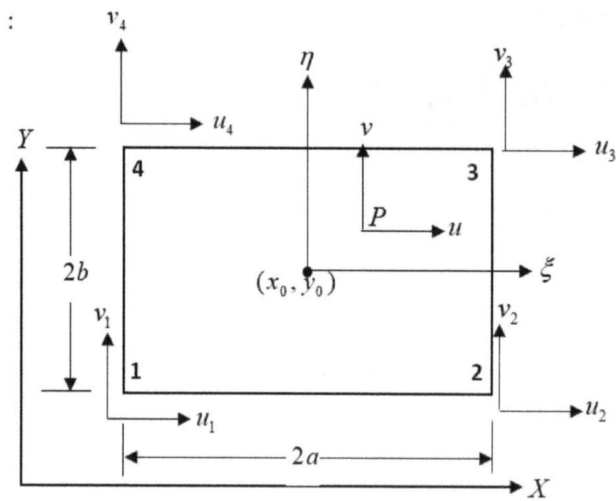

Figure 4.16 Four node rectangular element.

$$\xi = \frac{(x-x_0)}{a} \;\; ; \;\; \eta = \frac{(y-y_0)}{b}$$

.....(4.44)

Thus, the natural coordinates of nodes 1,2,3 and 4 are $(-1,-1)$, $(1,-1)$, $(1,1)$ and $(-1,1)$ respectively. The displacement u at any point within the element can be expressed as a polynomial bilinear function of these natural coordinates as follows:

$$u = a_0 + a_1\xi + a_2\eta + a_3\xi\eta = \lfloor 1 \;\; \xi \;\; \eta \;\; \xi\eta \rfloor \begin{Bmatrix} a_0 \\ a_1 \\ a_2 \\ a_3 \end{Bmatrix}$$

.....(4.45)

The nodal displacements along the X direction can be obtained by substituting the natural coordinates of the nodes in Eq. 4.44 as follows:

$$\begin{Bmatrix} u_1 \\ u_2 \\ u_3 \\ u_4 \end{Bmatrix} = \begin{bmatrix} 1 & -1 & -1 & 1 \\ 1 & 1 & -1 & -1 \\ 1 & 1 & 1 & 1 \\ 1 & -1 & 1 & -1 \end{bmatrix} \begin{Bmatrix} a_0 \\ a_1 \\ a_2 \\ a_3 \end{Bmatrix} \Rightarrow \{d_u\} = [A]\{\alpha\}$$

.....(4.46)

where $\{d_u\} = \lfloor u_1 \;\; u_2 \;\; u_3 \;\; u_4 \rfloor^T$

Eq. 4.46 can be rewritten as

$$\{\alpha\} = [A]^{-1}\{d_u\}$$

.....(4.47)

Substituting Eq. 4.47 in Eq. 4.45 leads to

$$\{u\} = \lfloor 1 \quad \xi \quad \eta \quad \xi\eta \rfloor [A]^{-1}\{d_u\} \Rightarrow \{u\} = \lfloor N_1 \quad N_2 \quad N_3 \quad N_4 \rfloor \{d_u\} = [N]\{d_u\} \quad .....(4.48)$$

Here $[N]$ represents the matrix of interpolation functions given as follows:

$$[N] = \left\lfloor \frac{(1-\xi)(1-\eta)}{4} \quad \frac{(1+\xi)(1-\eta)}{4} \quad \frac{(1+\xi)(1+\eta)}{4} \quad \frac{(1-\xi)(1+\eta)}{4} \right\rfloor \quad .....(4.49)$$

A similar approach can be followed to obtain the following:

$$v = \lfloor N_1 \quad N_2 \quad N_3 \quad N_4 \rfloor \{d_v\} \quad .....(4.50)$$

where

$$\{d_v\} = \lfloor v_1 \quad v_2 \quad v_3 \quad v_4 \rfloor^T \quad .....(4.51)$$

Eq. 4.48 and Eq. 4.51 can be combined to obtain

$$\begin{Bmatrix} u \\ v \end{Bmatrix} = \begin{bmatrix} N_1 & 0 & N_2 & 0 & N_3 & 0 & N_4 & 0 \\ 0 & N_1 & 0 & N_2 & 0 & N_3 & 0 & N_4 \end{bmatrix} \begin{Bmatrix} \{d_u\} \\ \{d_v\} \end{Bmatrix} = [N]\{d\} \quad .....(4.52)$$

The shape functions N_1, N_2, N_3 and N_4 can be sketched as shown in Figure 4.17:

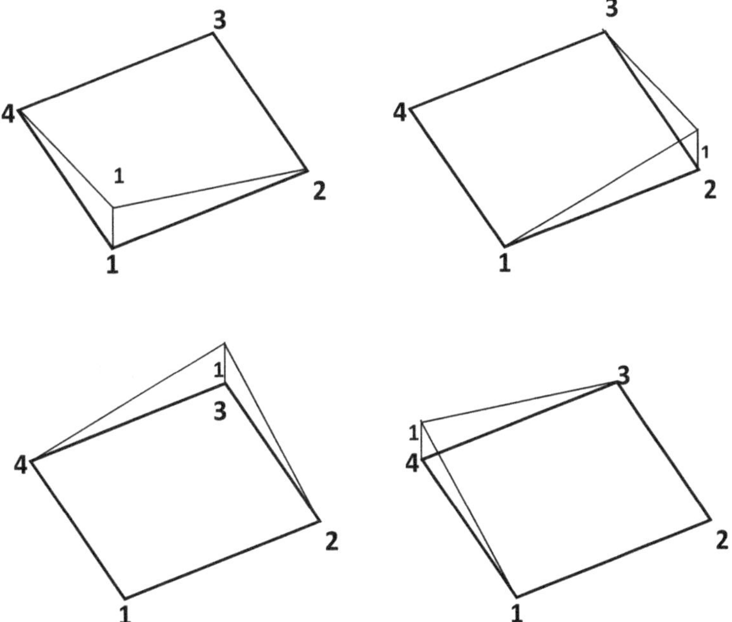

Figure 4.17 Interpolation functions for a four-node rectangular element.

4.4.5 Interpolation Functions for a 6-Node Triangular Element

Consider a triangular element with six nodes as shown in the Figure 4.18. Nodes 1,2 and 3 are corner node while nodes 4,5 and 6 are mid-side nodes.

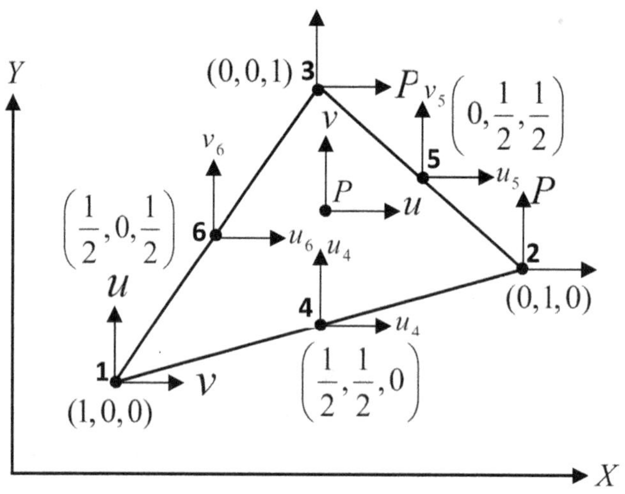

Figure 4.18 Six node triangular element.

The displacement u at any point P inside the element can be represented in terms of the nodal displacements in terms of natural coordinates as follows:

$$u = a_1\xi_1^2 + a_2\xi_2^2 + a_3\xi_3^2 + a_4\xi_1\xi_2 + a_5\xi_2\xi_3 + a_6\xi_3\xi_1 \qquad ....(4.53)$$

This can be written in the matrix form as follows

$$u = \lfloor \xi_1^2 \quad \xi_2^2 \quad \xi_3^2 \quad \xi_1\xi_2 \quad \xi_2\xi_3 \quad \xi_3\xi_1 \rfloor \{\alpha\} \qquad ......(4.54)$$

where $\{\alpha\} = \lfloor a_1 \quad a_2 \quad a_3 \quad a_4 \quad a_5 \quad a_6 \rfloor^T$. Using the nodal coordinates at each node i, $i = 1:6$ in Eq. 4.54, we obtain

$$\begin{Bmatrix} u_1 \\ u_2 \\ u_3 \\ u_4 \\ u_5 \\ u_6 \end{Bmatrix} = \begin{bmatrix} 1 & 0 & 0 & 0 & 0 & 0 \\ 0 & 1 & 0 & 0 & 0 & 0 \\ 0 & 0 & 1 & 0 & 0 & 0 \\ \dfrac{1}{4} & \dfrac{1}{4} & 0 & \dfrac{1}{4} & 0 & 0 \\ 0 & \dfrac{1}{4} & \dfrac{1}{4} & 0 & \dfrac{1}{4} & 0 \\ \dfrac{1}{4} & 0 & \dfrac{1}{4} & 0 & 0 & \dfrac{1}{4} \end{bmatrix} \{\alpha\} \qquad (4.55)$$

Eq.4.55 can be written as

$$\{d_u\} = [A]\{\alpha\} \qquad\qquad .....(4.56)$$

Eq. 4.56 can be inverted as $\{\alpha\} = [A]^{-1}\{d_u\}$ and substituted in Eq. 4.54 to obtain

$$u = \lfloor \xi_1^2 \quad \xi_2^2 \quad \xi_3^2 \quad \xi_1\xi_2 \quad \xi_2\xi_3 \quad \xi_3\xi_1 \rfloor [A]^{-1}\{d\} = [N]\{d_u\} \qquad\qquad .....(4.57)$$

where

$$N_1 = \xi_1^2 - \xi_1\xi_2 - \xi_3\xi_1 \ ; \ N_2 = \xi_2^2 - \xi_1\xi_2 - \xi_2\xi_3 \ ; \ N_3 = \xi_3^2 - \xi_2\xi_3 - \xi_3\xi_1$$
$$N_4 = 4\xi_1\xi_2 \ ; \ N_5 = 4\xi_2\xi_3 \ ; N_6 = 4\xi_3\xi_1 \qquad\qquad .....(4.58)$$

A relation like Eq. 4.57 can be obtained for the vertical displacement v as

$$v = [N]\{d_u\} \qquad\qquad .....(4.59)$$

Eq. 4.57 and Eq. 4.59 can be combined as follows:

$$\begin{Bmatrix} u \\ v \end{Bmatrix} = \begin{bmatrix} N_1 & N_2 & N_3 & N_4 & N_5 & N_6 & 0 & 0 & 0 & 0 & 0 & 0 \\ 0 & 0 & 0 & 0 & 0 & 0 & N_1 & N_2 & N_3 & N_4 & N_5 & N_6 \end{bmatrix} \begin{Bmatrix} d_u \\ d_v \end{Bmatrix} .....(4.60)$$

Eq. 4.60 can be written as

$$\{u\} = [N]\{d\} \qquad\qquad .....(4.61)$$

Shape functions can also be constructed using natural coordinates as follows:

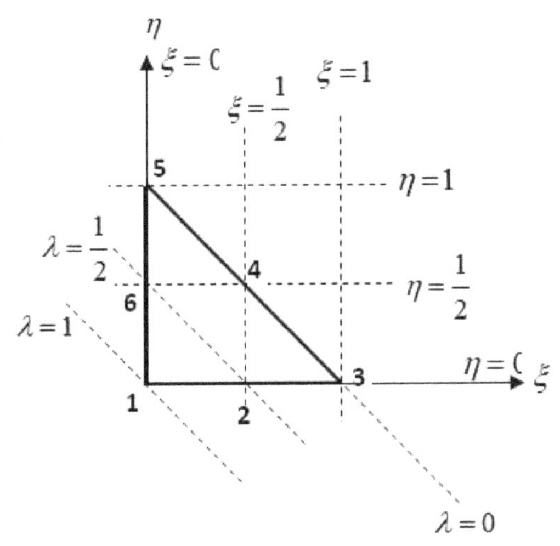

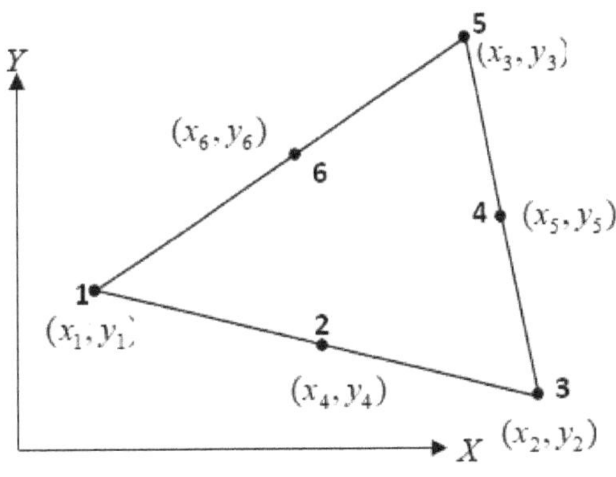

Equation of line passing the side 3-5 is $\lambda = 1 - \xi - \eta = 0$. A line parallel to side 3-5 and passing through node 1 has the equation $\lambda = 1$ while a line parallel to side 3-5 and passing through nodes 2 and 6 has the equation $\lambda = \dfrac{1}{2}$.

Similarly, line 1-5 has the equation $\xi = 0$, a line parallel to side 1-5 and passing through nodes 3 has the equation $\xi = 1$ while a line parallel to side 1-5 and passing through nodes 2 and 4 has the equation $\xi = \dfrac{1}{2}$.

Equation of line passing the side 1-3 is $\eta = 0$, a line parallel to side 1-3 and passing through node 5 has the equation $\eta = 1$ while a line parallel to side 1-3 and passing through nodes 4 and 6 has the equation $\eta = \dfrac{1}{2}$.

The shape function N_i corresponding to node i shall follow the following rules

$$N_j = 1 \quad \text{when } j = i$$
$$= 0 \quad \text{when } j \neq i \qquad \qquad \dots\dots(4.62)$$

From Figure 4.18, it is observed that node 1 lies on $\lambda = 1$, nodes 2 and 6 lie on $\lambda - \dfrac{1}{2} = 0$ while nodes 3,4,5 lie on $\lambda = 0$. Thus, the shape function for node 1 can be written as

$$N_1 = \psi \lambda \left(\lambda - \frac{1}{2} \right) \qquad \qquad \dots\dots(4.63)$$

At node 1, the value of N_1 shall be unity.

$$1 \equiv \psi \times 1 \times \left(1 - \frac{1}{2} \right) = \frac{1}{2}\psi \Rightarrow \psi = 2 \qquad \qquad \dots\dots(4.64)$$
$$N_1 = \lambda(2\lambda - 1)$$

Using a similar argument, shape functions for nodes 3 and 5 can be expressed as follows:

$$N_3 = \xi(2\xi - 1)$$
$$N_5 = \eta(2\eta - 1) \qquad \qquad \dots\dots(4.65)$$

Referring once again to Figure 4.18,

$$N_4 = \psi \xi \lambda \qquad \qquad \dots\dots(4.66)$$

At node 2, $\xi = \dfrac{1}{2}$ and $\lambda = \dfrac{1}{2}$. Substituting these values in Eq.4.66, we obtain

$$1 = \psi\left(\frac{1}{2}\right)\left(\frac{1}{2}\right) \Rightarrow \psi = 4$$

$$N_2 = 4\xi\lambda$$

.....(4.67)

By a similar argument, we obtain

$$N_4 = 4\xi\eta$$

$$N_6 = 4\eta\lambda$$

.....(4.68)

Thus, the shape functions for the 6-node triangular element using the above approach can be stated as follows as shown in Figure 4.19.

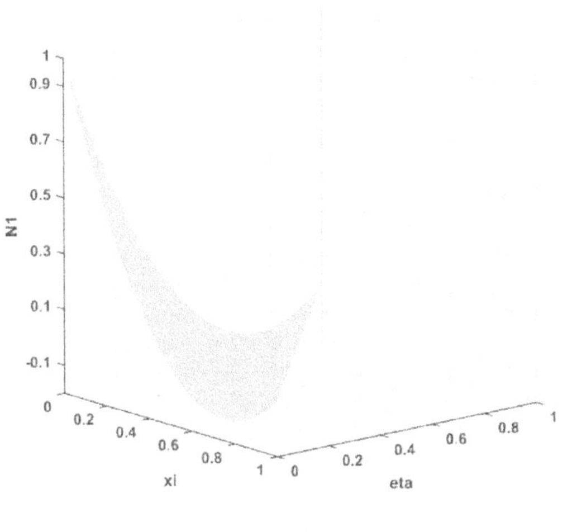

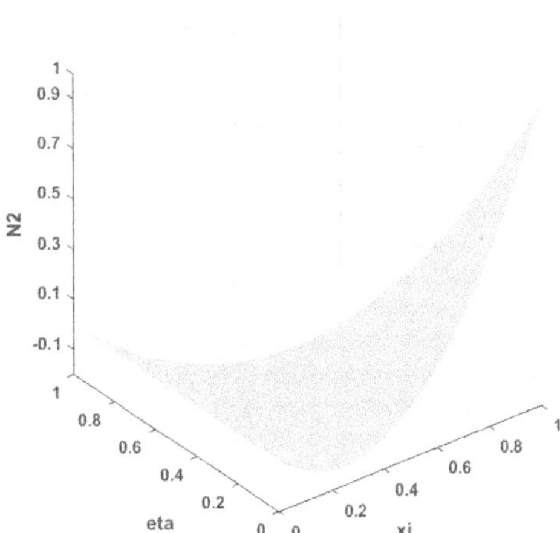

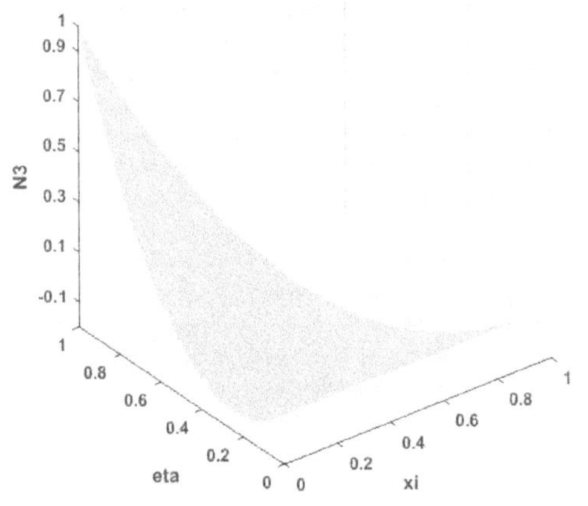

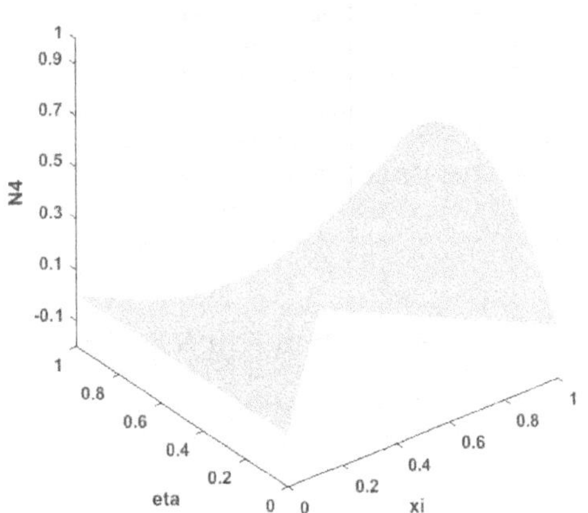

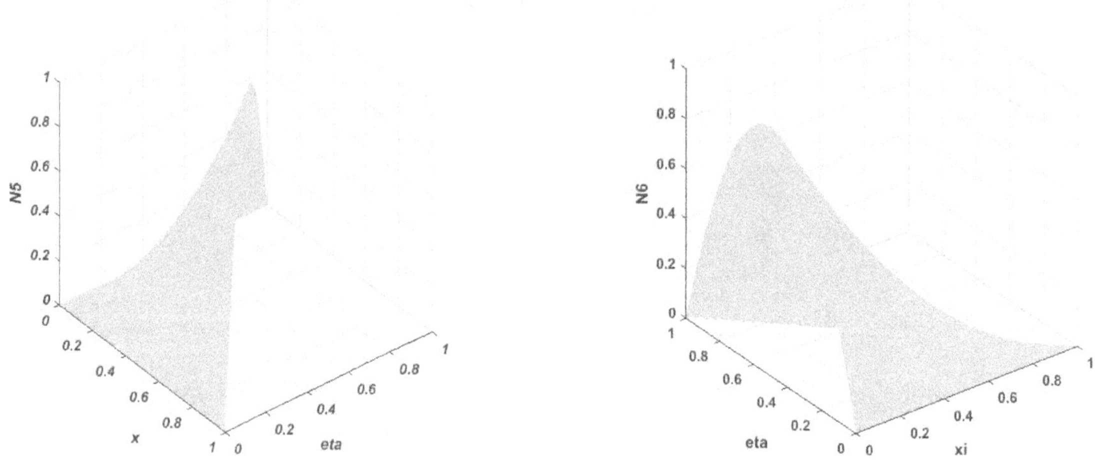

Figure 4.19 Shape functions of six node triangular element.

$$N_1 = \lambda(2\lambda - 1)\text{ where }\lambda = 1 - \xi - \eta$$
$$N_2 = 4\xi\lambda$$
$$N_3 = \xi(2\xi - 1)$$
$$N_4 = 4\xi\eta$$
$$N_5 = \eta(2\eta - 1)$$
$$N_6 = 4\eta\lambda$$

$$\text{.....}(4.69)$$

4.4.6 Interpolation Functions for a 4-node Tetrahedral Element

Consider the 4-node tetrahedral element shown in Figure 4.20. The displacements at each node along X, Y and Z directions are shown in Figure 4.20. The displacements at a point P inside the element along X, Y and Z directions are given as u, v and w respectively. The present task is the express these displacements in terms of the nodal displacements. This is done by choosing the natural coordinates defined by Eq. 4.19 as interpolation functions as follows:

$$u = u_1 N_1 + u_2 N_2 + u_3 N_3 + u_4 N_4$$
$$v = v_1 N_1 + v_2 N_2 + v_3 N_3 + v_4 N_4$$
$$w = w_1 N_1 + w_2 N_2 + w_3 N_3 + w_4 N_4$$

$$\text{.....}(4.70)$$

where
$$N_i = \frac{1}{6V}\left(a_i + b_i x + c_i y + d_i z\right), i = 1,4 \qquad \ldots\ldots(4.71)$$

also,
$$N_1 + N_2 + N_3 + N_4 = 1 \qquad \ldots\ldots(4.72)$$

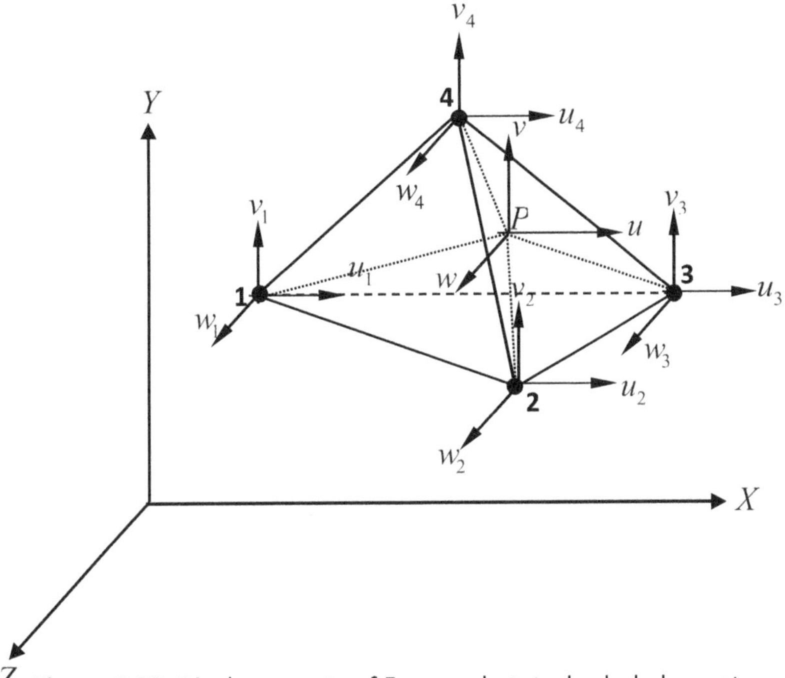

Figure 4.20 Displacements of Four node tetrahedral element.

4.5   LAGRANGIAN INTERPOLATION FUNCTIONS

Consider a line element oriented along ξ direction with N nodes as shown in the Figure 4.21:

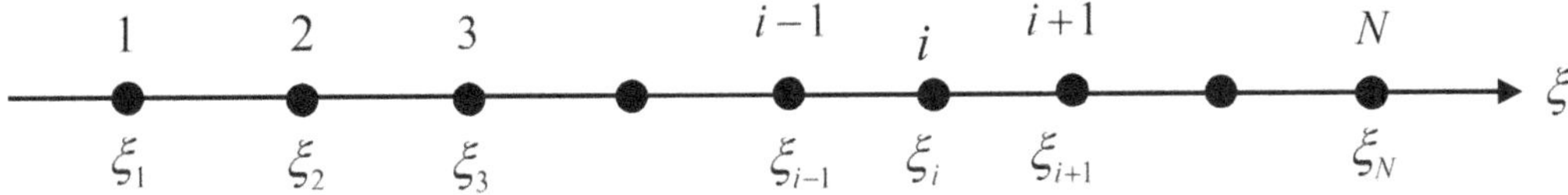

Figure 4.21 One dimensional $N-$node line element.

Lagrangian interpolation function $f(\xi_i)$ at node i can be defined as

$$f(\xi_i) = \frac{(\xi - \xi_1)(\xi - \xi_2)(\xi - \xi_3)\ldots(\xi - \xi_{i-1})(\xi - \xi_{i+1})\ldots(\xi - \xi_N)}{(\xi_i - \xi_1)(\xi_i - \xi_2)(\xi_i - \xi_3)\ldots(\xi_i - \xi_{i-1})(\xi_i - \xi_{i+1})\ldots(\xi_i - \xi_N)} \qquad \ldots\ldots(4.73)$$

From Eq. 4.73, it is observed that the function $f(\xi_i)$ contains product of $(N-1)$ terms each in numerator and denominator and excludes the term $\left(\xi - \xi_i\right)$ in the numerator. The terms in the denominator are then obtained by substituting $\xi = \xi_i$ in each corresponding term of the numerator. Thus, the function $f(\xi_i)$ has the order of $(N-1)$ in ξ.

In short form, $f(\xi_i)$ can be expressed as follows:

$$f_i(\xi_j) = \prod_{j=1}^{j=N, j \neq i} \frac{\left(\xi - \xi_j\right)}{\left(\xi_i - \xi_j\right)} \qquad .....(4.74)$$

Eq. 4.74 satisfies the property

$$\begin{aligned} f_i(\xi_j) &= 1 \ \ if \ \ i=j \\ &= 0 \ if \ \ i \neq j \end{aligned} \qquad (4.75)$$

For a one-dimensional element with N nodes, interpolation functions can be defined as

$$N_i \equiv f_i \qquad .....(4.76)$$

Lagrangian interpolation functions are derived for some elements in the succeeding sub-sections.

4.5.1 Two Node One Dimensional Element

Consider a two-node one-dimensional line element discussed previously in section 4.4.1, as shown in Figure 4.22.

4.22 Two-node one dimensional line element.

The cartesian coordinates corresponding to nodes 1 and 2 are x_1 and x_2 respectively. The length of the element is $L = x_2 - x_1$. These can be expressed in the natural coordinate system by using the following transformation:

$$\xi = \frac{2\left(x - x_1\right)}{L} - 1 \qquad .....(4.77)$$

Thus, at node 1, $\xi_1 = \dfrac{2\left(x_1 - x_1\right)}{L} - 1 = -1$ and $\xi_2 = \dfrac{2\left(x_2 - x_1\right)}{L} - 1 = 1$

Using Eq. 4.73, interpolation functions for the element can be written as follows:

$$N_1 = \frac{\left(\xi - \xi_2\right)}{\left(\xi_1 - \xi_2\right)} = \frac{\left(\xi - 1\right)}{\left(-1 - (1)\right)} = \frac{1}{2}(1 - \xi)$$

$$N_1 = \frac{\left(\xi - \xi_1\right)}{\left(\xi_2 - \xi_{1'}\right)} = \frac{\left(\xi - (-1)\right)}{\left(1 - (-1)\right)} = \frac{1}{2}(1 + \xi)$$

....(4.78)

The interpolation functions are linear and have the same shape as depicted in Figure 4.9.

4.5.2   Three Node One Dimensional Element

Consider a three-node one-dimensional line element as shown in Figure 4.23.

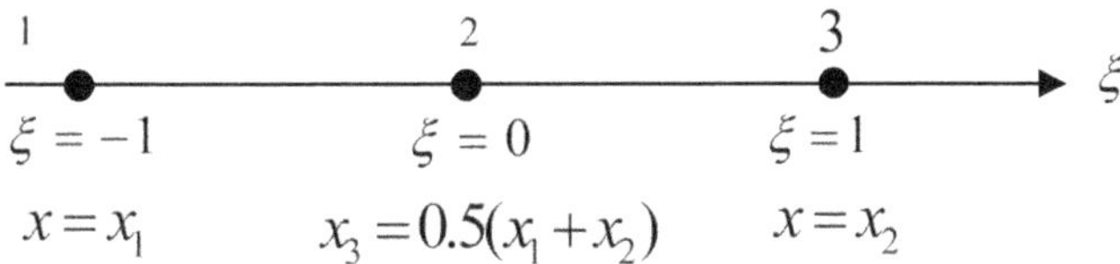

Figure 4.23 Three-node one dimensional line element.

The cartesian coordinates corresponding to nodes 1,2 and 3 are x_1, x_2 and x_3 respectively. The length of the element is $L = x_2 - x_1$. Thus, at node 1, $\xi = -1$; at node 2, $\xi = 0$; at node 3, $\xi = 1$. Using Eq. 4.73, interpolation functions for the element can be written as follows:

$$N_1 = \frac{\left(\xi - \xi_2\right)\left(\xi - \xi_3\right)}{\left(\xi_1 - \xi_2\right)\left(\xi_1 - \xi_3\right)} = \frac{(\xi - 1)(\xi - 0)}{(-1 - (1))(-1 - 0)} = \frac{1}{2}\xi(\xi - 1)$$

$$N_2 = \frac{\left(\xi - \xi_1\right)\left(\xi - \xi_3\right)}{\left(\xi_2 - \xi_{1'}\right)\left(\xi_2 - \xi_3\right)} = \frac{(\xi - (-1))(\xi - 1)}{(0 - (-1))(0 - 1)} = (1 - \xi^2)$$

.....(4.79)

$$N_3 = \frac{\left(\xi - \xi_1\right)\left(\xi - \xi_2\right)}{\left(\xi_3 - \xi_{1'}\right)\left(\xi_3 - \xi_2\right)} = \frac{(\xi - (-1))(\xi - 0)}{(1 - (-1))(1 - 0)} = \frac{1}{2}\xi(\xi + 1)$$

These interpolation functions are parabolic in shape and are plotted as shown in Figure 4.24:

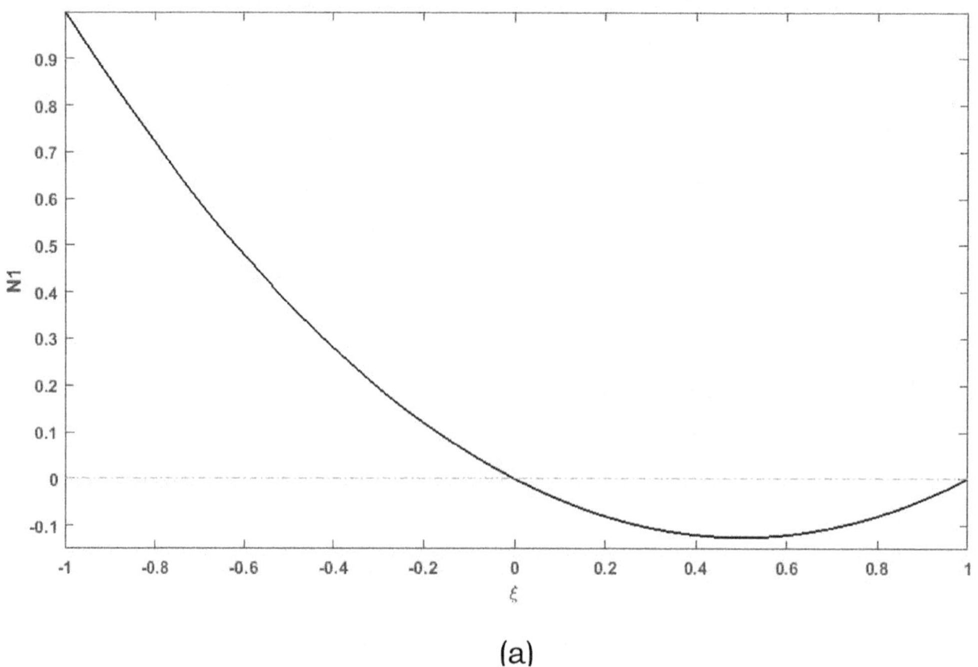

(a)

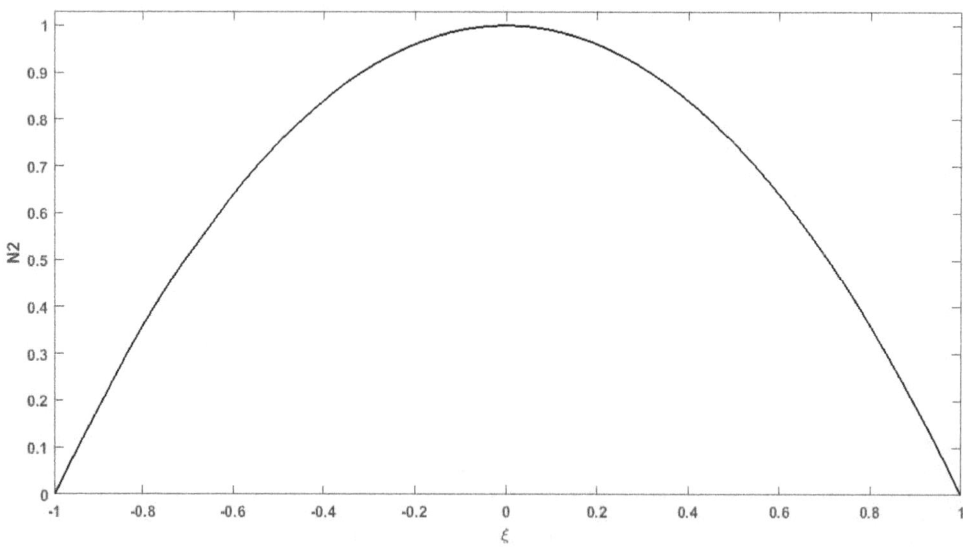

(b)

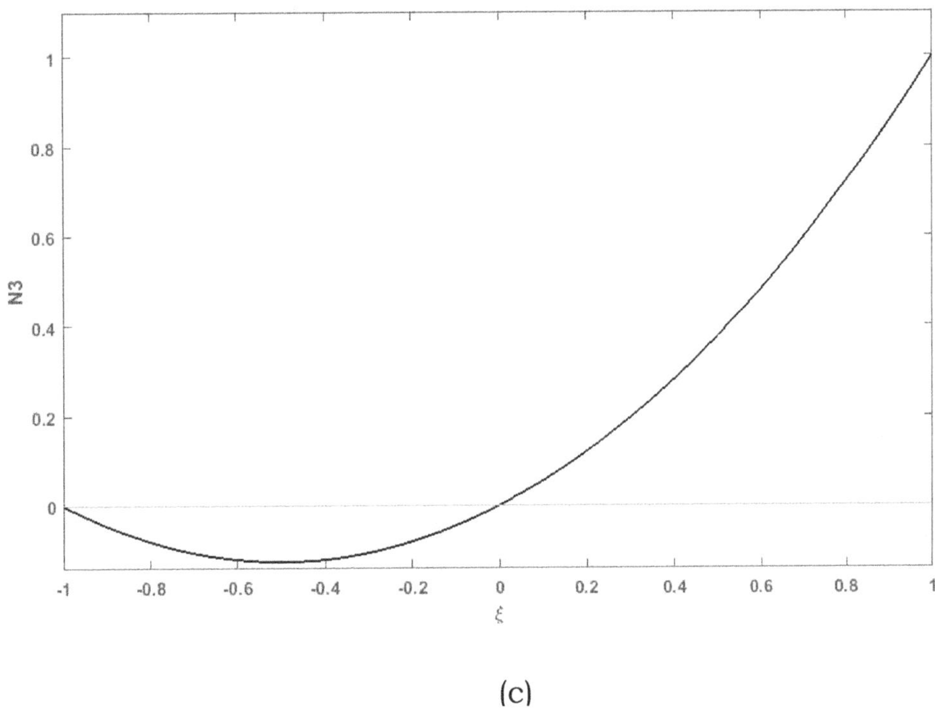

(c)

Figure 4.24 Shape functions for three node line element.

4.5.3   Four Node Rectangular Element

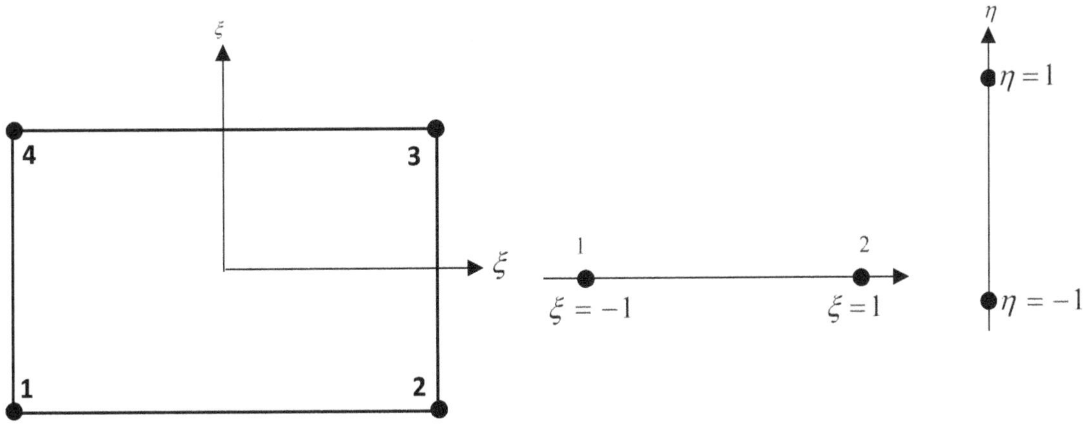

Figure 4.25 Interpolation functions of Four node rectangular element.

Consider the four-node rectangular element given in Figure 4.25. The shape functions for the element are given by

$$N_1(\xi,\eta) = f_1(\xi)f_1(\eta) = \frac{1}{4}(1-\xi)(1-\eta) \ ; \ N_2(\xi,\eta) = f_2(\xi)f_1(\eta) = \frac{1}{4}(1+\xi)(1-\eta)$$

$$N_3(\xi,\eta) = f_2(\xi)f_2(\eta) = \frac{1}{4}(1+\xi)(1+\eta) \ ; \ N_4(\xi,\eta) = f_1(\xi)f_2(\eta) = \frac{1}{4}(1-\xi)(1+\eta)$$

$$\dots\dots(4.80)$$

Sketches for the above functions are the same as those shown in Figure 4.16.

4.5.4 Nine Node Two Dimensional Lagrangian Element

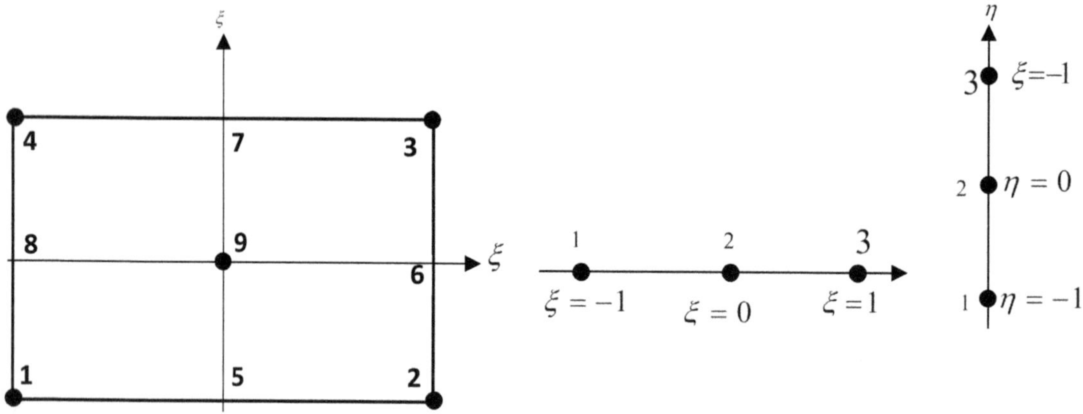

Figure 4.26 Interpolation functions of Nine-node rectangular element.

Consider the nine node two dimensional Lagrangian element given in Figure 4.26. The shape functions for the element can be computed using Eq. 4.79 as follows:

$$N_1(\xi,\eta) = f_1(\xi)f_1(\eta) = \frac{1}{2}\xi(\xi-1) \times \frac{1}{2}\eta(\eta-1) = \frac{1}{4}\xi\eta(\xi-1)(\eta-1)$$

$$N_2(\xi,\eta) = f_3(\xi)f_1(\eta) = \frac{1}{2}\xi(\xi+1) \times \frac{1}{2}\eta(\eta-1) = \frac{1}{4}\xi\eta(\xi+1)(\eta-1)$$

$$N_3(\xi,\eta) = f_3(\xi)f_3(\eta) = \frac{1}{2}\xi(\xi+1) \times \frac{1}{2}\eta(\eta+1) = \frac{1}{4}\xi\eta(\xi+1)(\eta+1)$$

$$N_4(\xi,\eta) = f_1(\xi)f_3(\eta) = \frac{1}{2}\xi(\xi-1) \times \frac{1}{2}\eta(\eta+1) = \frac{1}{4}\xi\eta(\xi-1)(\eta+1)$$

$$N_5(\xi,\eta) = f_2(\xi)f_1(\eta) = \left(1-\xi^2\right) \times \frac{1}{2}\eta(\eta-1) = \frac{1}{2}\eta\left(1-\xi^2\right)(\eta-1)$$

$$N_6(\xi,\eta) = f_3(\xi)f_2(\eta) = \frac{1}{2}\xi(\xi+1)\left(1-\eta^2\right)$$

$$N_7(\xi,\eta) = f_2(\xi)f_3(\eta) = \left(1-\xi^2\right) \times \frac{1}{2}\eta(\eta+1) = \frac{1}{2}\eta\left(1-\xi^2\right)(\eta+1)$$

$$N_8(\xi,\eta) = f_1(\xi)f_3(\eta) = \frac{1}{2}\xi(\xi-1)\left(1-\eta^2\right)$$

$$N_9(\xi,\eta) = f_2(\xi)f_2(\eta) = \left(1-\xi^2\right)\left(1-\eta^2\right) \qquad\qquad \dots\dots(4.81)$$

4.6   SERENDIPITY ELEMENTS

It is observed from Figure 4.26 that higher order Lagrangian elements have internal nodes that are not connected with the adjoining elements. The presence of these nodes increases the requirement of memory at no improvement in accuracy of solution. Being internal nodes, it is not possible to eliminate the contribution of these nodes using static condensation. This is a serious limitation on the efficiency of these elements. However, it is possible to develop elements with no internal nodes. Such elements are named as *serendipity elements.*

Serendipity refers "to the occurrence and development of events by chance in a happy or beneficial way". In the mid-1700s, English author Horace Walpole stumbled upon an interesting titbit of information while researching a coat of arms. He wrote: " I once read a silly fairy tale, called 'The Three Princes of Serendip': as their highnesses travelled, they were always making discoveries, by accidents and sagacity, of things they were not in quest of....".

The interpolation functions for serendipity elements cannot be obtained using tensor product of Lagrangian interpolation functions. Thus, an alternate approach is followed for obtaining the interpolation functions for serendipity elements. Figure 4.27 shows the Pascal's triangle for serendipity elements. It is observed that the middle terms (of the complete polynomial used for Lagrangian interpolation functions) are left out. This makes serendipity elements less accurate than Lagrangian elements as the completeness of a polynomial is indicative of the degree of accuracy of an element.

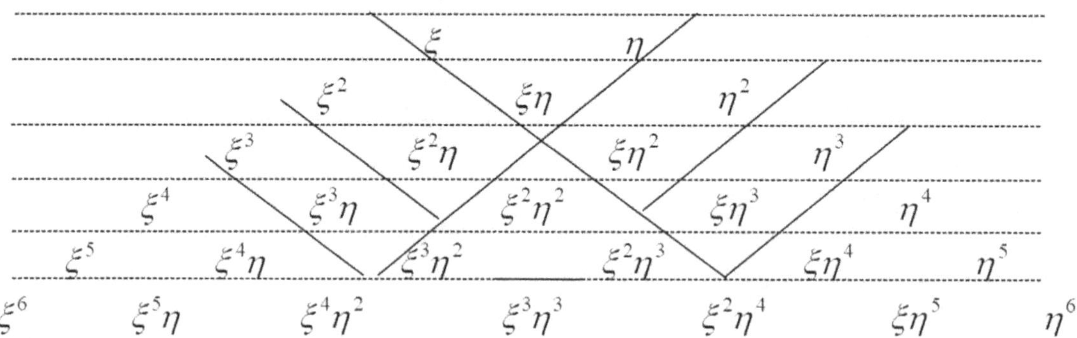

Figure 4.27 Pascal's triangle for Serendipity Elements.

Consider the 8-node two-dimensional serendipity element as shown in Figure 4.28. Nodes lying on lines shown in Figure 4.28 are given in Table 4.2.

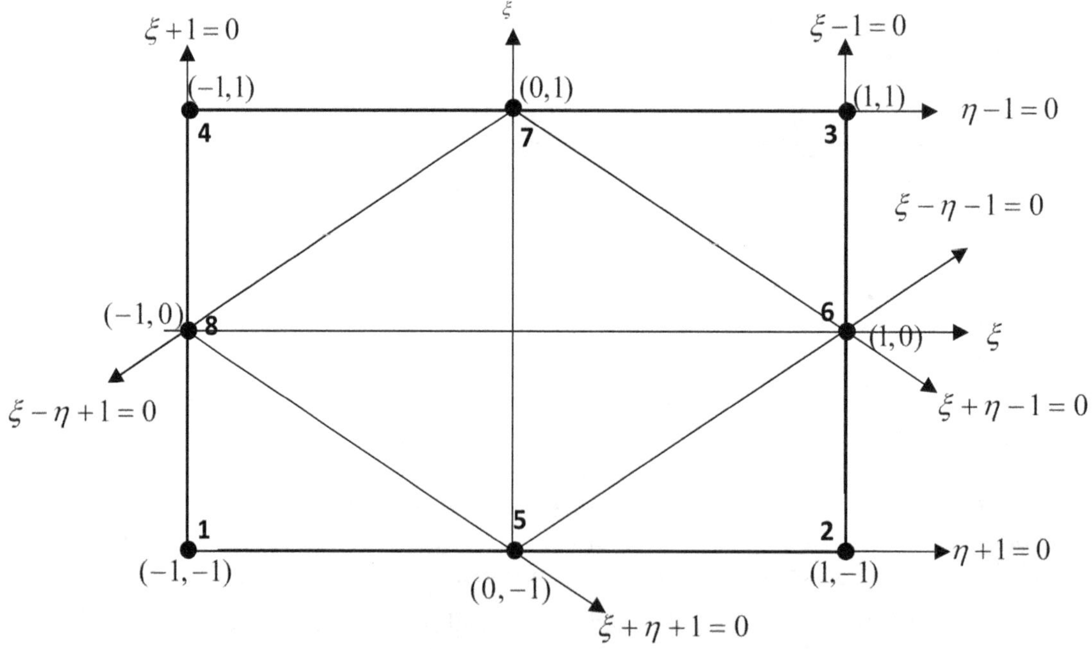

Figure 4.28 Interpolation functions of Eight node serendipity element.

Table 4.2 Nodes lying on lines.

Equation of lines	Nodes lying on the line
$(\xi-1)=0$	2,3,6
$(\xi+1)=0$	1,4,8
$(\eta-1)=0$	3,4,7
$(\eta+1)=0$	1,2,5
$\eta + 1 = 0$	5,8
$(-1,-1)$	6,7
$\xi+\eta+1=0$	5,6
$\xi-\eta+1=0$	7,8

The interpolation function for the first node is given as

$$N_1 \equiv \psi(\xi-1)(\eta-1)(\xi+\eta+1) \qquad\qquad .....(4.82)$$

satisfying the property,

$N_1 = 1$ *at node* 1

$\quad = 0$ *at remaining nodes*

It is observed from Eq.4.82 that, inclusion of expression $(\xi-1)$ makes value of N_1 zero at nodes 2,3,6 , inclusion of expression $(\eta-1)$ makes value of N_1 zero at nodes 3,4,7 while the expression $(\xi+\eta+1)$ makes value of N_1 zero at nodes 5,8. Thus the only non-zero value of N_1 is taken at node 1.

Thus, at node 1, $N_1 = 1 \Rightarrow \psi(-1-1)(-1-1)(-1-1+1) = 1 \Rightarrow \psi = -\dfrac{1}{4}$

$$N_1 \equiv -\frac{1}{4}(\xi-1)(\eta-1)(\xi+\eta+1)$$

On similar lines, interpolation functions for all nodes can be given as

$$N_1 \equiv \frac{1}{4}(\xi-1)(\eta-1)(\xi+\eta+1) \qquad N_2 \equiv -\frac{1}{4}(\xi+1)(\eta-1)(\xi-\eta-1)$$

$$N_3 \equiv \frac{1}{4}(\xi+1)(\eta+1)(\xi+\eta-1) \qquad N_4 \equiv -\frac{1}{4}(\xi-1)(\eta+1)(\xi-\eta+1)$$

$$N_5 \equiv \frac{1}{2}(\xi^2-1)(\eta-1) \qquad\qquad N_6 \equiv -\frac{1}{2}(\xi+1)(\eta^2-1) \qquad\qquad \dots\dots(4.83)$$

$$N_7 \equiv -\frac{1}{2}(\xi^2-1)(\eta+1) \qquad\qquad N_8 \equiv \frac{1}{2}(\xi-1)(\eta^2-1)$$

5 Formalisation of Finite Element Method

In this chapter, we present the procedure for obtaining the governing equations of finite element model of a structural system based on the concepts of strain energy and the principle of virtual displacement and Green's theorem.

5.1 PRINCIPLE OF VIRTUAL DISPLACEMENTS

Green's theorem, given by Eq.5.1, relates a line integral around a simple closed curve **C** to a double integral over the plane region D bounded by C.

$$\iint \left[\frac{\partial \phi}{\partial x}\frac{\partial \psi}{\partial x} + \frac{\partial \phi}{\partial y}\frac{\partial \psi}{\partial y} \right] dxdy = -\iint \phi \left(\frac{\partial^2 \psi}{\partial x^2} + \frac{\partial^2 \psi}{\partial y^2} \right) dxdy + \int \phi \left(\frac{\partial \psi}{\partial x}l + \frac{\partial \psi}{\partial y}m \right) dS \quad .....(5.1)$$

As outlined in Eq. 2.5, for a two-dimensional body held in a state of equilibrium, the equilibrium conditions for the body forces and surface tractions are given as follows:

$$\frac{\partial \sigma_x}{\partial x} + \frac{\partial \tau_{xy}}{\partial y} + X_b = 0$$

$$\frac{\partial \tau_{xy}}{\partial x} + \frac{\partial \sigma_y}{\partial y} + Y_b = 0$$

$$.....(5.2)$$

$$\sigma_x l + \tau_{xy} m = X_S$$

$$\tau_{xy} l + \sigma_y m = Y_S$$

$$.....(5.3)$$

Multiplying the components of equilibrium equations given by Eq. 5.2 along x and y directions by virtual displacements δu and δv respectively, and integrating over the entire domain, we obtain

$$\iint \left[\left(\frac{\partial \sigma_x}{\partial x} + \frac{\partial \tau_{xy}}{\partial y} + X_b \right) \delta u + \left(\frac{\partial \tau_{xy}}{\partial x} + \frac{\partial \sigma_y}{\partial y} + Y_b \right) \delta v \right] dxdy = 0 \qquad \text{.....(5.4)}$$

Now we apply Green's theorem to the above equation term-by-term and examine the nature of the equation.

Considering the first term of Eq. 5.4,

$$\iint \left(\frac{\partial \sigma_x}{\partial x} \right) \delta u \, dxdy = \iint \left[\frac{\partial \sigma_x}{\partial x} \delta u + \frac{\partial \sigma_x}{\partial y} \times 0 \right] dxdy \qquad \text{.....(5.5)}$$

Taking, $\qquad \phi = \sigma_x , \quad \dfrac{\partial \psi}{\partial x} = \delta u$ and $\dfrac{\partial \psi}{\partial y} = 0 \qquad$ .....(5.6)

The above equation is in the form

$$\iint \left[\frac{\partial \phi}{\partial x} \frac{\partial \psi}{\partial x} + \frac{\partial \phi}{\partial y} \frac{\partial \psi}{\partial y} \right] dxdy \qquad \text{.....(5.7)}$$

From Eq. 5.6,

$$\frac{\partial^2 \psi}{\partial x^2} = \frac{\partial}{\partial x}(\delta u) = \delta \left(\frac{\partial u}{\partial x} \right)$$

$$\frac{\partial^2 \psi}{\partial y^2} = \frac{\partial}{\partial y} \left(\frac{\partial \psi}{\partial y} \right) = 0 \qquad \text{.....(5.8)}$$

It is observed from Eq.5.8 that the variational operator is transposed with the partial derivative operator.

Using Eq. 5.6, Eq.5.5 can be recast as

$$\iint \left(\frac{\partial \sigma_x}{\partial x} \right) \delta u \, dxdy = \iint \left[\frac{\partial \sigma_x}{\partial x} \delta u + \frac{\partial \sigma_x}{\partial y} \times 0 \right] dxdy = -\iint \sigma_x \delta \left(\frac{\partial u}{\partial x} \right) dxdy + \int \sigma_x l \delta u \, dS \text{ .....(5.9)}$$

Taking,

$$\phi = \tau_{xy} \;\; ; \;\; \frac{\partial \psi}{\partial x} = \delta v \Rightarrow \frac{\partial^2 \psi}{\partial x^2} = \frac{\partial}{\partial x}(\delta v) = \delta \left(\frac{\partial \tau_{xy}}{\partial x} \right) ; \frac{\partial \psi}{\partial y} = 0 \Rightarrow \frac{\partial^2 \psi}{\partial x^2} = 0 \qquad \text{.....(5.10)}$$

$$\iint \left(\frac{\partial \tau_{xy}}{\partial x} \right) \delta v \, dxdy = \iint \left[\frac{\partial \tau_{xy}}{\partial x} \times \delta v + \frac{\partial \tau_{xy}}{\partial y} \times 0 \right] dxdy = -\iint \tau_{xy} \delta \left(\frac{\partial v}{\partial x} \right) dxdy + \int \tau_{xy} l \delta v \, dS$$

$$\text{.....(5.11)}$$

Taking

$$\phi = \sigma_y \ ; \ \frac{\partial \psi}{\partial x} = 0 \Rightarrow \frac{\partial^2 \psi}{\partial x^2} = 0; \frac{\partial \psi}{\partial y} = \delta v \ \Rightarrow \frac{\partial^2 \psi}{\partial x^2} = \frac{\partial}{\partial y}(\delta v) = \delta\left(\frac{\partial v}{\partial y}\right) \qquad .....(5.12)$$

$$\iint\left(\frac{\partial \sigma_y}{\partial y}\right)\delta v dx dy = \iint\left[\frac{\partial \sigma_x}{\partial x}\times 0 + \frac{\partial \sigma_y}{\partial y}\times \delta v\right]dx dy = -\iint \sigma_y \delta\left(\frac{\partial v}{\partial y}\right)dx dy + \int \sigma_y m \delta v dS$$

$$.....(5.13)$$

Taking

$$\phi = \tau_{xy} \ ; \ \frac{\partial \psi}{\partial x} = 0 \Rightarrow \frac{\partial^2 \psi}{\partial x^2} = 0; \frac{\partial \psi}{\partial y} = \delta u \ \Rightarrow \frac{\partial^2 \psi}{\partial x^2} = \frac{\partial}{\partial y}(\delta u) = \delta\left(\frac{\partial u}{\partial y}\right) \qquad .....(5.14)$$

$$\iint\left(\frac{\partial \tau_{xy}}{\partial y}\right)\delta u dx dy = \iint\left[\frac{\partial \tau_{xy}}{\partial x}\times 0 + \frac{\partial \tau_{xy}}{\partial y}\times \delta u\right]dx dy = -\iint \tau_{xy} \delta\left(\frac{\partial u}{\partial y}\right)dx dy + \int \tau_{xy} m \delta u dS$$

$$.....(5.15)$$

Combining terms from Eq. 5.9, 5.11, 5.13 and 5.15, we obtain

$$\iint\left[\left(\frac{\partial \sigma_x}{\partial x}+\frac{\partial \tau_{xy}}{\partial y}+X_b\right)\delta u + \left(\frac{\partial \tau_{xy}}{\partial x}+\frac{\partial \sigma_y}{\partial y}+Y_b\right)\delta v\right]dx dy =$$

$$-\iint \sigma_x \delta\left(\frac{\partial u}{\partial x}\right)dx dy - \iint \sigma_y \delta\left(\frac{\partial v}{\partial y}\right)dx dy - \iint \tau_{xy}\left[\delta\left(\frac{\partial v}{\partial x}\right)+\delta\left(\frac{\partial u}{\partial y}\right)\right]dx dy \qquad .....(5.16)$$

$$+\int\left(\sigma_x l + \tau_{xy} m\right)\delta u dS + \int\left(\tau_{xy} l + \sigma_y m\right)\delta v dS + \iint\left(X_b \delta u + Y_b \delta v\right)dx dy = 0$$

In the above equation,

$$\delta\left(\frac{\partial u}{\partial x}\right) = \delta\varepsilon_x$$

$$\delta\left(\frac{\partial v}{\partial y}\right) = \delta\varepsilon_y \qquad\qquad .....(5.17)$$

$$\delta\left(\frac{\partial u}{\partial y}\right)+\delta\left(\frac{\partial v}{\partial x}\right) = \delta\left(\frac{\partial u}{\partial y}+\frac{\partial v}{\partial x}\right) = \delta\gamma_{xy}$$

Substituting Eq. 5.17 in Eq. 5.16, we obtain

$$\iint \left[\left(\frac{\partial \sigma_x}{\partial x} + \frac{\partial \tau_{xy}}{\partial y} + X_b \right) \delta u + \left(\frac{\partial \tau_{xy}}{\partial x} + \frac{\partial \sigma_y}{\partial y} + Y_b \right) \delta v \right] dxdy =$$

$$-\iint \sigma_x \delta \varepsilon_x \, dxdy - \iint \sigma_y \delta \varepsilon_y \, dxdy - \iint \tau_{xy} \gamma_{xy} \, dxdy + \int (X_s \delta u + Y_s \delta v) dS \qquad .....(5.18)$$

$$+\iint (X_b \delta u + Y_b \delta v) \, dxdy = 0$$

Eq. 5.18 can be simplified as follows:

$$\iint (\sigma_x \delta \varepsilon_x + \sigma_y \delta \varepsilon_y + \tau_{xy} \delta \gamma_{xy}) dxdy = \iint (X_b \delta u + Y_b \delta v) dxdy + \int (X_s \delta u + Y_s \delta v) dS .....(5.19)$$

$$\iint \lfloor \delta \varepsilon_x \quad \delta \varepsilon_y \quad \delta \gamma_{xy} \rfloor \begin{Bmatrix} \sigma_x \\ \sigma_y \\ \tau_{xy} \end{Bmatrix} dxdy = \iint \lfloor \delta u \quad \delta v \rfloor \begin{Bmatrix} X_b \\ Y_b \end{Bmatrix} dxdy + \int \lfloor \delta u \quad \delta v \rfloor \begin{Bmatrix} X_s \\ Y_s \end{Bmatrix} dS$$

$$.....(5.20)$$

This can be written for a 3-dimensional solid as

$$\iiint \{\delta \varepsilon\}^T \{\sigma\} dV = \iiint \{\delta u\}^T \{X\} dV + \iint_{S_1} \{\delta u\}^T \{X_s\} dS \qquad .....(5.21)$$

Extracting the variational operator on both side of the equation, we obtain,

$$\delta \left(\iiint \{\varepsilon\}^T \{\sigma\} dV \right) = \delta \left(\iiint \{u\}^T \{X\} dV + \iint_{S_1} \{u\}^T \{X_s\} dS \right) \qquad .....(5.22)$$

$$\delta U = \delta W_e \qquad .....(5.23)$$

Eq. 5.23 implies that under the conditions of equilibrium, the total internal virtual energy stored inside the body is equal to the total external virtual work done on it, for every virtual displacement satisfying the kinematic boundary conditions.

5.2 STRAIN ENERGY STORED IN A DEFORMABLE BODY

The specific strain energy is given as

$$U_0 = \frac{1}{2} \{\varepsilon\}^T \{\sigma\} = \frac{1}{2} \{\varepsilon\}^T [C] \{\varepsilon\}$$

$$.....(5.24)$$

$$U = \iiint U_0 dV = \frac{1}{2} \iiint \{\varepsilon\}^T [C] \{\varepsilon\} dV \quad \text{is the total strain energy}$$

Work done by body forces and surface tractions is given as

$$W_e = -\iiint \{u^T\}\{X\}\,dV - \iint_{S_1} \{u^T\}\{X_s\}\,dS \qquad \qquad .....(5.25)$$

Total potential energy Π is defined as the sum of external work done and the internal strain energy stored. Thus,

$$\Pi = U + W_e = \frac{1}{2}\iiint \{\varepsilon\}^T [C]\{\varepsilon\}\,dV - \iiint \{u^T\}\{X\}\,dV - \iint_{S_1} \{u^T\}\{X_s\}\,dS \qquad ......(5.26)$$

Taking the first variation of total potential energy, we obtain

$$\delta\Pi = \iiint \{\delta\varepsilon\}^T \{\sigma\}\,dV - \iiint \{\delta u^T\}\{X\}\,dV - \iint_{S_1} \{\delta u^T\}\{X_s\}\,dS \qquad .....(5.27)$$

Substituting Eq. 5.23 in Eq. 5.27, we obtain

$$\delta\Pi = \delta U - \delta W_e = 0 \qquad \qquad .....(5.28)$$

Eq. 5.28 gives the *principle of stationary potential energy* which states that a system is in equilibrium if its potential energy is stationary viz, its first variation is zero. Thus, a system is in equilibrium if $\delta\Pi = 0$. Substituting Eq.5.28 in Eq. 5.27, we get

$$\iiint \{\delta\varepsilon\}^T \{\sigma\}\,dV = \iiint \{\delta u^T\}\{X\}\,dV + \iint_{S_1} \{\delta u^T\}\{X_s\}\,dS \qquad .....(5.29)$$

5.3   OBTAINING STIFFNESS MATRIX AND FORCE VECTOR

Eq. 5.29 is simplified by using the following relations

$$\{\varepsilon\} = [B]\{d\} \Rightarrow \{\delta\varepsilon\} = [B]\{\delta d\} \Rightarrow \{\delta\varepsilon\}^T = \{\delta d\}^T [B]^T$$

$$\{\sigma\} = [C]\{\varepsilon\} = [C][B]\{d\} \qquad \qquad .....(5.30)$$

$$\{u\} = [N]\{d\} \Rightarrow \{\delta u\} = [N]\{\delta d\} \Rightarrow \{\delta u\}^T = \{\delta d\}^T [N]^T$$

Substituting Eq.5.30 in Eq.5.29, we obtain

$$\iiint \{\delta d\}^T [B]^T [C][B]\{d\}\,dV - \iiint \{\delta d\}^T [N]^T \{X\}\,dV - \iint_{S_1} \{\delta d\}^T [N_s]^T \{X_s\}\,dS = 0$$

$$.....(5.31)$$

$$\{\delta d\}^T \left(\iiint [B]^T [C][B]\{d\}\, dV - \iiint [N]^T \{X\}\, dV - \iint_{S_1} [N_S]^T \{X_S\}\, dS \right) = 0 \quad .....(5.32)$$

$$\left(\iiint [B]^T [C][B]\, dV \right)\{d\} = \iiint [N]^T \{X\}\, dV + \iint_{S_1} [N_S]^T \{X_S\}\, dS \qquad .....(5.33)$$

Eq. 5.33 can be represented as

$$\left[K^{(e)} \right]\{d^{(e)}\} = \{Q^{(e)}\} \qquad\qquad .....(5.34)$$

In the above equation,

$$\left[K^{(e)} \right] = \iiint [B]^T [C][B]\, dV$$

$$\{Q^{(e)}\} = \iiint [N]^T \{X\}\, dV + \iint_{S_1} [N_S]^T \{X_S\}\, dS \qquad (5.35)$$

represent the element stiffness matrix and element force vector respectively. The first and second terms of $\{Q^{(e)}\}$ represent contribution due to the body forces and surface tractions respectively.

5.4   STEPS INVOLVED IN FINITE ELEMENT ANALYSIS

The following are the three major steps in any finite element software:

1. Pre-processing
2. Processing
3. Post-processing

These steps are briefly explained in the following sub-sections:

5.4.1   Pre-Processing

Pre-processing involves the definition of the following:

a) geometric domain of the problem.
b) element types to be used.
c) material properties of element.
d) geometric properties of elements.
e) element connectivity i.e., meshing the model.
f) develop equations for an element.
g) assemble the elements to present the entire problem.
h) construct the global stiffness matrix and apply boundary conditions, initial conditions, and loading.

Discretization (meshing) is the first step in solving a problem in Finite Element Analysis. Here the entire domain is discretized into finite number of divisions often called elements that are connected to each other by nodes. These elements form the building blocks on which the boundary conditions and external effects are specified later.

The number and type of elements chosen must be such that the variable distribution through the whole body is adequately approximated by the combined elemental representations. For example, if the mesh is too coarse, the resolution of the parametric distribution may be inadequate, whereas too fine a mesh is wasteful of computing time and possibly the user's time, and in some cases, won't even solve anyway. Part of the skill will be in designing and refining meshes in areas of high interest.

Next step is the selection of the interpolation functions (shape functions) to represent the physical behaviour of an element. A shape function is an approximate continuous function that is assumed to represent the solution of an element. Its purpose is to provide an approximation of the unknown solution within an element.

Often in commercially available FEA software, meshing techniques are used to create the model, generate an appropriate finite element grid, apply the appropriate boundary conditions, and view the total model.

5.4.2   Processing – Solution of Linear Matrix Equations

Processing involves the solution of the governing system of equations in matrix form and computes the unknown values of the primary variables. The results at nodes are typically displacement values at different nodes or temperature values at different nodes in a heat transfer problem.

5.4.3   Post-Processing of Solution

Post-processing involves computation of secondary variables such as forces, moments, stresses, strains etc., from the computed values of primary unknowns computed in the processing phase. Post-processing employs various sorting, printing, and plotting methods to show the output in the most intuitive way possible. It seeks to analyse, arrange, and illustrate output variables in the best way possible so that they can be interpreted effectively.

Post-processing also provides visualization of the computed results. In this phase, any other important information viz. values of principal stresses, heat fluxes, strains, etc. is obtained.

6 One Dimensional Elements

6.1 BAR ELEMENT

Consider the bar element as shown in the Figure 6.1. The length, cross sectional area and Young's modulus of the element are L, A and E respectively. The element has two nodes, located at the left and right ends. The axial stiffness of the element is $\dfrac{AE}{L}$.

The longitudinal axis of the element is oriented along the x-axis. The displacements at nodes 1 and 2 are u_1 and u_2 respectively, as shown in the figure.

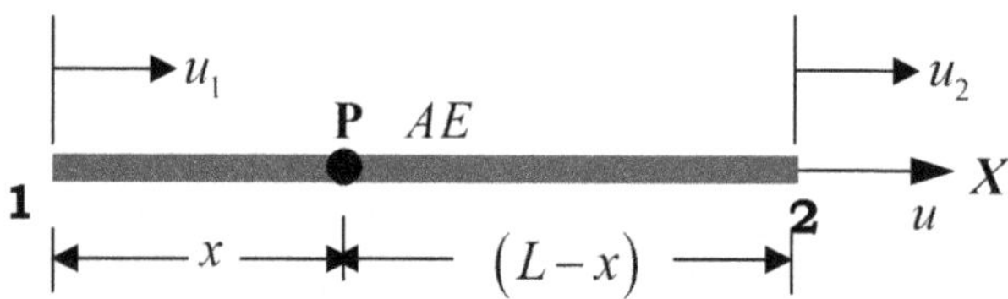

Figure 6.1 Bar Element.

The axial strain ε of the element is obtained by differentiating Eq. 4.25 as shown below:

$$\varepsilon = \frac{\partial u}{\partial x} = \left\lfloor \frac{\partial N_1}{\partial x} \quad \frac{\partial N_2}{\partial x} \right\rfloor \left\{ \begin{matrix} u_1 \\ u_2 \end{matrix} \right\} \qquad \qquad .....(6.1)$$

Using Eq. 4.23 and Eq. 6.1, axial strain ε of the element can be expressed as

$$\varepsilon = \frac{\partial u}{\partial x} = \left\lfloor -\frac{1}{L} \quad \frac{1}{L} \right\rfloor \left\{ \begin{matrix} u_1 \\ u_2 \end{matrix} \right\} = \lfloor B \rfloor \{d\} \qquad \qquad ....(6.2)$$

where $\lfloor B \rfloor$ is the *strain-displacement matrix* and $\{d\}$ is the *vector of nodal displacements*.

6.1.1 Stiffness Matrix for a Bar Element

The stiffness equation for the bar element can be given as

$$[K]\{d\} = \{Q\} \qquad\qquad .....(6.3)$$

$$[K] = \iiint \left[B^T\right][C][B]\,dV = A\int_0^L \left\{\begin{matrix} -\dfrac{1}{L} \\ \dfrac{1}{L} \end{matrix}\right\} E\left[-\dfrac{1}{L}\ \ \dfrac{1}{L}\right] dx = \dfrac{AE}{L}\begin{bmatrix} 1 & -1 \\ -1 & 1 \end{bmatrix} \qquad .....(6.4)$$

6.1.2 Force Vector for a Bar Element

Force vector for the bar element can be expressed as

$$\{Q\} = \iiint \left[N^T\right]\{X\}\,dV + \iint_{S_1} \left[N_S^T\right]\{f\}\,dx + \left\{\begin{matrix} Q_1 \\ Q_2 \end{matrix}\right\} \qquad ....(6.5)$$

The first term in the above equation corresponds to the *vector of body forces* and is evaluated as follows:

If γ is the unit weight (acting along the y-direction), the *vector of body forces* can be represented as $\{X_b\} = \gamma$.

$$\iiint \left[N^T\right]\{X_b\}\,dV = A\int_0^L \left\{\begin{matrix} 1 - \dfrac{x}{L} \\ \dfrac{x}{L} \end{matrix}\right\} \gamma\,dx = \dfrac{\gamma AL}{2}\left\{\begin{matrix} 1 \\ 1 \end{matrix}\right\} \qquad .....(6.6)$$

It is to be noted from the above equation that the term γAL, being the product of the unit weight and the volume of the bar is equal to the weight of the element. Thus Eq. 6.6 implies that the weight of the bar is equally distributed to both nodes. The second term in Eq. 6.5 corresponds to the vector of surface tractions and is evaluated as follows:

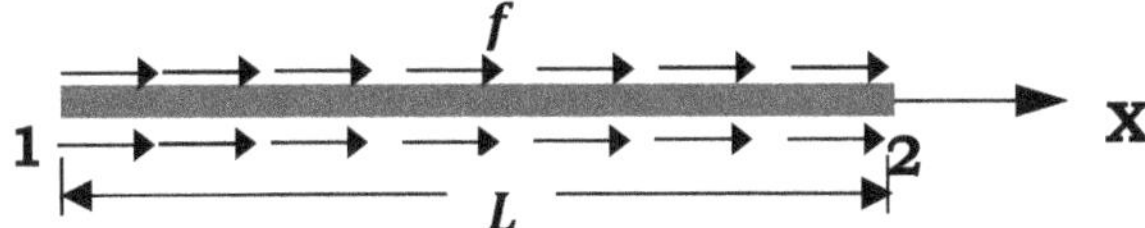

Figure 6.2 Surface traction on the bar element.

In the Figure 6.2, f represents the *surface traction per unit surface area*. The total surface area S is given by $S = pL$ where p is the perimeter of the bar. The total surface traction T is given by

$$T = fS = fpL \qquad \qquad .....(6.7)$$

Thus, the second term in Eq. 6.5 can be expressed as

$$\iint [N]^T f dS = \iint \begin{Bmatrix} 1 - \dfrac{x}{L} \\ \dfrac{x}{L} \end{Bmatrix} fp dx = \dfrac{fpL}{2} \begin{Bmatrix} 1 \\ 1 \end{Bmatrix} \qquad \qquad .....(6.8)$$

Here also it is observed that the surface traction force gets equally distributed to both nodes of the bar element. The third term in Eq. 6.5 represents vector of point loads acting directly at the nodes.

Eq. 6.3 is solved after application of boundary conditions.

Forces in an element can be computed using

$$F = AE\varepsilon = AE \lfloor B \rfloor \{d\} \qquad \qquad .....(6.9)$$

6.1.3 Numerical Examples

a) **Numerical example 6.1:** Determine the nodal displacement at nodes 2 and 3, stresses in each element and support reactions of the stepped bar-1 shown in Figure 6.3 due to an applied force of $P_1 = 250\ kN$ and $P_2 = 400\ kN$. Adopt $L_1 = 0.5\ m$, $L_2 = 0.4\ m$, $A_1 = 0.0026\ m^2$ and $A_2 = 0.0013\ m^2$. Note that $E = 2 \times 10^8\ kN/m^2$.

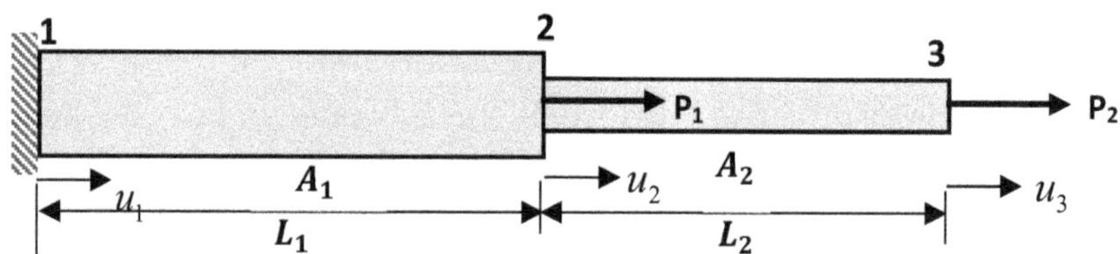

Figure 6.3 Stepped bar-1.

Solution:

Element 1: Nodes 1 and 2 of element 1 correspond to global degrees of freedom 1 and 2.

Element stiffness matrix is given by

$$\left[K_1^{(e)}\right] = \frac{A_1 E}{L_1}\begin{bmatrix} 1 & -1 \\ -1 & 1 \end{bmatrix} = \frac{0.0026 \times 2 \times 10^8}{0.5}\begin{bmatrix} 1 & -1 \\ -1 & 1 \end{bmatrix} = 10^6 \begin{bmatrix} 1.04 & -1.04 \\ -1.04 & 1.04 \end{bmatrix} \begin{matrix} 1 \\ 2 \end{matrix}$$

Element 2: Nodes 2 and 3 of element 1 correspond to global degrees of freedom 2 and 3.

Element stiffness matrix is given by

$$\left[K_2^{(e)}\right] = \frac{A_2 E}{L_2}\begin{bmatrix} 1 & -1 \\ -1 & 1 \end{bmatrix} = \frac{0.0013 \times 2 \times 10^8}{0.4}\begin{bmatrix} 1 & -1 \\ -1 & 1 \end{bmatrix} = 10^6 \begin{bmatrix} 0.65 & -0.65 \\ -0.65 & 0.65 \end{bmatrix} \begin{matrix} 2 \\ 3 \end{matrix}$$

Assembly

Global stiffness matrix is obtained by assembling the contributions made by the element stiffness matrices as follows:

$$[K] = 10^6 \begin{bmatrix} 1.04 & -1.04 & 0 \\ -1.04 & 1.04+0.65 & -0.65 \\ 0 & -0.65 & 0.65 \end{bmatrix} = 10^6 \begin{bmatrix} 1.04 & -1.04 & 0 \\ -1.04 & 1.69 & -0.65 \\ 0 & -0.65 & 0.65 \end{bmatrix} \begin{matrix} 1 \\ 2 \\ 3 \end{matrix}$$

Global force vector can be directly expressed as $\{Q\} = \begin{Bmatrix} 0 \\ 250 \\ 400 \end{Bmatrix}$

The stiffness equation for the structure can be expressed as

$$[K]\{d\} = \{Q\}$$

where $\{d\} = \begin{Bmatrix} u_1 \\ u_2 \\ u_3 \end{Bmatrix}$ is the vector of nodal displacements.

The stiffness equation for the stepped bar now can be written as:

$$10^6 \begin{bmatrix} 1.04 & -1.04 & 0 \\ -1.04 & 1.69 & -0.65 \\ 0 & -0.65 & 0.65 \end{bmatrix} \begin{Bmatrix} u_1 \\ u_2 \\ u_3 \end{Bmatrix} = \begin{Bmatrix} 0 \\ 250 \\ 400 \end{Bmatrix}$$

Boundary conditions: As the stepped bar is restrained at node 1, $u_1 = 0$.

This boundary condition can be easily incorporated by striking off the first row and first column in the stiffness matrix, and first row in the nodal displacement vector as well as the force vector. Thus, the stiffness equation for the stepped bar gets reduced to

$$10^6 \begin{bmatrix} 1.69 & -0.65 \\ -0.65 & 0.65 \end{bmatrix} \begin{Bmatrix} u_2 \\ u_3 \end{Bmatrix} = \begin{Bmatrix} 250 \\ 400 \end{Bmatrix}$$

Solving the above matrix equation leads to

$$\lfloor u_1 \ \ u_2 \ \ u_3 \rfloor = 10^{-3} \lfloor 0 \ \ 0.625 \ \ 1.2404 \rfloor \, m$$

From Eq. 6.9, axial force in elements 1 and 2 can be calculated as follows:

$$F_1 = A_1 E \varepsilon_1 = 0.0026 \times 2 \times 10^8 \left| -\frac{1}{0.5} \quad \frac{1}{0.5} \right| \begin{Bmatrix} u_1 \\ u_2 \end{Bmatrix} = 650 \ kN(Tension)$$

$$F_2 = A_2 E \varepsilon_2 = 0.0013 \times 2 \times 10^8 \left| -\frac{1}{0.4} \quad \frac{1}{0.4} \right| \begin{Bmatrix} u_2 \\ u_3 \end{Bmatrix} = 400 \ kN(Tension)$$

b) Numerical example 6.2: Determine the displacements at nodes 2 and 3, stresses in each element and support reactions of the stepped bar-2 shown in Figure 6.4 due to applied forces of $P_1 = 200 \ kN$ and $P_2 = 500 \ kN$. Adopt $L_1 = 0.5 \ m$, $L_2 = 0.4 \ m$, $L_3 = 0.6 \ m$, $A_1 = 0.002 \ m^2$, $A_2 = 0.001 \ m^2$ and $A_3 = 0.003 \ m^2$. Also adopt $E_1 = 2 \times 10^8 \ kN/m^2$, $E_2 = 0.7 \times 10^8 \ kN/m^2$ and $E_3 = 1 \times 10^8 \ kN/m^2$.

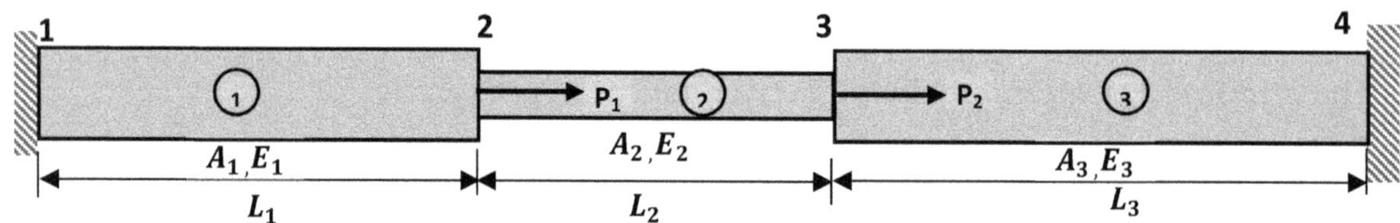

Figure 6.4 Stepped bar 2.

Element 1: Local nodes 1 and 2 of element 1 correspond to global degrees of freedom 1 and 2 respectively.

Element stiffness matrix is given by

$$\left[K_1^{(e)} \right] = \frac{A_1 E_1}{L_1} \begin{bmatrix} 1 & -1 \\ -1 & 1 \end{bmatrix} = \frac{0.002 \times 2 \times 10^8}{0.5} \begin{bmatrix} 1 & -1 \\ -1 & 1 \end{bmatrix} = 10^5 \begin{matrix} & 1 & 2 \\ \begin{bmatrix} 8 & -8 \\ -8 & 8 \end{bmatrix} & \begin{matrix} 1 \\ 2 \end{matrix} \end{matrix}$$

Element 2: Local nodes 1 and 2 of element 2 correspond to global degrees of freedom 2 and 3 respectively. Element stiffness matrix is given by

$$\left[K_2^{(e)}\right] = \frac{A_2 E_2}{L_2}\begin{bmatrix} 1 & -1 \\ -1 & 1 \end{bmatrix} = \frac{0.001\times 0.7\times 10^8}{0.4}\begin{bmatrix} 1 & -1 \\ -1 & 1 \end{bmatrix} = 10^6 \begin{bmatrix} 1.75 & -1.75 \\ -1.75 & 1.75 \end{bmatrix} \begin{matrix} 2 \\ 3 \end{matrix}$$

Element 3: Local nodes 1 and 2 of element 3 correspond to global degrees of freedom 3 and 4 respectively. Element stiffness matrix is given by

$$\left[K_3^{(e)}\right] = \frac{A_3 E_3}{L_3}\begin{bmatrix} 1 & -1 \\ -1 & 1 \end{bmatrix} = \frac{0.003\times 1\times 10^8}{0.6}\begin{bmatrix} 1 & -1 \\ -1 & 1 \end{bmatrix} = 10^5 \begin{bmatrix} 5 & -5 \\ -5 & 5 \end{bmatrix} \begin{matrix} 3 \\ 4 \end{matrix}$$

Assembly

Global stiffness matrix is obtained by assembling the contributions made by the element stiffness matrices as follows:

$$[K] = 10^5 \begin{bmatrix} 8 & -8 & 0 & 0 \\ -8 & 8+1.75 & -1.75 & 0 \\ 0 & -1.75 & 1.75+5 & -5 \\ 0 & 0 & -5 & 5 \end{bmatrix} = 10^5 \begin{bmatrix} 8 & -8 & 0 & 0 \\ -8 & 9.75 & -1.75 & 0 \\ 0 & -1.75 & 6.75 & -5 \\ 0 & 0 & -5 & 5 \end{bmatrix} \begin{matrix} 1 \\ 2 \\ 3 \\ 4 \end{matrix}$$

Global force vector can be directly expressed as $\{Q\} = \lfloor 0 \quad 200 \quad 500 \quad 0 \rfloor^T \, kN$

The stiffness equation for the stepped bar now can be written as:

$$10^5 \begin{bmatrix} 8 & -8 & 0 & 0 \\ -8 & 9.75 & -1.75 & 0 \\ 0 & -1.75 & 6.75 & -5 \\ 0 & 0 & -5 & 5 \end{bmatrix} \begin{Bmatrix} u_1 \\ u_2 \\ u_3 \\ u_4 \end{Bmatrix} = \begin{Bmatrix} 0 \\ 200 \\ 500 \\ 0 \end{Bmatrix}$$

Boundary conditions: As the stepped bar is restrained at nodes 1 and 4, $u_1 = 0$ and $u_4 = 0$. This boundary condition can be easily incorporated by striking off the first and fourth rows as well as first and fourth columns in the stiffness matrix, and first and fourth rows in the nodal displacement vector as well as the force vector. Thus, the stiffness equation for the stepped bar-2 gets reduced to

$$10^5 \begin{bmatrix} 9.75 & -1.75 \\ -1.75 & 6.75 \end{bmatrix} \begin{Bmatrix} u_2 \\ u_3 \end{Bmatrix} = \begin{Bmatrix} 200 \\ 500 \end{Bmatrix}$$

Solving the above matrix equation leads to

$$\begin{bmatrix} u_1 & u_2 & u_3 & u_4 \end{bmatrix} = 10^{-3} \begin{bmatrix} 0 & 0.3456 & 0.8327 & 0 \end{bmatrix} m$$

From Eq. 6.9, axial force in elements 1, 2 and 3 can be calculated as follows:

$$F_1 = A_1 E_1 \varepsilon_1 = 0.002 \times 2 \times 10^8 \begin{bmatrix} -\dfrac{1}{0.5} & \dfrac{1}{0.5} \end{bmatrix} \begin{Bmatrix} u_1 \\ u_2 \end{Bmatrix} = 283.67\,kN\,(Tension)$$

$$F_2 = A_2 E_2 \varepsilon_2 = 0.001 \times 0.7 \times 10^8 \begin{bmatrix} -\dfrac{1}{0.4} & \dfrac{1}{0.4} \end{bmatrix} \begin{Bmatrix} u_2 \\ u_3 \end{Bmatrix} = 83.67\,kN\,(Tension)$$

$$F_3 = A_3 E_3 \varepsilon_3 = 0.003 \times 1.0 \times 10^8 \begin{bmatrix} -\dfrac{1}{0.6} & \dfrac{1}{0.6} \end{bmatrix} \begin{Bmatrix} u_3 \\ u_4 \end{Bmatrix} = 416.33\,kN\,(Compression)$$

Numerical example 6.3 A stepped bar shown in the Figure 6.5 is subject to a load of $P = 500\,N$ at node 3 in addition to the self-weight. Find the stresses in each element and support reaction. The Young's modulus of the material and unit weight of the bar are $E = 2 \times 10^8\,kN/m^2$ and $\gamma = 0.785 \times 10^{-4}\,N/mm^3$ respectively. Adopt $L_1 = L_2 = 250\,mm$, $A_1 = 2500\,mm^2$, and $A_1 = 2000\,mm^2$.

Solution:

Element 1: Nodes 1 and 2 of element 1 correspond to global degrees of freedom 1 and 2.

Element stiffness matrix is given by

$$\left[K_1^{(e)} \right] = \frac{A_1 E}{L_1} \begin{bmatrix} 1 & -1 \\ -1 & 1 \end{bmatrix} = \frac{2500 \times 2 \times 10^5}{250} \begin{bmatrix} 1 & -1 \\ -1 & 1 \end{bmatrix} = 10^5 \begin{bmatrix} 20 & -20 \\ -20 & 20 \end{bmatrix} \begin{matrix} 1 \\ 2 \end{matrix}$$

Element force vector due to body forces can be expressed as

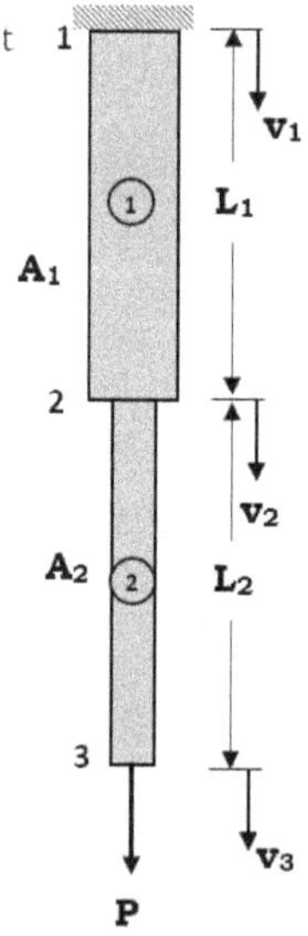

Figure 6.5
Stepped bar-3.

$$\left\{F_1^{(e)}\right\} = -\frac{\gamma A_1 L_1}{2}\begin{Bmatrix}1\\1\end{Bmatrix} = -\frac{0.785\times10^{-4}\times2500\times250}{2}\begin{Bmatrix}1\\1\end{Bmatrix} = \begin{Bmatrix}-24.531\\-24.531\end{Bmatrix}N$$

Element 2: Nodes 2 and 3 of element 1 correspond to global degrees of freedom 2 and 3.

Element stiffness matrix is given by

$$\left[K_2^{(e)}\right] = \frac{A_2 E}{L_2}\begin{bmatrix}1 & -1\\-1 & 1\end{bmatrix} = \frac{2000\times2\times10^5}{250}\begin{bmatrix}1 & -1\\-1 & 1\end{bmatrix} = 10^5\begin{matrix}\quad2\quad\quad3\\\begin{bmatrix}16 & -16\\-16 & 16\end{bmatrix}\end{matrix}\begin{matrix}2\\3\end{matrix}$$

Element force vector due to body forces can be expressed as

$$\left\{F_2^{(e)}\right\} = -\frac{\gamma A_2 L_2}{2}\begin{Bmatrix}1\\1\end{Bmatrix} = -\frac{0.785\times10^{-4}\times2000\times250}{2}\begin{Bmatrix}1\\1\end{Bmatrix} = \begin{Bmatrix}-19.625\\-19.625\end{Bmatrix}N$$

Assembly

Global stiffness matrix is obtained by assembling together the contributions made by the element stiffness matrices as follows:

$$[K] = 10^5\begin{bmatrix}20 & -20 & 0\\-20 & 20+16 & -16\\0 & -16 & 16\end{bmatrix} = 10^5\begin{bmatrix}20 & -20 & 0\\-20 & 36 & -16\\0 & -16 & 16\end{bmatrix}$$

Global force vector due to body forces can be assembled as

$$\{F\} = \begin{Bmatrix}-24.531\\-24.531-19.625\\-19.625\end{Bmatrix} = \begin{Bmatrix}-24.531\\-44.156\\-19.625\end{Bmatrix}N\begin{matrix}1\\2\\3\end{matrix}$$

Global force vector due to point loads can be expressed as $\{Q\} = \begin{matrix}1\\2\\3\end{matrix}\begin{Bmatrix}0\\0\\-500\end{Bmatrix}N$

The stiffness equation for the structure can be expressed as

$$[K]\{d\} = \{F\}+\{Q\}$$

The stiffness equation for the stepped bar now can be written as:

$$10^5 \begin{bmatrix} 20 & -20 & 0 \\ -20 & 36 & -16 \\ 0 & -16 & 16 \end{bmatrix} \begin{Bmatrix} v_1 \\ v_2 \\ v_3 \end{Bmatrix} = \begin{Bmatrix} -24.531 \\ -44.156 \\ -519.625 \end{Bmatrix} \begin{matrix} 1 \\ 2 \\ 3 \end{matrix}$$

Boundary conditions: As the stepped bar-3 is restrained at node 1, $v_1 = 0$. This boundary condition can be easily incorporated by striking off the first row and first column in the stiffness matrix, and first row in the nodal displacement vector as well as the force vector. Thus, the stiffness equation for the stepped bar-3 gets reduced to

$$10^5 \begin{bmatrix} 36 & -16 \\ -16 & 16 \end{bmatrix} \begin{Bmatrix} v_2 \\ v_3 \end{Bmatrix} = \begin{Bmatrix} -44.156 \\ -519.625 \end{Bmatrix}$$

The solution is given as $\begin{Bmatrix} v_1 \\ v_2 \\ v_3 \end{Bmatrix} = \begin{Bmatrix} 0 \\ -0.2819 \\ -0.6066 \end{Bmatrix} \times 10^{-3}\, m$

From Eq. 6.9, axial force in elements 1 and 2 can be calculated as follows:

$$F_1 = A_1 E \varepsilon_1 = 2500 \times 2 \times 10^5 \begin{bmatrix} -\dfrac{1}{250} & \dfrac{1}{250} \end{bmatrix} \begin{Bmatrix} v_1 \\ v_2 \end{Bmatrix} = 563.7\, N\,(Tension)$$

$$F_2 = A_2 E \varepsilon_2 = 2000 \times 2 \times 10^5 \begin{bmatrix} -\dfrac{1}{250} & \dfrac{1}{250} \end{bmatrix} \begin{Bmatrix} v_2 \\ v_3 \end{Bmatrix} = 519.6\, N\,(Tension)$$

6.2 PLANE TRUSS ELEMENT

Consider the triangular truss shown in the Figure 6.6. All the elements of the truss lie in the same vertical plane and hence the truss is a plane truss. The members of the truss meet at joints. Joints 1, 2 and 3 shown in the figure are pin jointed. In case of trusses, the loads are applied at the joints alone. Self-weight of the truss is usually much smaller in magnitude to the joint loads and thus is not usually considered in the analysis. The node numbers and element numbers are indicated in the figure. The element numbers are encircled to distinguish them from node numbers.

As elements of the truss are along different orientations, the element stiffness matrix needs to be transformed from the local axes of the elements to the structure axes. This requires specification of two separate sets of coordinate axes. Figure 6.7 shows coordinate axes X and Y for the entire planar truss structure.

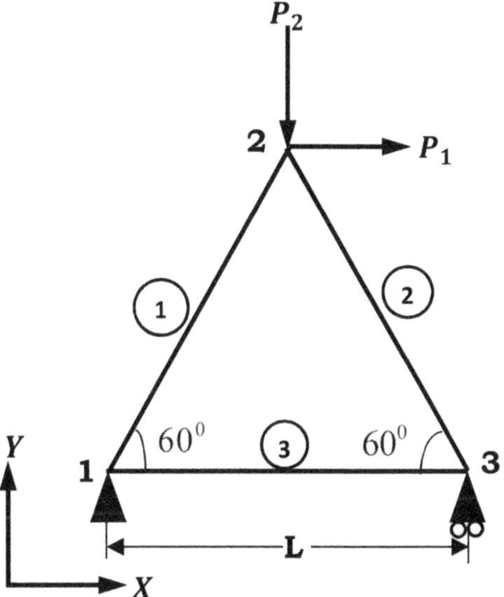

Figure 6.6 Triangular Truss.

Consider a plane truss element inclined to the positive direction of x-axis at an angle of θ as shown in the Figure 6.7. The displacements along x and y directions at node 1 are u_1 and v_1 respectively while the corresponding displacements at node 2 are u_2 and v_2 respectively. Here, X and Y are structure axes. X_M is the local x-axis of the member oriented along its longitudinal direction. Also, u_{1M} and u_{2M} are the displacements of nodes 1 and 2 *respectively along the* X_M *axis.*

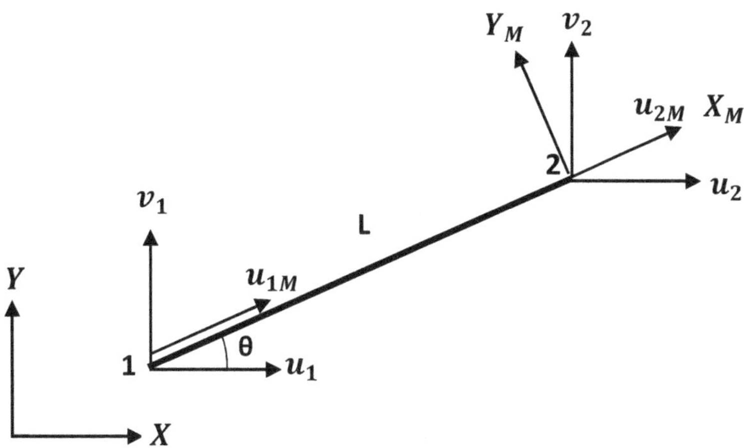

Figure 6.7 Plane Truss element.

In Figure 6.8 shown, (x_i, y_i) and (x_j, y_j) represent the coordinates of the node i and node j respectively. The node i corresponds to the degrees of freedom $(2i-1)$ and $2i$ along X and Y axes respectively. Node j corresponds to the degrees of freedom $(2j-1)$ and $2j$ along X and Y axes respectively.

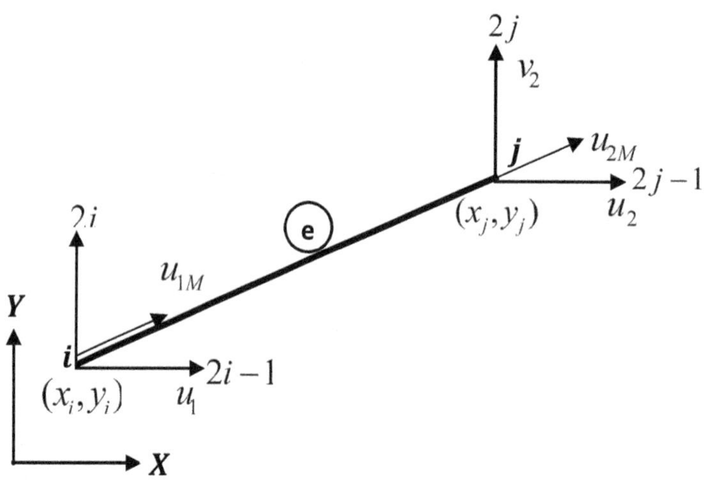

Figure 6.8   Degrees of freedom of
plane truss element along structure axes.

The length of L of the element is given as $L = \sqrt{(x_j - x_i)^2 + (y_j - y_i)^2}$. The direction cosines, which indicate the orientation of the element are defined as

$$C_x = \cos\theta = \frac{(x_j - x_i)}{L} \quad \text{and}$$

$$C_y = \sin\theta = \frac{(y_j - y_i)}{L} . \qquad \qquad .....(6.10)$$

The displacements, along the local X_M axis of the element, are expressed in terms of nodal displacements and direction cosines as

$$u_{1M} = C_x u_1 + C_y v_1$$
$$u_{2M} = C_x u_2 + C_y v_2 \qquad \qquad .....(6.11)$$

The longitudinal strain ε in the plane truss element can be expressed in terms of these nodal displacements (along the X_M axis) as follows:

$$\varepsilon = \frac{\left(u_{2M} - u_{1M}\right)}{L} = \frac{\left(-C_x u_1 - C_y v_1 + C_x u_2 + C_y v_2\right)}{L} \qquad .....(6.12)$$

$$\varepsilon = \frac{1}{L}\lfloor -C_x \quad -C_y \quad C_x \quad C_y \rfloor \begin{Bmatrix} u_1 \\ v_1 \\ u_2 \\ v_2 \end{Bmatrix} = [B]\{d\} \qquad .....(6.13)$$

where $[B] = \dfrac{1}{L}\lfloor -C_x \quad -C_y \quad C_x \quad C_y \rfloor$ is the strain displacement matrix and

$\{d\} = \lfloor u_1 \quad v_1 \quad u_2 \quad v_2 \rfloor^T$ is the vector of nodal displacement w.r.t structure axes.

6.2.1 Stiffness Matrix and Force Vector for Plane Truss Element

Stiffness matrix is computed as follows:

$$\left[K^{(e)}\right] = \iiint \left[B^T\right][C][B]\,dV \qquad .....(6.14)$$

Linear constitutive matrix in this case is $[C] = E$. Also, considering a prismatic element of cross-sectional area A , $dV = A\,dx$.

$$\left[K^{(e)}\right] = \int_{x=0}^{L} \frac{A}{L^2} \begin{Bmatrix} -C_x \\ -C_y \\ C_x \\ C_y \end{Bmatrix} E \lfloor -C_x \quad -C_y \quad C_x \quad C_y \rfloor dx \qquad .....(6.15)$$

$$\left[K^{(e)}\right] = \frac{AE}{L} \begin{bmatrix} C_x^2 & C_x C_y & -C_x^2 & -C_x C_y \\ C_x C_y & C_y^2 & -C_x C_y & -C_y^2 \\ -C_x^2 & -C_x C_y & C_x^2 & C_x C_y \\ -C_x C_y & -C_y^2 & C_x C_y & C_y^2 \end{bmatrix} \qquad .....(6.16)$$

Element stiffness matrix for each element with reference to structure axes is computed using Eq. 6.16.

Assembly of structure stiffness matrix is done as follows:

Element 1, connected between nodes 1 and 2 corresponds to degrees of freedom 1, 2, 3 and 4. Element 2, connected between nodes 2 and 3 corresponds to degrees of freedom 3,4,5 and 6. Element 3, connected between nodes 1 and 3 corresponds to degrees of freedom 1, 2, 5 and 6. The stiffness contributions of these elements will be reflected in the corresponding rows and columns of the structure stiffness matrix as per the scheme outlined below:

Suppose element stiffness matrices of elements 1, 2 and 3 are schematically represented as

$$
[K]_1^{(e)} = \begin{bmatrix} \times & \times & \times & \times \\ \times & \times & \times & \times \\ \times & \times & \times & \times \\ \times & \times & \times & \times \end{bmatrix} \begin{matrix} 1 \\ 2 \\ 3 \\ 4 \end{matrix} \qquad \text{.....(6.17)}
$$

$$
[K]_2^{(e)} = \begin{bmatrix} \otimes & \otimes & \otimes & \otimes \\ \otimes & \otimes & \otimes & \otimes \\ \otimes & \otimes & \otimes & \otimes \\ \otimes & \otimes & \otimes & \otimes \end{bmatrix} \begin{matrix} 1 \\ 2 \\ 3 \\ 4 \end{matrix} \qquad [K]_3^{(e)} = \begin{bmatrix} \oplus & \oplus & \oplus & \oplus \\ \oplus & \oplus & \oplus & \oplus \\ \oplus & \oplus & \oplus & \oplus \\ \oplus & \oplus & \oplus & \oplus \end{bmatrix} \begin{matrix} 1 \\ 2 \\ 3 \\ 4 \end{matrix}
$$

The structure has a total of 6 degrees of freedom, thus the stiffness matrix $[K]$ for the entire structure receives contribution from the three elements corresponding to these 6 degrees of freedom as illustrated below. Also, Force vector $\{Q\}$ is written directly from the joint loads as shown in Eq. 6.18.

$$
\begin{bmatrix} \times\oplus & \times\oplus & \times & \times & \oplus & \oplus \\ \times\oplus & \times\oplus & \times & \times & \oplus & \oplus \\ \times & \times & \times\otimes & \times\otimes & \otimes & \otimes \\ \times & \times & \times\otimes & \times\otimes & \otimes & \otimes \\ \oplus & \oplus & \otimes & \otimes & \otimes\oplus & \otimes\oplus \\ \oplus & \oplus & \otimes & \otimes & \otimes\oplus & \otimes\oplus \end{bmatrix} \begin{Bmatrix} u_1 \\ v_1 \\ u_2 \\ v_2 \\ u_3 \\ v_3 \end{Bmatrix} = \begin{Bmatrix} 0 \\ 0 \\ P_1 \\ -P_2 \\ 0 \\ 0 \end{Bmatrix} \qquad \text{.....(6.18)}
$$

Application of boundary conditions: Referring to the plane truss shown in Figure 6.8, the following are the boundary conditions applied on Eq. 6.18 are $u_1 = v_1 = v_3 = 0$. This amounts to striking off the first, second and sixth rows and columns in Eq. 6.18 leading to a reduced system of equations given by

$$\begin{bmatrix} \times\otimes & \times\otimes & \otimes \\ \times\otimes & \times\otimes & \otimes \\ \otimes & \otimes & \otimes \end{bmatrix} \begin{Bmatrix} u_2 \\ v_2 \\ u_3 \end{Bmatrix} = \begin{Bmatrix} P_1 \\ -P_2 \\ 0 \end{Bmatrix} \qquad ......(6.19)$$

Solving Eq. 6.19 yields the solution vector.

Obtaining member forces: Direction cosines C_X and C_Y are evaluated for each member. Using Eq. 6.13, member end forces can be written as follows:

$$F_1 = A_1 E_1 \varepsilon_1 = \frac{A_1 E_1}{L_1} \lfloor -C_X \quad -C_Y \quad C_X \quad C_Y \rfloor \lfloor u_1 \quad v_1 \quad u_2 \quad v_2 \rfloor^T$$

$$F_2 = A_2 E_2 \varepsilon_2 = \frac{A_2 E_2}{L_2} \lfloor -C_X \quad -C_Y \quad C_X \quad C_Y \rfloor \lfloor u_2 \quad v_2 \quad u_3 \quad v_3 \rfloor^T \qquad .....(6.20)$$

$$F_3 = A_3 E_3 \varepsilon_3 = \frac{A_3 E_3}{L_3} \lfloor -C_X \quad -C_Y \quad C_X \quad C_Y \rfloor \lfloor u_1 \quad v_1 \quad u_3 \quad v_3 \rfloor^T$$

6.2.2   Numerical Examples

Numerical examples are chosen to illustrate the applicability of the procedure outlined above.

Numerical example 6.4: Analyse the triangular truss shown in Figure 6.6. Length of each element is $L = 2m$, area of cross section is $A = 0.01\, m^2$ and $E = 2 \times 10^8\, kN/m^2$. $P_1 = 100\, kN$ and $P_2 = 150\, kN$.

Solution: The nodal data and element data are given in the Tables 6.1 and 6.2 respectively.

Table 6.1 Nodal data (Example 6.4).

Node	x (m)	y (m)
1	0.0	0.0
2	1.0	1.732
3	2.0	0.0

Table 6.2 Element data (Example 6.4).

Element	First Node	Second Node	Degrees of freedom
1	1	2	1,2,3,4
2	2	3	3,4,5,6
3	1	3	1,2,5,6

Element 1: $C_x = 0.5$, $C_y = 0.866$, $A = 0.01m^2$, $E = 2 \times 10^8 \, kN/m^2$, $L = 2.0 \, m$

$$[K]_1^{(e)} = 10^5 \begin{bmatrix} 2.5000 & 4.3301 & -2.5000 & -4.3301 \\ 4.3301 & 7.5000 & -4.3301 & -7.5000 \\ -2.5000 & -4.3301 & 2.5000 & 4.3301 \\ -4.3301 & -7.5000 & 4.3301 & 7.5000 \end{bmatrix}$$

Element 2: $C_x = 0.5$, $C_y = -0.866$, $A = 0.01m^2$, $E = 2 \times 10^8 \, kN/m^2$, $L = 2.0m$

$$[K]_2^{(e)} = 10^5 \begin{bmatrix} 2.5000 & -4.3301 & -2.5000 & 4.3301 \\ -4.3301 & 7.5000 & 4.3301 & -7.5000 \\ -2.5000 & 4.3301 & 2.5000 & -4.3301 \\ 4.3301 & -7.5000 & -4.3301 & 7.5000 \end{bmatrix}$$

Element 3: $C_x = 1.0$, $C_y = 0$, $A = 0.01 \, m^2$, $E = 2 \times 10^8 \, kN/m^2$, $L = 2.0 \, m$

$$[K]_3^{(e)} = 10^5 \begin{bmatrix} 1 & 0 & -1 & 0 \\ 0 & 0 & 0 & 0 \\ -1 & 0 & 1 & 0 \\ 0 & 0 & 0 & 0 \end{bmatrix}$$

Assembly of stiffness matrix and force vector for the structure:

$$10^5 \begin{bmatrix} 1.25 & 0.433 & -0.250 & -0.433 & -1.00 & 0.0 \\ 0.433 & 0.75 & -0.433 & -0.75 & 0.0 & 0.0 \\ -0.25 & -0.433 & 0.5 & 0.0 & -0.25 & 0.433 \\ -0.433 & -0.75 & 0.0 & 1.5 & 0.433 & -0.75 \\ -1.0 & 0.0 & -0.25 & 0.433 & 1.25 & -0.433 \\ 0.0 & 0.0 & 0.433 & -0.75 & -0.433 & 0.75 \end{bmatrix} \begin{Bmatrix} u_1 \\ v_1 \\ u_2 \\ v_2 \\ u_3 \\ v_3 \end{Bmatrix} = \begin{Bmatrix} 0 \\ 0 \\ 100 \\ -150 \\ 0 \\ 0 \end{Bmatrix}$$

Boundary conditions are $u_1 = v_1 = v_3 = 0$

Reduced system of equations after application of boundary conditions is

$$10^5 \begin{bmatrix} 0.5 & 0 & -0.25 \\ 0 & 1.5 & 0.433 \\ -0.25 & 0.433 & 1.25 \end{bmatrix} \begin{Bmatrix} u_2 \\ v_2 \\ u_3 \end{Bmatrix} = \begin{Bmatrix} 100 \\ -150 \\ 0 \end{Bmatrix}$$

Solution to the above system of equations is

$$\lfloor u_1 \quad v_1 \quad u_2 \quad v_2 \quad u_3 \quad v_3 \rfloor = 10^{-4} \lfloor 0 \quad 0 \quad 2.4665 \quad \text{-}1.2692 \quad 0.9329 \quad 0 \rfloor^T m$$

Obtaining member forces: Using Eq. 6.13, member end forces can be written as follows:

$$F_1 = A_1 E_1 \varepsilon_1 = \frac{A_1 E_1}{L_1} \lfloor -C_X \quad -C_Y \quad C_X \quad C_Y \rfloor \lfloor u_1 \quad v_1 \quad u_2 \quad v_2 \rfloor^T$$

$$= \left(\frac{0.01 \times 2 \times 10^5}{2} \right) \lfloor -0.5 \quad -0.866 \quad 0.5 \quad 0.86 \rfloor \begin{Bmatrix} 0 \\ 0 \\ 2.4665 \times 10^{-5} \\ -1.2692 \times 10^{-5} \end{Bmatrix} = 13.4122 \ kN(Tension)$$

$$F_2 = A_2 E_2 \varepsilon_2 = \frac{A_2 E_2}{L_2} \lfloor -C_X \quad -C_Y \quad C_X \quad C_Y \rfloor \lfloor u_2 \quad v_2 \quad u_3 \quad v_3 \rfloor^T$$

$$= \left(\frac{0.01 \times 2 \times 10^5}{2} \right) \lfloor -0.5 \quad 0.866 \quad 0.5 \quad -0.866 \rfloor \begin{Bmatrix} 2.4645 \times 10^{-4} \\ -1.2692 \times 10^{-4} \\ 0.9329 \times 10^{-4} \\ 0 \end{Bmatrix} = 186.59 \ kN(Comp)$$

$$F_3 = A_3 E_3 \varepsilon_3 = \frac{A_3 E_3}{L_3} \lfloor -C_X \quad -C_Y \quad C_X \quad C_Y \rfloor \lfloor u_1 \quad v_1 \quad u_3 \quad v_3 \rfloor^T$$

$$= \left(\frac{0.01 \times 2 \times 10^5}{2} \right) \lfloor -1 \quad 0 \quad 1 \quad 0 \rfloor \begin{Bmatrix} 0 \\ 0 \\ 9.3295 \times 10^{-5} \\ 0 \end{Bmatrix} = 93.295 \ kN(Tension)$$

Numerical example 6.5 Analyse the truss as shown in the Figure 6.9. For all members, area of cross section is $A = 0.01 \ m^2$ and $E = 2 \times 10^8 \ kN/m^2$.

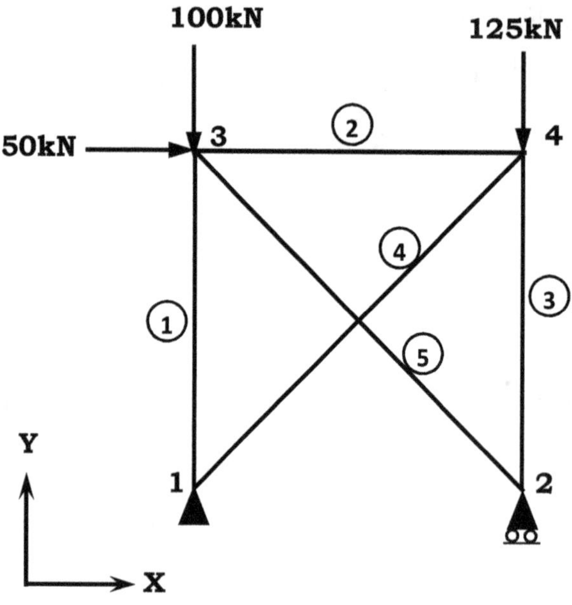

Figure 6.9 Plane truss (Example 6.5).

Table 6.3 Nodal data.

Node	x	y
1	0.0	0.0
2	2.0	0.0
3	0.0	2.0
4	2.0	2.0

Table 6.4 Element data.

Element	First Node	Second Node	Degrees of freedom
1	1	3	1,2,5,6
2	3	4	5,6,7,8
3	2	4	3,4,7,8
4	1	4	1,2,7,8
5	2	3	3,4,5,6

Element 1: $C_x = 0, C_Y = 1,\ A = 0.01m^2,\ E = 2\times10^8\,kN/m^2,\ L = 2.0\ m$

$$[K]_1^{(e)} = 10^6 \begin{bmatrix} 0 & 0 & 0 & 0 \\ 0 & 1 & 0 & -1 \\ 0 & 0 & 0 & 0 \\ 0 & -1 & 0 & 1 \end{bmatrix}$$

Element 2: $C_x = 1, C_Y = 0$, $A = 0.01\ m^2$, $E = 2 \times 10^8\ kN/m^2$, $L = 2.0m$

$$[K]_2^{(e)} = 10^6 \begin{bmatrix} 1 & 0 & -1 & 0 \\ 0 & 0 & 0 & 0 \\ -1 & 0 & 1 & 0 \\ 0 & 0 & 0 & 0 \end{bmatrix}$$

Element 3: $C_x = 1, C_Y = 0$, $A = 0.01\ m^2$, $E = 2 \times 10^8\ kN/m^2$, $L = 2.0\ m$

$$[K]_3^{(e)} = 10^6 \begin{bmatrix} 0 & 0 & 0 & 0 \\ 0 & 1 & 0 & -1 \\ 0 & 0 & 0 & 0 \\ 0 & -1 & 0 & 1 \end{bmatrix}$$

Element 4: $C_x = 0.707, C_Y = 0.707$, $A = 0.01\ m^2$, $E = 2 \times 10^8\ kN/m^2$, $L = 2.828\ m$

$$[K]_4^{(e)} = 10^6 \begin{bmatrix} 0.3536 & 0.3536 & -0.3536 & -0.3536 \\ 0.3536 & 0.3536 & -0.3536 & -0.3536 \\ -0.3536 & -0.3536 & 0.3536 & 0.3536 \\ -0.3536 & -0.3536 & 0.3536 & 0.3536 \end{bmatrix}$$

Element 5: $C_x = -0.707, C_Y = 0.707$, $A = 0.01\ m^2$, $E = 2 \times 10^8\ kN/m^2$, $L = 2.828\ m$

$$[K]_5^{(e)} = 10^6 \begin{bmatrix} 0.3536 & -0.3536 & -0.3536 & 0.3536 \\ -0.3536 & 0.3536 & 0.3536 & -0.3536 \\ -0.3536 & 0.3536 & 0.3536 & -0.3536 \\ 0.3536 & -0.3536 & -0.3536 & 0.3536 \end{bmatrix}$$

Assembly of stiffness matrix and force vector for the structure:

$$\begin{bmatrix} 0.3536 & 0.3536 & 0 & 0 & 0 & 0 & -0.3536 & -0.3536 \\ 0.3536 & 1.3536 & 0 & 0 & 0 & -1 & -0.3536 & -0.3536 \\ 0 & 0 & 0.3536 & -0.3536 & -0.3536 & 0.3536 & 0 & 0 \\ 0 & 0 & -0.3536 & 1.3536 & 0.3536 & -0.3536 & 0 & -1 \\ 0 & 0 & -0.3536 & 0.3536 & 1.3536 & -0.3536 & -1 & 0 \\ 0 & -1 & 0.3536 & -0.3536 & -0.3536 & 1.3536 & 0 & 0 \\ -0.3536 & -0.3536 & 0 & 0 & -1 & 0 & 1.3536 & 0.3536 \\ -0.3536 & -0.3536 & 0 & -1 & 0 & 0 & 0.3536 & 1.3536 \end{bmatrix} \begin{Bmatrix} u_1 \\ v_1 \\ u_2 \\ v_2 \\ u_3 \\ v_3 \\ u_4 \\ v_4 \end{Bmatrix} = \begin{Bmatrix} 0 \\ 0 \\ 0 \\ 0 \\ 50.0 \\ -100.0 \\ 0.0 \\ -125.0 \end{Bmatrix}$$

Boundary conditions are $u_1 = v_1 = v_2 = 0$

Reduced system of equations after application of boundary conditions is

$$10^6 \begin{bmatrix} 0.3536 & -0.3536 & 0.3536 & 0 & 0 \\ -0.3536 & 1.3536 & -0.3536 & -1 & 0 \\ 0.3536 & -0.3536 & 1.3536 & 0 & 0 \\ 0 & -1 & 0 & 1.3536 & 0.3536 \\ 0 & 0 & 0 & 0.3536 & 1.3536 \end{bmatrix} \begin{Bmatrix} u_2 \\ u_3 \\ v_3 \\ u_4 \\ v_4 \end{Bmatrix} = \begin{Bmatrix} 0 \\ 50 \\ -100 \\ 0 \\ -125 \end{Bmatrix}$$

The solution to the above system of equations is

$$\lfloor u_1 \ \ v_1 \ \ u_2 \ \ v_2 \ \ u_3 \ \ v_3 \ \ u_4 \ \ v_4 \rfloor = 10^{-4} \lfloor 0 \ \ 0 \ \ 4.6642 \ \ 0 \ \ 3.6642 \ \ -1.0 \ \ 3.1642 \ \ -1.75 \rfloor^T$$

Obtaining member forces: Using Eq. 6.13, member end forces can be written as follows:

$$F_1 = A_1 E_1 \varepsilon_1 = \frac{A_1 E_1}{L_1} \lfloor -C_X \ \ -C_Y \ \ C_X \ \ C_Y \rfloor [u_1 \ \ v_1 \ \ u_3 \ \ v_3]^T$$

$$= \left(\frac{0.01 \times 2 \times 10^8}{2} \right) \lfloor 0 \ \ -1 \ \ 0 \ \ 1 \rfloor 10^{-4} \lfloor 0 \ \ 0 \ \ 3.6642 \ \ -1.0 \rfloor^T = 100 \ kN (Comp)$$

$$F_2 = A_2 E_2 \varepsilon_2 = \frac{A_2 E_2}{L_2} \lfloor -C_X \ \ -C_Y \ \ C_X \ \ C_Y \rfloor [u_3 \ \ v_3 \ \ u_4 \ \ v_4]^T$$

$$= \left(\frac{0.01 \times 2 \times 10^8}{2} \right) \lfloor -1 \ \ 0 \ \ 1 \ \ 0 \rfloor 10^{-4} \lfloor 3.6642 \ \ -1.0 \ \ 3.1642 \ \ -1.75 \rfloor^T = 50.0 \ kN (Comp)$$

$$F_3 = A_3 E_3 \varepsilon_3 = \frac{A_3 E_3}{L_3} \lfloor -C_X \ \ -C_Y \ \ C_X \ \ C_Y \rfloor [u_2 \ \ v_2 \ \ u_4 \ \ v_4]^T$$

$$= \left(\frac{0.01 \times 2 \times 10^8}{2} \right) \lfloor -1 \ \ 0 \ \ 1 \ \ 0 \rfloor 10^{-4} \lfloor 0 \ \ 3.3642 \ \ 3.1642 \ \ -1.75 \rfloor^T = 175.0 \ kN (Comp)$$

$$F_4 = A_4 E_4 \varepsilon_4 = \frac{A_4 E_4}{L_4} \lfloor -C_X \ \ -C_Y \ \ C_X \ \ C_Y \rfloor [u_1 \ \ v_1 \ \ u_4 \ \ v_4]^T$$

$$= \left(\frac{0.01 \times 2 \times 10^8}{2\sqrt{2}} \right) \lfloor -0.707 \ \ -0.707 \ \ 0.707 \ \ 0.707 \rfloor 10^{-4} \lfloor 0 \ \ 0 \ \ 3.1642 \ \ -1.75 \rfloor^T = 70.71 \ kN (Tension)$$

$$F_5 = A_5 E_5 \varepsilon_5 = \frac{A_5 E_5}{L_5} \lfloor -C_X \quad -C_Y \quad C_X \quad C_Y \rfloor [u_2 \quad v_2 \quad u_3 \quad v_3]^T$$

$$= \left(\frac{0.01 \times 2 \times 10^8}{2\sqrt{2}} \right) \lfloor -0.707 \quad -0.707 \quad 0.707 \quad 0.707 \rfloor 10^{-4} \lfloor 4.6635 \quad 0 \quad 3.6638 \quad -0.9996 \rfloor^T = 0 \ kN$$

Numerical example 6.6: Analyse the Warren truss shown in the Figure 6.10.

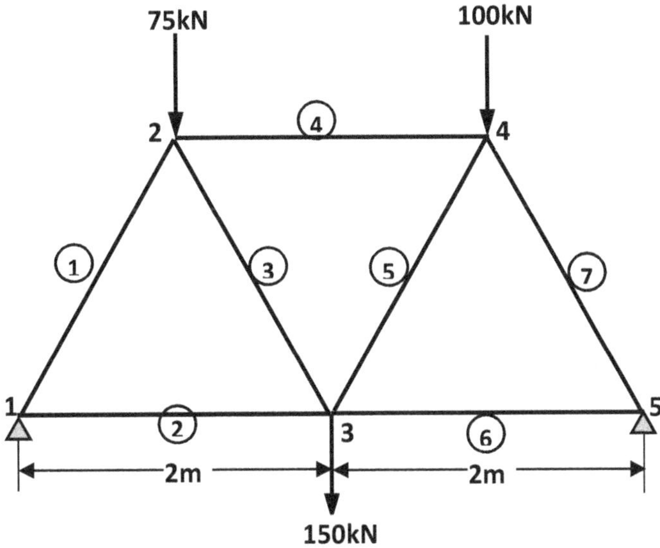

Figure 6.10 Warren truss (Example 6.6).

The nodal and element data for the truss are given in Table 6.5 and Table 6.6 respectively.

Table 6.5 Nodal data.

Node	x	y
1	0.0	0.0
2	1.0	1.732
3	2.0	0.0
4	3.0	1.732
5	4.0	0.0

Table 6.6 Element data.

Element	First Node	Second Node	Degrees of freedom
1	1	2	1,2,3,4
2	1	3	1,2,5,6
3	2	3	3,4,5,6
4	2	4	3,4,7,8
5	3	4	5,6,7,8
6	3	5	5,6,9,10
7	4	5	7,8,9,10

Elements 1 and 5: $C_X = 0.5$, $C_Y = 0.866$, $A = 0.01\ m^2$, $E = 2\times10^8\ kN/m^2$, $L = 2.0\ m$

$$[K]_1^{(e)} = [K]_5^{(e)} = 10^5 \begin{bmatrix} 2.5 & 4.33 & -2.5 & -4.33 \\ 4.33 & 7.5 & -4.33 & -7.5 \\ -2.5 & -4.33 & 2.5 & 4.33 \\ -4.33 & -7.5 & 4.33 & 7.5 \end{bmatrix}$$

Elements 3 and 7: $C_x = 0.5$, $C_Y = -0.866$, $A = 0.01\ m^2$, $E = 2\times10^8\ kN/m^2$, $L = 2.0\ m$

$$[K]_3^{(e)} = [K]_7^{(e)} = 10^5 \begin{bmatrix} 2.5 & -4.33 & -2.5 & 4.33 \\ -4.33 & 7.5 & 4.33 & -7.5 \\ -2.5 & 4.33 & 2.5 & -4.33 \\ 4.33 & -7.5 & -4.33 & 7.5 \end{bmatrix}$$

Elements 2, 4 and 6: $C_x = 1.0$, $C_Y = 0.0$, $A = 0.01 m^2$, $E = 2\times10^8\ kN/m^2$, $L = 2.0\ m$

$$[K]_{21}^{(e)} = [K]_4^{(e)} = [K]_6^{(e)} = 10^5 \begin{bmatrix} 10.0 & 0.0 & -10.0 & 0.0 \\ 0.0 & 0.0 & 0.0 & 0.0 \\ -10.0 & 0.0 & 10.0 & 0.0 \\ 0.0 & 0.0 & 0.0 & 10.0 \end{bmatrix}$$

Assembly of stiffness matrix and force vector for the structure:

$$10^6 \begin{bmatrix} 1.25 & 0.433 & -0.25 & -0.433 & -1 & 0 & 0 & 0 & 0 & 0 \\ 0.433 & 0.75 & -0.433 & -0.75 & 0 & 0 & 0 & 0 & 0 & 0 \\ -0.25 & -0.433 & 1.5 & 0 & -0.25 & 0.433 & -1 & 0 & 0 & 0 \\ -0.433 & -0.75 & 0 & 1.5 & 0.433 & -0.75 & 0 & 0 & 0 & 0 \\ -1 & 0 & -0.25 & 0.433 & 2.5 & 0 & -0.25 & -0.433 & 0 & 0 \\ 0 & 0 & 0.433 & -0.75 & 0 & 1.5 & -0.433 & -0.75 & 0 & 0 \\ 0 & 0 & -1 & 0 & -0.25 & -0.433 & 1.5 & 0 & 0 & 0 \\ 0 & 0 & 0 & 0 & -0.433 & -0.75 & 0 & 1.5 & 0.433 & -0.75 \\ 0 & 0 & 0 & 0 & 0 & 0 & 0 & 0.433 & 1.25 & -0.433 \\ 0 & 0 & 0 & 0 & 0 & 0 & 0 & -0.75 & -0.433 & 0.75 \end{bmatrix} \begin{Bmatrix} u_1 \\ v_1 \\ u_2 \\ v_2 \\ u_3 \\ v_3 \\ u_4 \\ v_4 \\ u_5 \\ v_5 \end{Bmatrix} = \begin{Bmatrix} 0 \\ 0 \\ 0 \\ -75 \\ 0 \\ -150 \\ 0 \\ -100 \\ 0 \\ 0 \end{Bmatrix}$$

Boundary conditions are $u_1 = v_1 = u_5 = v_5 = 0$

Reduced system of equations after application of boundary conditions is

$$10^6 \begin{bmatrix} 1.5 & 0 & -0.25 & 0.433 & -1 & 0 \\ 0 & 1.5 & 0.433 & -0.75 & 0 & 0 \\ -0.25 & 0.433 & 2.5 & 0 & -0.25 & -0.433 \\ 0.433 & -0.75 & 0 & 1.5 & -0.433 & -0.75 \\ -1 & 0 & -0.25 & -0.433 & 1.5 & 0 \\ 0 & 0 & -0.433 & -0.75 & 0 & 1.5 \end{bmatrix} \begin{Bmatrix} u_2 \\ v_2 \\ u_3 \\ v_3 \\ u_4 \\ v_4 \end{Bmatrix} = \begin{Bmatrix} 0 \\ -75 \\ 0 \\ -150 \\ 0 \\ -100 \end{Bmatrix}$$

Solution is given by

$$\lfloor u_1 \quad v_1 \quad u_2 \quad v_2 \quad u_3 \quad v_3 \quad u_4 \quad v_4 \quad u_5 \quad v_5 \rfloor$$

$$= 10^{-4} \lfloor 0 \quad 0 \quad 0.6132 \quad \text{-}2.186 \quad \text{-}0.0722 \quad \text{-}3.7488 \quad \text{-}0.6854 \quad \text{-}2.5617 \quad 0 \quad 0 \rfloor^T$$

Obtaining member forces: Using Eq. 6.13, member end forces can be written as follows:

$$F_1 = A_1 E_1 \varepsilon_1 = \frac{A_1 E_1}{L_1} \lfloor -C_X \quad -C_Y \quad C_X \quad C_Y \rfloor [u_1 \quad v_1 \quad u_2 \quad v_2]^T$$

$$= \left(\frac{0.01 \times 2 \times 10^8}{2} \right) \lfloor 0 \quad -1 \quad 0 \quad 1 \rfloor 10^{-4} \lfloor 0 \quad 0 \quad 3.6638 \quad -0.9996 \rfloor^T = 99.96 \; kN (Comp)$$

$$F_2 = A_2 E_2 \varepsilon_2 = \frac{A_2 E_2}{L_2} \lfloor -C_X \quad -C_Y \quad C_X \quad C_Y \rfloor [u_1 \quad v_1 \quad u_3 \quad v_3]^T$$

$$= \left(\frac{0.01 \times 2 \times 10^8}{2} \right) \lfloor -1 \quad 0 \quad 1 \quad 0 \rfloor 10^{-4} \lfloor 0 \quad 0 \quad -0.0072 \quad -3.7488 \rfloor^T = 7.216 \; kN (Comp)$$

$$F_3 = A_3 E_3 \varepsilon_3 = \frac{A_3 E_3}{L_3} \lfloor -C_X \quad -C_Y \quad C_X \quad C_Y \rfloor [u_2 \quad v_2 \quad u_3 \quad v_3]^T$$

$$= \left(\frac{0.01 \times 2 \times 10^8}{2} \right) \lfloor -0.5 \quad 0.866 \quad 0.5 \quad -0.866 \rfloor 10^{-4} \lfloor 0.6132 \quad -2.1867 \quad -0.6854 \quad -2.5617 \rfloor^T = 101.013 kN (Tension)$$

$$F_4 = A_4 E_4 \varepsilon_4 = \frac{A_4 E_4}{L_4} \lfloor -C_X \quad -C_Y \quad C_X \quad C_Y \rfloor [u_2 \quad v_2 \quad u_4 \quad v_4]^T$$

$$= \left(\frac{0.01 \times 2 \times 10^8}{2} \right) \lfloor -1 \quad 0 \quad 1 \quad 0 \rfloor 10^{-4} \lfloor 0.6132 \quad -2.1867 \quad -0.6854 \quad -2.5617 \rfloor^T = 129.863 \; kN (Comp)$$

$$F_5 = A_5 E_5 \varepsilon_5 = \frac{A_5 E_5}{L_5} \lfloor -C_X \quad -C_Y \quad C_X \quad C_Y \rfloor [u_3 \quad v_3 \quad u_4 \quad v_4]^T$$

$$= \left(\frac{0.01 \times 2 \times 10^8}{2} \right) \lfloor 0 \quad -1 \quad 0 \quad 1 \rfloor 10^{-4} \lfloor 0 \quad 0 \quad 0.6132 \quad -2.1867 \rfloor^T = 72.145 \ kN(Tension)$$

$$F_6 = A_6 E_6 \varepsilon_6 = \frac{A_5 E_5}{L_5} \lfloor -C_X \quad -C_Y \quad C_X \quad C_Y \rfloor [u_3 \quad v_3 \quad u_4 \quad v_4]^T$$

$$= \left(\frac{0.01 \times 2 \times 10^8}{2} \right) \lfloor 0 \quad -1 \quad 0 \quad 1 \rfloor 10^{-4} \lfloor 0 \quad 0 \quad 0.6132 \quad -2.1867 \rfloor^T = 7.216 \ kN(Tension)$$

$$F_7 = A_7 E_7 \varepsilon_7 = \frac{A_7 E_7}{L_7} \lfloor -C_X \quad -C_Y \quad C_X \quad C_Y \rfloor [u_4 \quad v_4 \quad u_5 \quad v_5]^T$$

$$= \left(\frac{0.01 \times 2 \times 10^8}{2} \right) \lfloor -0.5 \quad 0.866 \quad 0.5 \quad -0.866 \rfloor 10^{-4} \lfloor -0.6854 \quad -2.5617 \quad 0.0 \quad 0.0 \rfloor^T = 187.58 \ kN(Comp)$$

6.3 SPACE TRUSS ELEMENT

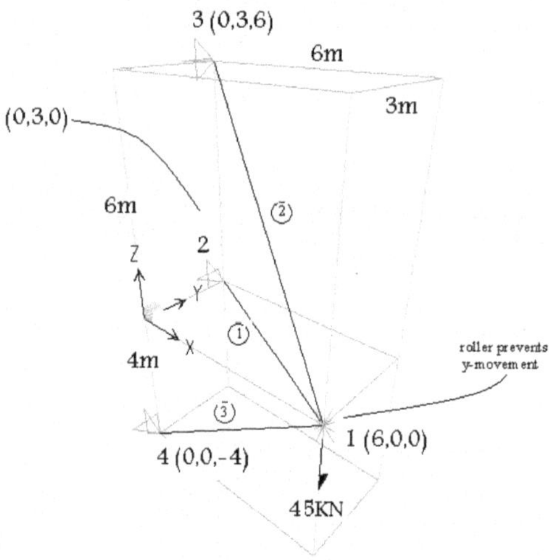

Figure 6.11 Three bar space truss.

Consider the three-bar space truss shown in the Figure 6.11. All the elements of the truss are oriented spatially and hence the truss is a space truss. The members of the truss meet at pin joints 1, 2 and 3 shown in the figure. The node numbers and element numbers are indicated in the figure. As elements of the truss are along different orientations, the element stiffness matrix needs to be transformed from the local axes of the elements to the structure axes. This requires specification of two separate sets of coordinate axes.

The coordinate axes X, Y and Z in Figure 6.12 are the structure axes, defined for the entire space truss structure. The space truss element is inclined to the positive direction of X, Y and Z axes at angles of θ_x, θ_y and θ_z as shown in the figure. The local X_M axis is along the longitudinal axis of the element. Y_M and Z_M axes are normal to the X_M axis. The displacements along X, Y and Z directions at node 1 are u_1, v_1 and w_1 respectively. The corresponding displacements are node 2 are u_2, v_2 and w_2 respectively. The displacements at node 1 along the local axes are u_{1M}, v_{1M} and w_{1M} respectively while the corresponding displacements at node 2 are u_{2M}, v_{2M} and w_{2M} respectively.

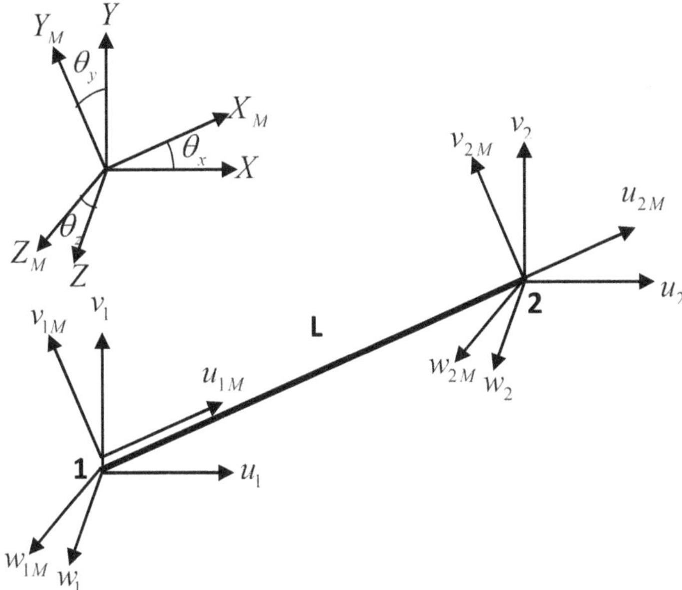

Figure 6.12 Space Truss element.

In Figure 6.13 shown (x_i, y_i, z_i) and (x_j, y_j, z_j) represent the coordinates of the nodes i and j respectively. The node i corresponds to the degrees of freedom $3i-2$, $3i-1$ and $3i$ along X, Y and Z axes respectively. The node j corresponds to the degrees of freedom $3j-2$, $3j-1$ and $3j$ along X, Y and Z axes respectively.

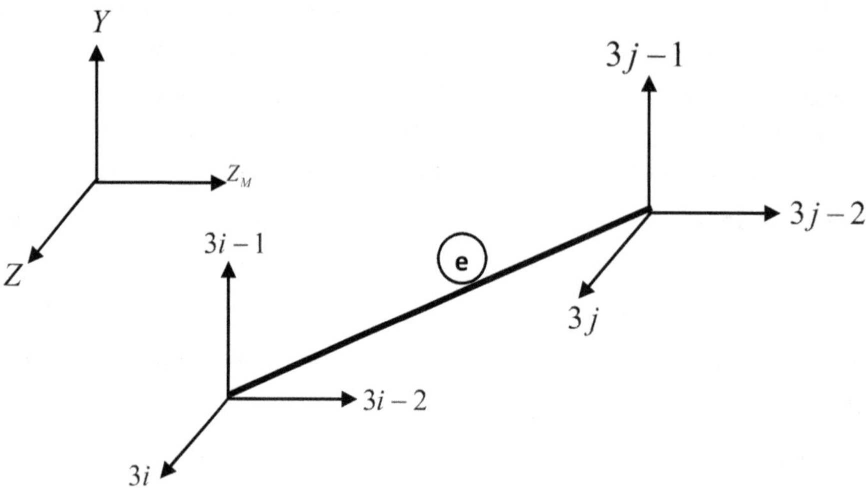

Figure 6.13 Degrees of freedom of space truss element.

The length of L of the element is given as $L = \sqrt{(x_j - x_i)^2 + (y_j - y_i)^2 + (z_j - z_i)^2}$. The direction cosines, which indicate the orientation of the element are defined as

$$C_x = \cos\theta_X = \frac{(x_j - x_i)}{L} \quad ; \quad C_y = \cos\theta_y = \frac{(y_j - y_i)}{L} \quad ; C_z = \cos\theta_z = \frac{(z_j - z_i)}{L} \quad .....(6.21)$$

The displacements along the local X_M axis of the element are expressed in terms of nodal displacements and direction cosines as

$$u_{1M} = C_x u_1 + C_y v_1 + C_z w_1$$
$$u_{2M} = C_x u_2 + C_y v_2 + C_z w_2 \qquad .....(6.22)$$

The longitudinal strain ε in the plane truss element can be expressed in terms of these nodal displacements (along the X_M axis) as follows:

$$\varepsilon = \frac{(u_{2M} - u_{1M})}{L} = \frac{\left(-C_x u_1 - C_y v_1 - C_z w_1 + C_x u_2 + C_y v_2 + C_z w_2\right)}{L} \qquad .....(6.23)$$

$$\varepsilon = \frac{1}{L}\left\lfloor -C_x \quad -C_y \quad -C_z \quad C_x \quad C_y \quad C_z \right\rfloor \left\lfloor u_1 \quad v_1 \quad w_1 \quad u_2 \quad v_2 \quad w_2 \right\rfloor^T = [B]\{d\}$$
$$.....(6.24)$$

where $[B] = \dfrac{1}{L}\lfloor -C_x \quad -C_y \quad -C_z \quad C_x \quad C_y \quad C_z \rfloor$ is the strain displacement matrix

and $\{d\} = \lfloor u_1 \quad v_1 \quad w_1 \quad u_2 \quad v_2 \quad w_2 \rfloor^T$ is the vector of nodal displacement w.r.t structure axes.

6.3.1 Stiffness Matrix for Space Truss Element

Stiffness matrix is computed as follows:

$$\left[K^{(e)}\right] = \iiint \left[B^T\right][C][B]\,dV \qquad \qquad .....(6.25)$$

Linear constitutive matrix in this case is $[C] = E$. Also, considering a prismatic element of cross-sectional area A, $dV = A\,dx$.

$$\left[K^{(e)}\right] = \int_{x=0}^{L} \frac{A}{L^2} \begin{Bmatrix} -C_x \\ -C_y \\ -C_z \\ C_x \\ C_y \\ C_z \end{Bmatrix} E\lfloor -C_x \quad -C_y \quad -C_z \quad C_x \quad C_y \quad C_z \rfloor dx \qquad .....(6.26)$$

Element stiffness matrix for each element with reference to structure axes is computed as follows:

$$\left[K^{(e)}\right] = \begin{bmatrix} C_x^2 & C_xC_y & C_xC_z & -C_x^2 & -C_xC_y & -C_xC_z \\ C_xC_y & C_y^2 & C_yC_z & -C_xC_y & -C_y^2 & -C_yC_z \\ C_xC_z & C_yC_z & C_z^2 & -C_xC_z & -C_yC_z & -C_z^2 \\ -C_x^2 & -C_xC_y & -C_xC_z & C_x^2 & C_xC_y & C_xC_z \\ -C_xC_y & -C_y^2 & -C_yC_z & C_xC_y & C_y^2 & C_yC_z \\ -C_xC_z & -C_yC_z & -C_z^2 & C_xC_z & C_yC_z & C_z^2 \end{bmatrix} \qquad .....(6.27)$$

6.3.2 Force Vector for Space Truss Element

Force vector for the space truss is directly assembled from the forces acting at its nodes.

6.3.3 Numerical Examples

Numerical example 6.7: Analyse the space truss shown in Figure 6.11. The three bars have cross sectional areas $0.0002\ m^2$, $0.0003\ m^2$ and $0.0001\ m^2$ respectively. Young's modulus $E = 2 \times 10^8\ kN/m^2$ for all members.

The structure has 3 nodes. The first node corresponds to degrees of freedom u_1, v_1 and w_1, second node corresponds to degrees of freedom u_2, v_2 and w_2 and third node corresponds to degrees of freedom u_3, v_3 and w_3.

The element stiffness matrices are computed as follows:

Element 1: $C_x = -0.8944$, $C_y = 0.4472$ and $C_z = 0$

$$\left[K_1^{(e)}\right] = 10^3 \begin{bmatrix} 4.770 & -2.385 & 0 & -4.770 & 2.385 & 0 \\ -2.385 & 1.1925 & 0 & 2.385 & -1.1925 & 0 \\ 0 & 0 & 0 & 0 & 0 & 0 \\ -4.770 & 2.385 & 0 & 4.770 & -2.385 & 0 \\ 2.385 & -1.1925 & 0 & -2.385 & 1.1925 & 0 \\ 0 & 0 & 0 & 0 & 0 & 0 \end{bmatrix}$$

Element 2: $C_x = -0.6666$, $C_y = 0.3333$ and $C_z = 0.6666$

$$\left[K_2^{(e)}\right] = 10^3 \begin{bmatrix} 4.9382 & -2.4691 & -4.9382 & -4.9382 & 2.4691 & 4.9382 \\ -2.4691 & 1.2345 & 2.4691 & 2.4691 & -1.2345 & -2.4691 \\ -4.9382 & 2.4691 & 4.9382 & 4.9382 & -2.4691 & -4.9382 \\ -4.9382 & 2.4691 & 4.9382 & 4.9382 & -2.4691 & -4.9382 \\ 2.4691 & 1.2345 & 2.4691 & -2.4691 & 1.2345 & 2.4691 \\ 4.9382 & 2.4691 & 4.9382 & -4.9382 & 2.4691 & 4.9382 \end{bmatrix}$$

Element 3: $C_x = -0.8320$, $C_y = 0$ and $C_z = -0.5547$

$$\left[K_3^{(e)}\right] = 10^3 \begin{bmatrix} 1.9201 & 0 & 1.2801 & -1.9201 & 0 & -1.2801 \\ 0 & 0 & 0 & 0 & 0 & 0 \\ 1.2801 & 0 & 0.8533 & -1.2801 & 0 & -0.8533 \\ -1.9201 & 0 & -1.2801 & 1.9201 & 0 & 1.2801 \\ 0 & 0 & 0 & 0 & 0 & 0 \\ -1.2801 & 0 & -0.8533 & 1.2801 & 0 & 0.8533 \end{bmatrix}$$

Assembly of stiffness equations leads to

$$10^3 \begin{bmatrix} 1.1628 & -0.4854 & -0.3658 & -0.4770 & 0.2385 & 0 & -0.4938 & 0.2469 & 0.4938 & -0.1920 & 0 & -0.1280 \\ -0.4854 & 0.2427 & 0.2469 & 0.2385 & -0.1192 & 0 & 0.2469 & -0.1234 & -0.2469 & 0 & 0 & 0 \\ -0.3658 & 0.2469 & 0.5791 & 0 & 0 & 0 & 0.4938 & -0.2469 & -0.4938 & -0.1280 & 0 & -0.0853 \\ -0.4770 & 0.2385 & 0 & 0.4770 & -0.2385 & 0 & 0 & 0 & 0 & 0 & 0 & 0 \\ 0.2385 & -0.1192 & 0 & -0.2385 & 0.1192 & 0 & 0 & 0 & 0 & 0 & 0 & 0 \\ 0 & 0 & 0 & 0 & 0 & 0 & 0 & 0 & 0 & 0 & 0 & 0 \\ -0.4938 & 0.2469 & 0.4938 & 0 & 0 & 0 & 0.4938 & -0.2469 & -0.4938 & 0 & 0 & 0 \\ 0.2469 & -0.1234 & -0.2469 & 0 & 0 & 0 & -0.2469 & 0.1234 & 0.2469 & 0 & 0 & 0 \\ 0.4938 & 0.2469 & -0.4938 & 0 & 0 & 0 & -0.4938 & 0.2469 & 0.4938 & 0 & 0 & 0 \\ -0.1920 & 0 & -0.1280 & 0 & 0 & 0 & 0 & 0 & 0 & 0.1920 & 0 & 0.1280 \\ 0 & 0 & 0 & 0 & 0 & 0 & 0 & 0 & 0 & 0 & 0 & 0 \\ -0.1280 & 0 & -0.0853 & 0 & 0 & 0 & 0 & 0 & 0 & 0.1280 & 0 & 0.0853 \end{bmatrix} \begin{Bmatrix} u_1 \\ v_1 \\ w_1 \\ u_2 \\ v_2 \\ w_2 \\ u_3 \\ v_3 \\ w_3 \\ u_4 \\ v_4 \\ w_4 \end{Bmatrix} = \begin{Bmatrix} 0 \\ 0 \\ -45 \\ 0 \\ 0 \\ 0 \\ 0 \\ 0 \\ 0 \\ 0 \\ 0 \\ 0 \end{Bmatrix}$$

Boundary conditions are $v_1 = u_2 = v_2 = w_2 = u_3 = v_3 = w_3 = u_4 = v_4 = w_4 = 0$

$$10^4 \begin{bmatrix} 1.1628 & -0.3658 \\ -0.3658 & 0.5791 \end{bmatrix} \begin{Bmatrix} u_2 \\ w_2 \end{Bmatrix} = \begin{Bmatrix} 0 \\ -45 \end{Bmatrix}$$

The solution vector is $\lfloor u_1 \quad w_1 \rfloor = 10^{-3} \lfloor -3.0503 \quad -9.6965 \rfloor$

Postprocessing: Computation of axial forces in elements

Axial force is given by $F^{(e)} = AE\varepsilon = \dfrac{AE}{L} \lfloor -C_x \quad -C_y \quad -C_z \quad C_x \quad C_y \quad C_z \rfloor \{d^{(e)}\}$

$$F_1 = \frac{0.0002 \times 2 \times 10^8}{6.7082} \lfloor 0.8944 \quad -0.4472 \quad 0 \quad -0.8944 \quad 0.4472 \quad 0 \rfloor 10^{-3} \begin{Bmatrix} -3.0503 \\ 0 \\ -9.6965 \\ 0 \\ 0 \\ 0 \end{Bmatrix}$$

$$= 16.268 \, kN \,(Compression)$$

$$F_2 = \frac{0.0005 \times 2 \times 10^8}{9.0}$$

$$\begin{bmatrix} 0.6666 & -0.3333 & -0.6666 & -0.6666 & 0.3333 & 0.6666 \end{bmatrix} 10^{-3} \begin{Bmatrix} -3.0503 \\ 0 \\ -9.6965 \\ 0 \\ 0 \\ 0 \end{Bmatrix}$$

$$= 49.23 \; kN \, (Tension)$$

$$F_3 = \frac{0.0001 \times 2 \times 10^8}{7.211}$$

$$\begin{bmatrix} 0.8320 & 0 & -0.6666 & -0.8320 & 0 & -0.5547 \end{bmatrix} 10^{-3} \begin{Bmatrix} -3.0503 \\ 0 \\ -9.6965 \\ 0 \\ 0 \\ 0 \end{Bmatrix}$$

$$= 21.957 kN \, (Compression)$$

Numerical example 6.8: Analyse microwave tower as shown in Figure 6.14. This tower is square in shape with a side of 1.5 m and a height of 15 m. The tower is subdivided into 5 panels each of height 3 m. Each body is provided with X-bracing on all four faces and horizontal members at the top level. The Indian standard equal angle section *ISA* 110×110×10 is used for all members of the section.

The finite element model of the structure has 24 nodes and 80 members. Each member has two equal legs each of 110 mm with thickness of 10 mm with a cross sectional area of 21.06×10⁻⁴ m².

The tower is subjected to point loads of 5 kN acting normal to the left face at nodes 5, 8, 9, 12, 13, 16, 17, 20, 21 and 24 along X-axis. Young's modulus $E = 2 \times 10^8 \; kN/m^2$ for all members.

Solution: Detailed calculations cannot be shown owing to the large size of the structure. However, selected nodal displacements and member forces are shown in Table 6.7 and Table 6.8 respectively.

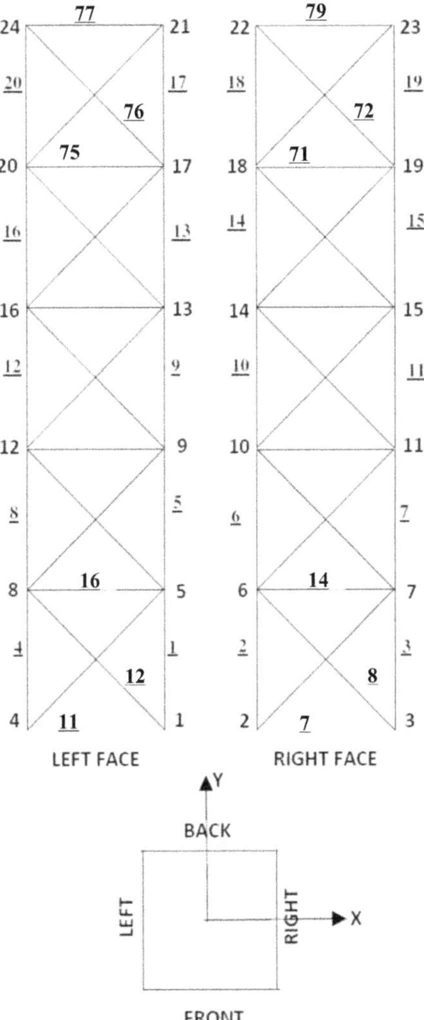

Figure 6.14 Finite element model of microwave tower.

Table 6.7 Nodal displacements of selected nodes.

Node	X-displacement $\times 10^{-3}\,m$	Y-displacement $\times 10^{-3}\,m$	Z-displacement $\times 10^{-3}\,m$
5	4.703	0.0456	0.8836
13	26.525	0.0143	0.1776
21	56.734	0.0011	1.9563

Table 6.8 Forces in elements.

Element	Force (kN)
4	124.058
16	-48.068
76	4.531

6.4   BEAM ELEMENT

Consider the beam element as shown in the Figure 6.15. The element has two nodes. At each node, there is a displacement v along the positive y-direction and an in-plane rotation θ (anticlockwise), about the z-axis.

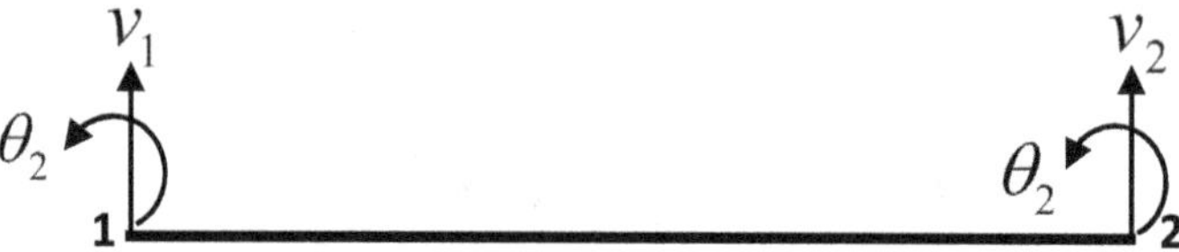

Figure 6.15 Beam Element.

The vertical displacement at any point on the beam can be expressed in terms of nodal displacements by using the Hermite Cubic polynomial functions defined by Eq. 4.31 as shown below:

$$v(x) = \lfloor N_1(x) \quad N_2(x) \quad N_3(x) \quad N_4(x) \rfloor \begin{Bmatrix} v_1 \\ \theta_1 \\ v_2 \\ \theta_2 \end{Bmatrix} = \lfloor N(x) \rfloor \{d\} \qquad \ldots\ldots(6.28)$$

where $\{d\}$ is the vector of nodal displacements.

For small deformations under conditions of linear elasticity, the curvature of the beam element is given by the second derivative of $v(x)$ as

$$v''(x) = \lfloor N''(x) \rfloor \{d\} \qquad \ldots\ldots(6.29)$$

For a prismatic Euler-Bernoulli beam, the moment curvature relation can now be written as

$$\sigma_x = Ey\frac{d^2v}{dx^2} \qquad \ldots\ldots(6.30)$$

$$M_x = \int_0^h \sigma \, y \, b \, dy = EI \frac{d^2 v}{dx^2} \qquad\qquad .....(6.31)$$

where $I = \int_0^h y^2 b \, dy$ is the area moment of inertia about the x-axis. It is to be noted here that b is the width of the beam at a distance of y from the neutral axis and h is the depth of the beam. Thus, the above relations are valid for prismatic beam of any cross-section.

Bending moment M_x can now be expressed as

$$M_x = EIv'' = EI\lfloor N''(x) \rfloor \{d\} \qquad\qquad .....(6.32)$$

By analogy, M_x is treated similar to σ and v'' is treated similar to ε. Thus,

$$[B] = \lfloor N''(x) \rfloor \text{ and } [C] = EI$$

Here $\lfloor N''(x) \rfloor$ is given by

$$\lfloor N''(x) \rfloor = \left\lfloor \frac{1}{L^2}(-6+12\eta) \quad \frac{1}{L}(-4+6\eta) \quad \frac{1}{L^2}(6-12\eta) \quad \frac{1}{L}(-2+6\eta) \right\rfloor \qquad .....(6.33)$$

6.4.1 Stiffness Matrix for Beam Element

The stiffness matrix for the beam element is evaluated as

$$[K] = \int_0^L [B^T][C][B] dx = \int_0^L \lfloor N''(x) \rfloor^T EI \lfloor N''(x) \rfloor dx$$

$$K_{11} = \int_{x=0}^L N_1'' EI N_1'' dx = EI \int_{x=0}^L \left(\frac{1}{L^2}(-6+12\eta) \right)^2 (L d\eta) = \frac{EI}{L^3} \int_{\eta=0}^1 (-6+12\eta) d\eta = \frac{12EI}{L^3}$$

$$K_{21} = K_{12} = \int_{x=0}^L N_1'' EI N_2'' dx = EI \int_{x=0}^L \left(\frac{1}{L^2}(-6+12\eta) \times \frac{1}{L}(-4+6\eta) \right)(L d\eta) = \frac{EI}{L^2} \int_{\eta=0}^1 (-6+12\eta)(-4+6\eta) d\eta = \frac{6EI}{L^2}$$

$$K_{31} = K_{13} = \int_{x=0}^L N_1'' EI N_3'' dx = EI \int_{x=0}^L \left(\frac{1}{L^2}(-6+12\eta) \times \frac{1}{L^2}(6-12\eta) \right)(L d\eta) = \frac{EI}{L^3} \int_{\eta=0}^1 (-6+12\eta)^2 d\eta = -\frac{12EI}{L^3}$$

$$K_{41} = K_{14} = \int_{x=0}^L N_1'' EI N_4'' dx = EI \int_{x=0}^L \left(\frac{1}{L^2}(-6+12\eta) \times \frac{1}{L}(-2+6\eta) \right)(L d\eta) = \frac{EI}{L^2} \int_{\eta=0}^1 (-6+12\eta)(-2+6\eta) d\eta = \frac{6EI}{L^2}$$

$$K_{22} = \int_{x=0}^{L} N_2'' EI N_2'' dx = EI \int_{x=0}^{L} \left(\frac{1}{L}(-4+6\eta) \right)^2 (Ld\eta) = \frac{EI}{L} \int_{\eta=0}^{1} (-4+6\eta)^2 d\eta = \frac{4EI}{L}$$

$$K_{32} = K_{23} = \int_{x=0}^{L} N_2'' EI N_3'' dx = EI \int_{x=0}^{L} \left(\frac{1}{L}(-4+6\eta) \times \frac{1}{L^2}(6-12\eta) \right)(Ld\eta) = \frac{EI}{L^2} \int_{\eta=0}^{1} \frac{1}{L}(-4+6\eta)(-6+12\eta)d\eta = -\frac{6EI}{L^2}$$

$$K_{42} = K_{24} = \int_{x=0}^{L} N_2'' EI N_4'' dx = EI \int_{x=0}^{L} \left(\frac{1}{L}(-4+6\eta) \times \frac{1}{L}(-2+6\eta) \right)(Ld\eta) = \frac{EI}{L^2} \int_{\eta=0}^{1} (-4+6\eta)(-2+6\eta)d\eta = \frac{2EI}{L}$$

$$K_{44} = \int_{x=0}^{L} N_4'' EI N_4'' dx = EI \int_{x=0}^{L} \left(\frac{1}{L}(-2+6\eta) \times \frac{1}{L}(-2+6\eta) \right)(Ld\eta) = \frac{EI}{L^2} \int_{\eta=0}^{1} (-2+6\eta)^2 d\eta = \frac{4EI}{L}$$

$$[K] = \begin{bmatrix} \dfrac{12EI}{L^3} & \dfrac{6EI}{L^2} & -\dfrac{12EI}{L^3} & \dfrac{6EI}{L^2} \\[2ex] \dfrac{6EI}{L^2} & \dfrac{4EI}{L} & -\dfrac{6EI}{L^2} & \dfrac{2EI}{L} \\[2ex] -\dfrac{12EI}{L^3} & -\dfrac{6EI}{L^2} & \dfrac{12EI}{L^3} & \dfrac{6EI}{L^2} \\[2ex] \dfrac{6EI}{L^2} & \dfrac{2EI}{L} & -\dfrac{6EI}{L^2} & \dfrac{4EI}{L} \end{bmatrix} \qquad \qquad .....(6.34)$$

6.4.2　Force Vector for Beam Element

Following the analogy of Eq. 6.32, element force vector for the beam element can be written as

$$\{Q^{(e)}\} = \int_{0}^{L} \left[N^T \right] f(x) dx \qquad\qquad .....(6.35)$$

where $f(x)$ defines the distribution of load along the length of the beam. For gravity loading, as load acts along negative-Y direction, negative sign is adopted. In case of an element subjected to a uniformly distributed load of p in the *vertically upward direction* and adopting $\eta = \dfrac{x}{L}$, Eq. (6.35) can be written as

$$\{Q\} = L \int_{0}^{1} \left[N^T \right] p\, d\eta \qquad\qquad .....(6.36)$$

$$\{Q^{(e)}\} = L\int_0^1 p \begin{Bmatrix} (1-3\eta^2+2\eta^3) \\ L(\eta-2\eta^2+\eta^3) \\ (3\eta^2-2\eta^3) \\ L(-\eta^2+\eta^3) \end{Bmatrix} d\eta = \begin{vmatrix} \dfrac{pL}{2} & \dfrac{pL^2}{12} & \dfrac{pL}{2} & -\dfrac{pL^2}{12} \end{vmatrix}^T \qquad .....(6.37)$$

In case of an element subjected to a point load acting at the midpoint of an element $(\eta = 0.5)$ in *the vertically upward direction*, the point load W can be expressed as

$$p(x) = W\delta(\eta - 0.5) \qquad\qquad .....(6.38)$$

Where δ is the Dirac-delta function. Substituting Eq. 6.38 in Eq. 6.36, the following is obtained

$$\{Q^{(e)}\} = L\int_0^1 W\delta(\eta-0.5) \begin{Bmatrix} (1-3\eta^2+2\eta^3) \\ L(\eta-2\eta^2+\eta^3) \\ (3\eta^2-2\eta^3) \\ L(-\eta^2+\eta^3) \end{Bmatrix} d\eta = \begin{vmatrix} \dfrac{W}{2} & \dfrac{WL}{8} & \dfrac{W}{2} & -\dfrac{WL}{8} \end{vmatrix}^T \qquad .....(6.39)$$

In case of an eccentric point load W *acting vertically upwards* as a distance of a from the left end and b from the right end.

$$\{Q^{(e)}\} = \begin{vmatrix} \dfrac{Wb}{L} & \dfrac{Wab^2}{L^2} & \dfrac{Wa}{L} & -\dfrac{Wa^2b}{L^2} \end{vmatrix}^T \qquad .....(6.40)$$

Consider a three-span continuous beam as shown in the Figure 6.16:

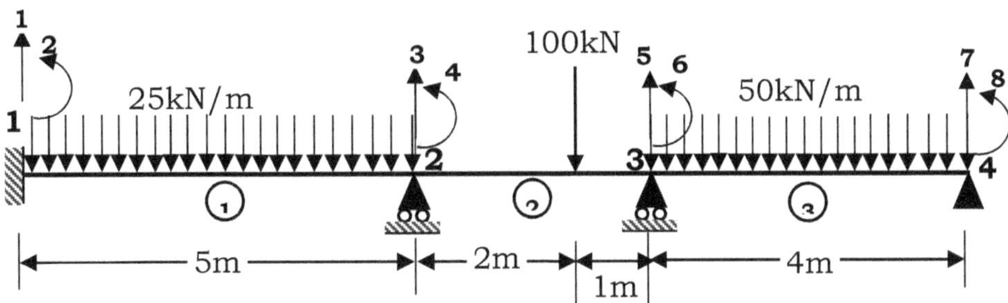

Figure 6.16 Three span continuous beam.

The beam has 3 elements and 4 nodes. The total degrees of freedom of the structure are 8. Thus, the structure has a stiffness matrix of size 8. Assembly of structure stiffness matrix is done as follows:

Element 1, connected between nodes 1 and 2 corresponds to degrees of freedom 1, 2, 3 and 4. Element 2, connected between nodes 2 and 3 corresponds to degrees of freedom 3,4,5 and 6. Element 3, connected between nodes 3 and

4 corresponds to degrees of freedom 5, 6, 7 and 8. The stiffness contributions of these elements will be reflected in the corresponding rows and columns of the structure stiffness matrix as per the scheme outlined below:

Element stiffness matrices of elements 1, 2 and 3 are schematically represented as follows:

$$
K_1^{(e)} =
\begin{array}{c@{}c}
\begin{array}{cccc} 1 & 2 & 3 & 4 \end{array} & \\
\begin{bmatrix}
\times & \times & \times & \times \\
\times & \times & \times & \times \\
\times & \times & \times & \times \\
\times & \times & \times & \times
\end{bmatrix} &
\begin{array}{c} 1 \\ 2 \\ 3 \\ 4 \end{array}
\end{array}
\qquad
K_2^{(e)} =
\begin{array}{c@{}c}
\begin{array}{cccc} 3 & 4 & 5 & 6 \end{array} & \\
\begin{bmatrix}
\otimes & \otimes & \otimes & \otimes \\
\otimes & \otimes & \otimes & \otimes \\
\otimes & \otimes & \otimes & \otimes \\
\otimes & \otimes & \otimes & \otimes
\end{bmatrix} &
\begin{array}{c} 3 \\ 4 \\ 5 \\ 6 \end{array}
\end{array}
\qquad
K_3^{(e)} =
\begin{array}{c@{}c}
\begin{array}{cccc} 5 & 6 & 7 & 8 \end{array} & \\
\begin{bmatrix}
\oplus & \oplus & \oplus & \oplus \\
\oplus & \oplus & \oplus & \oplus \\
\oplus & \oplus & \oplus & \oplus \\
\oplus & \oplus & \oplus & \oplus
\end{bmatrix} &
\begin{array}{c} 5 \\ 6 \\ 7 \\ 8 \end{array}
\end{array}
$$

The structure has a total of 8 degrees of freedom, thus the stiffness matrix $[K]$ for the entire structure receives contribution from the three elements corresponding to these 8 degrees of freedom as illustrated below. The structure force vector receives contribution from the element 1 at the degrees of freedom 1, 2, 3 and 4, from element 2 at the degrees of freedom 3, 4, 5 and 6 and from element 3 at the degrees of freedom 5, 6, 7 and 8 respectively. This is indicated in Eq. 6.41 below. Any loads acting at the global degrees of freedom are directly included in the global force vector $\{Q\}$.

$$
\begin{array}{c}
\begin{array}{cccccccc} 1 & 2 & 3 & 4 & 5 & 6 & 7 & 8 \end{array} \\
\begin{bmatrix}
\times & \times & \times & \times & 0 & 0 & 0 & 0 \\
\times & \times & \times & \times & 0 & 0 & 0 & 0 \\
\times & \times & \times\otimes & \times\otimes & \otimes & \otimes & 0 & 0 \\
\times & \times & \times\otimes & \times\otimes & \otimes & \otimes & 0 & 0 \\
0 & 0 & \otimes & \otimes & \otimes\oplus & \otimes\oplus & \oplus & \oplus \\
0 & 0 & \otimes & \otimes & \otimes\oplus & \otimes\oplus & \oplus & \oplus \\
0 & 0 & 0 & 0 & \oplus & \oplus & \oplus & \oplus \\
0 & 0 & 0 & 0 & \oplus & \oplus & \oplus & \oplus
\end{bmatrix}
\end{array}
\begin{Bmatrix}
u_1 \\ v_1 \\ u_2 \\ v_2 \\ u_3 \\ v_3 \\ u_4 \\ v_4
\end{Bmatrix}
=
\begin{Bmatrix}
\times \\ \times \\ \times\otimes \\ \times\otimes \\ \otimes\oplus \\ \otimes\oplus \\ \oplus \\ \oplus
\end{Bmatrix}
+
\begin{Bmatrix} Q \end{Bmatrix}
$$

$$\dots\dots(6.41)$$

Boundary conditions are given by $v_1 = \theta_1 = v_2 = v_3 = v_4 = 0$. Thus, the first, second, third, fifth, seventh rows and columns are struck off from stiffness matrix and force vector of the structure. Thus, the reduced system of matrix equations is given by

$$
\begin{bmatrix} \times \otimes & \otimes & 0 \\ \otimes & \otimes \oplus & \oplus \\ 0 & \oplus & \oplus \end{bmatrix} \begin{Bmatrix} \theta_2 \\ \theta_3 \\ \theta_4 \end{Bmatrix} = \begin{Bmatrix} \times \otimes \\ \otimes \oplus \\ \oplus \end{Bmatrix}
\qquad \dots\text{(6.42)}
$$

Solving Eq. 6.42, we obtain the vector of unknowns $\lfloor \theta_1 \quad \theta_2 \quad \theta_3 \rfloor^T$

Post-processing:

Using Eq. 6.32 and Eq.6.33, the following is the bending moment obtained

$$
M_x = EI \left\lfloor \frac{1}{L^2}(-6+12\eta) \quad \frac{1}{L}(-4+6\eta) \quad \frac{1}{L^2}(6-12\eta) \quad \frac{1}{L}(-2+6\eta) \right\rfloor \{d\}
\qquad \dots\text{(6.43)}
$$

Differentiating we obtain the shear force as

$$
V_x = \frac{dM_x}{dx} = \frac{1}{L}\frac{dM_x}{d\eta} = EI \left\lfloor \frac{12}{L^3} \quad \frac{6}{L^2} \quad \frac{-12}{L^3} \quad \frac{6}{L^2} \right\rfloor \{d\}
\qquad \dots\text{(6.44)}
$$

To obtain the moment and shear at the end of the element, superimpose fixed end reactions to the above moment and shear. However, these values are exact only when point loads at nodes of the structure. In case of elements being subjected to loads, Eq. 6.43 and Eq.6.44 do not yield an exact result and thus need to be corrected. The correction to be applied to the bending moment and shear force is computed as shown below:

Consider an element acted upon by uniformly distributed load per metre run over the entire span in the upward direction. The fixed end reactions are shown in Figure 6.17:

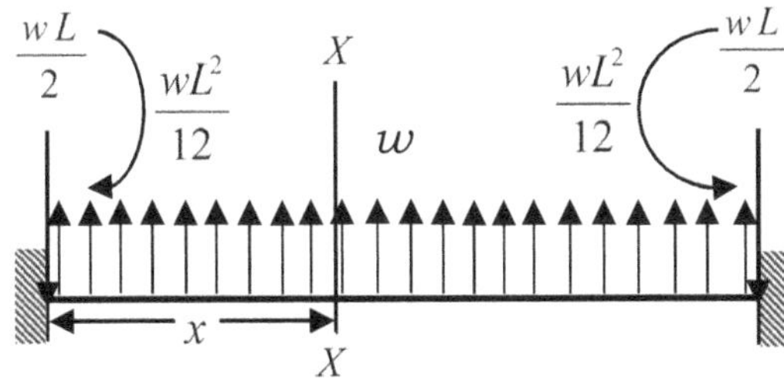

Figure 6.17 Fixed beam subjected to a uniformly distributed load over entire span.

Bending moment at any section at a distance of x can be obtained as

$$M_{1x} = -\frac{wL}{2}x + \frac{wL^2}{12} + \frac{wx^2}{2} = \frac{wL^2}{12}\left(1 - \frac{6x}{L} + \frac{6x^2}{L^2}\right) = \frac{wL^2}{12}\left(1 - 6\eta + 6\eta^2\right) \quad \dots\dots(6.45)$$

$$V_{1x} = -\frac{wL}{2} + wx = -\frac{wL}{2}\left(1 - \frac{2x}{L}\right) = -\frac{wL}{2}\left(1 - 2\eta\right) \quad \dots\dots(6.46)$$

Thus, the overall bending moment and shear force are computed by using Eq.6.43 *through* Eq.6.46 as follows:

$$M_x = EI\left\lfloor \frac{1}{L^2}(-6+12\eta) \quad \frac{1}{L}(-4+6\eta) \quad \frac{1}{L^2}(6-12\eta) \quad \frac{1}{L}(-2+6\eta)\right\rfloor\{d\} + \frac{wL^2}{12}\left(1 - 6\eta + 6\eta^2\right)$$

$$\dots\dots(6.47)$$

$$V_x = EI\left\lfloor \frac{12}{L^3} \quad \frac{6}{L^2} \quad \frac{-12}{L^3} \quad \frac{6}{L^2}\right\rfloor\{d\} - \frac{wL}{2}\left(1 - 2\eta\right) \quad \dots\dots(6.48)$$

Substitute $\eta = 0$ and 1 respectively in Eq. 6.47 and Eq. 6.48 for obtaining expressions for M_x and V_x at nodes 1 and 2.

At node 1, $\quad M_x = EI\left\lfloor -\frac{6}{L^2} \quad \frac{-4}{L} \quad \frac{6}{L^2} \quad \frac{-2}{L}\right\rfloor\{d\} + \frac{wL^2}{12} \quad \dots\dots(6.49)$

and $\quad V_x = EI\left\lfloor \frac{12}{L^3} \quad \frac{6}{L^2} \quad -\frac{12}{L^3} \quad \frac{6}{L^2}\right\rfloor\{d\} - \frac{wL}{2}$

At node 2, $M_x = EI\left\lfloor \frac{6}{L^2} \quad \frac{2}{L} \quad -\frac{6}{L^2} \quad \frac{4}{L}\right\rfloor\{d\} + \frac{wL^2}{12} \quad \dots\dots(6.50)$

and $\quad V_x = EI\left\lfloor \frac{12}{L^3} \quad \frac{6}{L^2} \quad -\frac{12}{L^3} \quad \frac{6}{L^2}\right\rfloor\{d\} + \frac{wL}{2}$

Consider an element subjected to a point load W acting in the upward direction at a distance a from the left end (and at a distance b from the right end) as shown in the Figure 6.18. The bending moment and shear force at a distance $x = \eta L$ from the left end are written as:

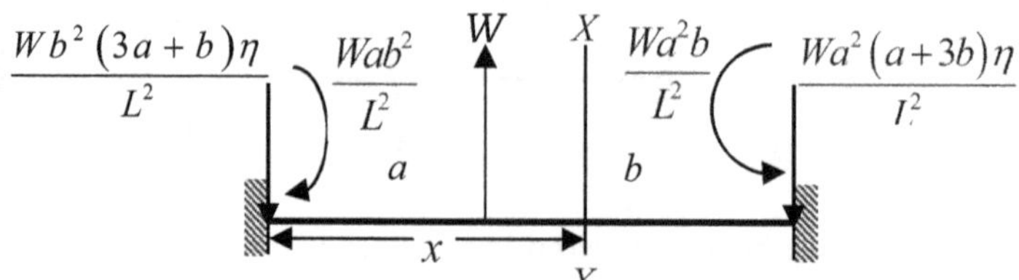

Figure 6.18 Fixed beam subjected to an eccentric point load.

$$M_{1x} = -\frac{Wb^2(3a+b)}{L^2}\eta + \frac{Wab^2}{L^2} + WL\left(\eta - \frac{a}{L}\right)H\left(\eta - \frac{a}{L}\right) \quad\quad .....(6.51)$$

$$V_{1x} = -\frac{Wb^2(3a+b)}{L^3} + WH\left(\eta - \frac{a}{L}\right) \quad\quad .....(6.52)$$

These are superimposed with the values of bending moment and shear force obtained using Finite element analysis as shown below:

$$M_x = EI\left\lfloor \frac{1}{L^2}(-6+12\eta) \quad \frac{1}{L}(-4+6\eta) \quad \frac{1}{L^2}(6-12\eta) \quad \frac{1}{L}(-2+6\eta)\right\rfloor\{d\}$$

$$-\frac{Wb^2(3a+b)}{L^2}\eta + \frac{Wab^2}{L^2} + WL\left(\eta - \frac{a}{L}\right)H\left(\eta - \frac{a}{L}\right) \quad\quad .....(6.53)$$

$$V_x = EI\left\lfloor \frac{12}{L^3} \quad \frac{6}{L^2} \quad \frac{-12}{L^3} \quad \frac{6}{L^2}\right\rfloor\{d\} - \frac{Wb^2(3a+b)}{L^3} + WH\left(\eta - \frac{a}{L}\right) \quad\quad .....(6.54)$$

where $H(x-a)$ is the Heaviside function defined as follows:

$$H(x-a) = 0 \ \ if \ x < a$$
$$= 1 \ if \ x > a \quad\quad .....(6.55)$$
$$= \frac{1}{2} \ if \ x = a$$

Substitute $\eta = 0$ and 1 respectively in Eq. 6.53 and Eq. 6.54 for obtaining expressions for M_x and V_x at nodes 1 and 2. At node 1,

$$M_x = EI\left\lfloor -\frac{6}{L^2} \quad \frac{-4}{L} \quad \frac{6}{L^2} \quad \frac{-2}{L}\right\rfloor\{d\} + \frac{Wab^2}{L^2} \quad\quad .....(6.56)$$

$$V_x = EI\left\lfloor \frac{12}{L^3} \quad \frac{6}{L^2} \quad \frac{-12}{L^3} \quad \frac{6}{L^2}\right\rfloor\{d\} - \frac{Wb^2(3a+b)}{L^3}$$

At node 2:

$$M_x = EI\left\lfloor \frac{6}{L^2} \quad \frac{2}{L} \quad \frac{-6}{L^2} \quad \frac{4}{L}\right\rfloor\{d\} - \frac{Wb^2(3a+b)}{L^2} + \frac{Wab^2}{L^2} + Wb \quad\quad .....(6.57)$$

$$V_x = EI\left\lfloor \frac{12}{L^3} \quad \frac{6}{L^2} \quad \frac{-12}{L^3} \quad \frac{6}{L^2}\right\rfloor\{d\} + \frac{Wa^2(a+3b)}{L^3}$$

The detailed procedure for the analysis of a beam is illustrated by the numerical examples presented in the next section.

6.4.3 Numerical Examples

Numerical Example 6.9: Analyse the three-span continuous beam shown in Figure 6.16. Adopt $A = 0.06\ m^2$, $I = 4.5 \times 10^{-4}\ m^4$ and $E = 2 \times 10^8\ kN/m^2$ for all members.

Solution:

Element 1:

$$\left[K_1^{(e)}\right] = 10^4 \begin{bmatrix} 0.864 & 2.16 & -0.864 & 2.16 \\ 2.16 & 7.2 & -2.16 & 3.6 \\ -0.864 & -2.16 & 0.864 & -2.160 \\ 2.16 & 3.6 & -2.16 & 7.2 \end{bmatrix}; \ \left\{Q_1^{(e)}\right\} = \begin{Bmatrix} -62.5 \\ -52.0833 \\ -62.5 \\ 52.0833 \end{Bmatrix}$$

$$\left[K_2^{(e)}\right] = 10^4 \begin{bmatrix} 4.0 & 6.0 & -4.0 & 6.0 \\ 6.0 & 12.0 & -6.0 & 6.0 \\ -4.0 & -6.0 & 4.0 & -6.0 \\ 6.0 & 6.0 & -6.0 & 12.0 \end{bmatrix}; \ \left\{Q_2^{(e)}\right\} = \begin{Bmatrix} -33.333 \\ -22.222 \\ -33.334 \\ 44.444 \end{Bmatrix}$$

$$\left[K_3^{(e)}\right] = 10^4 \begin{bmatrix} 1.6875 & 3.375 & -1.6875 & 3.375 \\ 3.375 & 9.0 & -3.375 & 4.5 \\ -1.6875 & -3.3.75 & 1.6875 & -3.375 \\ 3.375 & 4.5 & -3.375 & 9.0 \end{bmatrix}; \ \left\{Q_3^{(e)}\right\} = \begin{Bmatrix} -100 \\ -66.667 \\ -100 \\ 66.667 \end{Bmatrix}$$

The overall structure stiffness matrix and force vector are obtained as follows:

$$10^4 \begin{bmatrix} 0.864 & 2.16 & -0.864 & -2.16 & 0 & 0 & 0 & 0 \\ 2.16 & 7.2 & -2.16 & 3.6 & 0 & 0 & 0 & 0 \\ -0.864 & -2.16 & 0.864+4.0 & 2.16+6.0 & -4.0 & 6.0 & 0 & 0 \\ 2.16 & 3.6 & -2.16+6.0 & 7.2+12.0 & -6.0 & 6.0 & 0 & 0 \\ 0 & 0 & -4.0 & -6.0 & 4.0+1.6875 & -6.0+3.375 & -1.6875 & 3.375 \\ 0 & 0 & 6.0 & 6.0 & -6.0+3.375 & 12.0+9.0 & -3.375 & 4.5 \\ 0 & 0 & 0 & 0 & -1.6875 & -3.375 & 1.6875 & -3.375 \\ 0 & 0 & 0 & 0 & 3.375 & 4.5 & -3.375 & 9.0 \end{bmatrix} \begin{Bmatrix} v_1 \\ \theta_1 \\ v_2 \\ \theta_2 \\ v_2 \\ \theta_3 \\ v_4 \\ \theta_4 \end{Bmatrix} = \begin{Bmatrix} -62.5 \\ -52.0833 \\ -62.5-33.333 \\ 52.0833-22.222 \\ -33.333-100.0 \\ 44.444-66.667 \\ -100.0 \\ 66.667 \end{Bmatrix}$$

Boundary conditions: $v_1 = \theta_1 = v_2 = v_3 = 0$

Reduced system of equations after application of boundary conditions is

$$10^4 \begin{bmatrix} 19.2 & 6.0 & 0.0 \\ 6.0 & 21.0 & 4.5 \\ 0.0 & 4.5 & 9.0 \end{bmatrix} \begin{Bmatrix} \theta_2 \\ \theta_3 \\ \theta_4 \end{Bmatrix} = \begin{Bmatrix} 29.8613 \\ -22.222 \\ 66.667 \end{Bmatrix}$$

Solution is $\lfloor \theta_2 \quad \theta_3 \quad \theta_4 \rfloor = 10^{-4} \lfloor 2.7568 \quad -3.8451 \quad 9.3299 \rfloor$

Post Processing:

Element 1: Using Eq. 6.49 and Eq. 6.50, bending moments and shear forces at nodes 1 and 2 are computed as follows:

$$M_x = EI \left\lfloor -\frac{6}{5^2} \quad \frac{-4}{5} \quad \frac{6}{5^2} \quad \frac{-2}{5} \right\rfloor \times 10^{-4} \lfloor 0 \quad 0 \quad 0 \quad 2.75688 \rfloor^T + \frac{(-25 \times 5^2)}{12} = -62.008 \ kNm$$

$$V_x = EI \left\lfloor \frac{12}{5^3} \quad \frac{6}{5^2} \quad -\frac{12}{5^3} \quad \frac{6}{5^2} \right\rfloor \times 10^{-4} \lfloor 0 \quad 0 \quad 0 \quad 2.75688 \rfloor - \frac{(-25 \times 5)}{2} = 68.455 \ kN$$

At node 2, $M_x = EI \left\lfloor \frac{6}{L^2} \quad \frac{2}{L} \quad -\frac{6}{L^2} \quad \frac{4}{L} \right\rfloor \{d\} + \frac{wL^2}{12}$

$$V_x = EI \left\lfloor \frac{12}{L^3} \quad \frac{6}{L^2} \quad -\frac{12}{L^3} \quad \frac{6}{L^2} \right\rfloor + \frac{wL}{2}$$

$$M_x = EI \left\lfloor \frac{6}{5^2} \quad \frac{2}{5} \quad -\frac{6}{5^2} \quad \frac{4}{5} \right\rfloor \times 10^{-4} \lfloor 0 \quad 0 \quad 0 \quad 2.75688 \rfloor + \frac{(-25 \times 5^2)}{12} = -32.233 \ kNm$$

$$V_x = EI \left\lfloor \frac{12}{5^3} \quad \frac{6}{5^2} \quad -\frac{12}{5^3} \quad \frac{6}{5^2} \right\rfloor \times 10^{-4} \lfloor 0 \quad 0 \quad 0 \quad 2.75688 \rfloor + \frac{(-25 \times 5)}{2} = -56.545 \ kN$$

Element 2: Using Eq. 6.56 and Eq. 6.57, bending moment and shear force are obtained at nodes 1 and 2 as follows:

Node 1:

$$M_x = EI \left\lfloor -\frac{6}{3^2} \quad \frac{-4}{3} \quad \frac{6}{3^2} \quad \frac{-2}{3} \right\rfloor \times 10^{-4} \lfloor 0 \quad 2.75688 \quad 0 \quad -3.845165 \rfloor^T + \frac{(-100 \times 2 \times 1^2)}{3^2} = -32.234 \ kNm$$

$$V_X = EI \left\lfloor \frac{12}{3^3} \quad \frac{6}{3^2} \quad \frac{-12}{3^3} \quad \frac{6}{3^2} \right\rfloor 10^{-4} \lfloor 0 \quad 2.75688 \quad 0 \quad -3.845165 \rfloor^T - \frac{(-100) \times 1^2 \times (3 \times 2 + 1)}{3^3} = 19.396 \ kN$$

Node 2:

$$M_x = EI \left| \frac{6}{3^2} \quad \frac{2}{3} \quad -\frac{6}{3^2} \quad \frac{4}{3} \right| \times 10^{-4} \lfloor 0 \quad 2.75688 \quad 0 \quad \text{-}3.845165 \rfloor^T$$

$$-\frac{(100 \times 1^2 (3 \times 2 + 1))}{3^2} + \frac{-100 \times 2 \times 1^2}{3^2} + (-100 \times 1) = -74.045 \ kNm$$

$$V_X = EI \left| \frac{12}{3^3} \quad \frac{6}{3^2} \quad \frac{-12}{3^3} \quad \frac{6}{3^2} \right| 10^{-4} \lfloor 0 \quad 2.75688 \quad 0 \quad \text{-}3.845165 \rfloor^T + \frac{(-100) \times 2^2 (2 + 3 \times 1)}{3^3} = -80.604 \ kN$$

Element 3: Using Eq. 6.49 and Eq. 6.50, bending moments and shear forces at nodes 1 and 2 are computed as follows:

At node 1,

$$M_x = EI \left| -\frac{6}{4^2} \quad \frac{-4}{4} \quad \frac{6}{4^2} \quad \frac{-2}{4} \right| \times 10^{-4} \lfloor 0 \quad \text{-}3.84516 \quad 0 \quad 9.32998 \rfloor^T + \frac{(-50 \times 4^2)}{12} = -74.075 \ kNm$$

$$V_x = EI \left| \frac{12}{4^3} \quad \frac{6}{4^2} \quad -\frac{12}{4^3} \quad \frac{6}{4^2} \right| \times 10^{-4} \lfloor 0 \quad \text{-}3.84516 \quad 0 \quad 9.32998 \rfloor - \frac{(-50 \times 4)}{2} = 118.511 \ kN$$

At node 2,

$$M_x = EI \left| \frac{6}{4^2} \quad \frac{2}{4} \quad -\frac{6}{4^2} \quad \frac{4}{4} \right| \times 10^{-4} \lfloor 0 \quad \text{-}3.84516 \quad 0 \quad 9.32998 \rfloor + \frac{(-50 \times 4^2)}{12} = 0.0 \ kNm$$

$$V_x = EI \left| \frac{12}{4^3} \quad \frac{6}{4^2} \quad -\frac{12}{4^3} \quad \frac{6}{4^2} \right| \times 10^{-4} \lfloor 0 \quad \text{-}3.84516 \quad 0 \quad 9.32998 \rfloor + \frac{(-50 \times 4)}{2} = -81.489 \ kN$$

Numerical Example 6.10: Analyse the two-span continuous beam shown in Figure 6.19. Adopt $A = 0.09 m^2$, $I = 6 \times 10^{-4} m^4$ and $E = 2 \times 10^8 kN/m^2$ for all members.

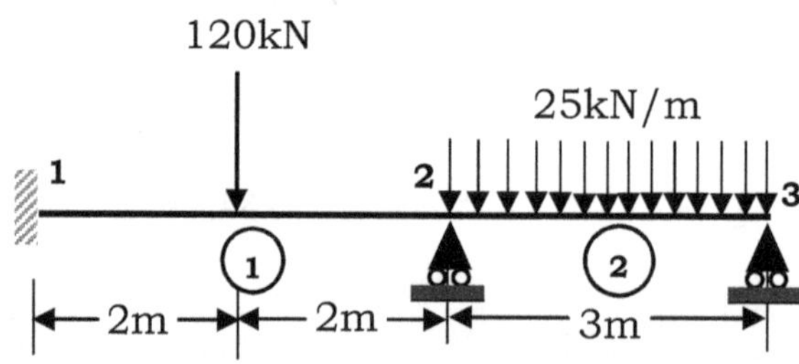

Figure 6.19 Two span continuous beam.

Element 1:

Element stiffness matrix $\left[K_1^{(e)}\right] = 10^4 \begin{bmatrix} 2.25 & 4.5 & -2.25 & 4.5 \\ 4.5 & 12.0 & -4.5 & 6.0 \\ -2.25 & -4.5 & 2.25 & -4.5 \\ 4.0 & 6.0 & -4.5 & 12.0 \end{bmatrix}$

Element force vector $\left\{Q_1^{(e)}\right\} = \begin{bmatrix} -60.0 & -60.0 & -60.0 & 60.0 \end{bmatrix}^T$

Element 2:

Element stiffness matrix $\left[K_2^{(e)}\right] = 10^4 \begin{bmatrix} 5.333 & 8.0 & -5.333 & 8.0 \\ 8.0 & 16.0 & -8.0 & 8.0 \\ -5.333 & -8.0 & 5.333 & -8.0 \\ 8.0 & 8.0 & -8.0 & 16.0 \end{bmatrix}$

Element force vector $\left\{Q_2^{(e)}\right\} = \begin{bmatrix} -37.5 & -18.75 & -37.5 & 18.75 \end{bmatrix}^T$

The overall structure stiffness matrix and force vector are obtained as follows:

$$10^4 \begin{bmatrix} 2.25 & 4.5 & -2.25 & 4.5 & 0 & 0 \\ 4.5 & 12.0 & -4.5 & 6.0 & 0 & 0 \\ -2.25 & -4.5 & 2.25+5.333 & -4.5+8.0 & -5.333 & 8.0 \\ 4.5 & 6.0 & -4.5+8.0 & 12.0+16.0 & -8.0 & 8.0 \\ 0 & 0 & -5.333 & -8.0 & 5.333 & -8.0 \\ 0 & 0 & 8.0 & 8.0 & -8.0 & 16.0 \end{bmatrix} \begin{Bmatrix} v_1 \\ \theta_1 \\ v_2 \\ \theta_2 \\ v_3 \\ \theta_3 \end{Bmatrix} = \begin{Bmatrix} -60.0 \\ -60.0 \\ -60.0-37.5 \\ 60.0-18.75 \\ -37.5 \\ 18.75 \end{Bmatrix}$$

Boundary conditions: $v_1 = \theta_1 = v_2 = v_3 = 0$. The reduced system of equations after application of boundary conditions is

$$10^4 \begin{bmatrix} 28.0 & 8.0 \\ 8.0 & 16.0 \end{bmatrix} \begin{Bmatrix} \theta_2 \\ \theta_3 \end{Bmatrix} = \begin{Bmatrix} 41.25 \\ 18.75 \end{Bmatrix}$$

The solution is given by $\lfloor \theta_2 \quad \theta_3 \rfloor = 10^{-4} \lfloor 1.328125 \quad 0.507812 \rfloor$

Computation of bending moment and shear force:

Element 1: Using Eq. 6.56 and Eq. 6.57, bending moment and shear force are obtained at nodes 1 and 2 as follows:

Node 1:

$$M_x = EI \left\lfloor -\frac{6}{4^2} \quad \frac{-4}{4} \quad \frac{6}{4^2} \quad \frac{-2}{4} \right\rfloor \times 10^{-4} \lfloor 0 \quad 0 \quad 0 \quad 1.328125 \rfloor^T + \frac{(-120 \times 2 \times 2^2)}{4^2} = -67.969 \ kNm$$

$$V_X = EI \left\lfloor \frac{12}{4^3} \quad \frac{6}{4^2} \quad \frac{-12}{4^3} \quad \frac{6}{4^2} \right\rfloor 10^{-4} \lfloor 0 \quad 0 \quad 0 \quad 1.328125 \rfloor^T - \frac{(-120) \times 2^2 \times (3 \times 2 + 2)}{4^3} = 65.977 \ kN$$

Node 2:

$$M_x = EI \left\lfloor \frac{6}{4^2} \quad \frac{2}{4} \quad -\frac{6}{4^2} \quad \frac{4}{4} \right\rfloor \times 10^{-4} \lfloor 0 \quad 0 \quad 0 \quad 1.328125 \rfloor^T$$

$$-\frac{(-120) \times 2^2 (3 \times 2 + 2)}{4^2} + \frac{(-120) \times 2 \times 2^2}{4^2} + (-120) \times 2 = -44.063 \ kNm$$

$$V_X = EI \left\lfloor \frac{12}{4^3} \quad \frac{6}{4^2} \quad \frac{-12}{4^3} \quad \frac{6}{4^2} \right\rfloor 10^{-4} \lfloor 0 \quad 0 \quad 0 \quad 1.328125 \rfloor^T + \frac{(-120) \times 2^2 (2 + 3 \times 2)}{4^3} = -54.023 \ kN$$

Element 2: Using Eq. 5.49 and Eq. 5.50, bending moments and shear forces at nodes 1 and 2 are computed as follows:

At node 1,

$$M_x = EI \left\lfloor -\frac{6}{3^2} \quad \frac{-4}{3} \quad \frac{6}{3^2} \quad \frac{-2}{3} \right\rfloor \times 10^{-4} \lfloor 0 \quad 1.328125 \quad 0 \quad 0.507812 \rfloor^T + \frac{(-25 \times 3^2)}{12} = -44.063 \ kNm$$

$$V_x = EI \left\lfloor \frac{12}{3^3} \quad \frac{6}{3^2} \quad -\frac{12}{3^3} \quad \frac{6}{3^2} \right\rfloor \times 10^{-4} \lfloor 0 \quad 1.328125 \quad 0 \quad 0.507812 \rfloor - \frac{(-25 \times 3)}{2} = 52.188 \ kN$$

At node 2,

$$M_x = EI \left\lfloor \frac{6}{3^2} \quad \frac{2}{3} \quad -\frac{6}{3^2} \quad \frac{4}{3} \right\rfloor \times 10^{-4} \lfloor 0 \quad 1.328125 \quad 0 \quad 0.507812 \rfloor^T + \frac{(-25 \times 3^2)}{12} = 0.0 \ kNm$$

$$V_x = V_x = EI \left\lfloor \frac{12}{3^3} \quad \frac{6}{3^2} \quad -\frac{12}{3^3} \quad \frac{6}{3^2} \right\rfloor \times 10^{-4} \lfloor 0 \quad 1.328125 \quad 0 \quad 0.507812 \rfloor + \frac{(-25 \times 3)}{2} = -22.813 \ kN$$

6.5 PLANE FRAME ELEMENT

Consider a beam element with three degrees of freedom as shown in the Figure 6.20. The degrees of freedom at the first node at u_1, v_1 and θ_1 respectively corresponding to translation along x and y directions and in-plane rotation about the z axis. The corresponding degrees of freedom at the second node are u_2, v_2 and θ_2 respectively.

Figure 6.20 Beam element with 3 d.o.f at each node.

The stiffness matrix for this element can simply be obtained by collating the stiffness contributions corresponding to the bar and beam elements discussed earlier. Thus, the element stiffness matrix can be written as follows:

$$
\left[K^{(e)}\right] =
\begin{bmatrix}
\dfrac{AE}{L} & 0 & 0 & -\dfrac{AE}{L} & 0 & 0 \\[2ex]
0 & \dfrac{12EI}{L^3} & \dfrac{6EI}{L^2} & 0 & -\dfrac{12EI}{L^3} & \dfrac{6EI}{L^2} \\[2ex]
0 & \dfrac{6EI}{L^2} & \dfrac{4EI}{L} & 0 & -\dfrac{6EI}{L^2} & \dfrac{2EI}{L} \\[2ex]
-\dfrac{AE}{L} & 0 & 0 & \dfrac{AE}{L} & 0 & 0 \\[2ex]
0 & -\dfrac{12EI}{L^3} & -\dfrac{6EI}{L^2} & 0 & \dfrac{12EI}{L^3} & -\dfrac{6EI}{L^2} \\[2ex]
0 & \dfrac{6EI}{L^2} & \dfrac{2EI}{L} & 0 & -\dfrac{6EI}{L^2} & \dfrac{4EI}{L}
\end{bmatrix}
\qquad \ldots\ldots(6.58)
$$

The stiffness matrix of the plane frame element is obtained from the above beam element by a process of transformation as outlined below:

6.5.1 Stiffness Matrix for a Plane Frame Element

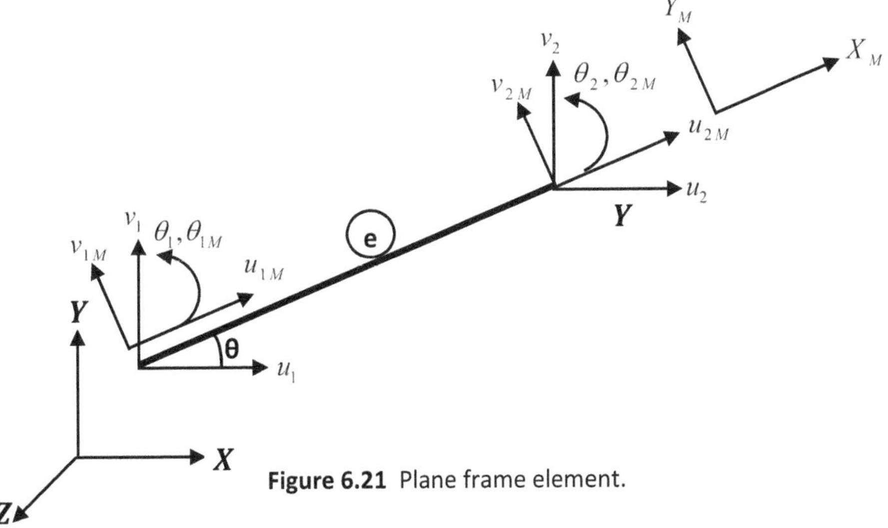

Figure 6.21 Plane frame element.

The plane frame element is shown in the Figure 6.21. The degrees of freedom at the first node are u_1, v_1 and θ_1 (horizontal translation, vertical translation, and in-plane rotation respectively). The corresponding degrees of freedom at the second node are u_2, v_2 and θ_2 respectively.

X, Y and Z axes refer to the structure axes while the corresponding member reference axes are X_M, Y_M and Z_M respectively. The corresponding degrees of freedom along the member axes are u_{1M}, v_{1M} and θ_{1M} at node 1 and u_{2M}, v_{2M} and θ_{2M} at node 2 respectively. Now the displacements with reference to the member axes can be expressed in terms of the corresponding displacements with reference to structure axes as follows:

$$u_{1M} = u_1 \cos\theta + v_1 \sin\theta = u_1 C_X + v_1 C_Y$$

$$v_{1M} = -u_1 \sin\theta + v_1 \cos\theta = -u_1 C_Y + v_1 C_X \qquad \dots\dots (6.59)$$

$$\theta_{1M} = \theta_1$$

Eq. 6.59 can be cast in the matrix form as follows:

$$\begin{Bmatrix} u_{1M} \\ v_{1M} \\ \theta_{1M} \end{Bmatrix} = \begin{bmatrix} C_x & C_y & 0 \\ -C_y & C_x & 0 \\ 0 & 0 & 1 \end{bmatrix} \begin{Bmatrix} u_1 \\ v_1 \\ \theta_1 \end{Bmatrix} \qquad \dots\dots(6.60)$$

A similar relation exists at the node 2 of the element also as follows:

$$\begin{Bmatrix} u_{2M} \\ v_{2M} \\ \theta_{2M} \end{Bmatrix} = \begin{bmatrix} C_x & C_y & 0 \\ -C_y & C_x & 0 \\ 0 & 0 & 1 \end{bmatrix} \begin{Bmatrix} u_2 \\ v_2 \\ \theta_2 \end{Bmatrix} \qquad \dots\dots (6.61)$$

Eq. 6.60 and Eq. 6.61 can be combined as follows:

$$\begin{Bmatrix} u_{1M} \\ v_{1M} \\ \theta_{1M} \\ u_{2M} \\ v_{2M} \\ \theta_{2M} \end{Bmatrix} = \begin{bmatrix} C_x & C_y & 0 & 0 & 0 & 0 \\ -C_y & C_x & 0 & 0 & 0 & 0 \\ 0 & 0 & 1 & 0 & 0 & 0 \\ 0 & 0 & 0 & C_x & C_y & 0 \\ 0 & 0 & 0 & -C_y & C_x & 0 \\ 0 & 0 & 0 & 0 & 0 & 1 \end{bmatrix} \begin{Bmatrix} u_1 \\ v_1 \\ \theta_1 \\ u_2 \\ v_2 \\ \theta_2 \end{Bmatrix} \qquad \dots\dots(6.62)$$

Eq. 6.62 can be expressed as

$$\{d_M\} = [T]\{d\} \qquad .....(6.63)$$

where $[T]$ is the rotation transformation matrix while $\{d\}$ and $\{d_M\}$ are the vectors of nodal displacement with reference to the structure axes and member axes respectively. A similar relationship exists between force components along structure and member axes as shown below:

$$\{Q_M\} = [T]\{Q\} \qquad (6.64)$$

$$\{Q\} = [T]^{-1}\{Q_M\} = [T]^{T}\{Q_M\} \qquad (6.65)$$

Here $\{Q_M\} = \lfloor Q_{1M} \quad Q_{2M} \quad Q_{3M} \quad Q_{4M} \quad Q_{5M} \quad Q_{6M} \rfloor^{T}$ is the vector of force components with reference to the member axes along and $\{Q\} = \lfloor Q_1 \quad Q_2 \quad Q_3 \quad Q_4 \quad Q_5 \quad Q_6 \rfloor^{T}$ along the structure axes. Here it is to be noted that the rotation transformation matrix is orthonormal. Thus $[T]^{T}[T] = [I]$ and $[T]$ represents a *similarity transformation.*

The stiffness equation for the plane frame element with reference to the member axes can be written as follows:

$$\left[K_M^{(e)} \right]\{Q_M\} = \{d_M\} \qquad .....(6.66)$$

Using Eq. 6.62 and Eq. 6.64 in Eq. 6.66 we get

$$\left[K_M^{(e)} \right][T]\{Q\} = [T]\{d\} \qquad .....(6.67)$$

Pre-multiplying on both sides by $[T]^{T}$ yields

$$[T]^{T}\left[K_M^{(e)} \right][T]\{Q\} = \{d\} \Rightarrow \left[K^{(e)} \right]\{Q\} = \{d\} \qquad ......(6.68)$$

where $\left[K^{(e)} \right]$ and $\{d\}$ are the stiffness matrix and vector of nodal displacements of the element with reference to the structure axes.

The structure stiffness matrix $[K]$ is assembled from the stiffness matrices of individual elements.

6.5.2   Force Vector for Plane Frame Element

Forces can be applied on a plane frame either directly at the nodes of the structure or over the body of the elements themselves. Thus, the overall force vector is obtained as

$$\{Q\} = \{F\} + \{Q\} \qquad \qquad .....(6.69)$$

Here $\{F\}$ is the force vector due to loads directly acting at the nodes of the structure. The vector $\{Q_{eq}\}$ is obtained by combining the element load vectors owing to loads acting on the body of each element.

For an element acted upon by a uniformly distributed load over the entire span, Eq. 6.37 is modified to give the element force vector with reference to the member axes as

$$\{Q_M^{(e)}\} = \left[0 \quad \frac{pL}{2} \quad \frac{pL^2}{12} \quad 0 \quad \frac{pL}{2} \quad -\frac{pL^2}{12} \right]^T \qquad .....(6.70)$$

For an element acted upon by a point load at the mid span, Eq. 6.39 is modified to give the element force vector with reference to the member axes as

$$\{Q_M^{(e)}\} = \left[0 \quad \frac{W}{2} \quad \frac{WL}{8} \quad 0 \quad \frac{W}{2} \quad -\frac{WL}{8} \right]^T \qquad .....(6.71)$$

For an element acted upon by an eccentric point load at a distance of a from the left end of the element and b from the right end of the element, Eq. 6.40 is modified to give the element force vector with reference to the member axes as

$$\{Q_M^{(e)}\} = \left[0 \quad \frac{Wb}{L} \quad \frac{Wab^2}{L^2} \quad 0 \quad \frac{Wa}{L} \quad -\frac{Wa^2b}{L^2} \right]^T \qquad .....(6.72)$$

Contributions from the element force vectors are assembled to form the equivalent force vector due to element forces $\{Q_{eq}\}$. Force vector $\{F\}$ is assembled directly from the nodal forces and moments. The combined force vector for the structure $\{Q\}$ is then obtained using Eq. 6.69. This results in the stiffness equation for the entire structure as $[K]\{d\} = \{Q\}$. Boundary conditions are applied and the solution vector $\{d\}$ is obtained by the solution of the reduced system of stiffness equations resulting out of application of boundary conditions.

Post-processing

Axial force is calculated using Eq.6.2 as follows:

$$F = AE\frac{\partial u}{\partial x} = AE\left[-\frac{1}{L} \quad \frac{1}{L} \right]\left\{ \begin{array}{c} u_1 \\ u_2 \end{array} \right\} \qquad (6.73)$$

Bending moment and shear force at nodes 1 and 2 are computed using Eq. 6.8, Eq. 6.49, Eq.6.55 and Eq. 6.56.

6.5.3 Numerical Examples

Numerical Example 6.11: Analyse the single bay single storey portal frame shown in Figure 6.22. Adopt

$$A = 0.06m^2, I = 4.5\times10^{-4}m^4, E = 2\times10^8 \; kN/m^2$$

for all members

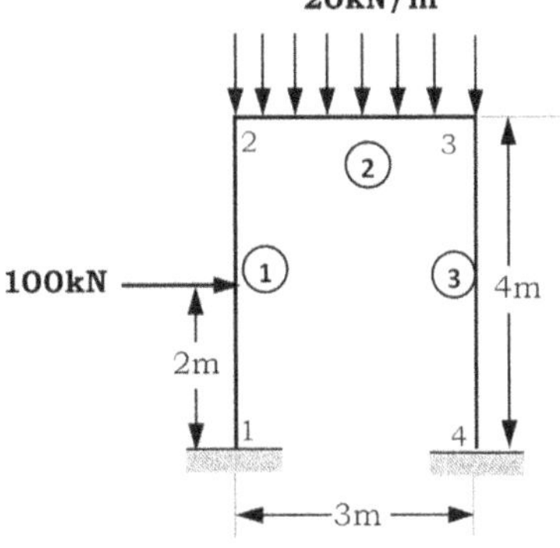

Figure 6.22 Portal Frame.

Solution:

Element 1: Element 1 is connected between nodes 1 and 2. The corresponding degrees of freedom are 1,2,3,4,5 and 6. The direction cosines for the element are $C_x = 0$ $C_y = 1$.

Element stiffness matrix with reference to member axes is

$$\left[K_{1M}^{(e)}\right] = 10^5 \begin{bmatrix} 30 & 0 & 0 & -30 & 0 & 0 \\ 0 & 0.16875 & 0.3375 & 0 & -0.16875 & 0.3375 \\ 0 & 0.3375 & 0.9 & 0 & -0.3375 & 0.45 \\ -30 & 0 & 0 & 30 & 0 & 0 \\ 0 & -0.16875 & -0.3375 & 0 & 0.16875 & -0.3375 \\ 0 & 0.3375 & 0.45 & 0 & -0.3375 & 0.9 \end{bmatrix}$$

The rotation transformation matrix is given by

$$\left[T_1^{(e)}\right] = \begin{bmatrix} 0 & 1 & 0 & 0 & 0 & 0 \\ -1 & 0 & 0 & 0 & 0 & 0 \\ 0 & 0 & 0 & 0 & 0 & 0 \\ 0 & 0 & 0 & 0 & 1 & 0 \\ 0 & 0 & 0 & -1 & 0 & 0 \\ 0 & 0 & 0 & 0 & 0 & 0 \end{bmatrix}$$

The element stiffness matrix with reference to structure axes is given by

$$\left[K_1^{(e)}\right] = \left[T^T\right]\left[K_{1M}^{(e)}\right][T] = \begin{bmatrix} 0.16875 & 0 & -0.3375 & -0.16875 & 0 & -0.3375 \\ 0 & 30 & 0 & 0 & -30 & 0 \\ -0.3375 & 0 & 0.9 & 0.3375 & 0 & 0.45 \\ -0.16875 & 0 & 0.3375 & 0.16875 & 0 & 0.3375 \\ 0 & -30 & 0 & 0 & 30 & 0 \\ -0.3375 & 0 & 0.45 & 0.3375 & 0 & 0.9 \end{bmatrix}$$

The element force vector with reference to member axes is given by

$$\left\{Q_{1M}^{(e)}\right\} = \lfloor 0 \quad -50 \quad -50 \quad 0 \quad -50 \quad 50 \rfloor^T$$

The element force vector with reference to structure axes is given by

$$\left\{Q_1^{(e)}\right\} = \lfloor 50 \quad 0 \quad -50 \quad 50 \quad 0 \quad 50 \rfloor^T$$

Element 2: Element 2 is connected between nodes 2 and 3. The corresponding degrees of freedom are 4, 5, 6, 7, 8 and 9. The direction cosines of the element are $C_x = 1 \quad C_y = 0$

Element stiffness matrix with reference to local axes is

$$\left[K_{2M}^{(e)}\right] = 10^5 \begin{bmatrix} 40 & 0 & 0 & -40 & 0 & 0 \\ 0 & 0.4 & 0.6 & 0 & -0.4 & 0.6 \\ 0 & 0.6 & 1.2 & 0 & -0.6 & 0.6 \\ -40 & 0 & 0 & 40 & 0 & 0 \\ 0 & -0.4 & -0.6 & 0 & 0.4 & -0.6 \\ 0 & 0.6 & 0.6 & 0 & -0.6 & 1.2 \end{bmatrix}$$

The rotation transformation matrix is given by

$$\left[T_2^{(e)}\right] = \begin{bmatrix} 1 & 0 & 0 & 0 & 0 & 0 \\ 0 & 1 & 0 & 0 & 0 & 0 \\ 0 & 0 & 1 & 0 & 0 & 0 \\ 0 & 0 & 0 & 1 & 0 & 0 \\ 0 & 0 & 0 & 0 & 1 & 0 \\ 0 & 0 & 0 & 0 & 0 & 1 \end{bmatrix}$$

The element stiffness matrix with reference to structure axes is given by

$$\left[K_2^{(e)}\right]=\left[T_2^T\right]\left[K_2^{(e)}\right]\left[T_2\right]=10^5\begin{bmatrix} 40 & 0 & 0 & -40 & 0 & 0 \\ 0 & 0.4 & 0.6 & 0 & -0.4 & 0.6 \\ 0 & 0.6 & 1.2 & 0 & -0.6 & 0.6 \\ -40 & 0 & 0 & 40 & 0 & 0 \\ 0 & -0.4 & -0.6 & 0 & 0.4 & -0.6 \\ 0 & 0.6 & 0.6 & 0 & -0.6 & 1.2 \end{bmatrix}$$

The element force vector with reference to member axes is given by

$$\left\{Q_{1M}^{(e)}\right\}=\lfloor 0 \quad -30 \quad -15 \quad 0 \quad -30 \quad 15\rfloor^T$$

The element force vector with reference to structure axes is given by

$$\left\{Q_1^{(e)}\right\}=\lfloor 0 \quad -30 \quad -15 \quad 0 \quad -30 \quad 15\rfloor^T$$

Element 3: Element 3 is connected between nodes 3 and 4. The corresponding degrees of freedom are 7, 8, 9, 10, 11 and 12. The direction cosines for the element are $C_x=0$ $C_y=1$. Element stiffness matrix with reference to member axes is

$$\left[K_{2M}^{(e)}\right]=10^5\begin{bmatrix} 30 & 0 & 0 & -30 & 0 & 0 \\ 0 & 0.16875 & 0.3375 & 0 & -0.16875 & 0.3375 \\ 0 & 0.3375 & 0.9 & 0 & -0.3375 & 0.45 \\ -30 & 0 & 0 & 30 & 0 & 0 \\ 0 & -0.16875 & -0.3375 & 0 & 0.16875 & -0.3375 \\ 0 & 0.3375 & 0.45 & 0 & -0.3375 & 0.9 \end{bmatrix}$$

The rotation transformation matrix is given by

$$\left[T_1^{(e)}\right]=\left[\begin{array}{ccc|ccc} 0 & -1 & 0 & 0 & 0 & 0 \\ 1 & 0 & 0 & 0 & 0 & 0 \\ 0 & 0 & 0 & 0 & 0 & 0 \\ \hline 0 & 0 & 0 & 0 & -1 & 0 \\ 0 & 0 & 0 & 1 & 0 & 0 \\ 0 & 0 & 0 & 0 & 0 & 0 \end{array}\right]$$

The element stiffness matrix with reference to structure axes is given by

$$\left[K_3^{(e)}\right] = \left[T_3^T\right]\left[K_3^{(e)}\right]\left[T_3\right] = 10^5 \begin{bmatrix} 40 & 0 & 0 & -40 & 0 & 0 \\ 0 & 0.4 & 0.6 & 0 & -0.4 & 0.6 \\ 0 & 0.6 & 1.2 & 0 & -0.6 & 0.6 \\ -40 & 0 & 0 & 40 & 0 & 0 \\ 0 & -0.4 & -0.6 & 0 & 0.4 & -0.6 \\ 0 & 0.6 & 0.6 & 0 & -0.6 & 1.2 \end{bmatrix}$$

Element 3 is not subjected to any loading, $\{Q_3^{(e)}\} = \lfloor 0 \ \ 0 \ \ 0 \ \ 0 \ \ 0 \ \ 0 \rfloor^T$.

Assembly: The structure stiffness matrix and force vector are assembled as follows:

$$\begin{bmatrix} 0.16875 & 0 & -0.3375 & -0.16875 & 0 & -0.3375 & 0 & 0 & 0 & 0 & 0 & 0 \\ 0 & 30 & 0 & 0 & -30 & 0 & 0 & 0 & 0 & 0 & 0 & 0 \\ -0.3375 & 0 & 0.9 & 0.3375 & 0 & 0.45 & 0 & 0 & 0 & 0 & 0 & 0 \\ -0.16875 & 0 & 0.3375 & 40.16875 & 0 & 0.3375 & -40 & 0 & 0 & 0 & 0 & 0 \\ 0 & -30 & 0 & 0 & 30.4 & 0.6 & 0 & -0.4 & 0.6 & 0 & 0 & 0 \\ -0.3375 & 0 & 0.45 & 0.3375 & 0.6 & 2.1 & 0 & -0.6 & 0.6 & 0 & 0 & 0 \\ 0 & 0 & 0 & -40 & 0 & 0 & 40.16875 & 0 & 0.3375 & -0.16875 & 0 & 0.3375 \\ 0 & 0 & 0 & 0 & -0.4 & -0.6 & 0 & 30.4 & -0.6 & 0 & -30 & 0 \\ 0 & 0 & 0 & 0 & 0.6 & 0.6 & 0.3375 & -0.6 & 2.1 & -0.3375 & 0 & 0.45 \\ 0 & 0 & 0 & 0 & 0 & 0 & -0.16875 & 0 & -0.3375 & 0.16875 & 0 & -0.3375 \\ 0 & 0 & 0 & 0 & 0 & 0 & 0 & -30 & 0 & 0 & 30 & 0 \\ 0 & 0 & 0 & 0 & 0 & 0 & 0.3375 & 0 & 0.45 & -0.3375 & 0 & 0.9 \end{bmatrix} \begin{Bmatrix} u_1 \\ v_1 \\ \theta_1 \\ u_2 \\ v_2 \\ \theta_2 \\ u_3 \\ v_3 \\ \theta_3 \\ u_4 \\ v_4 \\ \theta_4 \end{Bmatrix} = \begin{Bmatrix} 50 \\ 0 \\ -50 \\ 50 \\ -30 \\ 35 \\ 0 \\ -30 \\ 15 \\ 0 \\ 0 \\ 0 \end{Bmatrix}$$

Boundary conditions:

$$u_1 = v_1 = \theta_1 = u_4 = v_4 = \theta_4 = 0$$

The reduced system of equations after application of boundary conditions is

$$\begin{bmatrix} 40.16875 & 0 & 0.3375 & -40 & 0 & 0 \\ 0 & 30.4 & 0.6 & 0 & -0.4 & 0.6 \\ 0.3375 & 0.6 & 2.1 & 0 & -0.6 & 0.6 \\ -40 & 0 & 0 & 40.16875 & 0 & 0.3375 \\ 0 & -0.4 & -0.6 & 0 & 30.4 & -0.6 \\ 0 & 0.6 & 0.6 & 0.3375 & -0.6 & 2.1 \end{bmatrix} \begin{Bmatrix} u_2 \\ v_2 \\ \theta_2 \\ u_3 \\ v_3 \\ \theta_3 \end{Bmatrix} = \begin{Bmatrix} 50 \\ -30 \\ 35 \\ 0 \\ -30 \\ 15 \end{Bmatrix}$$

The solution vector is given as $\lfloor u_2 \ \ v_2 \ \ \theta_2 \ \ u_3 \ \ v_3 \ \ \theta_3 \rfloor = 10^{-3} \lfloor 1.73707 \ \ -0.000507 \ \ 0.006034 \ \ 1.73139 \ \ -0.001492 \ \ 0.0192402 \rfloor$

Forces and Moments in Elements

Axial force, shear force and bending moment at nodes 1 and 2 of elements 1,2 and 3 are tabulated (Table 6.9):

Table 6.9 Forces and Moments.

Element	At Node 1			At Node 2		
	Axial	**Shear**	**Bending**	**Axial**	**Shear**	**Bending**
1	15.229	76.276	-105.911	-15.229	-22.724	3.195
2	22.724	15.229	3.195	-22.724	-44.771	-41.118
3	44.771	22.724	-41.118	-44.771	22.724	49.776

6.6 SPACE FRAME ELEMENT

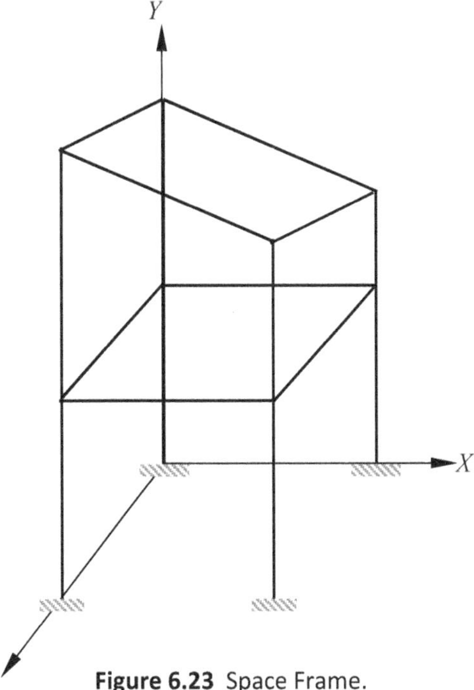

Figure 6.23 Space Frame.

Consider a space frame as shown in the Figure 6.23. Members of the frame are oriented in space. An element of this frame is known as a space frame element. The objective of the present discussion is to develop the stiffness matrix and force vector for the space frame element.

122 | **Finite Element Analysis of Structures using MATLAB**

6.6.1 Stiffness Matrix for the Space Frame Element

A space frame element exhibits axial deformation in addition to shear and flexural deformations. The stiffness matrix of this element with reference to the local axes is given as the follows:

$$\left[K_M^{(e)}\right] =$$

$$
\begin{bmatrix}
\dfrac{EA}{L} & 0 & 0 & 0 & 0 & 0 & -\dfrac{EA}{L} & 0 & 0 & 0 & 0 & 0 \\[2ex]
0 & \dfrac{12EI_z}{L^3} & 0 & 0 & 0 & \dfrac{6EI_z}{L^2} & 0 & -\dfrac{12EI_z}{L^3} & 0 & 0 & 0 & \dfrac{6EI_z}{L^2} \\[2ex]
0 & 0 & \dfrac{12EI_y}{L^3} & 0 & -\dfrac{6EI_y}{L^2} & 0 & 0 & 0 & -\dfrac{12EI_y}{L^3} & 0 & -\dfrac{6EI_y}{L^2} & 0 \\[2ex]
0 & 0 & 0 & \dfrac{GI_x}{L} & 0 & 0 & 0 & 0 & 0 & \dfrac{GI_x}{L} & 0 & 0 \\[2ex]
0 & 0 & -\dfrac{6EI_y}{L^2} & 0 & \dfrac{4EI_y}{L} & 0 & 0 & 0 & \dfrac{6EI_y}{L^2} & 0 & \dfrac{2EI_y}{L} & 0 \\[2ex]
0 & \dfrac{6EI_z}{L^2} & 0 & 0 & 0 & \dfrac{4EI_z}{L} & 0 & -\dfrac{6EI_z}{L^2} & 0 & 0 & 0 & \dfrac{2EI_z}{L} \\[2ex]
-\dfrac{EA}{L} & 0 & 0 & 0 & 0 & 0 & \dfrac{EA}{L} & 0 & 0 & 0 & 0 & 0 \\[2ex]
0 & -\dfrac{12EI_z}{L^3} & 0 & 0 & 0 & -\dfrac{6EI_z}{L^2} & 0 & \dfrac{12EI_z}{L^3} & 0 & 0 & 0 & -\dfrac{6EI_z}{L^2} \\[2ex]
0 & 0 & -\dfrac{12EI_y}{L^3} & 0 & \dfrac{6EI_y}{L^2} & 0 & 0 & 0 & \dfrac{12EI_y}{L^3} & 0 & \dfrac{6EI_y}{L^2} & 0 \\[2ex]
0 & 0 & 0 & -\dfrac{GI_x}{L} & 0 & 0 & 0 & 0 & 0 & \dfrac{GI_x}{L} & 0 & 0 \\[2ex]
0 & 0 & -\dfrac{6EI_y}{L^2} & 0 & \dfrac{2EI_y}{L} & 0 & 0 & 0 & \dfrac{6EI_y}{L^2} & 0 & \dfrac{4EI_y}{L} & 0 \\[2ex]
0 & \dfrac{6EI_z}{L^2} & 0 & 0 & 0 & \dfrac{2EI_z}{L} & 0 & -\dfrac{6EI_z}{L^2} & 0 & 0 & 0 & \dfrac{4EI_z}{L}
\end{bmatrix}
$$

$$.....(6.74)$$

6.6.2 Rotation Transformation Matrix for the Space Frame Element

Consider a space frame element as shown in Figure 6.24. The element has two nodes and has six degrees of freedom at each node, making it a total of 12 degrees of freedom for the element. The local degrees of freedom for the element and the corresponding force components are shown in Figure 6.24. Also, global degrees of freedom for the element and the corresponding force components are shown in Figure 6.25.

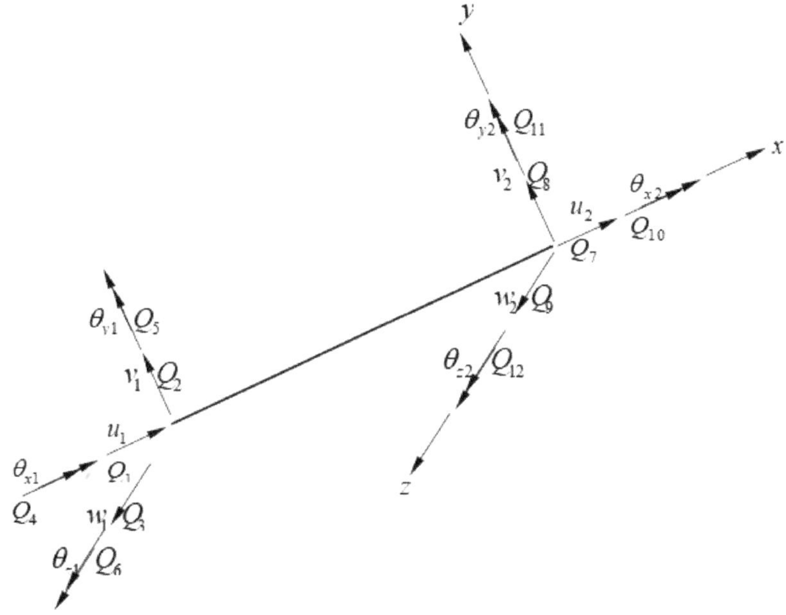

Figure 6.24 Space Frame element- Local degrees of Freedom.

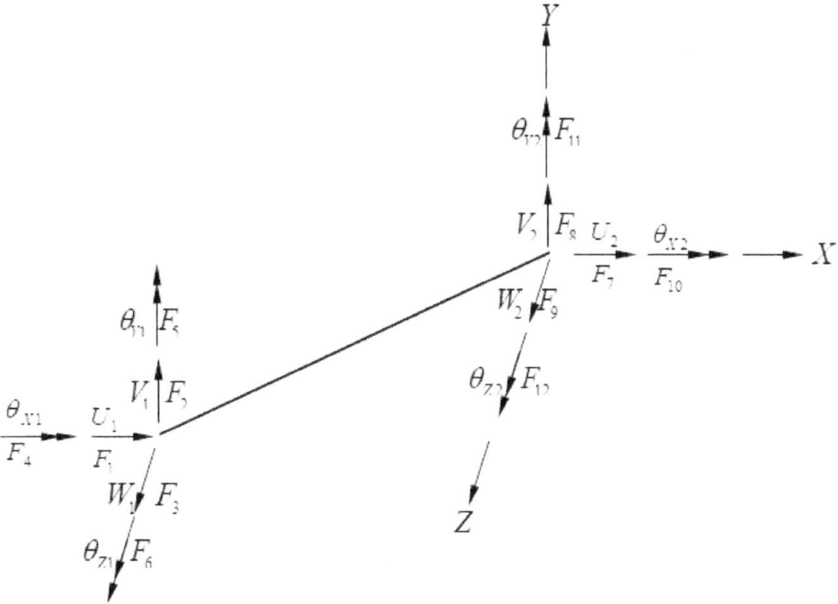

Figure 6.25 Space Frame element- Global degrees of Freedom.

The inclinations made by the local x, y and z axes with the global X, Y and Z axes are shown in Figures 6.26(a), (b) and (c) respectively.

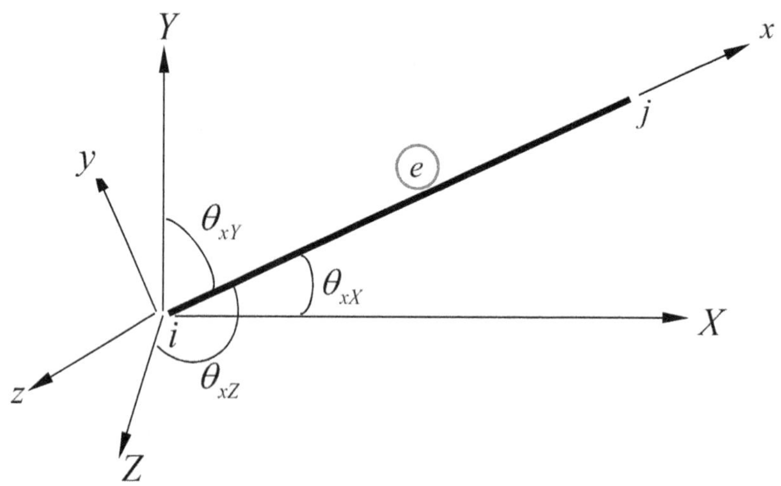

Figure 6.26(a) Space Frame element- Local and Global axes.

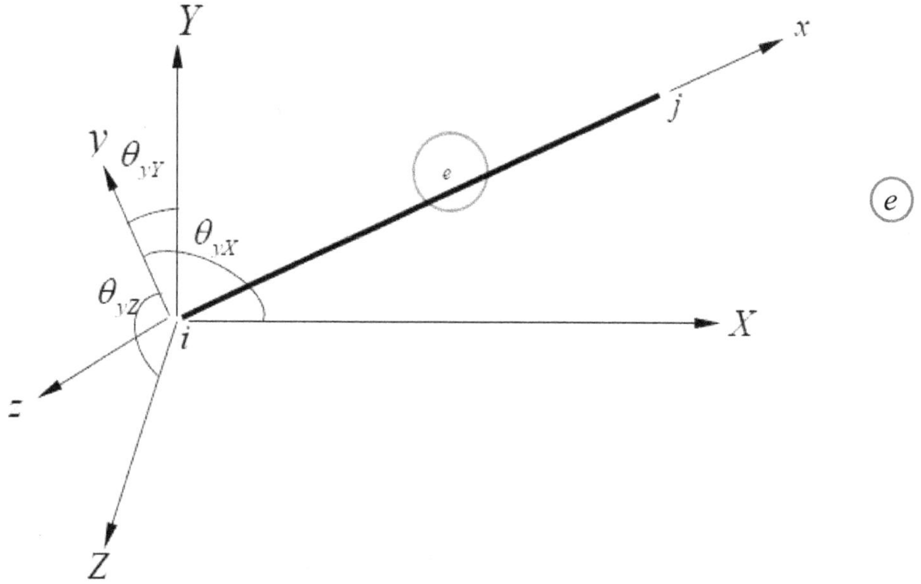

Figure 6.26(b) Space Frame element- Local and Global axes.

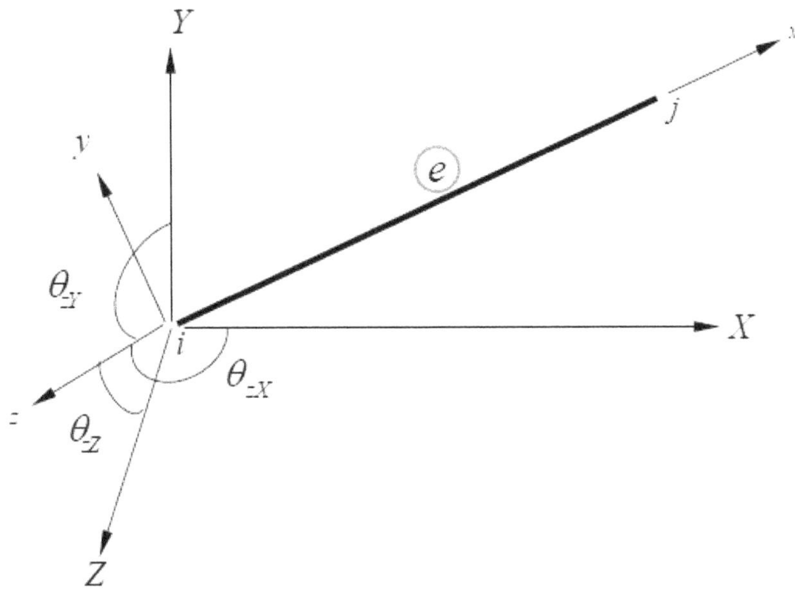

Figure 6.26(c) Space Frame element- Local and Global axes.

If Q_1, Q_2 and Q_3 are the forces along the local x, y and z axes respectively and F_1, F_2 and F_3 are the forces along the global X, Y and Z axes respectively, the relationship between them can be stated in terms of the direction cosines of the local axes with reference to the global axes as follows:

$$\begin{Bmatrix} Q_1 \\ Q_2 \\ Q_3 \end{Bmatrix} = \begin{bmatrix} C_{xX} & C_{xY} & C_{xZ} \\ C_{yX} & C_{yY} & C_{yZ} \\ C_{zX} & C_{zY} & C_{zZ} \end{bmatrix} \begin{Bmatrix} F_1 \\ F_2 \\ F_3 \end{Bmatrix} = [T_1] \begin{Bmatrix} F_1 \\ F_2 \\ F_3 \end{Bmatrix} \qquad .....(6.75a)$$

Similar relationship can be obtained for the remaining forces components as well as follows:

$$\begin{Bmatrix} Q_4 \\ Q_5 \\ Q_6 \end{Bmatrix} = [T_1] \begin{Bmatrix} F_4 \\ F_5 \\ F_6 \end{Bmatrix} ; \begin{Bmatrix} Q_7 \\ Q_8 \\ Q_9 \end{Bmatrix} = [T_1] \begin{Bmatrix} F_7 \\ F_8 \\ F_9 \end{Bmatrix} \qquad .....(6.75b)$$

Eq. 6.75 (a) and (b) can be combined as

$$\{Q\} = [T]\{F\}$$
$$\{F\} = [T]^T \{Q\} \qquad .....(6.76)$$

where

$$\{Q\} = \lfloor Q_1 \quad Q_2 \quad Q_3 \quad Q_4 \quad Q_5 \quad Q_6 \quad Q_7 \quad Q_8 \quad Q_9 \quad Q_{10} \quad Q_{11} \quad Q_{12} \rfloor^T$$

$$\{F\} = \lfloor F_1 \quad F_2 \quad F_3 \quad F_4 \quad F_5 \quad F_6 \quad F_7 \quad F_8 \quad F_9 \quad F_{10} \quad F_{11} \quad F_{12} \rfloor^T \qquad .....(6.77)$$

and

$$[T] = \begin{bmatrix} T_1 & O & O & O \\ O & T_1 & O & O \\ O & O & T_1 & O \\ O & O & O & T_1 \end{bmatrix}; \quad [O] = \begin{bmatrix} 0 & 0 & 0 & 0 \\ 0 & 0 & 0 & 0 \\ 0 & 0 & 0 & 0 \\ 0 & 0 & 0 & 0 \end{bmatrix} \qquad .....(6.78)$$

Since member end displacements are also vector quantities, the relationship between local displacements and global displacements can be stated as follows:

$$\{u\} = [T]\{U\} \qquad .....(6.79)$$

where

$$\{u\} = \lfloor u_1 \quad u_2 \quad u_3 \quad u_4 \quad u_5 \quad u_6 \quad u_7 \quad u_8 \quad u_9 \quad u_{10} \quad u_{11} \quad u_{12} \rfloor^T$$

$$\{U\} = \lfloor U_1 \quad U_2 \quad U_3 \quad U_4 \quad U_5 \quad U_6 \quad U_7 \quad U_8 \quad U_9 \quad U_{10} \quad U_{11} \quad U_{12} \rfloor^T \qquad .....(6.80)$$

Once the rotation transformation matrix $[T_1]$ is obtained, $[T]$ can be obtained using Eq. 6.77 to obtain the global stiffness matrix as $\left[K^{(e)} \right] = [T]^T \left[K_M^{(e)} \right] [T]$.

The next step is to obtain the rotation transformation matrix $[T_1]$. Consider the element e as shown in the Figure 6.26(a). The coordinates of nodes i and j are used to compute the direction cosines in the first row of $[T_1]$ as follows:

$$C_{xX} = \cos\theta_{xX} = \frac{\left(x_j - x_i\right)}{L}$$

$$C_{xY} = \cos\theta_{xY} = \frac{\left(y_j - y_i\right)}{L} \qquad .....(6.81)$$

$$C_{xZ} = \cos\theta_{xZ} = \frac{\left(z_j - z_i\right)}{L}$$

$$L = \sqrt{\left(x_j - x_i\right)^2 + \left(y_j - y_i\right)^2 + \left(z_j - z_i\right)^2}$$

Now, the direction cosines in the second and third rows of $[T_1]$ need to be calculated.

Let I_X, I_Y and I_Z be the unit vectors along X, Y and Z axes respectively. Also, let i_x, i_y and i_z be the unit vectors along x, y and z axes respectively. The relationship between these two sets of unit vectors is given by

$$\begin{Bmatrix} i_x \\ i_y \\ i_z \end{Bmatrix} = [T_1] \begin{Bmatrix} I_X \\ I_Y \\ I_Z \end{Bmatrix} \qquad .....(6.82)$$

The member's actual orientation, as seen in Figure 6.27, is reached in two steps:

(a) In the first step, the x-axis of the member is allowed be inclined in the desired direction as per design. The y and z axes of the member are so oriented that xy plane is vertical while the z axis of the member lies in a horizontal plane. The y and z axes of the member, in the present (imaginary) orientation, as designated as $\bar{y}$ and $\bar{z}$ axes to indicate that these axes have not yet reached their final orientation. Thus $x\bar{y}$ plane is vertical just as XY plane is vertical. As the $\bar{z}$ axis is perpendicular to the $x\bar{y}$ plane, a vector $\bar{z}$ oriented along the $\bar{z}$ axis can be expressed as a cross product of the unit vectors i_X and I_Y as follows:

$$\bar{z} = i_x \times I_Y = \begin{vmatrix} I_X & I_Y & I_Z \\ C_{xX} & C_{xY} & C_{xZ} \\ 0 & 1 & 0 \end{vmatrix} = -C_{xZ} I_X + 0.I_Y + C_{xX} I_Z$$

A unit vector along the direction of vector $\bar{z}$ is given by

$$i_{\bar{z}} = \frac{1}{|\bar{z}|} \bar{z} = \frac{-C_{xZ}}{\sqrt{C_{xX}^2 + C_{xZ}^2}} I_X + 0 I_Y + \frac{C_{xX}}{\sqrt{C_{xX}^2 + C_{xZ}^2}} I_Z \qquad .....(6.83)$$

The unit vector $i_{\bar{y}}$ can be expressed as product of unit vectors i_x and $i_{\bar{z}}$ as

$$i_{\bar{y}} = i_x \times i_{\bar{z}}$$

$$i_{\bar{y}} = \begin{vmatrix} I_X & I_Y & I_Z \\[2mm] \dfrac{-C_{xZ}}{\sqrt{C_{xX}^2 + C_{xZ}^2}} & 0 & \dfrac{C_{xX}}{\sqrt{C_{xX}^2 + C_{xZ}^2}} \\[2mm] C_{xX} & C_{xY} & C_{xZ} \end{vmatrix}$$

$$= \left(\dfrac{C_{xX} C_{xY}}{\sqrt{C_{xX}^2 + C_{xZ}^2}} \right) I_X + \left(\sqrt{C_{xX}^2 + C_{xZ}^2} \right) I_Y + \left(\dfrac{-C_{xY} C_{xZ}}{\sqrt{C_{xX}^2 + C_{xZ}^2}} \right) I_Z$$

$$.....(6.84)$$

The relationship between unit vectors $i_x, i_{\bar{y}}$ and $i_{\bar{z}}$ along $x, \bar{y}, \bar{z}$ axes and corresponding unit vectors I_X, I_Y and I_Z along X, Y, Z axes can be represented as

$$\left\{ \begin{array}{c} i_x \\ i_{\bar{y}} \\ i_{\bar{z}} \end{array} \right\} = \begin{bmatrix} C_{xX} & C_{xY} & C_{xZ} \\[2mm] \dfrac{C_{xX} C_{xY}}{\sqrt{C_{xX}^2 + C_{xZ}^2}} & \sqrt{C_{xX}^2 + C_{xZ}^2} & \dfrac{-C_{xY} C_{xZ}}{\sqrt{C_{xX}^2 + C_{xZ}^2}} \\[2mm] \dfrac{-C_{xZ}}{\sqrt{C_{xX}^2 + C_{xZ}^2}} & 0 & \dfrac{C_{xX}}{\sqrt{C_{xX}^2 + C_{xZ}^2}} \end{bmatrix} \left\{ \begin{array}{c} I_X \\ I_Y \\ I_Z \end{array} \right\} \qquad .....(6.85)$$

(b) Now we rotate the $\bar{y}$ and $\bar{z}$ axes about the x axis, in anti-clockwise sense, through an angle α (called the *angle of roll*) so that the principal axes of the member are in their desired orientations. After rotating anti-clockwise through an angle α, the $\bar{y}$ and $\bar{z}$ axes now reach their final orientation and are now called y, z axes. The unit vectors along $x, \bar{y}, \bar{z}$ axes can now be expressed in terms of corresponding unit vectors along x, y, z axes as follows:

$$\left\{ \begin{array}{c} i_x \\ i_y \\ i_z \end{array} \right\} = \begin{bmatrix} 1 & 0 & 0 \\ 0 & \cos\alpha & \sin\alpha \\ 0 & -\sin\alpha & \cos\alpha \end{bmatrix} \left\{ \begin{array}{c} i_x \\ i_{\bar{y}} \\ i_{\bar{z}} \end{array} \right\} \qquad .....(6.86)$$

Substituting Eq. 6.85 into the right-hand side of Eq.6.85, we obtain

$$\begin{Bmatrix} i_x \\ i_y \\ i_z \end{Bmatrix} = \begin{bmatrix} 1 & 0 & 0 \\ 0 & \cos\alpha & \sin\alpha \\ 0 & -\sin\alpha & \cos\alpha \end{bmatrix} \begin{bmatrix} C_{xX} & C_{xY} & C_{xZ} \\ \dfrac{C_{xX}C_{xY}}{\sqrt{C_{xX}^2 + C_{xZ}^2}} & \sqrt{C_{xX}^2 + C_{xZ}^2} & \dfrac{-C_{xY}C_{xZ}}{\sqrt{C_{xX}^2 + C_{xZ}^2}} \\ \dfrac{-C_{xZ}}{\sqrt{C_{xX}^2 + C_{xZ}^2}} & 0 & \dfrac{C_{xX}}{\sqrt{C_{xX}^2 + C_{xZ}^2}} \end{bmatrix} \begin{Bmatrix} I_X \\ I_Y \\ I_Z \end{Bmatrix} \quad .....(6.87)$$

Comparing Eq. 6.87 with Eq. 6.82, we obtain

$$[T_1] = \begin{bmatrix} 1 & 0 & 0 \\ 0 & \cos\alpha & \sin\alpha \\ 0 & -\sin\alpha & \cos\alpha \end{bmatrix} \begin{bmatrix} C_{xX} & C_{xY} & C_{xZ} \\ \dfrac{C_{xX}C_{xY}}{\sqrt{C_{xX}^2 + C_{xZ}^2}} & \sqrt{C_{xX}^2 + C_{xZ}^2} & \dfrac{-C_{xY}C_{xZ}}{\sqrt{C_{xX}^2 + C_{xZ}^2}} \\ \dfrac{-C_{xZ}}{\sqrt{C_{xX}^2 + C_{xZ}^2}} & 0 & \dfrac{C_{xX}}{\sqrt{C_{xX}^2 + C_{xZ}^2}} \end{bmatrix} \quad .....(6.88)$$

Simplifying, we obtain

$$[T_1] = \begin{bmatrix} C_{xX} & C_{xY} & C_{xZ} \\ \dfrac{-C_{xX}C_{xY}\cos\alpha - C_{xZ}\sin\alpha}{\sqrt{C_{xX}^2 + C_{xZ}^2}} & \sqrt{C_{xX}^2 + C_{xZ}^2}\cos\alpha & \dfrac{-C_{xY}C_{xZ}\cos\alpha + C_{xX}\sin\alpha}{\sqrt{C_{xX}^2 + C_{xZ}^2}} \\ \dfrac{C_{xX}C_{xY}\sin\alpha - C_{xZ}\cos\alpha}{\sqrt{C_{xX}^2 + C_{xZ}^2}} & -\sqrt{C_{xX}^2 + C_{xZ}^2}\sin\alpha & \dfrac{C_{xY}C_{xZ}\sin\alpha + C_{xX}\cos\alpha}{\sqrt{C_{xX}^2 + C_{xZ}^2}} \end{bmatrix}$$

$$.....(6.89)$$

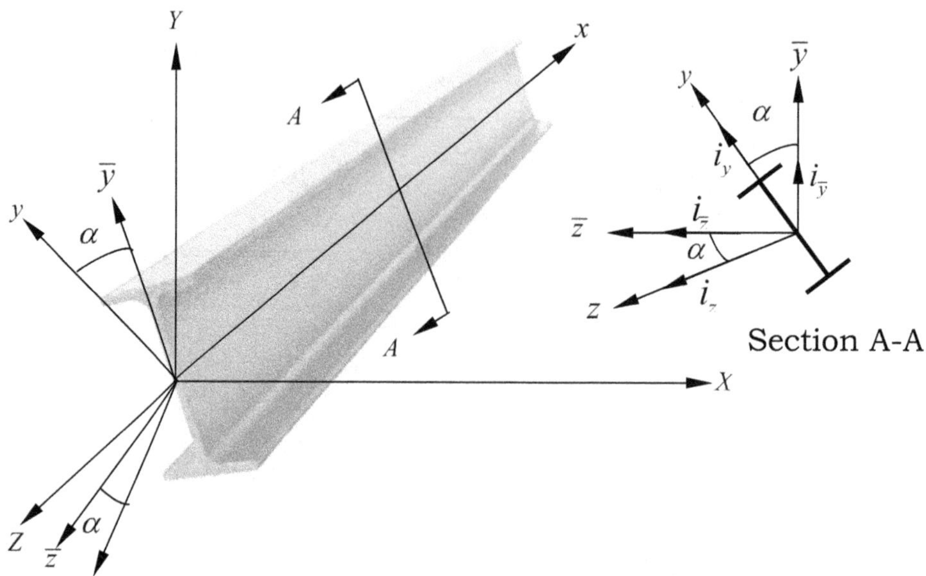

Figure 6.27 Actual Member Orientation.

The *angle of roll α is defined as the angle, taken positive when looking in the negative x direction, through which local xyz coordinate system shall be rotated about its x axis, so that the xy plane becomes vertical with the y axis pointing upwards, in the positive direction of global Y axis.* $[T_1]$ obtained from Eq. 6.89 then can be substituted in Eq. 6.78, to obtain the rotation transformation matrix $[T]$.

However, this rotation transformation matrix is not valid for a vertical member as the value of C_{xX} and C_{xZ} are zero. In such a case, some elements of the rotation transformation matrix $[T_1]$ given by Eq. 6.89 become undefined. However, for a vertical member, the rotation transformation matrix can be defined using the following approach:

We imagine that the vertical member's desired (or actual design) orientation is reached in two steps, as shown in Figure 6.28. In the first step, while the member's x axis is oriented in the desired (vertical) direction, its y and z axes are oriented so that the local z axis is parallel to the global Z axis, as shown in Figure 6.28. As indicated the member's principal axes in this (imaginary) orientation are designated as $\bar{y}$ and $\bar{z}$ (instead of y and z, respectively). The direction of the local x axis (known from the global coordinates of the member ends) is represented by the unit vector $i_x = C_{xY} I_Y$, while the direction of the $\bar{z}$

axis is given by the unit vector. The unit vector $i_{\bar{y}}$, directed along the $\bar{y}$ axis, can therefore be conveniently established using the cross product $i_{\bar{z}} \times i_x$, as

$$i_{\bar{y}} = i_{\bar{z}} \times i_x = \begin{vmatrix} I_X & I_Y & I_Z \\ 0 & 0 & 1 \\ 0 & C_{xY} & 0 \end{vmatrix} = -C_{xY} I_X \qquad \qquad .....(6.90)$$

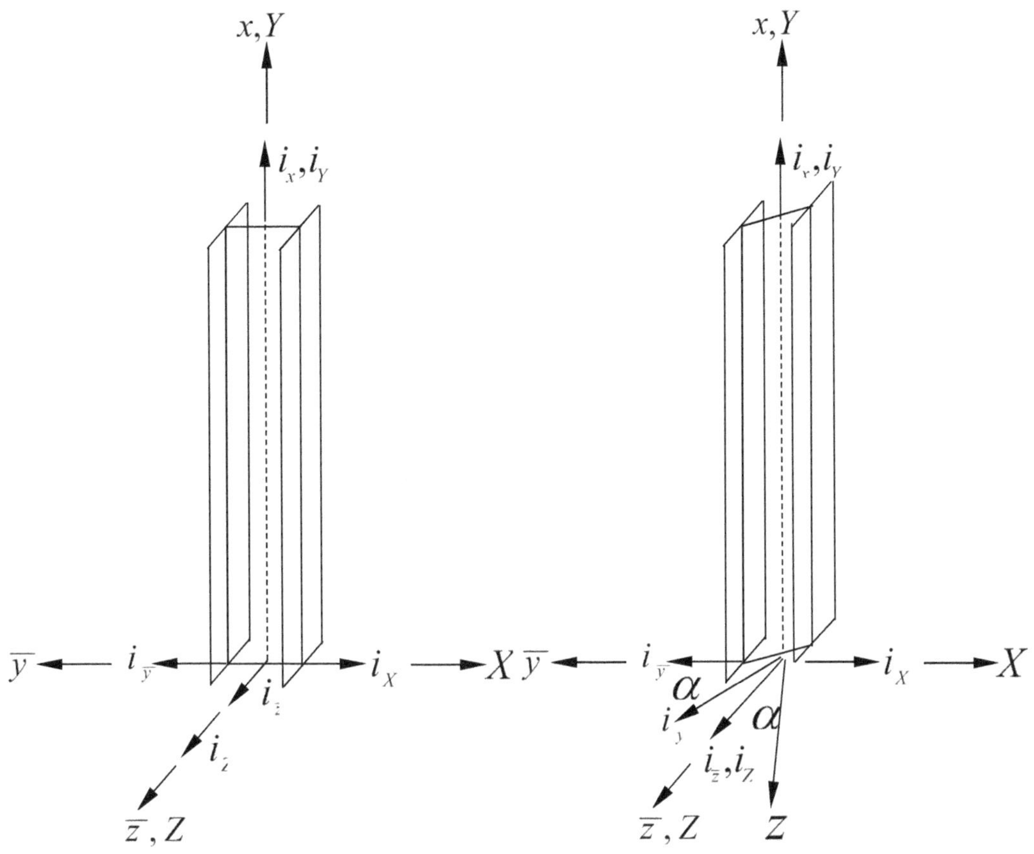

a) z-axis parallel to Z axis

b) Actual orientation of vertical

Figure 6.28 Vertical member.

Thus, the transformation relationship between the global XYZ and the auxiliary local $\overline{xyz}$ coordinate system is given by

$$\begin{Bmatrix} i_x \\ i_{\overline{y}} \\ i_{\overline{z}} \end{Bmatrix} = \begin{bmatrix} 0 & C_{xY} & 0 \\ -C_{xY} & 0 & 1 \\ 0 & 0 & 1 \end{bmatrix} \begin{Bmatrix} I_X \\ I_Y \\ I_Z \end{Bmatrix} \tag{6.91}$$

In the next step, we rotate the auxiliary $\overline{xyz}$ coordinate system about its x axis, in a counter-clockwise sense, by the angle of roll α, until the member's principal axes are in their desired orientations. This final orientation of the member is depicted in Fig. 6.28(b), in which the member principal axes are now designated as the y and z axes. From this figure, we can see that the transformation relationship between the auxiliary $\overline{xyz}$ and the actual xyz coordinate systems is the same as given previously in Eq. 6.86. Thus, the desired transformation from the global XYZ coordinate system to the local xyz coordinate system can be obtained by substituting Eq. 6.91 into Eq. 6.86 and performing the required matrix multiplication. This yields,

$$\begin{Bmatrix} i_x \\ i_y \\ i_z \end{Bmatrix} = \begin{bmatrix} 0 & C_{xY} & 0 \\ -C_{xY}\cos\alpha & 0 & \sin\alpha \\ C_{xY}\sin\alpha & 0 & \cos\alpha \end{bmatrix} \begin{Bmatrix} I_X \\ I_Y \\ I_Z \end{Bmatrix} \quad .....(6.92)$$

To define angle of roll α in terms of a point P, we define the position vector $\vec{p}$ directed from node i of the member to the point P as follows:

$$\vec{p} = (x_P - x_i)I_X + (y_P - y_i)I_Y + (z_P - z_i)I_Z \quad .....(6.93)$$

Also, point P and node i lie in the local xy plane, the position vector $\vec{p}$ in the same plane. Since the direction cosines of the local x axis are already known from the global coordinates of the member ends, the direction cosines of the local z axis can be conveniently established using the following relationship.

$$i_z = \frac{i_x \times \vec{p}}{\left| i_x \times \vec{p} \right|} \quad .....(6.94)$$

With both i_x and i_z now known, the direction cosines of the local y axis can be obtained via the cross product,

$$i_y = i_x \times i_z \quad .....(6.95)$$

In the case that the reference point P is specified in the local xz plane of the member, the direction cosines of the local y axis need to be determined first using the relationship

$$i_z = \frac{\vec{p} \times i_x}{\left| \vec{p} \times i_x \right|} \qquad \qquad .....(6.96)$$

and then the direction cosines of the local z axis are obtained via the cross product

$$i_z = i_x \times i_y \qquad \qquad .....(6.97)$$

6.6.3 Stiffness Matrix of Space Frame Element with w.r.t Global Axes

Stiffness matrix with reference to the global axes can be obtained as

$$\left[K^{(e)} \right] = \left[T^T \right] \left[K_M^{(e)} \right] [T] \qquad \qquad (6.98)$$

The global stiffness matrix and global force vector are assembled from local stiffness matrix and force vector by assembly.

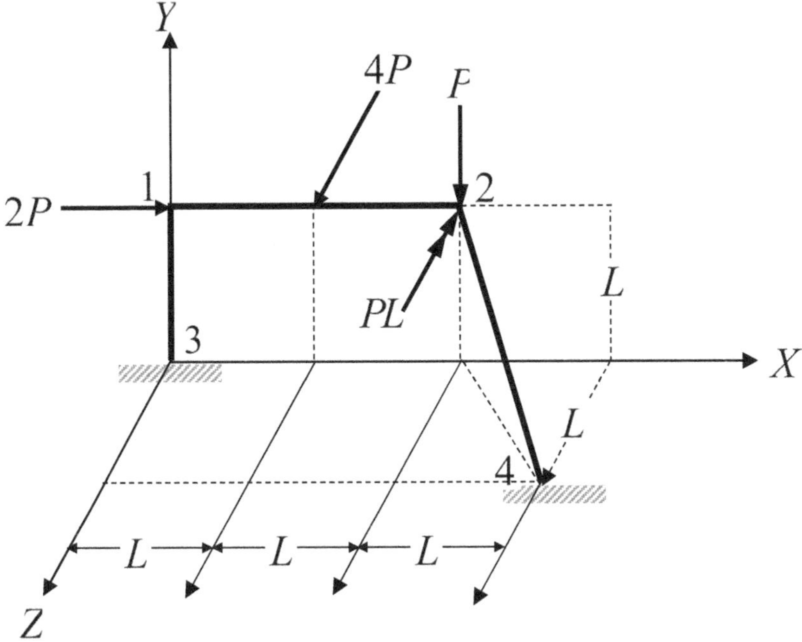

Figure 6.29 Space Frame

6.6.4   Numerical Examples

Numerical Example 6.12

Analyse the space frame given in Figure 6.29. Given $E = 2 \times 10^8 \, kN/m^2$, $C = 0.8 \times 10^8 \, kN/m^2$, $L = 3m$, $A = 0.01 \, m^2$, $I_X = 2 \times 10^{-3} m^4$, $I_Y = I_Y = 1 \times 10^{-3} m^4$ and $P = 60 \, kN$.

Solution:

Nodal displacements

Node	Translation ($\times 10^{-3}$ m)			Rotation ($\times 10^{-3}$ radians)		
	X	Y	Z	θ_X	θ_Y	θ_Z
1	-0.8594	0.0577	5.0076	2.3933	-1.6231	0.6813
2	-1.1760	3.2531	5.2555	1.2884	1.7209	-0.7715
3	0.0	0.0	0.0	0.0	0.0	0.0
4	0.0	0.0	0.0	0.0	0.0	0.0

FORCES AND MOMENTS IN ELEMENTS

Element	Node	Forces (kN)			Moments (kNm)		
		Fx	Fy	Fz	Mx	My	Mz
1	1	105.548	-38.509	-126.013	29.464	86.569	-66.100
	2	-105.548	38.509	-113.987	-29.464	-50.491	-163.954
2	3	-38.509	14.452	-126.013	86.569	348.575	-23.744
	1	38.509	-14.452	126.013	-86.569	29.464	66.100
3	2	183.623	9.192	5.967	-21.404	46.703	-32.181
	4	-183.623	-9.192	-5.967	21.404	-76.711	79.946

Numerical Example 6.13

A two-storey frame subjected to lateral loads as shown in Figure 6.30. The section properties of beams and columns are given below. The 'k' nodes for various members are referred to the node numbers and are given below:

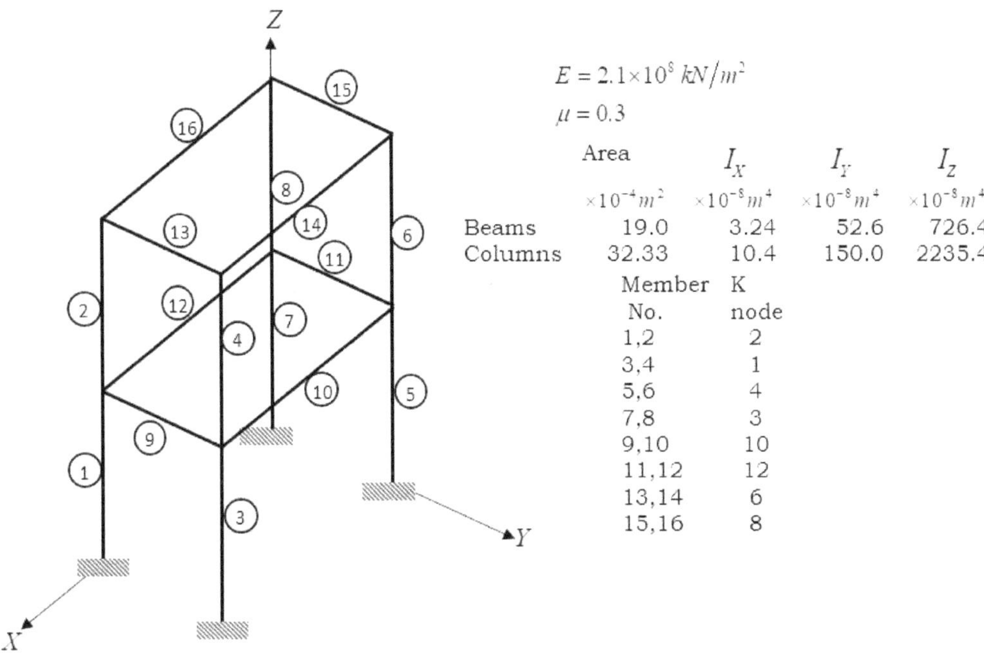

Figure 6.30 Two-bay two storey portal frame

Selected displacements and moments are given below:

Nodal Displacements

Node	Translation ($\times 10^{-3}$ m)			Rotation ($\times 10^{-3}$ radians)		
	X	Y	Z	θ_X	θ_Y	θ_Z
5	-0.3972	5.3809	0.0205	-1.6427	-0.0474	-0.1305
6	0.3972	5.3363	-0.0205	-1.6378	0.0474	-0.1239
11	0.9846	13.466	-0.0548	-1.9983	0.0242	-1.1161
12	-0.9846	13.494	0.0548	-2.0086	-0.0242	-1.1161

Forces and Moments in Elements

Element	Node	Forces (kN)			Moments (kNm)		
		Fx	Fy	Fz	Mx	My	Mz
1	1	-4.6509	-6.0847	-0.0457	0.0003	0.0735	-11.697
	5	4.6509	6.0847	0.0457	-0.0003	0.0635	-6.5570
2	5	-1.2339	-0.0481	-0.0672	0.0027	0.0984	1.5914
	9	1.2339	0.0481	0.0672	-0.0027	0.1032	-1.7357
11	7	0.0134	4.6522	-0.0215	-0.0001	0.0319	6.9763
	8	-0.0134	-4.6522	0.0215	0.0001	0.0327	6.9805

6.7 EXERCISE PROBLEMS

1. Compute the nodal displacement at node 2, stresses in each element and support reactions of the stepped bar-1 shown in Figure due to the applied forces. Adopt $L_1 = 0.5\ m$, $L_2 = 0.4\ m$, $A_1 = 0.0024\ m^2$ and $A_2 = 0.0012\ m^2$. Adopt $E_1 = 2 \times 10^8\ kN/m^2$ and $E_2 = 1 \times 10^8\ kN/m^2$.

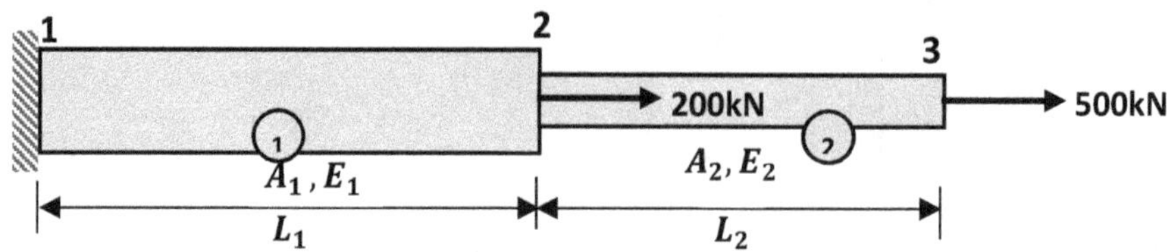

2. Compute the displacements at nodes 2 and 3, stresses in each element and support reactions of the stepped bar shown in Figure due to the applied forces. Adopt $L_1 = 0.4\ m$, $L_2 = 0.3\ m$, $L_3 = 0.5\ m$, $A_1 = 0.002\ m^2$, $A_2 = 0.001\ m^2$ and $A_3 = 0.003\ m^2$. Note that $E_1 = 2 \times 10^8\ kN/m^2$, $E_2 = 0.7 \times 10^8\ kN/m^2$ and $E_3 = 1 \times 10^8\ kN/m^2$.

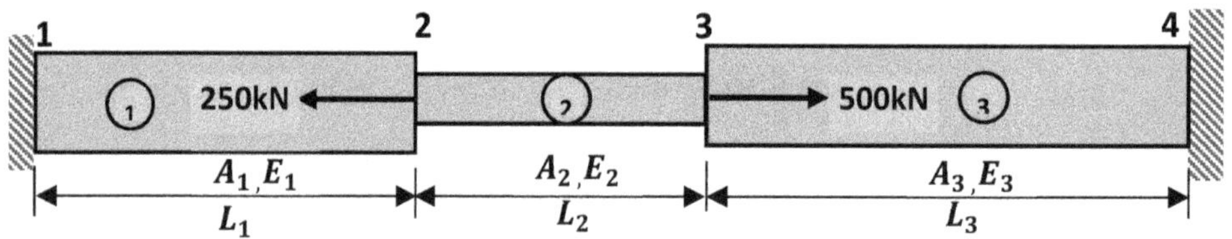

3. A stepped bar shown in the Figure is subject to a load of $1000\ N$ at node 3 in addition to the self-weight. Find the displacements at nodes 2 and 3, stresses in each element and support reaction. The Young's modulus of the material of the bar is $E = 2 \times 10^5\ N/mm^2$ and $\gamma = 0.785 \times 10^{-4}\ N/mm^3$.

Adopt $L_1 = 600\ mm$, $L_2 = 300\ mm$, $A_1 = 3000\ mm^2$ and $A_2 = 2000\ mm^2$.

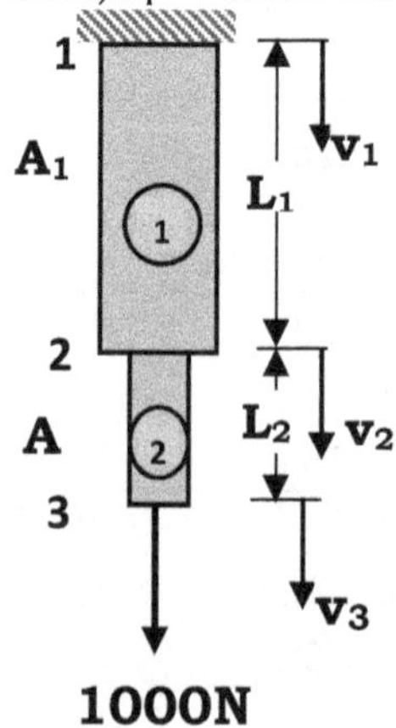

4. Analyse the triangular truss shown in the figure below. Adopt
   $A_1 = 0.02\ m^2\ A_2 = 0.01\ m^2\ A_3 = 0.03\ m^2\ E = 2 \times 10^8\ kN/m^2$

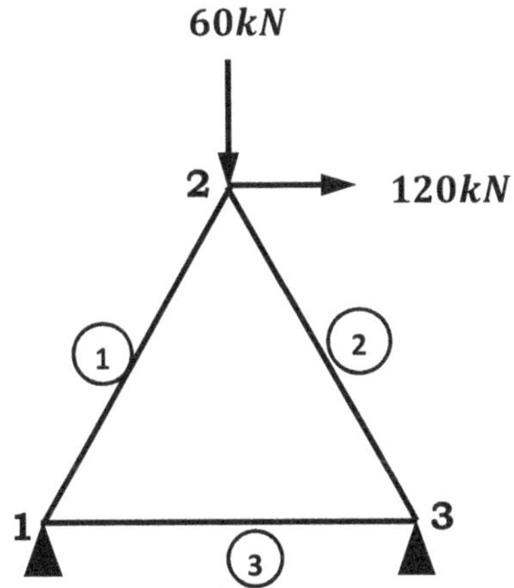

5. Analyse the Warren truss shown below. Adopt $A = 0.02 m^2$ and
   $E = 2 \times 10^8\ kN/m^2$ for all members

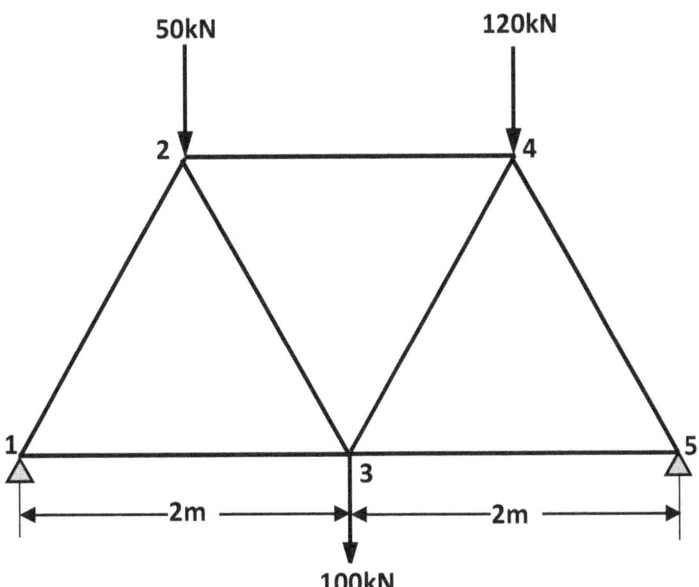

6. Analyse the Pratt truss shown in the figure. Adopt $A = 0.02m^2$ and $E = 2 \times 10^8\, kN / m^2$ for all members.

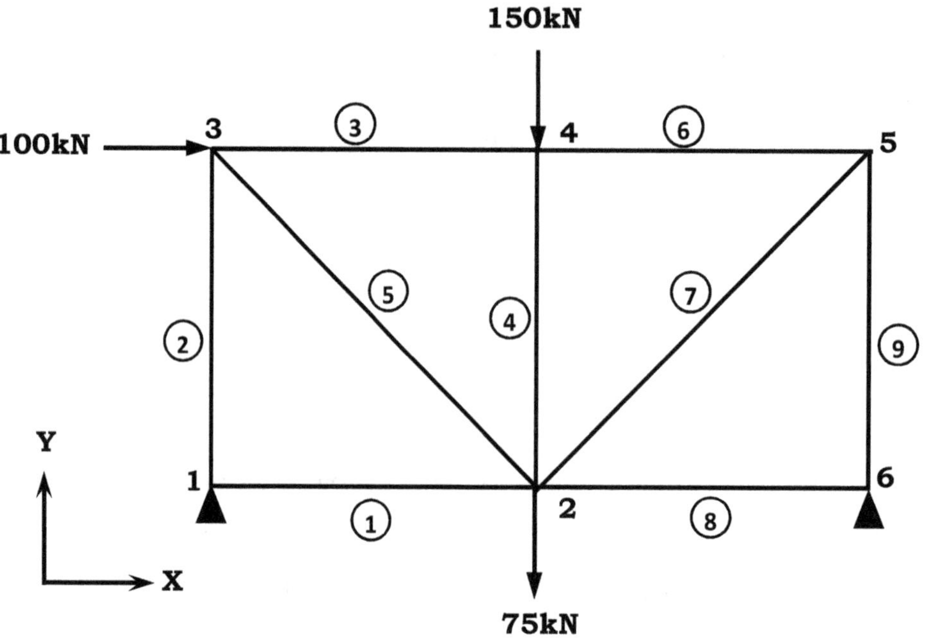

7. Analyse the space truss shown below: Adopt $A = 0.04\, m^2$ and $E = 2 \times 10^8\, kN / m^2$.

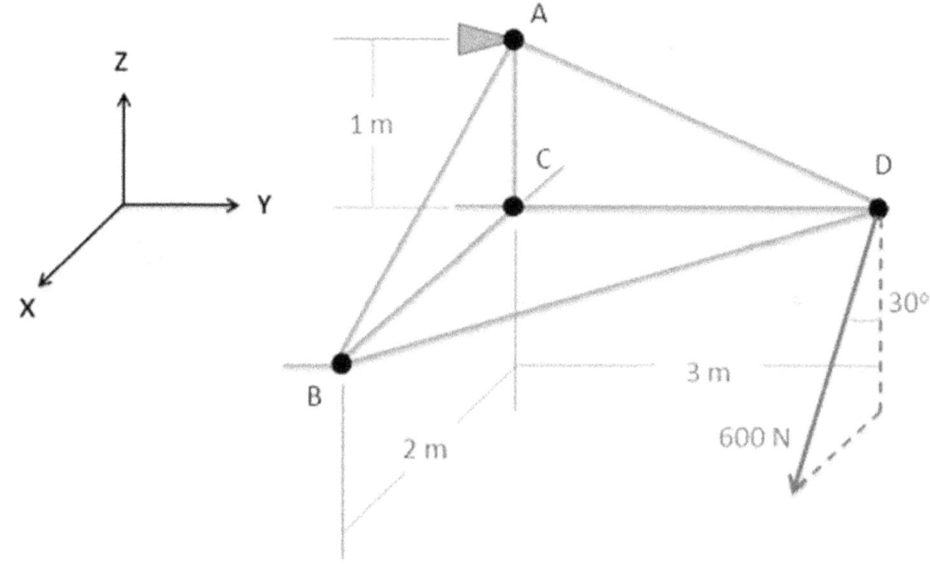

8. Analyse the two-span continuous beam as shown in the figure. Adopt $A = 0.04 \ m^2$, $I = 5 \times 10^{-4} m^4$ and $E = 2 \times 10^8 \ kN \, / \, m^2$ for all members.

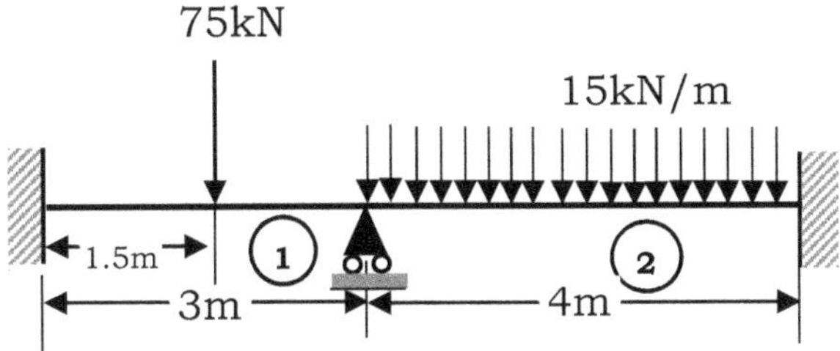

9. Analyse the two-span continuous beam as shown in the figure below. Adopt $A_1 = 0.06 \ m^2$, $I_1 = 4.5 \times 10^{-4} m^4$ for the first element and $A_2 = 0.09 \ m^2$, $I_2 = 4.5 \times 10^{-4} m^4$ for the second element. $E = 2 \times 10^8 \ kN \, / \, m^2$ for both elements.

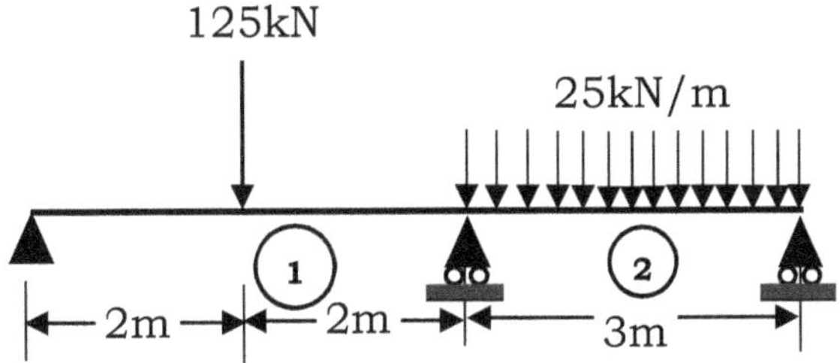

10. Analyse the three-span continuous beam as shown in the figure below. Adopt $A = 0.06 \ m^2$, $I = 4.5 \times 10^{-4} m^4$ and $E = 2 \times 10^8 \ kN \, / \, m^2$ for all members.

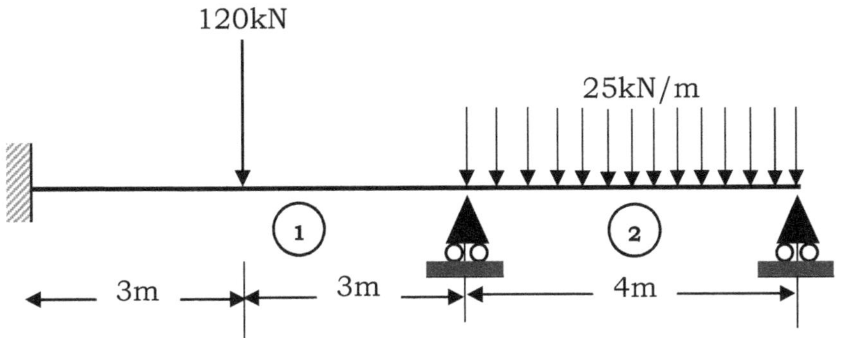

11. Analyse the three-span continuous beam as shown in the figure below. Adopt $A = 0.09\ m^2$, $I = 7.5 \times 10^{-4} m^4$ and $E = 2 \times 10^8\ kN/m^2$ for all members.

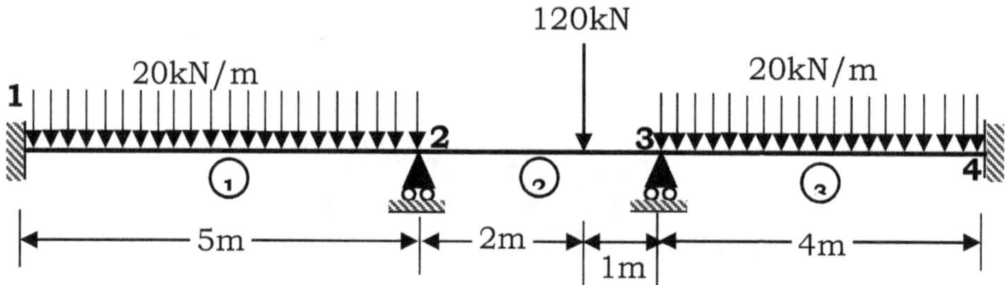

12. Analyse the portal frame shown in the figure. Adopt $A_1 = 0.06m^2$, $I_1 = 5 \times 10^{-4} m^4$, $A_2 = 0.04m^2$, $I_2 = 4.5 \times 10^{-4} m^4$, $A_2 = 0.05m^2$, $I_3 = 5 \times 10^{-4} m^4$, $A_2 = 0.03m^2$.

Adopt $E = 2 \times 10^8\ kN/m^2$ for all members.

13. Analyse the plane frame shown in the figure. Adopt $A_1 = 0.03m^2$, $I_1 = 4 \times 10^{-4} m^4$, $A_2 = 0.05m^2$, $I_2 = 6 \times 10^{-4} m^4$. Adopt $E = 2 \times 10^8\ kN/m^2$ for all members.

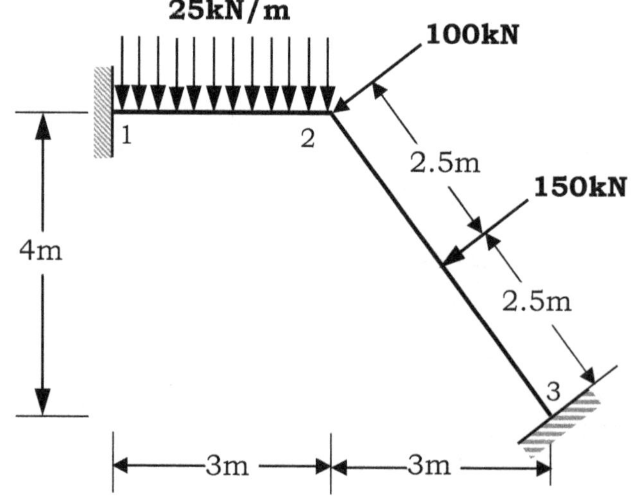

14. Analyse the space frame shown below. Given $E = 2 \times 10^8 \, kN/m^2$, $G = 0.8 \times 10^8 \, kN/m^2$, $L = 2m$, $A = 0.015 \, m^2$, $I_X = 2.5 \times 10^{-3} \, m^4$ $I_Y = I_Z = 1.5 \times 10^{-3} \, m^4$ and $P = 50kN$.

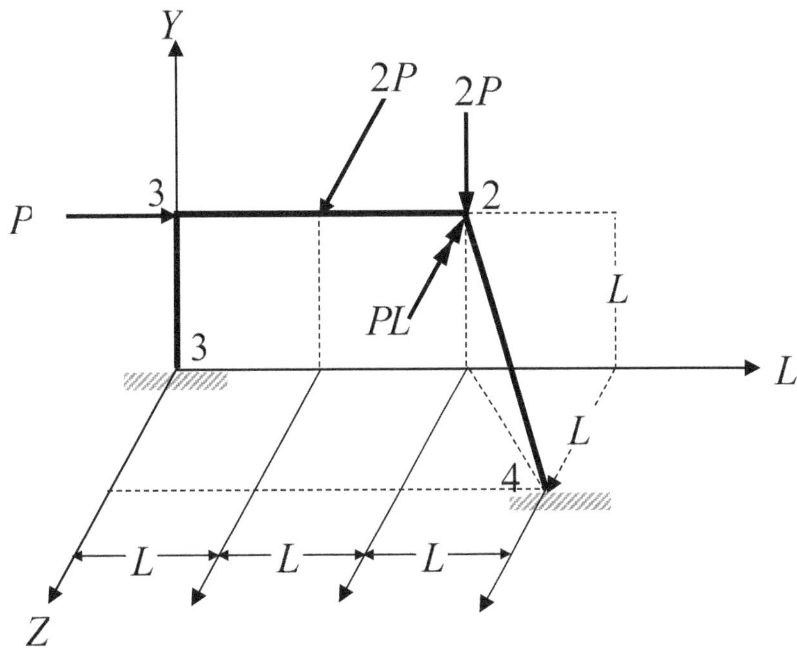

CHAPTER 7

Two Dimensional Elements

In finite element analysis, a mechanical structure is discretized into various types of finite elements, according to the shape and geometry of the analysed system.

Two dimensional elements are defined by three or more nodes in a plane and are used for structural problems where both the loading and the corresponding deformation, occur within a plane. Two-dimensional finite elements are widely used in finite element analysis and using them is generally easy and fast. The physical domain considered for the analysis is an area with uniform thickness.

7.1 SIMPLE TWO-DIMENSIONAL ELEMENTS

Some simple and popular two-dimensional elements are triangular and rectangular elements. The triangular element was the first and simplest element developed for 2D solids. Triangular element is used for many practical engineering problems owing to its adaptivity to complex geometry. Triangular elements are normally used when we want to mesh a 2D model involving complex geometry with acute corners. In addition, the triangular configuration with the simplest topological features makes it easier to develop meshing processors. Triangular elements are usually less accurate than rectangular elements. The strain-displacement matrix of the linear triangular elements is constant, accounting for the inaccuracy.

However, in case of the rectangular element, the strain-displacement matrix is not a constant. This provides for a more realistic presentation in strain, and hence the stress distribution, across the structure. The formulation of the equations for the rectangular elements is simpler compared to the triangular elements, because the shape functions can be formulated very easily due to the regularity in the shape of the rectangular element.

7.2 ELEMENTS OF ARBITRARY SHAPE

The triangular and rectangular elements discussed in the previous section, though popular and easy to use, have their limitations in case of analysis of structures with complicated geometry and displacement domains. This necessitated the development of elements of more arbitrary shape.

To define a finite element, one needs to define the geometry as well as the displacement domain of the element. Consider an element whose nodal coordinates are (x_i, y_i, z_i), $i = 1, n$. An arbitrary point (x, y, z) inside the element can be defined in terms of these nodal coordinates as,

$$\left. \begin{aligned} x &= N_1' x_1 + N_2' x_2 + N_3' x_3 + ... + N_n' x_n \\ y &= N_1' y_1 + N_2' y_2 + N_3' y_3 + ... + N_n' y_m \\ z &= N_1' z_1 + N_2' z_2 + N_3' z_3 + ... + N_n' z_n \end{aligned} \right\} \Rightarrow \{x\} = \lfloor N'(x) \rfloor \{x_m\} \qquad .....(7.1)$$

The nodal displacements are (u_i, v_i, w_i), $i = 1, m_2$. Displacements (u, v, w) at an arbitrary point inside the element can be defined in terms of these nodal displacements as,

$$\left. \begin{aligned} u &= N_1 u_1 + N_2 u_2 + N_3 u_3 + ... + N_m u_m \\ v &= N_1 v_1 + N_2 v_2 + N_3 v_3 + ... + N_m v_m \\ w &= N_1 w_1 + N_2 w_2 + N_3 w_3 + ... + N_m w_m \end{aligned} \right\} \Rightarrow \{u\} = \lfloor N(x) \rfloor \{u_m\} \qquad .....(7.2)$$

In this context, the following types of elements are defined:

(a) *Sub-parametric elements*: If the interpolation functions are used to model geometry are *less* than the interpolation function used to model displacements, such elements are called sub-parametric elements. For sub-parametric elements, $\lfloor N'(x) \rfloor < \lfloor N(x) \rfloor$. Such elements possess simpler geometry and complicated displacement domain.

(b) *Super-parametric elements*: If the interpolation functions are used to model geometry are *more* than the interpolation function used to model displacements, such elements are called super-parametric elements. For super-parametric elements, $\lfloor N'(x) \rfloor > \lfloor N(x) \rfloor$. Such elements possess complicated geometry and simpler displacement domain.

(c) *Iso-parametric elements*: If the same set of interpolation functions are used to model geometry as well as displacements, such elements are called iso-parametric elements. For iso-parametric elements, $\lfloor N'(x) \rfloor \equiv \lfloor N'(x) \rfloor$. For such elements geometry and displacement domains have the same level of complexity. Iso-parametric elements are widely used in two dimensional and three-dimensional stress analysis of plates and shells. e.g., linear strain triangular (LST) element, four node quadrilateral element and eight-node parabolic elements are iso-parametric elements.

Some two-dimensional elements are discussed in the succeeding sections.

7.3 CONSTANT STRAIN TRIANGULAR (CST) ELEMENT

Consider the three-node triangular element shown in Figure 7.1. The interpolation functions for this element are discussed in Section 4.4.3.

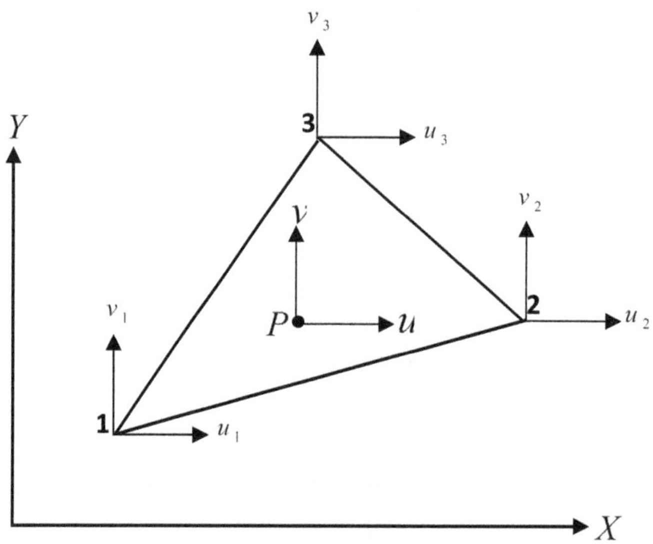

Figure 7.1 Three-node triangular element.

The natural coordinates ξ_1, ξ_2 and ξ_3 defined in Eq. 4.10 are used as the linear interpolation functions N_1, N_2 and N_3.

$$N_i = \frac{1}{2A}\left(c_i + b_i x + a_i y\right), \quad i = 1, 2, 3 \qquad \ldots\ldots(7.3)$$

Thus, the displacements at any point inside the element are expressed in terms of nodal displacements at the three nodes of the element as given by Eq. 4.34 as

$$\begin{Bmatrix} u \\ v \end{Bmatrix} = \begin{bmatrix} N_1 & 0 & N_2 & 0 & N_3 & 0 \\ 0 & N_1 & 0 & N_2 & 0 & N_3 \end{bmatrix}\{d\} = [N]\{d\} \qquad \ldots\ldots(7.4)$$

where $\{d\}$ is the vector of nodal displacements and $[N]$ is the shape function matrix.

The strain-displacement relationship for the element is now expressed as

$$\varepsilon_x = \frac{\partial u}{\partial x}$$

$$\varepsilon_y = \frac{\partial v}{\partial y}$$

$$\gamma_{xy} = \frac{\partial u}{\partial y} + \frac{\partial v}{\partial x}$$

.....(7.5)

Eq. 7.5 can be written in matrix form as follows:

$$\{\varepsilon\} = \left\{\begin{array}{c} \varepsilon_x \\ \varepsilon_y \\ \gamma_{xy} \end{array}\right\} = \begin{bmatrix} \dfrac{\partial}{\partial x} & 0 \\[2ex] 0 & \dfrac{\partial}{\partial y} \\[2ex] \dfrac{\partial}{\partial y} & \dfrac{\partial}{\partial x} \end{bmatrix} \left\{\begin{array}{c} u \\ v \end{array}\right\}$$

.....(7.6)

Combining Eq. 7.6 with Eq.7.4 yields

$$\{\varepsilon\} = \left\{\begin{array}{c} \varepsilon_x \\ \varepsilon_y \\ \gamma_{xy} \end{array}\right\} = \begin{bmatrix} \dfrac{\partial}{\partial x} & 0 \\[2ex] 0 & \dfrac{\partial}{\partial y} \\[2ex] \dfrac{\partial}{\partial y} & \dfrac{\partial}{\partial x} \end{bmatrix} \begin{bmatrix} N_1 & 0 & N_2 & 0 & N_3 & 0 \\ 0 & N_1 & 0 & N_2 & 0 & N_3 \end{bmatrix} \{d\}$$

.....(7.7)

Eq. 7.7 can be simplified as

$$\{\varepsilon\} = \begin{bmatrix} \dfrac{\partial N_1}{\partial x} & 0 & \dfrac{\partial N_2}{\partial x} & 0 & \dfrac{\partial N_3}{\partial x} & 0 \\[2ex] 0 & \dfrac{\partial N_1}{\partial y} & 0 & \dfrac{\partial N_2}{\partial y} & 0 & \dfrac{\partial N_3}{\partial y} \\[2ex] \dfrac{\partial N_1}{\partial y} & \dfrac{\partial N_1}{\partial x} & \dfrac{\partial N_2}{\partial y} & \dfrac{\partial N_2}{\partial x} & \dfrac{\partial N_3}{\partial y} & \dfrac{\partial N_3}{\partial x} \end{bmatrix} \{d\}$$

.....(7.8)

The partial derivatives of interpolation functions N_1, N_2 and N_3 with reference to x and y can be obtained using Eq. 7.3 as follows:

$$\frac{\partial N_i}{\partial x} = \frac{b_i}{2A}$$
$$\frac{\partial N_i}{\partial y} = \frac{a_i}{2A}$$

$$.....(7.9)$$

Substituting Eq. 7.9 in Eq. 7.8, we obtain

$$\{\varepsilon\} = \frac{1}{2A}\begin{bmatrix} b_1 & 0 & b_2 & 0 & b_3 & 0 \\ 0 & a_1 & 0 & a_2 & 0 & a_3 \\ a_1 & b_1 & a_2 & b_2 & a_3 & b_3 \end{bmatrix}\{d\} = [B]\{d\} \qquad .....(7.10)$$

where $[B]$ is the strain-displacement matrix having constant terms, making the strain $\{\varepsilon\}$ over the element constant. $[B]$ matrix is independent of the x and y coordinates and depends solely on the element nodal coordinates, as seen from Eq. 7.10. Thus, the element is called a *Constant Strain Triangular (CST) Element.*

The stiffness matrix of the CST element is given as

$$\left[K^{(e)}\right] = \iiint [B]^T [C][B]\,dV \qquad ....(7.11)$$

Considering an element dA of uniform thickness h, $dV = hdA$, Eq. 7.11 can be rewritten as

$$\left[K^{(e)}\right] = h\int [B]^T [C][B]\,dA \qquad .....(7.12)$$

In Eq. 7.12, the integrand $[B]^T [C][B]$ is not a function of x or y for a CST element. Thus Eq. 7.12 can be rewritten as

$$\left[K^{(e)}\right] = hA[B]^T [C][B] \qquad ....(7.13)$$

Here $[C]$ matrix is the constitutive matrix for two-dimensional continuum. For plane stress and plane strain problems, $[C]$ matrix is given by Eq. 2.23 and Eq.2.29 respectively.

Thus, for plane stress problems

$$[C] = \frac{E}{\left(1-\mu^2\right)} \begin{bmatrix} 1 & \mu & 0 \\ \mu & 1 & 0 \\ 0 & 0 & 1 \end{bmatrix} \qquad .....(7.14)$$

And for plane strain problems

$$[C] = \frac{E}{\left(1+\mu\right)\left(1-2\mu\right)} \begin{bmatrix} \left(1-\mu\right) & \mu & 0 \\ \mu & 1-\mu & 0 \\ 0 & 0 & \dfrac{\left(1-2\mu\right)}{2} \end{bmatrix} \qquad ......(7.15)$$

For the CST element with constant thickness, solution converges when the element size is small.

In the formulation of the element stiffness matrix using Eq. 7.13, the nodal displacements have been along the global axes. Thus, the element stiffness matrix derived is with reference to the global axes and thus does not require any rotation transformation.

Force vector for the CST element can be expressed as

$$\left\{Q^{(e)}\right\} = \iiint \left[N^T\right]\{X\}\,dV + \iint_{S_1}\left[N_S^T\right]\{f\}\,dA + \begin{Bmatrix} Q_1 \\ Q_2 \end{Bmatrix} \qquad .....(7.16)$$

The first, second and third terms in the above equation correspond to the *vector of body forces, vector of surface tractions and point loads.* First, the vector of body forces is evaluated as follows.

$$\{X\} = \begin{Bmatrix} X_b \\ Y_b \end{Bmatrix} = \begin{Bmatrix} 0 \\ -\rho g \end{Bmatrix} \qquad .....(7.17)$$

Here, X_b and Y_b are body forces per unit volume along the x and y directions respectively. Substituting Eq. 7.17 and $[N]$ from Eq.7.4 in the first term of Eq.7.16 yields

$$\iiint \left[N^T \right]\{X\}\,dV = h \iint \begin{bmatrix} N_1 & 0 \\ 0 & N_1 \\ N_2 & 0 \\ 0 & N_2 \\ N_3 & 0 \\ 0 & N_3 \end{bmatrix} \begin{Bmatrix} 0 \\ -\rho g \end{Bmatrix} dA = h \begin{Bmatrix} 0 \\ -\iint N_1 \rho g\, dA \\ 0 \\ -\iint N_2 \rho g\, dA \\ 0 \\ -\iint N_3 \rho g\, dA \end{Bmatrix} = -\frac{h}{3} \begin{Bmatrix} 0 \\ \rho g A \\ 0 \\ \rho g A \\ 0 \\ \rho g A \end{Bmatrix} \quad \dots\dots(7.18)$$

It is observed from Eq. 7.18 that the self-weight gets uniformly shared to the three nodes of the element.

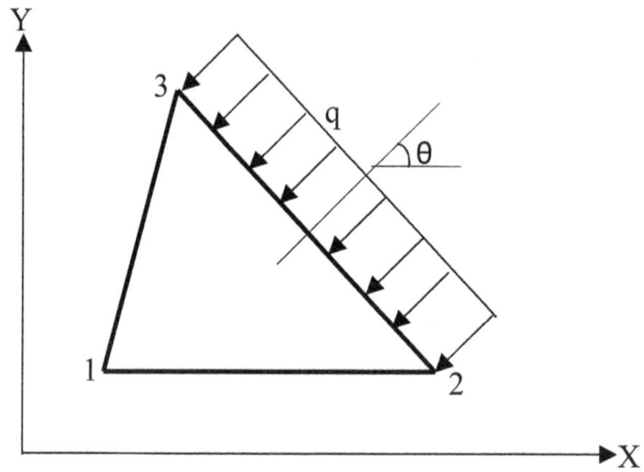

Figure 7.2 Surface traction on an edge.

Considering an element subjected to a uniformly distributed load of magnitude q on the edge 2-3 as shown in Figure 7.2, vector of traction forces can be written as

$$\{f\} = \begin{Bmatrix} -q\cos\theta \\ -q\sin\theta \end{Bmatrix} \qquad \dots(7.19)$$

Substituting in the second term of Eq. 7.16, and considering that N_1 is zero on side 2-3(denoted by $L_{2\text{-}3}$), we obtain

$$\iint_{S_1}\left[N_S^T\right]\{f\}dA = \int\begin{bmatrix}0 & 0\\ 0 & 0\\ N_2 & 0\\ 0 & N_2\\ N_3 & 0\\ 0 & N_3\end{bmatrix}\begin{Bmatrix}-q\cos\theta\\ -q\sin\theta\end{Bmatrix}hdl = h\int\begin{Bmatrix}0\\ 0\\ -N_2 q\cos\theta\\ -N_2 q\sin\theta\\ -N_3 q\cos\theta\\ -N_3 q\sin\theta\end{Bmatrix}dl = \begin{Bmatrix}0\\ 0\\ 0.5qL_{2-3}\cos\theta\\ 0.5qL_{2-3}\sin\theta\\ 0.5qL_{2-3}\cos\theta\\ 0.5qL_{2-3}\sin\theta\end{Bmatrix}$$

$$.....(7.20)$$

The third term of Eq. 7.16 represents the vector of point loads acting on the element.

7.3.1   Numerical Examples

Numerical examples are taken up to illustrate the procedure outlined in the preceding discussion.

Numerical Example 7.1: Consider a thin plate as shown in the Figure 7.3 below. Given $E=2.1\times10^8\, kN/m^2$, $\mu=0.3$ and $t=0.025\, m$. Analyse the plate under conditions of plane stress.

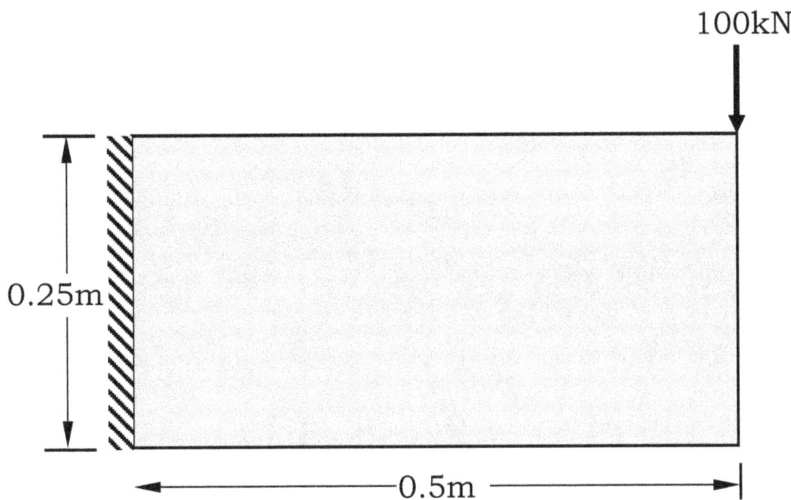

Figure 7.3 Rectangular plate fixed at the edge.

The plate is divided into two linear triangular elements as shown in Figure 7.4, *only for illustration purposes.*

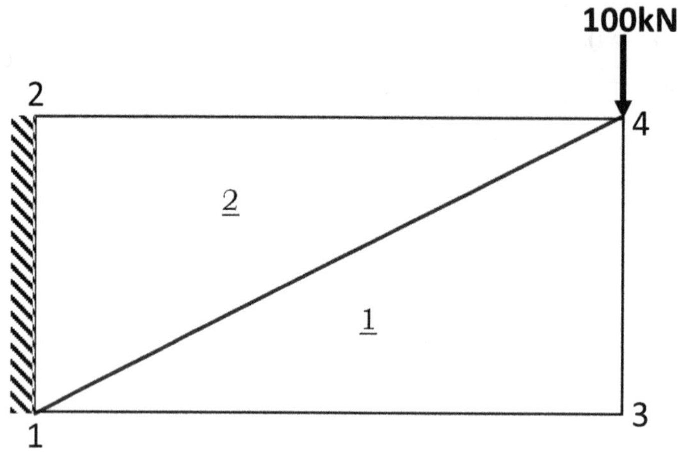

Figure 7.4 Rectangular plate fixed at the edge.

Nodal Information		
Node	x (m)	Y (m)
1	0.0	0.0
2	0.0	0.25
3	0.5	0.0
4	0.5	0.25

Element Information			
Element	Node1	Node 2	Node 3
1	1	3	4
2	2	1	4

Element stiffness matrices are computed for elements 1 and 2 as follows:

Element 1:

$$A = \frac{1}{2}\begin{vmatrix} 1 & 0.0 & 0.0 \\ 1 & 0.5 & 0.0 \\ 1 & 0.5 & 0.25 \end{vmatrix} = 0.0625$$

$$[B] = \begin{bmatrix} -2 & 0 & 0 & 0 & 2 & 0 \\ 0 & 4 & 0 & -4 & 0 & 0 \\ 4 & -2 & 4 & 0 & 0 & 2 \end{bmatrix}$$

$$[C] = 10^8 \begin{bmatrix} 2.30769 & 0.69231 & 0 \\ 0.69231 & 2.307692 & 0 \\ 0 & 0 & 0.80769 \end{bmatrix}$$

$$\left[K_1^{(e)} \right] = 10^6 \begin{bmatrix} 3.4615 & -1.875 & -2.0192 & 0.8653 & -1.442 & 1.009 \\ -1.875 & 6.274 & 1.009 & -5.769 & 0.865 & -0.504 \\ -2.0192 & 1.009 & 2.0192 & 0 & 0 & -1.009 \\ 0.8653 & -5.769 & 0 & 5.769 & -0.865 & 0 \\ -1.442 & 0.865 & 0 & -0.865 & 1.442 & 0 \\ 1.009 & -0.504 & -1.009 & 0 & 0 & 0.504 \end{bmatrix}$$

Element 2:

$$A = \frac{1}{2} \begin{vmatrix} 1 & 0.5 & 0.0 \\ 1 & 0.5 & 0.25 \\ 1 & 0.0 & 0.0 \end{vmatrix} = 0.0625$$

$$[B] = \begin{bmatrix} 2 & 0 & 0 & 0 & -2 & 0 \\ 0 & -4 & 0 & 4 & 0 & 0 \\ -4 & 2 & 4 & 0 & 0 & -2 \end{bmatrix}$$

$$[C] = 10^8 \begin{bmatrix} 2.30769 & 0.69231 & 0 \\ 0.69231 & 2.307692 & 0 \\ 0 & 0 & 0.80769 \end{bmatrix}$$

$$\left[K_2^{(e)} \right] = 10^6 \begin{bmatrix} 3.4615 & -1.875 & -2.0192 & 0.8653 & -1.442 & 1.009 \\ -1.875 & 6.274 & 1.009 & -5.769 & 0.865 & -0.504 \\ -2.0192 & 1.009 & 2.0192 & 0 & 0 & -1.009 \\ 0.8653 & -5.769 & 0 & 5.769 & -0.865 & 0 \\ -1.442 & 0.865 & 0 & -0.865 & 1.442 & 0 \\ 1.009 & -0.504 & -1.009 & 0 & 0 & 0.504 \end{bmatrix}$$

The structure stiffness matrix is assembled from element stiffness matrices is

$$[K] = 10^6 \begin{bmatrix} 3.461 & 0 & -2.0192 & 1.0096 & -1.442 & 0.865 & 0 & -1.875 \\ 0 & 6.274 & 0.865 & -5.769 & 1.009 & -0.504 & -1.875 & 0 \\ -2.0192 & 0.865 & 3.461 & -1.875 & 0 & 0 & -1.442 & 1.009 \\ 1.0096 & -5.769 & -1.875 & 6.274 & 0 & 0 & 0.865 & -0.504 \\ -1.442 & 1.009 & 0 & 0 & 3.461 & -1.875 & -2.019 & 0.865 \\ 0.865 & -0.504 & 0 & 0 & -1.875 & 6.274 & 1.009 & -5.769 \\ 0 & -1.875 & -1.442 & 0.865 & -2.019 & 1.009 & 3.461 & 0 \\ -1.875 & 0 & 1.009 & -0.504 & 0.865 & -5.769 & 0 & 6.274 \end{bmatrix}$$

Force vector for the structure is

$$\{F\} = \begin{bmatrix} 0 & 0 & 0 & 0 & 0 & 0 & 0 & -100 \end{bmatrix}^T$$

Boundary conditions are: $u_1 = v_1 = u_2 = v_2 = 0$

After applying the boundary conditions, matrix equation becomes:

$$10^6 \begin{bmatrix} 3.461 & -1.875 & -2.019 & 0.865 \\ -1.875 & 6.274 & 1.009 & -5.769 \\ -2.019 & 1.009 & 3.461 & 0 \\ 0.865 & -5.769 & 0 & 6.274 \end{bmatrix} \begin{Bmatrix} u_3 \\ v_3 \\ u_4 \\ v_4 \end{Bmatrix} = \begin{Bmatrix} 0 \\ 0 \\ 0 \\ -100 \end{Bmatrix}$$

Solution is

$$\begin{Bmatrix} u_3 \\ v_3 \\ u_4 \\ v_4 \end{Bmatrix} = \begin{Bmatrix} -0.02934 \\ -0.15765 \\ 0.02886 \\ -0.15685 \end{Bmatrix} m$$

However, the values of displacements obtained above are much smaller than the actual deformations.

This is because

(a) Only two elements are used, thus the mesh is very coarse. The assumption of constant strain across the entire element is not valid in the case of a coarse mesh.

(b) Constant strain triangular elements are very stiff with reference to bending.

However, accuracy of the results can be improved by increasing the number of elements. Calculations are once again performed by dividing the plate given in Figure 7.2 into a finer mesh of 100 elements as shown in figure 7.5:

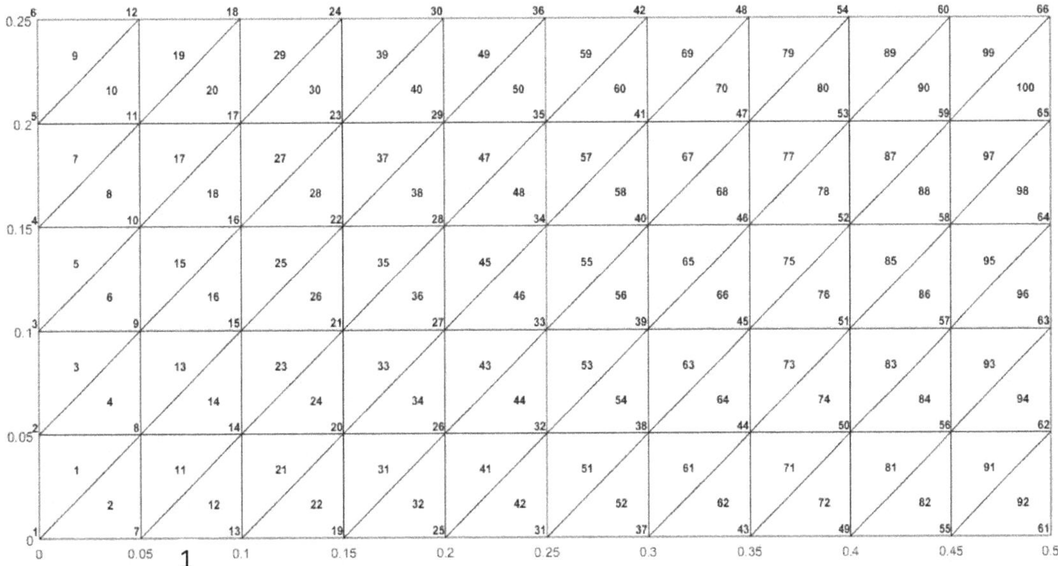

Figure 7.5 Finer mesh for rectangular plate.

A program listed in section A.5 is utilized for obtaining the results. The following is the input file used:

Input:
```
0.5   0.25   0.025   10   5   210e6 7850   0.3    0.0   1 0
4     0.0    -100.0
```

The following is the dialog generated during execution of the program to input for other quantities. Note that the responses are marked in bold font:

```
Enter 1 for Plane Stress or 2 for Plane Strain:1
Point loads are present at specified nodes
Enter 1 for Edge restraint OR 2 for Corner restraint:1
enter edge restraints in the order [Bottom Top Left Right]
enter 0 for restrained edge and 1 for unrestrained edge
Bottom Edge(1 or 0): 1
Top Edge    (1 or 0): 1
```

```
Left Edge  (1 or 0): 0

Right Edge (1 or 0): 1

Do you want plots on deformed configuration (1 for Yes 0 for
No): 0
```

Deformations obtained are compared with those obtained using ABAQUS *with the same level of discretization*. The values of selected displacements and stresses are listed in Tables 7.1 and 7.2:

Table 7.1 Comparison of displacements obtained with results of ABAQUS.

Node Location	Horizontal displacement ×10⁻⁴ m		Vertical displacement ×10⁻⁴ m	
	Present	ABAQUS	Present	ABAQUS
Top right corner	2.13099	2.13100	–6.804009	–6.80401
Bottom right corner	–1.93550	–1.93550	–6.21791	–6.21791

Table 7.2 Comparison of stresses obtained with results of ABAQUS.

Element	$\sigma_X(\times10^5\,kN/m^2)$		$\sigma_Y(\times10^5\,kN/m^2)$		$\tau_{XY}(\times10^5\,kN/m^2)$	
	Present	ABAQUS	Present	ABAQUS	Present	ABAQUS
2	–1.7384	–1.73844	–0.1821	–0.18212	–0.0786	0.07862
19	1.4986	1.4986	0.1713	0.1713	–0.2420	–0.2421
79	0.5711	0.5711	0.0427	0.0427	–0.1013	–0.1013

It is observed from the above tables that the program outlined in section A.5 gives accurate results. Figures 7.6 and 7.7 show the plots of horizontal and vertical displacements of various nodes of the mesh (given in Figure 7.5). Figure 7.8, Figure 7.9, and Figure 7.10 show the variation of in-plane stresses σ_X, σ_Y and τ_{XY} respectively.

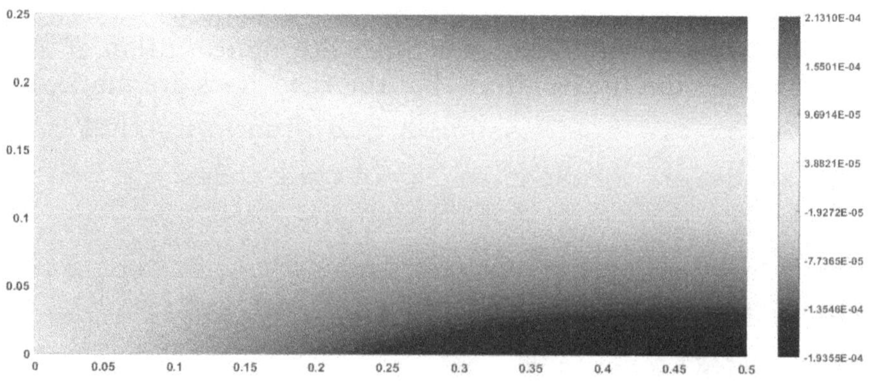

Figure 7.6 Horizontal displacement (m).

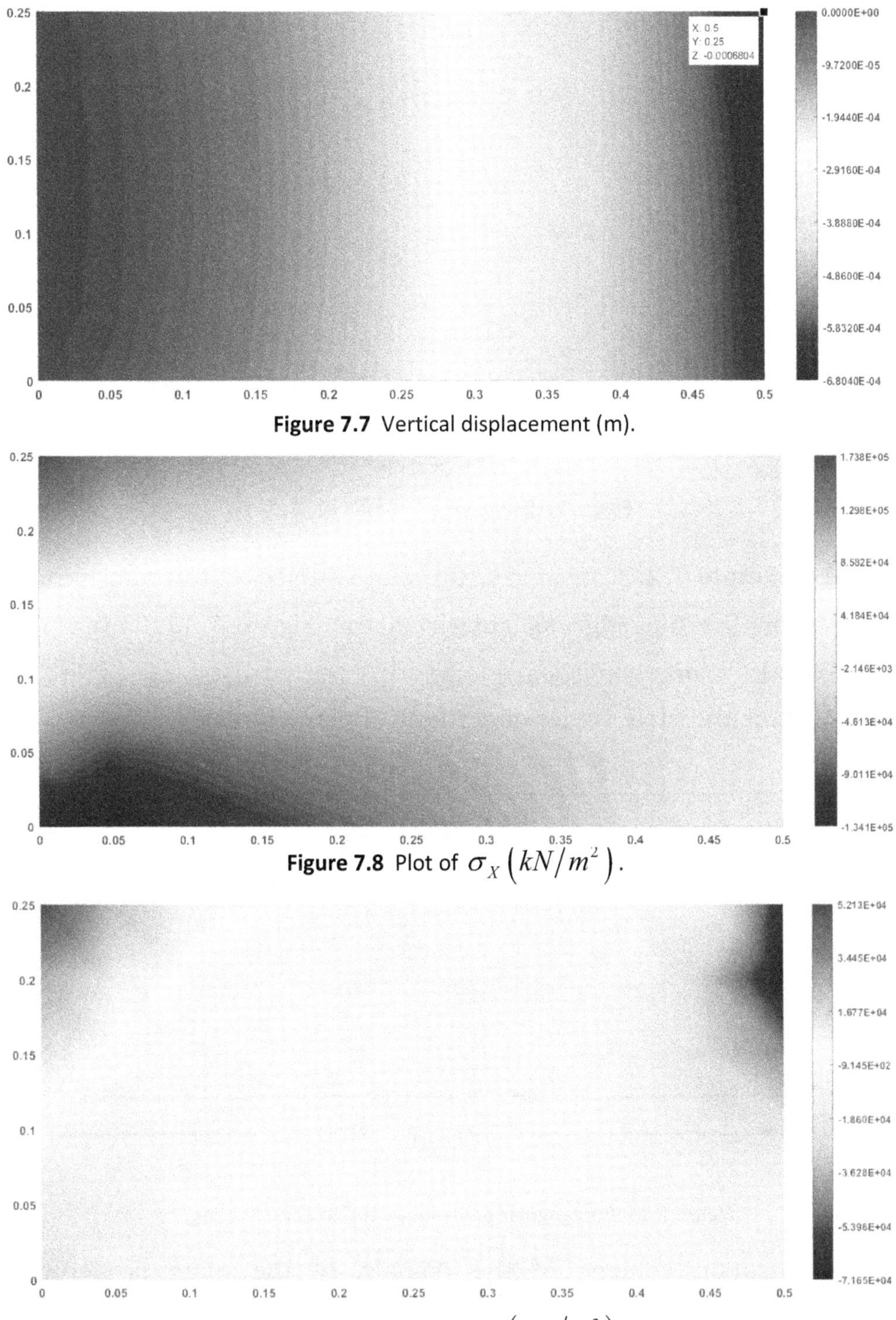

Figure 7.7 Vertical displacement (m).

Figure 7.8 Plot of $\sigma_X \left(kN/m^2 \right)$.

Figure 7.9 Plot of $\sigma_Y \left(kN/m^2 \right)$.

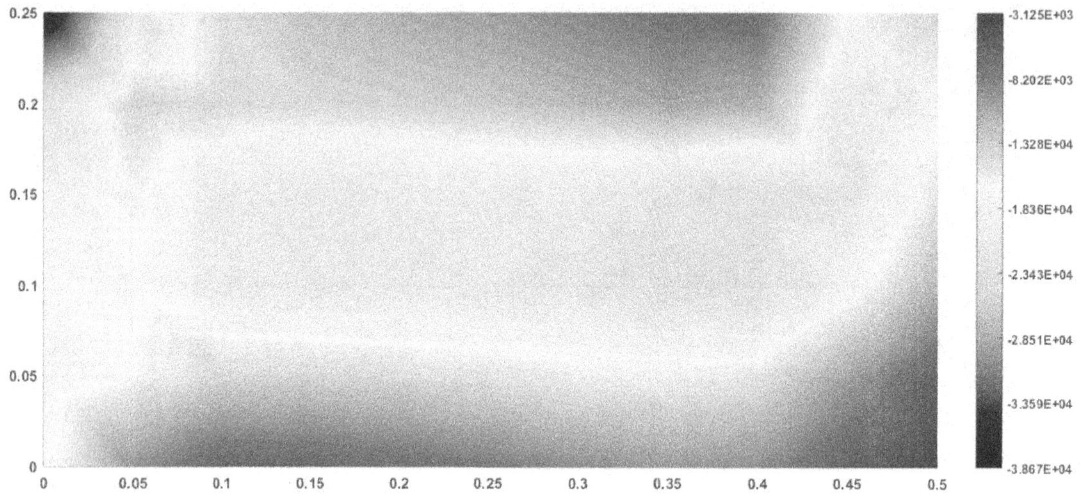

Figure 7.10 Plot of $\tau_{XY}\left(kN/m^2\right)$.

Numerical Example 7.2: Consider a thin plate subjected to uniform pressure of $100\,N/mm^2$ on the top edge as shown in the Figure 7.11. Given Young's modulus $E=2\times10^5\,N/mm^2$, Poisson's ratio $\mu=0.3$. Thickness of the plate $t=5\,mm$. Analyse the plate under conditions of plane stress.

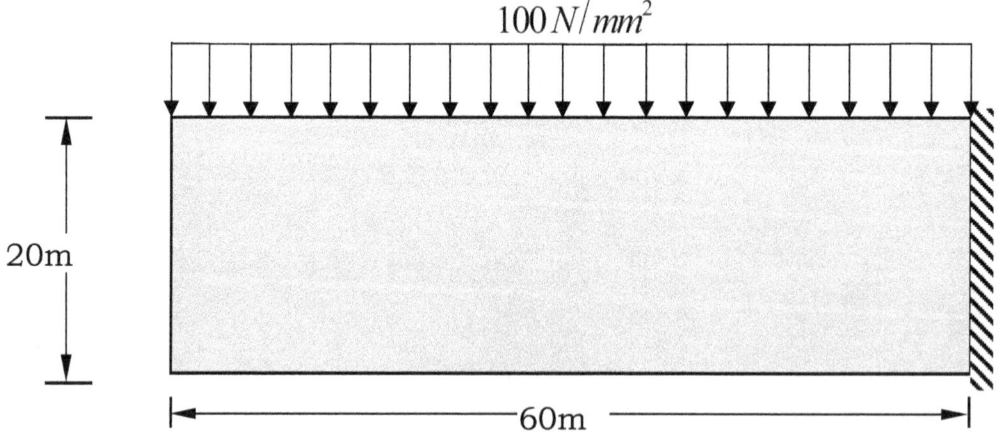

Figure 7.11 Rectangular plate subjected to edge loading.

The discretization scheme of the domain of the plate is depicted in Figure 7.12. Using Eq. 7.20, the uniform pressure acting on the top edge of the plate is expressed as equivalent static loads applied at nodes.

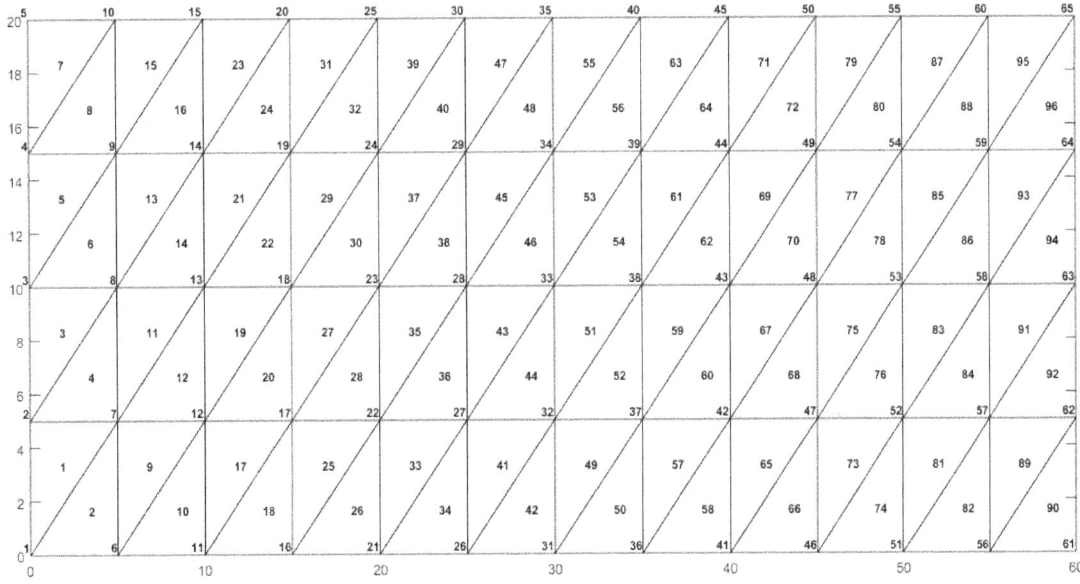

Figure 7.12 Discretization of rectangular plate.

Once again, the program listed in section A.5 is utilized for obtaining the results. The following is the input file used:

Input:

```
60.0   20.0 5.0 12 4 2.0e5    77.5      0.30     0.0  1  0
2         0.0  -1000.0
```

The following is the dialog generated during execution of the program to input for other quantities. Note that the responses are marked in bold font:

```
Enter 1 for Plane Stress or 2 for Plane Strain:1
Distributed load is present on the edges
Top edge is loaded
Enter 1 for Edge restraint OR 2 for Corner restraint:1
enter edge restraints in the order [Bottom Top Left Right]
enter 0 for restrained edge and 1 for unrestrained edge
Bottom Edge(1 or 0): 1
Top Edge(1 or 0): 1
Left Edge(1 or 0): 1
Right Edge(1 or 0): 0
Do you want plots on deformed configuration (1 for Yes 0 for No):0
```

Deformations obtained are compared with those obtained using ABAQUS with the same level of discretization. The values of selected displacements and stresses are listed in Tables 7.3 and 7.4:

Table 7.3 Comparison of displacements obtained with results of ABAQUS.

Node Location	Horizontal displacement $\times 10^{-4}$ m		Vertical displacement $\times 10^{-4}$ m	
	Present	ABAQUS	Present	ABAQUS
Top left corner	−0.23013	−0.23013	−1.12708	−1.12709
Bottom left corner	0.21919	0.21919	−1.12198	−1.12199

Table 7.4 Comparison of stresses obtained at integration points with results of ABAQUS.

Element	$\sigma_X(N/mm^2)$		$\sigma_Y(\times 10^5 N/mm^2)$		$\tau_{XY}(N/mm^2)$	
	Present	ABAQUS	Present	ABAQUS	Present	ABAQUS
2	−2.6515	−2.6515	−6.6120	−6.6120	3.7498	3.7498
19	7.6254	7.6254	−31.3286	−31.3286	69.8309	69.8309
50	−655.076	−655.076	−54.220	−54.220	164.925	164.925

It is observed from the above tables that the results obtained using the MATLAB program outlined in Section A.5 compare very well with the results obtained using ABAQUS.

Numerical Example 7.3: Analyze the plate shown in Figure 7.13 under conditions of plane stress. The thickness of the plate is 10mm, Young's modulus is $2.08 \times 10^5 N/mm^2$ and $\mu = 0.32$. Body force is neglected in comparison with externally applied load.

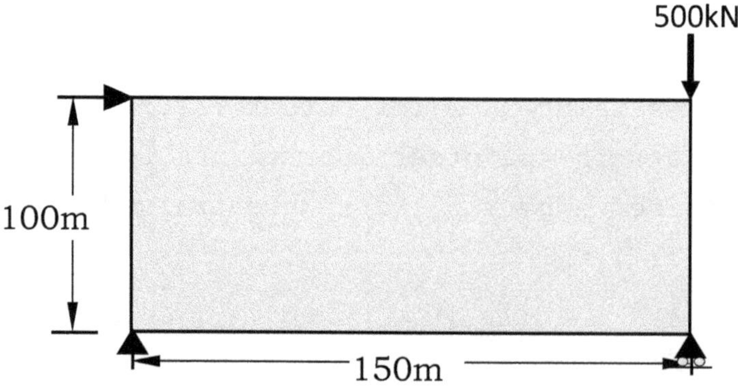

Figure 7.13 Rectangular plate.

Once again, the program listed in section A.5 is utilized for obtaining the results. The following is the input file used for analysis:

```
150.0   100.0 10.0 10 7 2.08e5     77.5       0.32     0.0   1   0

4   0.0   -500.0e3

%point load at top-right corner
```

The following is the dialog generated during execution of the program to input for other quantities. Note that the responses are marked in bold font:

```
Enter 1 for Plane Stress or  2 for Plane Strain:1

Point loads are present at specified nodes

Enter 1 for Edge restraint OR 2 for Corner restraint:2

enter corner restraints

enter 1 for Free  2 for hinge   3 for roller

Top-Left Corner(1/2/3)     : 2

Bottom-Left Corner(1/2/3) : 2

Top-Right Corner(1/2/3)    : 1

Bottom-Right Corner(1/2/3): 3

Do you want plots on deformed configuration (1 for Yes 0 for No):0
```

The mesh generated is shown in the Figure 7.14

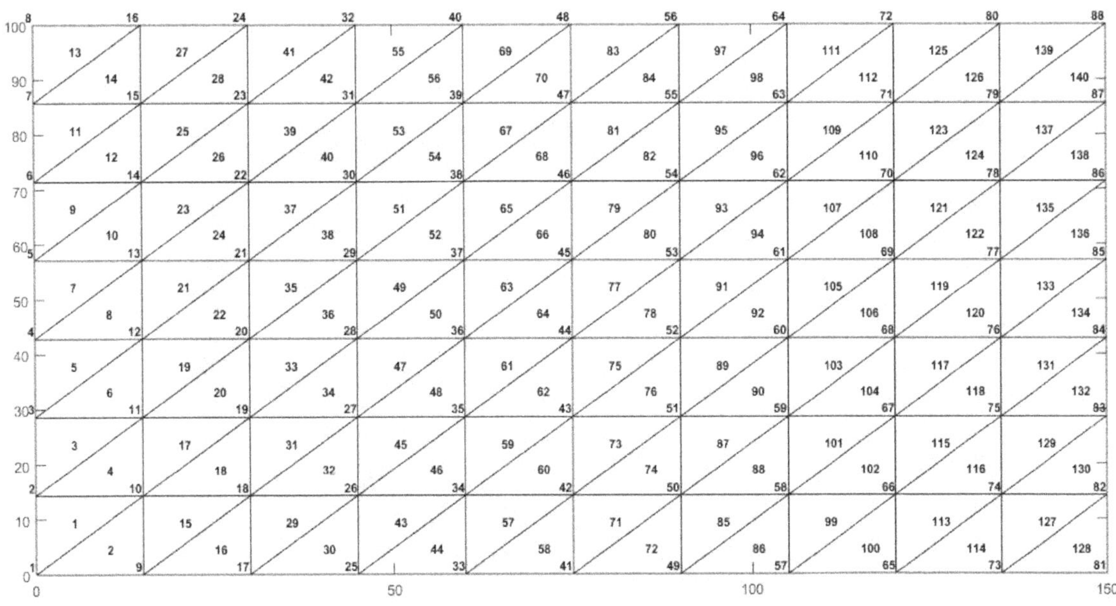

Figure 7.14 Mesh for the rectangular plate.

Deformations obtained are compared with those obtained using ABAQUS with the same level of discretization. The values of selected displacements and stresses are listed in Tables 7.5 and 7.6 below:

Table 7.5 Comparison of displacements obtained with results of ABAQUS.

Node Location	Horizontal displacement(mm)		Vertical displacement (mm)	
	Present	ABAQUS	Present	ABAQUS
Top right corner	0.64128	0.6413	−1.70198	−1.702
Bottom right corner	0.13592	0.13592	0.0	0.0

Table 7.6 Comparison of stresses obtained at integration points with results of ABAQUS.

Element	$\sigma_X(N/mm^2)$		$\sigma_Y(\times 10^5 N/mm^2)$		$\tau_{XY}(N/mm^2)$	
	Present	ABAQUS	Present	ABAQUS	Present	ABAQUS
1	−234.031	−234.031	−277.642	−277.642	−466.934	−466.934
13	1291.01	1291.01	231.881	231.881	−487.171	−487.172
7	7.78024	7.78024	95.3002	95.3002	32.1473	32.1473

It is observed from the above tables that the results obtained using the presented program compare very well with the results obtained using ABAQUS. Figures 7.15 and 7.16 show the variation of horizontal displacement across various nodes of the plate obtained using ABAQUS and the presented program. Similarly, Figures 7.17 and 7.18 show the variation of horizontal displacement across various nodes of the plate obtained using ABAQUS and the presented program. It is observed from these figures that the results obtained using the presented program match very well with the results obtained using MATLAB.

Figure 7.15 Plot of horizontal displacement using ABAQUS.

Figure 7.16 Plot of horizontal displacement- presented program.

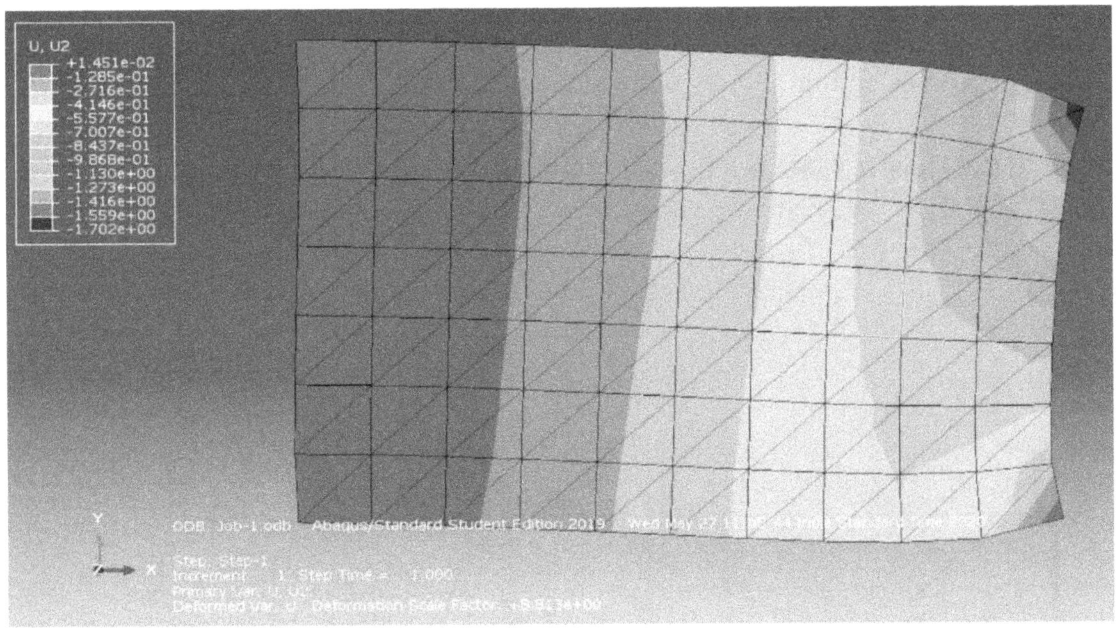

Figure 7.17 Plot of vertical displacement using ABAQUS.

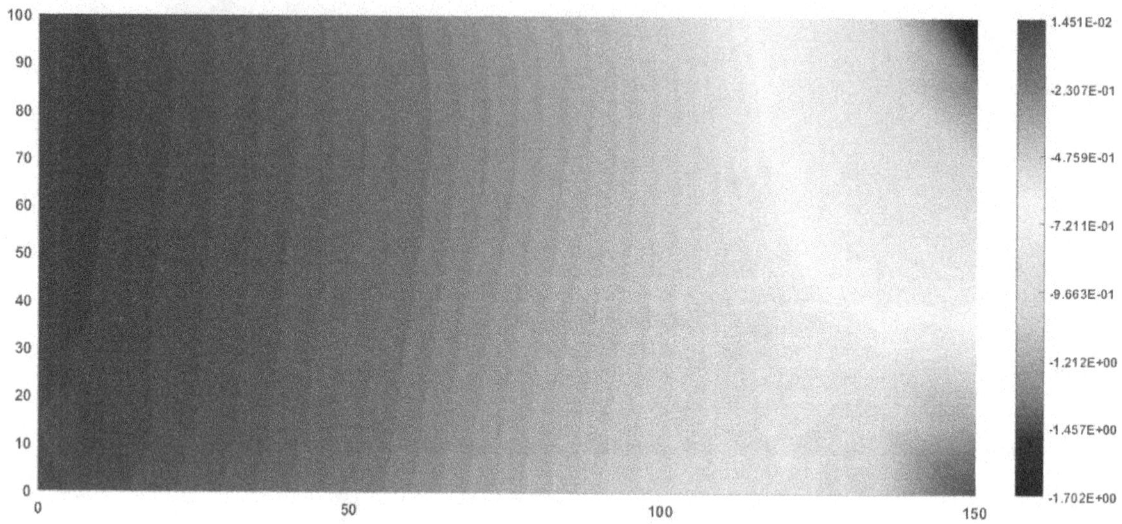

Figure 7.18 Plot of horizontal displacement- presented program.

Problems involving plane strain can also be solved with results matching the results obtained using ABAQUS.

7.4   SIX-NODE (LINEAR STRAIN) TRIANGULAR ELEMENT (LST)

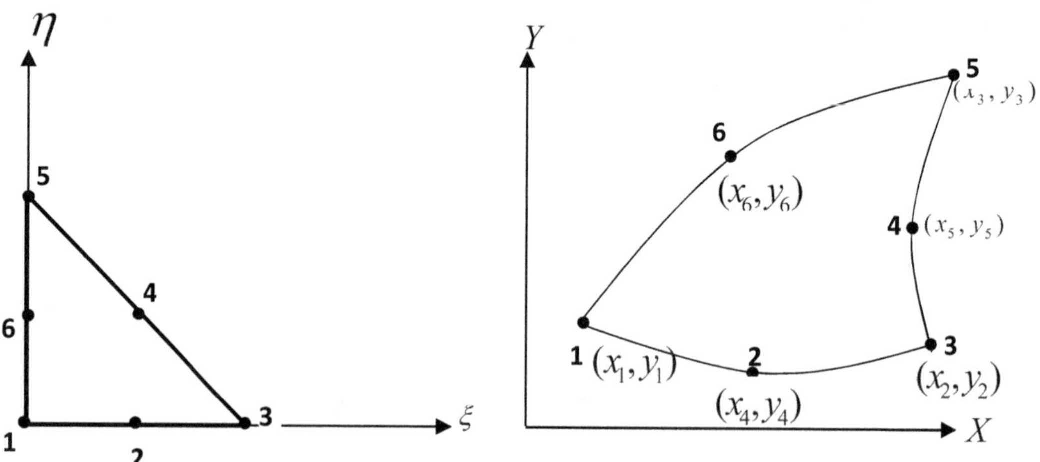

Figure 7.19 Six-node triangular element.

Consider the six-node triangular element shown in Figure 7.19. The interpolation functions for this element are discussed in Section 4.4.5.

The interpolation functions for the element are presented in Eq.4.60 can be used to model displacement of the element too. Thus, if u and v are the displacements of a point inside the element, they can be related to the nodal displacements as follows:

$$u = u_1 N_1 + u_2 N_2 + u_3 N_3 + u_4 N_4 + u_5 N_5 + u_6 N_6$$
$$v = v_1 N_1 + v_2 N_2 + v_3 N_3 + v_4 N_4 + v_5 N_5 + v_6 N_6$$

$$.....(7.21)$$

This can be written in the matrix form as follows:

$$\begin{Bmatrix} u \\ v \end{Bmatrix} = \begin{bmatrix} N_1 & 0 & N_2 & 0 & N_3 & 0 & N_4 & 0 & N_5 & 0 & N_6 & 0 \\ 0 & N_1 & 0 & N_2 & 0 & N_3 & 0 & N_4 & 0 & N_5 & 0 & N_6 \end{bmatrix} \{d\} \quad .....(7.22)$$

where

$$\{d\} = \begin{bmatrix} u_1 & v_1 & u_2 & v_2 & u_3 & v_3 & u_4 & v_4 & u_5 & v_5 & u_6 & v_6 \end{bmatrix}^T \qquad .....(7.23)$$

This can be represented as

$$\{u\} = [N]\{d\} \qquad\qquad .....(7.24)$$

Here $[N]$ is the shape function matrix and $\{d\}$ is the vector of nodal displacements. On similar lines, the x and y coordinates of any point inside the element are related to the nodal coordinates as

$$x = N_1 x_1 + N_2 x_2 + N_3 x_3 + N_4 x_4 + N_5 x_5 + N_6 x_6$$
$$y = N_1 y_1 + N_2 y_2 + N_3 y_3 + N_4 y_4 + N_5 y_5 + N_6 y_6$$

$$.....(7.25)$$

The partial derivatives are given by

$$\frac{\partial x}{\partial \xi} = \sum_{i=1}^{6} \frac{\partial N_i}{\partial \xi} x_i \quad ; \quad \frac{\partial y}{\partial \xi} = \sum_{i=1}^{6} \frac{\partial N_i}{\partial \xi} y_i$$

$$\frac{\partial x}{\partial \eta} = \sum_{i=1}^{6} \frac{\partial N_i}{\partial \eta} x_i \quad ; \quad \frac{\partial y}{\partial \eta} = \sum_{i=1}^{6} \frac{\partial N_i}{\partial \eta} y_i$$

$$.....(7.26)$$

The Jacobian matrix $[J]$ is given by

$$[J] = \begin{bmatrix} \dfrac{\partial x}{\partial \xi} & \dfrac{\partial y}{\partial \xi} \\[2ex] \dfrac{\partial x}{\partial \eta} & \dfrac{\partial y}{\partial \eta} \end{bmatrix} = \begin{bmatrix} \displaystyle\sum_{i=1}^{6} \frac{\partial N_i}{\partial \xi} x_i & \displaystyle\sum_{i=1}^{6} \frac{\partial N_i}{\partial \xi} y_i \\[3ex] \displaystyle\sum_{i=1}^{6} \frac{\partial N_i}{\partial \eta} x_i & \displaystyle\sum_{i=1}^{6} \frac{\partial N_i}{\partial \eta} y_i \end{bmatrix} \qquad .....(7.27)$$

$$[J] = \begin{bmatrix} \dfrac{\partial N_1}{\partial \xi} & \dfrac{\partial N_2}{\partial \xi} & \dfrac{\partial N_3}{\partial \xi} & \dfrac{\partial N_4}{\partial \xi} & \dfrac{\partial N_5}{\partial \xi} & \dfrac{\partial N_6}{\partial \xi} \\[2mm] \dfrac{\partial N_1}{\partial \eta} & \dfrac{\partial N_2}{\partial \eta} & \dfrac{\partial N_3}{\partial \eta} & \dfrac{\partial N_4}{\partial \eta} & \dfrac{\partial N_5}{\partial \eta} & \dfrac{\partial N_6}{\partial \eta} \end{bmatrix} \begin{bmatrix} x_1 & y_1 \\ x_2 & y_2 \\ x_3 & y_3 \\ x_4 & y_4 \\ x_5 & y_5 \\ x_6 & y_6 \end{bmatrix} \qquad \text{.....(7.28)}$$

Using the interpolation functions for the element listed in Eq. 4.60, we can compute the Jacobian matrix as follows:

$$[J] = \begin{bmatrix} 1-4\lambda & 4(\lambda-\xi) & 4\xi-1 & 4\eta & 0 & -4\eta \\ 1-4\lambda & -4\xi & 0 & 4\xi & 4\eta-1 & 4(\lambda-\eta) \end{bmatrix} \begin{bmatrix} x_1 & y_1 \\ x_2 & y_2 \\ x_3 & y_3 \\ x_4 & y_4 \\ x_5 & y_5 \\ x_6 & y_6 \end{bmatrix} \qquad \text{.....(7.29)}$$

where $\lambda = 1 - \xi - \eta$.

The strain vector can be computed as follows:

$$\{\varepsilon\} = \left\{ \begin{array}{c} \varepsilon_x \\ \varepsilon_y \\ \gamma_{xy} \end{array} \right\} = \begin{bmatrix} \dfrac{\partial}{\partial x} & 0 \\[2mm] 0 & \dfrac{\partial}{\partial y} \\[2mm] \dfrac{\partial}{\partial y} & \dfrac{\partial}{\partial x} \end{bmatrix} \begin{bmatrix} N_1 & 0 & N_2 & 0 & N_3 & 0 & N_4 & 0 & N_5 & 0 & N_6 & 0 \\ 0 & N_1 & 0 & N_2 & 0 & N_3 & 0 & N_4 & 0 & N_5 & 0 & N_6 \end{bmatrix} \{d\}$$

$$\text{.....(7.30)}$$

$$\{\varepsilon\} = \begin{bmatrix} \dfrac{\partial N_1}{\partial x} & 0 & \dfrac{\partial N_2}{\partial x} & 0 & \dfrac{\partial N_3}{\partial x} & 0 & \dfrac{\partial N_4}{\partial x} & 0 & \dfrac{\partial N_5}{\partial x} & 0 & \dfrac{\partial N_6}{\partial x} & 0 \\[2mm] 0 & \dfrac{\partial N_1}{\partial y} & 0 & \dfrac{\partial N_2}{\partial y} & 0 & \dfrac{\partial N_3}{\partial y} & 0 & \dfrac{\partial N_4}{\partial y} & 0 & \dfrac{\partial N_5}{\partial y} & 0 & \dfrac{\partial N_6}{\partial y} \\[2mm] \dfrac{\partial N_1}{\partial y} & \dfrac{\partial N_1}{\partial x} & \dfrac{\partial N_2}{\partial y} & \dfrac{\partial N_2}{\partial x} & \dfrac{\partial N_3}{\partial y} & \dfrac{\partial N_3}{\partial x} & \dfrac{\partial N_4}{\partial y} & \dfrac{\partial N_4}{\partial x} & \dfrac{\partial N_5}{\partial y} & \dfrac{\partial N_5}{\partial x} & \dfrac{\partial N_6}{\partial y} & \dfrac{\partial N_6}{\partial x} \end{bmatrix} \{d\}$$

$$\text{....(7.31)}$$

$$\{\varepsilon\} = [B]\{d\} \qquad \text{.....(7.32)}$$

where $[B]$ is the strain-displacement matrix.

In order to compute the $[B]$ matrix, it is necessary to compute partial derivatives of the interpolation functions with reference to x and y coordinates. This is achieved as follows:

$$\frac{\partial N_i}{\partial \xi} = \frac{\partial N_i}{\partial x}\frac{\partial x}{\partial \xi} + \frac{\partial N_i}{\partial y}\frac{\partial y}{\partial \xi}$$

$$\frac{\partial N_i}{\partial \eta} = \frac{\partial N_i}{\partial x}\frac{\partial x}{\partial \eta} + \frac{\partial N_i}{\partial y}\frac{\partial y}{\partial \eta}$$

$$\ldots\ldots(7.33)$$

Eq. 7.30 can be rewritten in the matrix form as follows:

$$\begin{Bmatrix} \dfrac{\partial N_i}{\partial \xi} \\ \dfrac{\partial N_i}{\partial \eta} \end{Bmatrix} = \begin{bmatrix} \dfrac{\partial x}{\partial \xi} & \dfrac{\partial y}{\partial \xi} \\ \dfrac{\partial x}{\partial \eta} & \dfrac{\partial y}{\partial \eta} \end{bmatrix} \begin{Bmatrix} \dfrac{\partial N_i}{\partial x} \\ \dfrac{\partial N_i}{\partial y} \end{Bmatrix} = [J] \begin{Bmatrix} \dfrac{\partial N_i}{\partial x} \\ \dfrac{\partial N_i}{\partial y} \end{Bmatrix} \quad \ldots\ldots(7.34)$$

For a bijective relation between set of points in the reference element and the corresponding set of points in the parent element, an inverse relation to Eq. 7.34 exists and can be expressed as

$$\begin{Bmatrix} \dfrac{\partial N_i}{\partial x} \\ \dfrac{\partial N_i}{\partial y} \end{Bmatrix} = [J]^{-1} \begin{Bmatrix} \dfrac{\partial N_i}{\partial \xi} \\ \dfrac{\partial N_i}{\partial \eta} \end{Bmatrix} \quad \ldots\ldots(7.35)$$

Eq. 7.35 can be expanded as

$$\begin{bmatrix} \dfrac{\partial N_1}{\partial x} & \dfrac{\partial N_2}{\partial x} & \dfrac{\partial N_3}{\partial x} & \dfrac{\partial N_4}{\partial x} & \dfrac{\partial N_5}{\partial x} & \dfrac{\partial N_6}{\partial x} \\ \dfrac{\partial N_1}{\partial y} & \dfrac{\partial N_2}{\partial y} & \dfrac{\partial N_3}{\partial y} & \dfrac{\partial N_4}{\partial y} & \dfrac{\partial N_5}{\partial y} & \dfrac{\partial N_6}{\partial y} \end{bmatrix} = [J^{-1}] \begin{bmatrix} \dfrac{\partial N_1}{\partial \xi} & \dfrac{\partial N_2}{\partial \xi} & \dfrac{\partial N_3}{\partial \xi} & \dfrac{\partial N_4}{\partial \xi} & \dfrac{\partial N_5}{\partial \xi} & \dfrac{\partial N_6}{\partial \xi} \\ \dfrac{\partial N_1}{\partial \eta} & \dfrac{\partial N_2}{\partial \eta} & \dfrac{\partial N_3}{\partial \eta} & \dfrac{\partial N_4}{\partial \eta} & \dfrac{\partial N_5}{\partial \eta} & \dfrac{\partial N_6}{\partial \eta} \end{bmatrix}$$

$$\ldots\ldots(7.36)$$

The partial derivatives obtained using Eq. 7.36 are utilised in assembling the B matrix given in Eq. 7.31.

It is observed from Eq.7.31 and Eq.7.36 that strain $\{\varepsilon\}$ computed is a linear function of ξ and η. Thus, strain varies linearly from node to node across the element and the element is called *Linear Strain Triangular (LST) element*.

The stiffness matrix for the LST element can be expressed as

$$\left[K^{(e)} \right] = h\int \left[B^T \right][D][B]\,dA \qquad \qquad .....(7.37)$$

$$\left[K^{(e)} \right] = h\int_{0}^{1}\int_{0}^{1-\xi} \left[B(\xi,\eta)^T \right][D]\left[B(\xi,\eta) \right]|J|\,d\xi d\eta$$

$$= h\sum_{i=1}^{nhp} W_i \left[B(\xi_i,\eta_i)^T \right][D]\left[B(\xi_i,\eta_i) \right]|J| \qquad \qquad ......(7.38)$$

where *nhp* represents the number of Hammer points.

Force vector for the LST element can be expressed as

$$\{Q^{(e)}\} = \iiint \left[N^T \right]\{X\}\,dV + \iint_{S_1} \left[N_S^T \right]\{f\}\,dA + \begin{Bmatrix} Q_1 \\ Q_2 \end{Bmatrix} \qquad .....(7.39)$$

The first, second and third terms in the above equation correspond to the *vector of body forces, vector of surface tractions and point loads*. First, the vector of surface tractions is evaluated as follows.

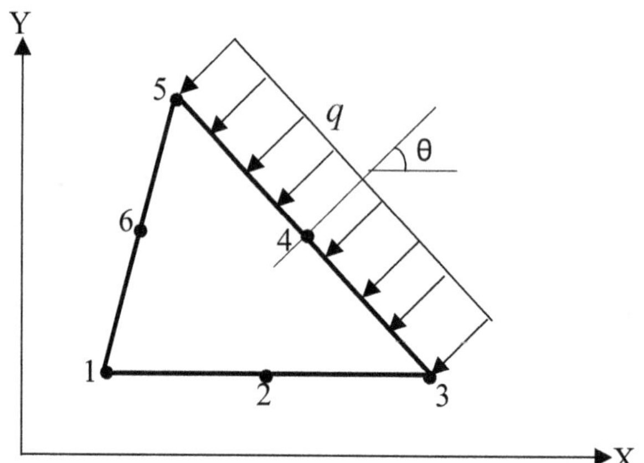

Figure 7.20 Six-node triangular element -Surface traction on an edge.

Considering an element subjected to a normal pressure of magnitude q on the edge 3-4-5 as shown in Figure 7.20, vector of traction forces can be written as

$$\{f\} = \begin{Bmatrix} q\cos\theta \\ q\sin\theta \end{Bmatrix} \qquad \qquad .....(7.40)$$

Substituting Eq.7.40 in the second term of Eq. 7.39, and considering that N_1, N_2 and N_6 are zero on side 3-4-5 (denoted by $L_{3\text{-}4\text{-}5}$), we obtain

$$\iint_{S_1}\left[N_S^T\right]\{f\}\,dA = \int \begin{bmatrix} 0 & 0 \\ 0 & 0 \\ 0 & 0 \\ 0 & 0 \\ N_3 & 0 \\ 0 & N_3 \\ N_4 & 0 \\ 0 & N_4 \\ N_5 & 0 \\ 0 & N_5 \\ 0 & 0 \\ 0 & 0 \end{bmatrix} \begin{Bmatrix} q\cos\theta \\ q\sin\theta \end{Bmatrix} h\,dl = h\int \begin{Bmatrix} 0 \\ 0 \\ 0 \\ 0 \\ N_3 q\cos\theta \\ N_3 q\sin\theta \\ N_4 q\cos\theta \\ N_4 q\sin\theta \\ N_5 q\cos\theta \\ N_5 q\sin\theta \\ 0 \\ 0 \end{Bmatrix} dL = \begin{Bmatrix} 0 \\ 0 \\ 0 \\ 0 \\ \dfrac{1}{6}qL_{3-5}\cos\theta \\ \dfrac{1}{6}qL_{3-5}\sin\theta \\ \dfrac{1}{3}qL_{3-5}\cos\theta \\ \dfrac{1}{3}qL_{3-5}\sin\theta \\ \dfrac{1}{6}qL_{3-5}\cos\theta \\ \dfrac{1}{6}qL_{3-5}\sin\theta \\ 0 \\ 0 \end{Bmatrix}$$

$$....(7.41)$$

The third term of Eq. 7.39 represents the vector of point loads acting on the element.

7.4.1   Numerical Example

Consider a thin plate subjected to a point load of $1000\ N$ at the centre of the left edge as shown in the Figure 7.21. Adopt Young's modulus $E=2\times10^5\ N/mm^2$, Poisson's ratio $\mu = 0.3$. Thickness of the plate $t = 5\ mm$. Analyse the plate under conditions of plane stress using LST elements.

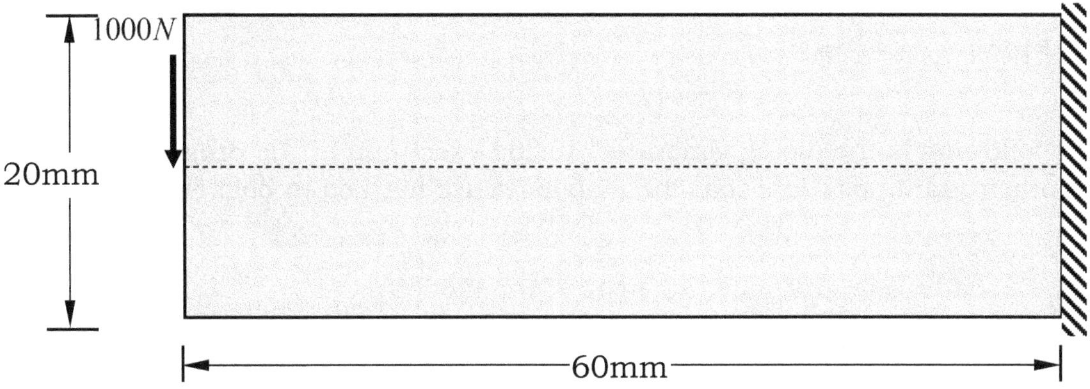

Figure 7.21 Rectangular plate subjected to point load at the left end at mid-depth.

The program listed in section A.6 is utilized for obtaining the results. The following is the input file used for analysis:

```
60.0     20.0     5.0 10  4  2.0e5 77.5   0.30    0.0  1  0
2  0.0  -1000.0
%point load at center-left node
%Lx      Ly      t  nx ny  E     rho     mue     pressure edge corner
% Lx, Ly,t in metres
% E in kN/m^2
pressure kN/m^2
mass density kg/m^3
Corner codes
1: TL
2: CL
3: BL
4: TR
5: CR
6: BR
Edges
1: Top
2. Left
3. Bottom
4. Right
```

The following is the dialog generated during execution of the program to input for other quantities. Note that the responses are marked in bold font:

```
Enter 1 for Plane Stress or  2 for Plane Strain:1
Point loads are present at specified nodes
Enter 1 for Edge restraint or 2 for Corner restraint:1
enter edge restraints in the order [Bottom Top Left Right]
enter 0 for restrained edge and 1 for unrestrained edge
```

```
Bottom Edge(1 or 0): 1

Top Edge(1 or 0): 1

Left Edge(1 or 0): 1

Right Edge(1 or 0): 0

Do you want plots on deformed configuration (1 for Yes 0 for No):0
```

The mesh generated is shown in Figure 7.22. The plate is also analysed using ABAQUS software. *CPS6 elements are used in the analysis.* The results obtained using the present program match exactly with the results obtained using ABAQUS. For example, Figures 7.23 and 7.25 depict the variation of horizontal and vertical displacements obtained using ABAQUS. Similar plots for horizontal and vertical displacements using the present program are presented in Figures 7.24 and 7.26. A comparison of Figures 7.23 and 7.25 for horizontal displacement and Figures 7.24 and 7.26 for vertical displacements reveals that the results obtained using the present MATLAB program are accurate.

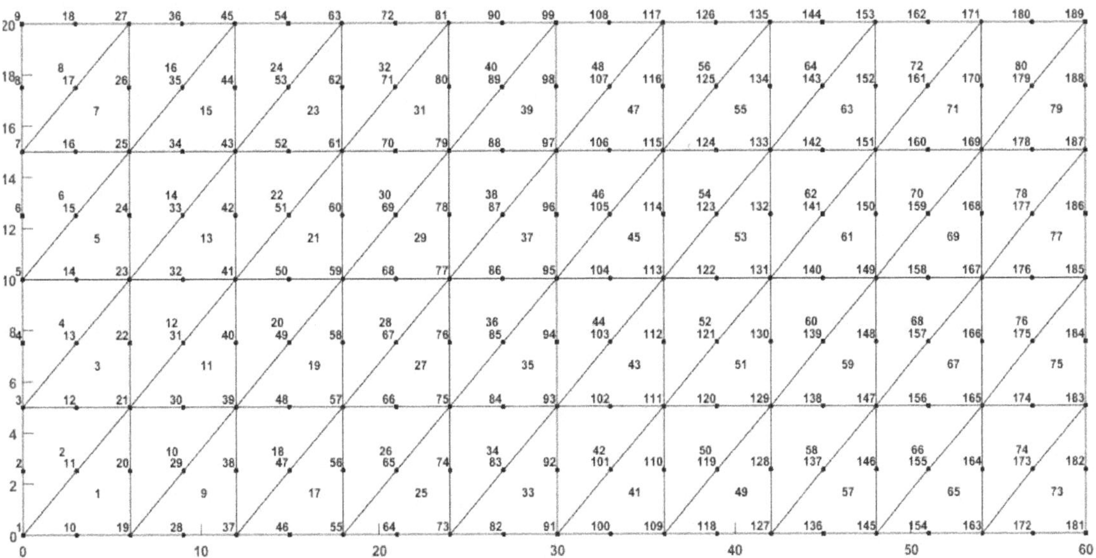

Figure 7.22 Mesh for the rectangular plate with LST elements.

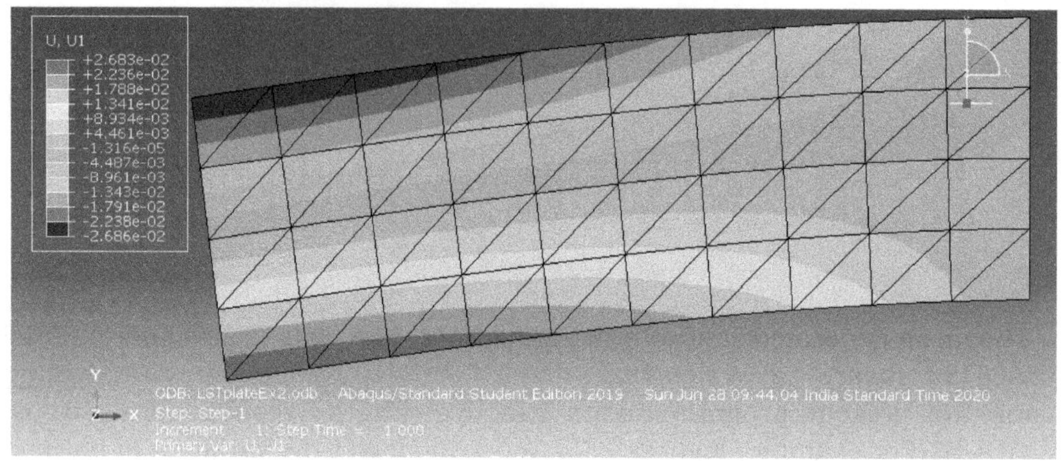

Figure 7.23 Plot of horizontal displacement (ABAQUS).

Figure 7.24 Plot of horizontal displacement (Present).

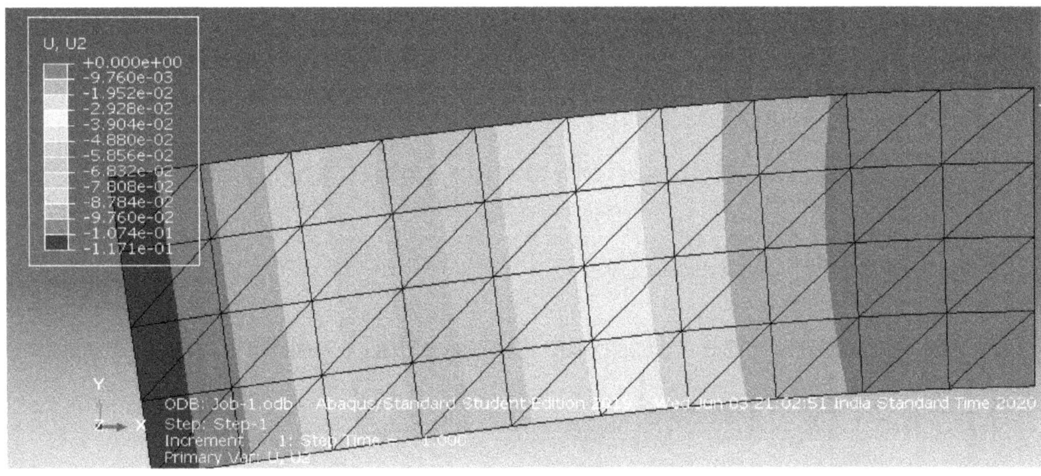

Figure 7.25 Plot of vertical displacement (ABACUS).

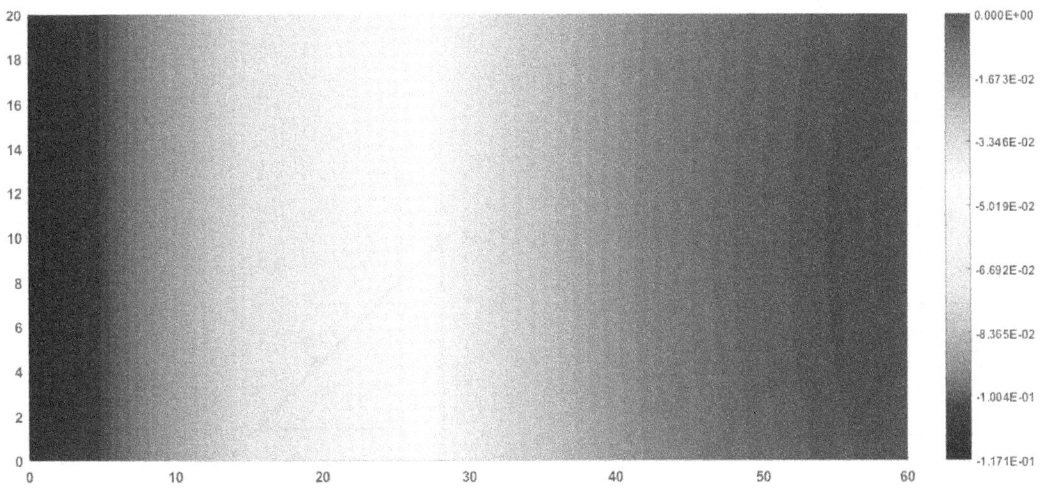

Figure 7.26 Plot of vertical displacement (Present).

7.4.2   Nodal Stress Recovery

The values of stresses σ_X, σ_Y and τ_{XY} are obtained at integration points of each element during post processing. However, it is often required to plot the variation of there stresses across the plate. For this purpose, stresses are extrapolated from the integration points of the element to the element nodes by the process of *nodal averging*. Each node receives contributions from elements connected to it. Nodal averaging is performed to compute the average of contributions of stress received at a node from elements connected to it.

For example, consider a parent triangular element with six nodes as shown in Figure 7.27. The natural coordinates of nodes of this element are given in Table 7.7. Two options for the Hammer integration points is available:

a) In the Case-A, the Hammer integration points are located at $1'\left(\dfrac{1}{2},\dfrac{1}{2}\right)$, $2'\left(0,\dfrac{1}{2}\right)$ and $3'\left(\dfrac{1}{2},0\right)$ respectively.

b) In the Case-B, the Hammer integration points are located at $1'\left(\dfrac{1}{6},\dfrac{1}{6}\right)$, $2'\left(\dfrac{2}{3},\dfrac{1}{6}\right)$ and $3'\left(\dfrac{1}{6},\dfrac{2}{3}\right)$ respectively.

The triangle that is obtained by joining these three points is called *Hammer element*. The Hammer element can be visualized as an ' *three node triangular element*' inside the parent triangular element with six nodes. Coordinate axes ξ' and η' are passed with the node $1'$ as origin as shown in the figure .

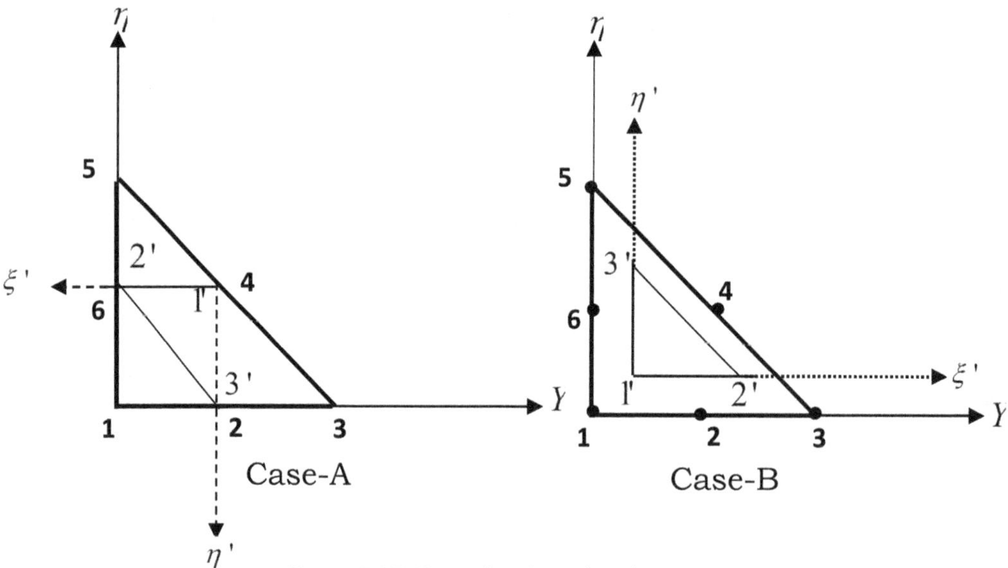

Figure 7.27 Six node triangular element.

The natural coordinates of nodes 1 through 6 with reference to the new natural coordinate system $\xi'-\eta'$ are given in columns 4 and 5 of Table 7.7 for case-A. Corresponding coordinates for case-B are given in Table 7.7. These coordinates are substituted in the shape functions N_1', N_2' and N_3' of the 3 node

Hammer element. These shape functions are the same as those given in Eq. 4.39 and are reproduced here for convenience.

$$N_1' = 1 - \xi - \eta$$
$$N_2' = \xi \qquad \qquad \qquad ....(7.42)$$
$$N_3' = \eta$$

Any scalar quantity w whose values w_i' at the corners of Hammer element are known can be interpolated through the above shape functions now expressed in terms of ξ and η as follows:

$$w(\xi',\eta') = \lfloor w_1' \quad w_2' \quad w_3' \rfloor \begin{Bmatrix} N_1^{(e')} \\ N_2^{(e')} \\ N_3^{(e')} \end{Bmatrix} = \lfloor N_1^{(e')} \quad N_2^{(e')} \quad N_3^{(e')} \rfloor^T \begin{Bmatrix} w_1' \\ w_2' \\ w_3' \end{Bmatrix} \qquad .....(7.43)$$

The natural coordinates of the six nodes of original element can be substituted in Eq.7.40, to obtain the following relationship for extrapolation of nodal stresses.

$$\lfloor w_1 \quad w_2 \quad w_3 \quad w_4 \quad w_5 \quad w_6 \rfloor^T = [T] \lfloor w_1' \quad w_2' \quad w_3' \rfloor^T \qquad (7.44)$$

Here $\lfloor w_1' \quad w_2' \quad w_3' \rfloor$ represents the each of the stress values (σ_{XX}, σ_{YY} or τ_{XY}) at the Hammer integration points whereas $\lfloor w_1 \quad w_2 \quad w_3 \quad w_4 \quad w_5 \quad w_6 \rfloor$ represents the corresponding values at the six nodes of the original element.

Table 7.7 Natural coordinates of six-noded triangular element for Case – A.

Corner node	ξ	η	ξ'	η'	Hammer node	ξ	η	ξ'	η'
1	0	0	1	1	1'	$\frac{1}{2}$	$\frac{1}{2}$	0	0
2	$\frac{1}{2}$	0	0	1	2'	0	$\frac{1}{2}$	1	0
3	1	0	-1	1	3'	$\frac{1}{2}$	0	0	1
4	$\frac{1}{2}$	$\frac{1}{2}$	0	0					
5	0	1	1	-1					
6	0	$\frac{1}{2}$	1	0					

Table 7.8 Natural coordinates of six-noded triangular element-Case-B.

Corner node	ξ	η	ξ'	η'	Hammer node	ξ	η	ξ'	η'
1	0	0	$-\dfrac{1}{6}$	$-\dfrac{1}{6}$	1'	$\dfrac{1}{6}$	$\dfrac{1}{6}$	0	0
2	$\dfrac{1}{2}$	0	$\dfrac{1}{3}$	$-\dfrac{1}{6}$	2'	$\dfrac{2}{3}$	$\dfrac{1}{6}$	1	0
3	1	0	$\dfrac{5}{6}$	$-\dfrac{1}{6}$	3'	$\dfrac{1}{6}$	$\dfrac{2}{3}$	0	1
4	$\dfrac{1}{2}$	$\dfrac{1}{2}$	$\dfrac{1}{3}$	$\dfrac{1}{3}$					
5	0	1	$-\dfrac{1}{6}$	$\dfrac{5}{6}$					
6	0	$\dfrac{1}{2}$	$-\dfrac{1}{6}$	$\dfrac{1}{3}$					

Figure 7.28, Figure 7.29 and Figure 7.30 represent the variation of the in-plane stresses σ_{XX}, σ_{YY} and τ_{XY} across the plate.

Figure 7.28 Plot of variation of σ_X across the plate.

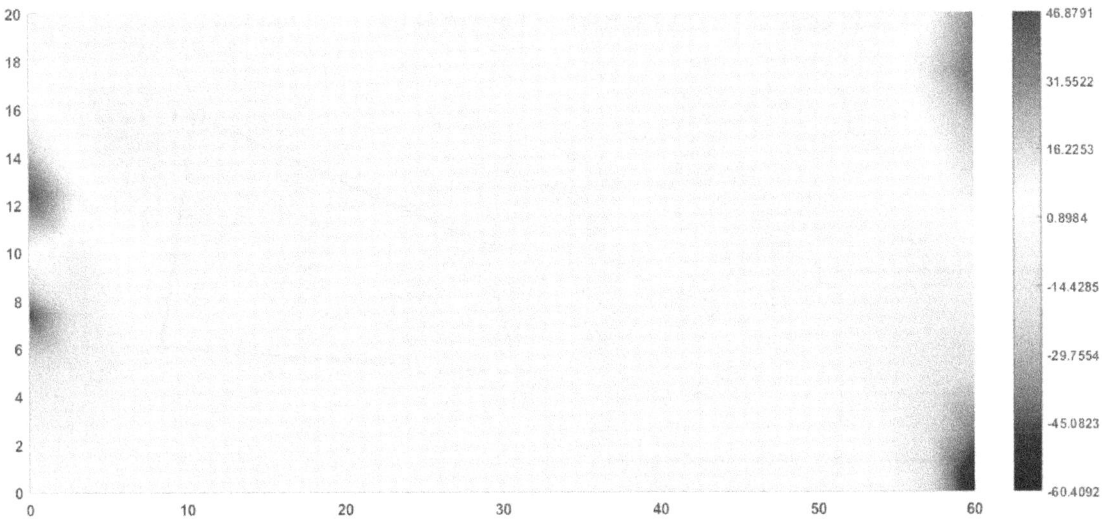

Figure 7.29 Plot of variation of σ_Y across the plate.

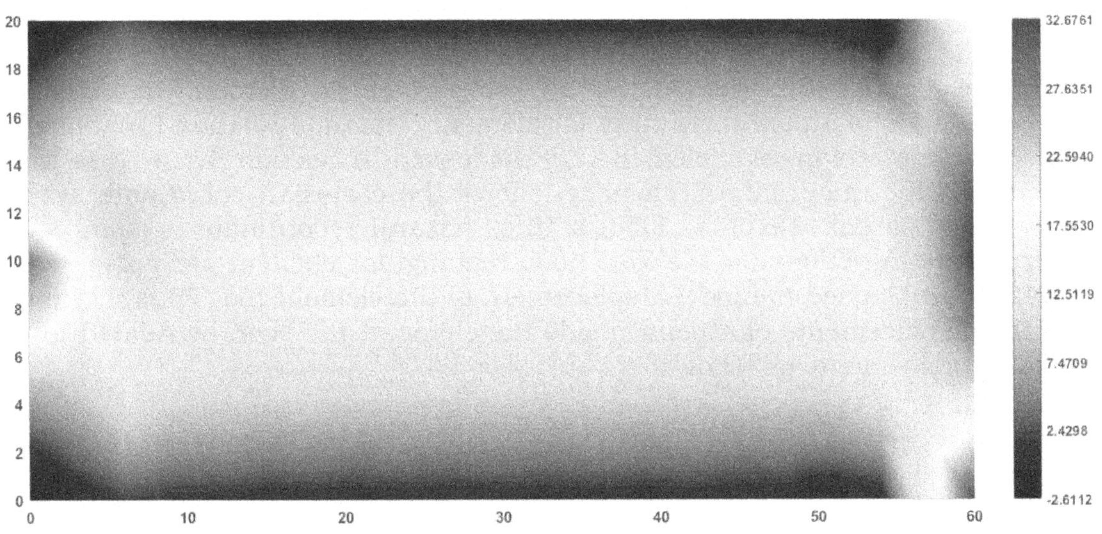

Figure 7.30 Plot of variation of τ_{XY} across the plate.

7.5 FOUR-NODE QUADRILATERAL ELEMENT

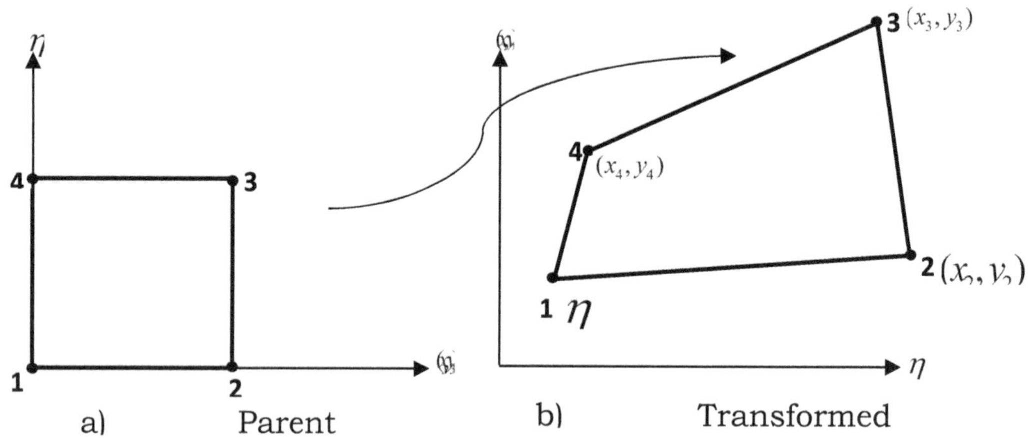

Figure 7.31 Four-node quadrilateral element.

Consider the four-node quadrilateral element shown in Figure 7.31. As seen from the figure, this element is an iso-parametric element obtained by transformation of a parent rectangular element. The interpolation functions for the four-node rectangular element are discussed in Section 4.4.4. It is to be noted that the quadrilateral element follows the cartesian coordinate system while the parent element follows the natural coordinate system. The interpolation functions for the four-node rectangular element are presented in Eq.4.49 can be used to model displacement of the element too. Thus, if u and v are the displacements of a point inside the element, they can be related to the nodal displacements as follows:

$$u = u_1 N_1 + u_2 N_2 + u_3 N_3 + u_4 N_4$$
$$v = v_1 N_1 + v_2 N_2 + v_3 N_3 + v_4 N_4$$

......(7.45)

This can be written in the matrix form as follows:

$$\left\{ \begin{matrix} u \\ v \end{matrix} \right\} = \begin{bmatrix} N_1 & 0 & N_2 & 0 & N_3 & 0 & N_4 & 0 \\ 0 & N_1 & 0 & N_2 & 0 & N_3 & 0 & N_4 \end{bmatrix} \{d\}$$

.....(7.46)

where

$$\{d\} = \begin{bmatrix} u_1 & v_1 & u_2 & v_2 & u_3 & v_3 & u_4 & v_4 \end{bmatrix}^T$$

.....(7.47)

This can be represented as

$$\{u\} = [N]\{d\} \qquad \qquad .....(7.48)$$

Here $[N]$ is the shape function matrix and $\{d\}$ is the vector of nodal displacements.

Partial derivatives w.r.t Cartesian coordinates x and y can be expressed in terms of partial derivatives w.r.t natural coordinates ξ and η as

$$\frac{\partial}{\partial \xi} = \frac{\partial}{\partial x}\frac{\partial x}{\partial \xi} + \frac{\partial}{\partial y}\frac{\partial y}{\partial \xi}$$

$$\frac{\partial}{\partial \eta} = \frac{\partial}{\partial x}\frac{\partial x}{\partial \eta} + \frac{\partial}{\partial y}\frac{\partial y}{\partial \eta}$$

$$.....(7.49)$$

This can be expressed in matrix form as

$$\left\{\begin{matrix} \dfrac{\partial}{\partial \xi} \\[2mm] \dfrac{\partial}{\partial \eta} \end{matrix}\right\} = \begin{bmatrix} \dfrac{\partial x}{\partial \xi} & \dfrac{\partial y}{\partial \xi} \\[2mm] \dfrac{\partial x}{\partial \eta} & \dfrac{\partial y}{\partial \eta} \end{bmatrix} \left\{\begin{matrix} \dfrac{\partial}{\partial x} \\[2mm] \dfrac{\partial}{\partial y} \end{matrix}\right\} = [J]\left\{\begin{matrix} \dfrac{\partial}{\partial x} \\[2mm] \dfrac{\partial}{\partial y} \end{matrix}\right\} \qquad .....(7.50)$$

where $[J]$, the Jacobian matrix is given by

$$[J] = \begin{bmatrix} \dfrac{\partial x}{\partial \xi} & \dfrac{\partial y}{\partial \xi} \\[2mm] \dfrac{\partial x}{\partial \eta} & \dfrac{\partial y}{\partial \eta} \end{bmatrix} \qquad \qquad .....(7.51)$$

The x and y coordinates of any point inside the element are related to the nodal coordinates as

$$x = N_1 x_1 + N_2 x_2 + N_3 x_3 + N_4 x_4$$

$$y = N_1 y_1 + N_2 y_2 + N_3 y_3 + N_4 y_4$$

$$.....(7.52)$$

Using Eq. 7.51, Jacobian matrix is expressed as

$$[J] = \begin{bmatrix} \dfrac{\partial x}{\partial \xi} & \dfrac{\partial y}{\partial \xi} \\[2mm] \dfrac{\partial x}{\partial \eta} & \dfrac{\partial y}{\partial \eta} \end{bmatrix} = \begin{bmatrix} \sum\limits_{i=1}^{4}\dfrac{\partial N_i}{\partial \xi}x_i & \sum\limits_{i=1}^{4}\dfrac{\partial N_i}{\partial \xi}y_i \\[4mm] \sum\limits_{i=1}^{4}\dfrac{\partial N_i}{\partial \eta}x_i & \sum\limits_{i=1}^{4}\dfrac{\partial N_i}{\partial \eta}y_i \end{bmatrix} = \begin{bmatrix} \dfrac{\partial N_1}{\partial \xi} & \dfrac{\partial N_2}{\partial \xi} & \dfrac{\partial N_3}{\partial \xi} & \dfrac{\partial N_4}{\partial \xi} \\[2mm] \dfrac{\partial N_1}{\partial \eta} & \dfrac{\partial N_2}{\partial \eta} & \dfrac{\partial N_3}{\partial \eta} & \dfrac{\partial N_4}{\partial \eta} \end{bmatrix}\begin{bmatrix} x_1 & y_1 \\ x_2 & y_2 \\ x_3 & y_3 \\ x_4 & y_4 \end{bmatrix}$$

$$.....(7.53)$$

Using the interpolation functions for the element listed in Eq.4.49, we can compute J matrix as follows:

$$[J] = \frac{1}{4}\begin{bmatrix} \eta-1 & -\eta+1 & \eta+1 & -\eta-1 \\ \xi-1 & -\xi+1 & \xi+1 & -\xi-1 \end{bmatrix}\begin{bmatrix} x_1 & y_1 \\ x_2 & y_2 \\ x_3 & y_3 \\ x_4 & y_4 \end{bmatrix} \qquad .....(7.54)$$

The Jacobian matrix $[J]$ defines the transformation between natural and cartesian coordinates and is given by

Following Eq. 7.50, the vector of partial derivatives with reference to the cartesian coordinates can be expressed in terms of the vector of partial derivatives with reference to natural coordinates as follows:

$$\begin{Bmatrix} \dfrac{\partial}{\partial x} \\ \dfrac{\partial}{\partial y} \end{Bmatrix} = [J^{-1}]\begin{Bmatrix} \dfrac{\partial}{\partial \xi} \\ \dfrac{\partial}{\partial \eta} \end{Bmatrix} \qquad .....(7.55)$$

For this inverse transformation to exist, the Jacobian matrix $[J]$ shall be a non-singular matrix.

$$|J| = \frac{\partial x}{\partial \xi}\frac{\partial y}{\partial \eta} - \frac{\partial x}{\partial \eta}\frac{\partial y}{\partial \xi} \neq 0 \qquad .....(7.56)$$

The strain vector can be computed as follows:

$$\{\varepsilon\} = \begin{Bmatrix} \varepsilon_x \\ \varepsilon_y \\ \gamma_{xy} \end{Bmatrix} = \begin{Bmatrix} \dfrac{\partial u}{\partial x} \\ \dfrac{\partial v}{\partial y} \\ \dfrac{\partial u}{\partial y}+\dfrac{\partial v}{\partial x} \end{Bmatrix} = \begin{bmatrix} \dfrac{\partial}{\partial x} & 0 \\ 0 & \dfrac{\partial}{\partial y} \\ \dfrac{\partial}{\partial y} & \dfrac{\partial}{\partial x} \end{bmatrix}\begin{Bmatrix} u \\ v \end{Bmatrix} = \begin{bmatrix} \dfrac{\partial}{\partial x} & 0 \\ 0 & \dfrac{\partial}{\partial y} \\ \dfrac{\partial}{\partial y} & \dfrac{\partial}{\partial x} \end{bmatrix}[N]\{d\} \qquad .....(7.57)$$

$$\{\varepsilon\} = \begin{bmatrix} \dfrac{\partial N_1}{\partial x} & 0 & \dfrac{\partial N_2}{\partial x} & 0 & \dfrac{\partial N_3}{\partial x} & 0 & \dfrac{\partial N_4}{\partial x} & 0 \\ 0 & \dfrac{\partial N_1}{\partial x} & 0 & \dfrac{\partial N_2}{\partial x} & 0 & \dfrac{\partial N_3}{\partial x} & 0 & \dfrac{\partial N_4}{\partial x} \end{bmatrix}\{d\} \qquad .....(7.58)$$

$$\{\varepsilon\} = [B]\{d\} \qquad .....(7.59)$$

where $[B]$ is the strain-displacement matrix.

Eq. 7.55 can be used to write

$$\begin{bmatrix} \dfrac{\partial N_1}{\partial x} & \dfrac{\partial N_2}{\partial x} & \dfrac{\partial N_3}{\partial x} & \dfrac{\partial N_4}{\partial x} \\[2mm] \dfrac{\partial N_1}{\partial y} & \dfrac{\partial N_2}{\partial y} & \dfrac{\partial N_3}{\partial y} & \dfrac{\partial N_4}{\partial y} \end{bmatrix} = \left[J^{-1} \right] \begin{bmatrix} \dfrac{\partial N_1}{\partial \xi} & \dfrac{\partial N_2}{\partial \xi} & \dfrac{\partial N_3}{\partial \xi} & \dfrac{\partial N_4}{\partial \xi} \\[2mm] \dfrac{\partial N_1}{\partial \eta} & \dfrac{\partial N_2}{\partial \eta} & \dfrac{\partial N_3}{\partial \eta} & \dfrac{\partial N_4}{\partial \eta} \end{bmatrix} \qquad \text{.....(7.60)}$$

B matrix can be assembled by substituting the partial derivatives obtained from Eq. 7.60.in Eq. 7.58.

Thus, the stiffness matrix for the four-node quadrilateral element can be expressed as

$$\left[K^{(e)} \right] = h \iint \left[B^T \right] [D][B]\, dA \qquad \text{.....(7.61)}$$

$$\left[K^{(e)} \right] = h \int_{\xi=-1}^{1} \int_{\eta=-1}^{1} \left[B(\xi,\eta)^T \right] [D]\left[B(\xi,\eta) \right] |J|\, d\xi\, d\eta \qquad \text{.....(7.62)}$$

$$\left[K^{(e)} \right] = h \sum_{i=1}^{ngp} \sum_{j=1}^{ngp} W_i W_j \left[B(\xi_i,\eta_j)^T \right] [D]\left[B(\xi_i,\eta_j) \right] |J| \qquad \text{.....(7.63)}$$

where ngp represents the number of Gauss integration points.

7.5.1   Numerical Example

Analyse the skew plate shown in Figure 7.32, using four-node quadrilateral elements under conditions of plane stress. The thickness of the plate is 10 mm, Young's modulus is $2\times10^5 \, N/mm^2$ and $\mu = 0.3$. Body force is neglected in comparison with externally applied load.

Angle of skew $\alpha = 5^0$ with horizontal direction.

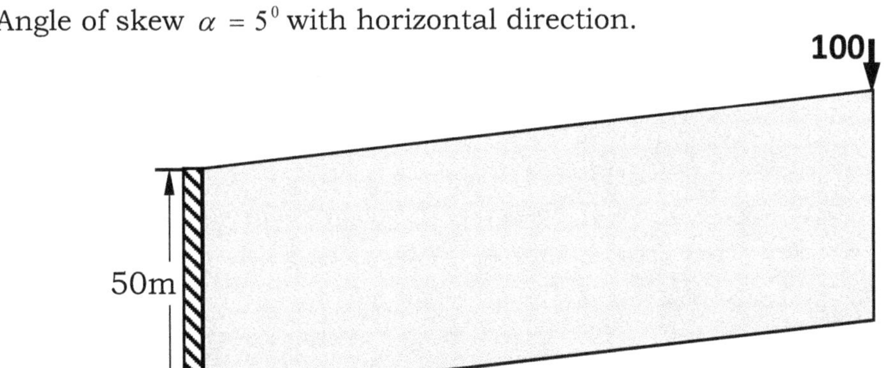

Figure 7.32 Skew plate.

The program listed in section A.7 is utilized for obtaining the results. The following is the input file used for analysis:

```
150.0   50.0 10.0 10 4 2.0e5    77.5      0.3    0.0  1  0
4   0.0   -100.0e3
%point load at top-right corner
```

The following is the dialog generated during execution of the program to input for other quantities. Note that the responses are marked in bold font:

```
Enter 1 for Plane Stress or  2 for Plane Strain:1
Point loads are present at specified nodes
Enter 1 for Edge restraint or 2 for Corner restraint:1
enter edge restraints in the order [Bottom Top Left Right]
enter 0 for restrained edge and 1 for unrestrained edge
Bottom Edge(1 or 0): 1
Top Edge(1 or 0): 1
Left Edge(1 or 0): 0
Right Edge(1 or 0): 1
```

```
Do you want plots on deformed configuration (1 for Yes 0 for No):1
```

The mesh generated is shown in Figure 7.33. The plate is also analysed using ABAQUS software *using CPS6 elements*. The results obtained using the present program match exactly with the results obtained using ABAQUS. For example, Figures 7.34 and 7.35 depict the variation of horizontal and vertical displacements obtained using ABAQUS. Similar plots for horizontal and vertical displacements using the present program are presented in Figures 7.36 and 7.37. A comparison of Figures 7.34 and 7.36 for horizontal displacement and Figures 7.35 and 7.37 for vertical displacements reveals that the results obtained using the present MATLAB program are accurate.

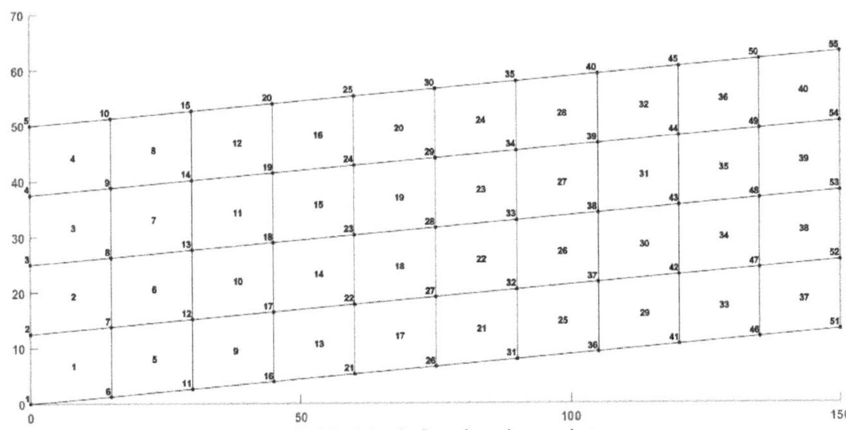

Figure 7.33 Mesh for the skew plate.

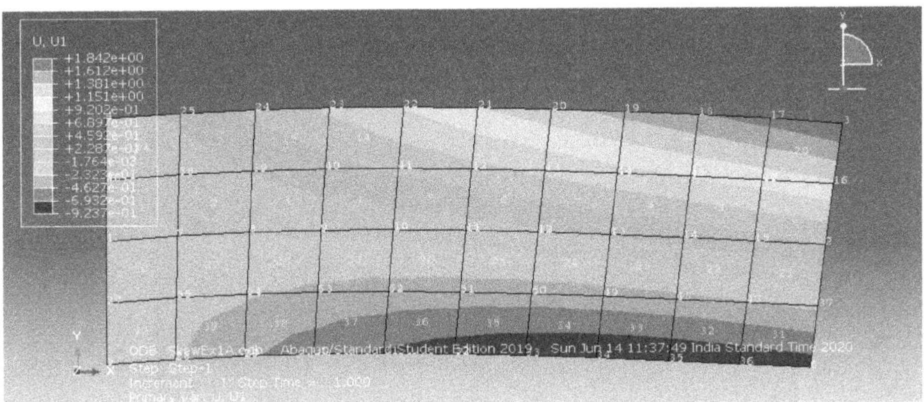

Figure 7.34 Plot of horizontal displacement (ABAQUS).

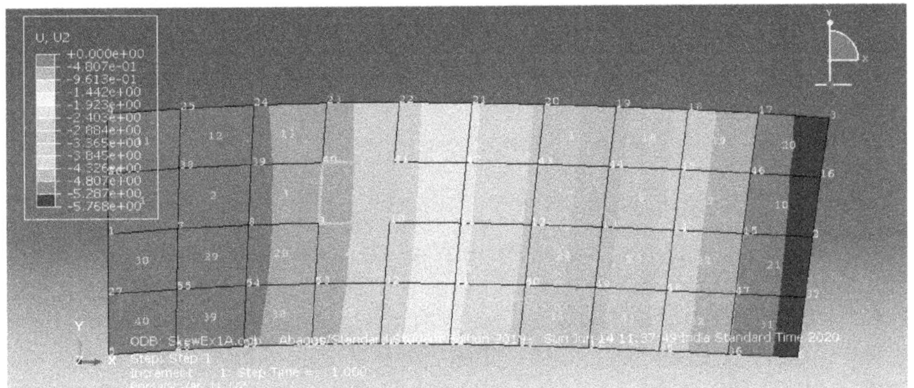

Figure 7.35 Plot of vertical displacement (ABAQUS).

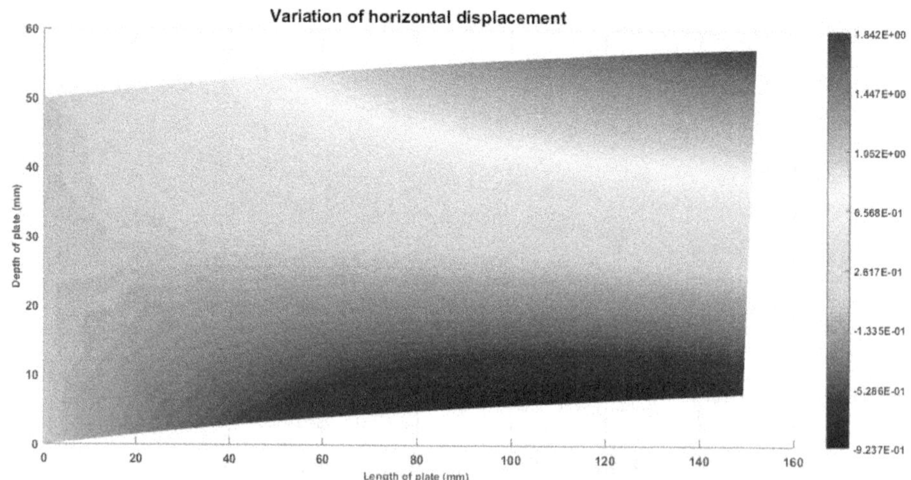

Figure 7.36 Plot of horizontal displacement (Present).

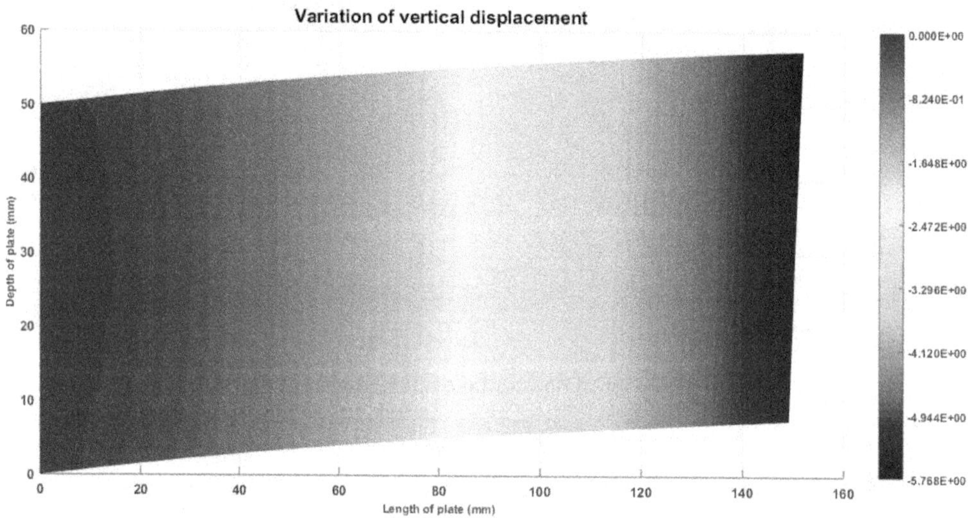

Fig. 7.37 Plot of vertical displacement (Present).

7.5.2 Nodal Stress Recovery

Consider the parent rectangular element with four nodes as shown in Figure 7.38. The natural coordinates of nodes of this element are given in Table Natural coordinates for the Gauss integration points for this element are also given in this table. The rectangle that is obtained by joining these Gauss integration points is called Gauss element. Thus the Gauss element can be visualized as an ' *four-node rectangular element*' inside the parent rectangular element with four nodes. Coordinate axes ξ' and η' have the centre of the element as origin as shown in the Figure 7.38. The natural coordinates of nodes

1 through 4 with reference to the new natural coordinate system $\xi'-\eta'$ are given in columns 4 and 5 of Table 7.7. These coordinates are substituted in the shape functions of the 4 noded Gauss element. These shape functions are the same as those given in Eq. 4.49 and are reproduced here for convenience.

$$[N'] = \left| \frac{(1-\xi')(1-\eta')}{4} \quad \frac{(1+\xi')(1-\eta')}{4} \quad \frac{(1+\xi')(1+\eta')}{4} \quad \frac{(1-\xi')(1+\eta')}{4} \right| \quad .....(7.64)$$

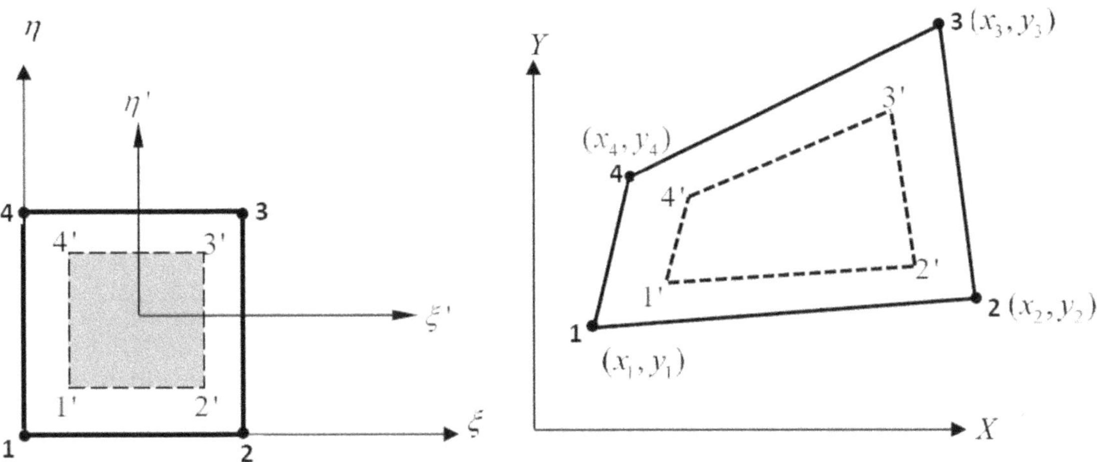

Figure 7.38 Gauss Element inside four-node quadrilateral element.

Any scalar quantity w whose values w'_i at the corners of Gauss element are known can be interpolated through the above shape functions expressed in terms of ξ' and η' as follows:

$$w(\xi',\eta') = \lfloor w'_1 \quad w'_2 \quad w'_3 \quad w'_4 \rfloor \begin{Bmatrix} N_1^{(e')} \\ N_2^{(e')} \\ N_3^{(e')} \\ N_4^{(e')} \end{Bmatrix} \quad .....(7.65)$$

The natural coordinates of the four nodes of parent element are substituted in Eq.7.65, to obtain the following relationship for extrapolation of nodal stresses.

$$\lfloor w_1 \quad w_2 \quad w_3 \quad w_4 \rfloor^T = [T] \begin{Bmatrix} w'_1 \\ w'_2 \\ w'_3 \\ w'_4 \end{Bmatrix} \quad .....(7.66)$$

Here $\{w'\}$ represents the each of the stress values (σ_{XX}, σ_{YY} or τ_{XY}) at the Gauss integration points.

Table 7.9 Natural coordinates of four-noded quadrilateral element.

Corner node	ξ	η	ξ'	η'	Gauss node	ξ	η	ξ'	η'
1	-1	-1	$-\sqrt{3}$	$-\sqrt{3}$	1'	$\dfrac{-1}{\sqrt{3}}$	$\dfrac{-1}{\sqrt{3}}$	-1	-1
2	1	-1	$\sqrt{3}$	$-\sqrt{3}$	2'	$\dfrac{1}{\sqrt{3}}$	$\dfrac{-1}{\sqrt{3}}$	1	-1
3	1	1	$\sqrt{3}$	$\sqrt{3}$	3'	$\dfrac{1}{\sqrt{3}}$	$\dfrac{1}{\sqrt{3}}$	1	1
4	-1	1	$-\sqrt{3}$	$\sqrt{3}$	4'	$\dfrac{-1}{\sqrt{3}}$	$\dfrac{1}{\sqrt{3}}$	-1	1

Figure 7.39, Figure 7.40 and Figure 7.41 represent the variation of the in-plane stresses σ_{XX}, σ_{YY} and τ_{XY} across the plate. These stresses are obtained using the nodal stress recovery outlined above. The stresses obtained from using the developed MATLAB program match exactly with the results obtained using ABAQUS.

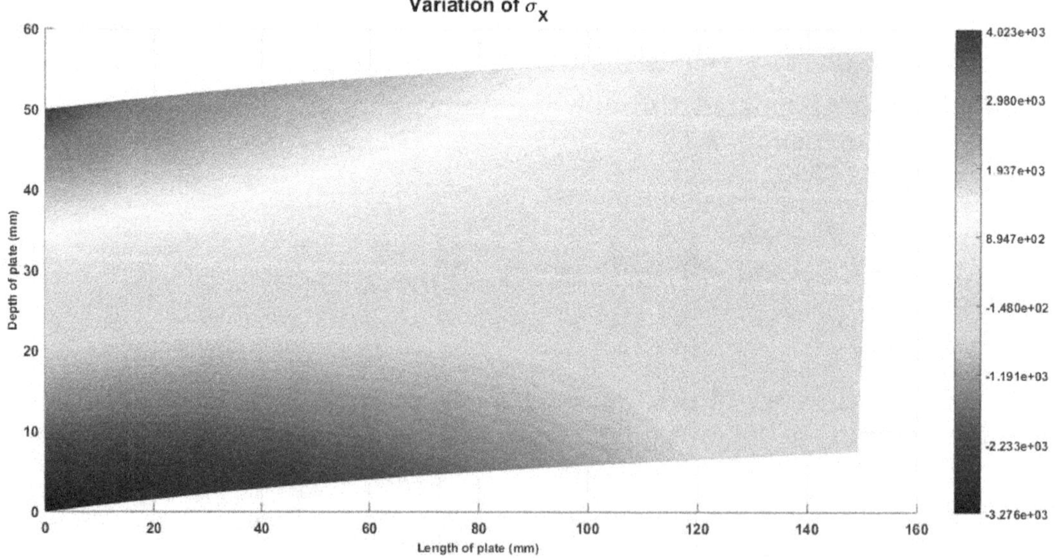

Figure 7.39 Plot of variation of σ_X across the plate.

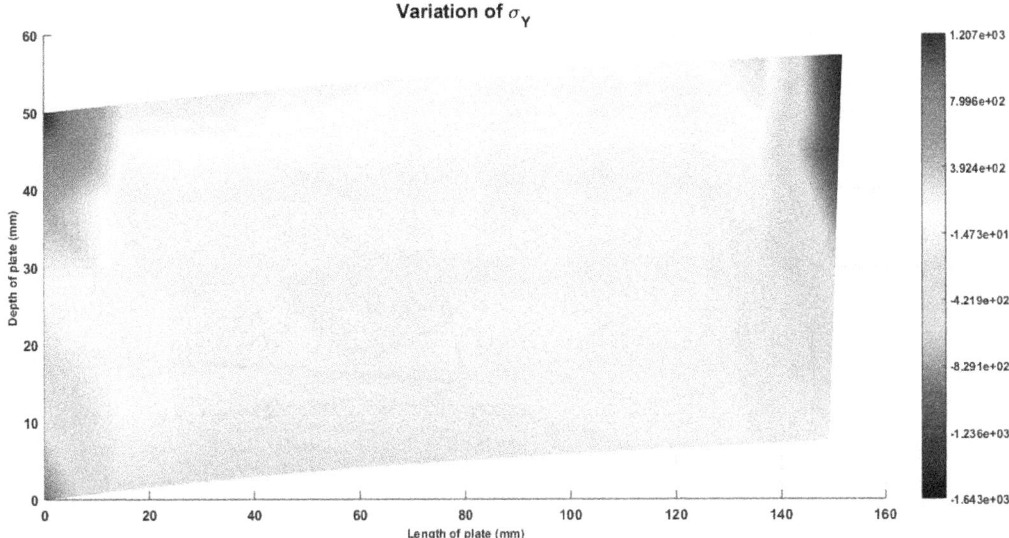

Figure 7.40 Plot of variation of σ_Y across the plate.

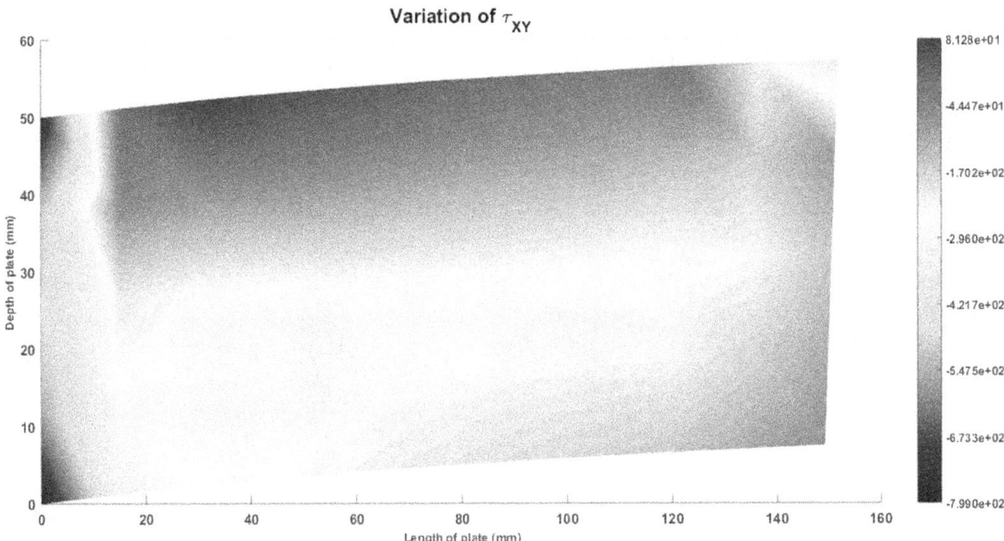

Figure 7.41 Plot of variation of τ_{XY} across the plate.

7.6 EIGHT-NODE QUADRILATERAL ELEMENT

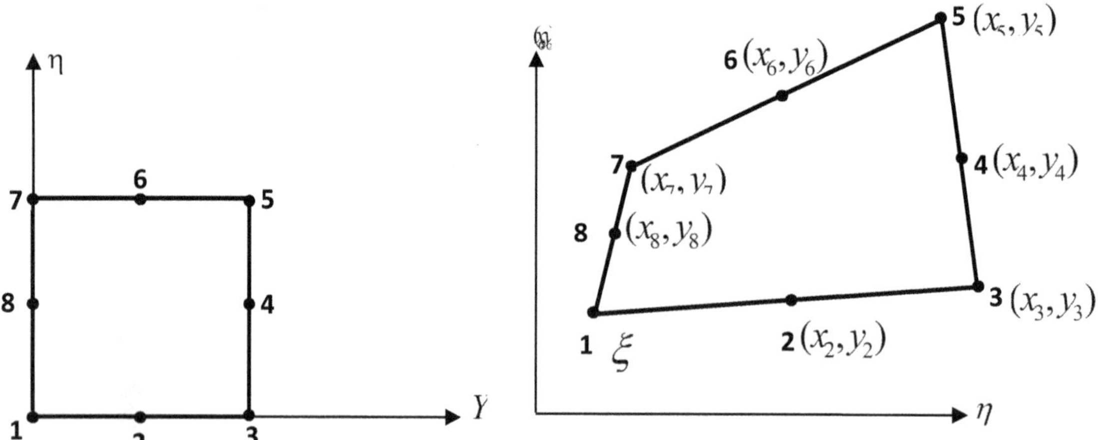

Figure 7.42 Eight-node quadrilateral element.

Consider the eight-node quadrilateral element shown in Figure 7.42. As seen from the figure, this element is an *iso-parametric element* obtained by transformation of a parent eight-node rectangular serendipity element. The interpolation functions for the eight-node rectangular serendipity element are discussed in Section 4.6. It is to be noted that the present eight-node quadrilateral element follows the cartesian coordinate system while the parent element follows the natural coordinate system. The interpolation functions for the eight-node rectangular element are presented in Eq.4.74 are used to model displacement of the element too. These shape functions are now presented in a revised order as follows:

$$N_1 \equiv -\frac{1}{4}(\xi-1)(\eta-1)(\xi+\eta+1) \qquad N_2 \equiv \frac{1}{2}(\xi^2-1)(\eta-1)$$

$$N_3 \equiv -\frac{1}{4}(\xi+1)(\eta-1)(\xi-\eta-1) \qquad N_4 \equiv -\frac{1}{2}(\xi+1)(\eta^2-1)$$

$$N_5 \equiv \frac{1}{4}(\xi+1)(\eta+1)(\xi+\eta-1) \qquad N_6 \equiv -\frac{1}{2}(\xi^2-1)(\eta+1) \qquad \dots(7.67)$$

$$N_7 \equiv -\frac{1}{4}(\xi-1)(\eta+1)(\xi-\eta+1) \qquad N_8 \equiv \frac{1}{2}(\xi-1)(\eta^2-1)$$

Thus, if u and v are the displacements of a point inside the element, they can be related to the nodal displacements as follows:

$$u = u_1 N_1 + u_2 N_2 + u_3 N_3 + u_4 N_4 + u_5 N_5 + u_6 N_6 + u_7 N_7 + u_8 N_8$$
$$v = v_1 N_1 + v_2 N_2 + v_3 N_3 + v_4 N_4 + v_5 N_5 + v_6 N_6 + v_7 N_7 + v_8 N_8$$
$$\qquad\qquad .....(7.68)$$

This can be written in the matrix form as follows:

$$\begin{Bmatrix} u \\ v \end{Bmatrix} = \begin{bmatrix} N_1 & 0 & N_2 & 0 & N_3 & 0 & N_4 & 0 & N_5 & 0 & N_6 & 0 & N_7 & 0 & N_8 & 0 \\ 0 & N_1 & 0 & N_2 & 0 & N_3 & 0 & N_4 & 0 & N_5 & 0 & N_6 & 0 & N_7 & 0 & N_8 \end{bmatrix} \{d\}$$
$$\qquad\qquad .....(7.69)$$

where

$$\{d\} = \begin{bmatrix} u_1 & v_1 & u_2 & v_2 & u_3 & v_3 & u_4 & v_4 & u_5 & v_5 & u_6 & v_6 & u_7 & v_7 & u_8 & v_8 \end{bmatrix}^T \quad .....(7.70)$$

This can be represented as

$$\{u\} = [N]\{d\} \qquad\qquad .....(7.71)$$

Here $[N]$ is the shape function matrix and $\{d\}$ is the vector of nodal displacements. On similar lines, the x and y coordinates of any point inside the element are related to the nodal coordinates as

$$x = N_1 x_1 + N_2 x_2 + N_3 x_3 + N_4 x_4 + N_5 x_5 + N_6 x_6 + N_7 x_7 + N_8 x_8$$
$$y = N_1 y_1 + N_2 y_2 + N_3 y_3 + N_4 y_4 + N_5 y_5 + N_6 y_6 + N_7 y_7 + N_8 y_8$$
$$\qquad\qquad .....(7.72)$$

The partial derivatives are given by

$$\frac{\partial x}{\partial \xi} = \sum_{i=1}^{8} \frac{\partial N_i}{\partial \xi} x_i \quad ; \quad \frac{\partial y}{\partial \xi} = \sum_{i=1}^{8} \frac{\partial N_i}{\partial \xi} y_i$$

$$\frac{\partial x}{\partial \eta} = \sum_{i=1}^{8} \frac{\partial N_i}{\partial \eta} x_i \quad ; \quad \frac{\partial y}{\partial \eta} = \sum_{i=1}^{8} \frac{\partial N_i}{\partial \eta} y_i$$
$$\qquad\qquad .....(7.73)$$

The Jacobian matrix $[J]$ is given by

$$[J] = \begin{bmatrix} \dfrac{\partial x}{\partial \xi} & \dfrac{\partial y}{\partial \xi} \\[2ex] \dfrac{\partial x}{\partial \eta} & \dfrac{\partial y}{\partial \eta} \end{bmatrix} = \begin{bmatrix} \displaystyle\sum_{i=1}^{8} \frac{\partial N_i}{\partial \xi} x_i & \displaystyle\sum_{i=1}^{8} \frac{\partial N_i}{\partial \xi} y_i \\[3ex] \displaystyle\sum_{i=1}^{8} \frac{\partial N_i}{\partial \eta} x_i & \displaystyle\sum_{i=1}^{8} \frac{\partial N_i}{\partial \eta} y_i \end{bmatrix} \qquad .....(7.74)$$

$$[J] = \begin{bmatrix} \dfrac{\partial N_1}{\partial \xi} & \dfrac{\partial N_2}{\partial \xi} & \dfrac{\partial N_3}{\partial \xi} & \dfrac{\partial N_4}{\partial \xi} & \dfrac{\partial N_5}{\partial \xi} & \dfrac{\partial N_6}{\partial \xi} & \dfrac{\partial N_7}{\partial \xi} & \dfrac{\partial N_8}{\partial \xi} \\[3mm] \dfrac{\partial N_1}{\partial \eta} & \dfrac{\partial N_2}{\partial \eta} & \dfrac{\partial N_3}{\partial \eta} & \dfrac{\partial N_4}{\partial \eta} & \dfrac{\partial N_5}{\partial \eta} & \dfrac{\partial N_6}{\partial \eta} & \dfrac{\partial N_7}{\partial \eta} & \dfrac{\partial N_8}{\partial \eta} \end{bmatrix} \begin{bmatrix} x_1 & y_1 \\ x_2 & y_2 \\ x_3 & y_3 \\ x_4 & y_4 \\ x_5 & y_5 \\ x_6 & y_6 \\ x_7 & y_7 \\ x_8 & y_8 \end{bmatrix} \qquad \ldots\ldots(7.75)$$

The strain vector can be computed as follows:

$$\{\varepsilon\} = \begin{Bmatrix} \varepsilon_x \\ \varepsilon_y \\ \gamma_{xy} \end{Bmatrix} = \begin{bmatrix} \dfrac{\partial}{\partial x} & 0 \\[3mm] 0 & \dfrac{\partial}{\partial y} \\[3mm] \dfrac{\partial}{\partial y} & \dfrac{\partial}{\partial x} \end{bmatrix} \begin{bmatrix} N_1 & 0 & N_2 & 0 & N_3 & 0 & N_4 & 0 & N_5 & 0 & N_6 & 0 & N_7 & 0 & N_8 & 0 \\ 0 & N_1 & 0 & N_2 & 0 & N_3 & 0 & N_4 & 0 & N_5 & 0 & N_6 & 0 & N_7 & 0 & N_8 \end{bmatrix} \{d\}$$

$$\ldots\ldots(7.76)$$

$$\{\varepsilon\} = [B]\{d\} \qquad \ldots\ldots(7.77)$$

$$[B] = \begin{Bmatrix} \dfrac{\partial N_1}{\partial x} & 0 & \dfrac{\partial N_2}{\partial x} & 0 & \dfrac{\partial N_3}{\partial x} & 0 & \dfrac{\partial N_4}{\partial x} & 0 & \dfrac{\partial N_5}{\partial x} & 0 & \dfrac{\partial N_6}{\partial x} & 0 & \dfrac{\partial N_7}{\partial x} & 0 & \dfrac{\partial N_8}{\partial x} & 0 \\[3mm] 0 & \dfrac{\partial N_1}{\partial y} & 0 & \dfrac{\partial N_2}{\partial y} & 0 & \dfrac{\partial N_3}{\partial y} & 0 & \dfrac{\partial N_4}{\partial y} & 0 & \dfrac{\partial N_5}{\partial y} & 0 & \dfrac{\partial N_6}{\partial y} & 0 & \dfrac{\partial N_7}{\partial y} & 0 & \dfrac{\partial N_8}{\partial y} \\[3mm] \dfrac{\partial N_1}{\partial y} & \dfrac{\partial N_1}{\partial x} & \dfrac{\partial N_2}{\partial y} & \dfrac{\partial N_2}{\partial x} & \dfrac{\partial N_3}{\partial y} & \dfrac{\partial N_3}{\partial x} & \dfrac{\partial N_4}{\partial y} & \dfrac{\partial N_4}{\partial x} & \dfrac{\partial N_5}{\partial y} & \dfrac{\partial N_5}{\partial x} & \dfrac{\partial N_6}{\partial y} & \dfrac{\partial N_6}{\partial x} & \dfrac{\partial N_7}{\partial y} & \dfrac{\partial N_7}{\partial x} & \dfrac{\partial N_8}{\partial y} & \dfrac{\partial N_8}{\partial x} \end{Bmatrix}$$

$$\ldots\ldots(7.78)$$

where $[B]$ is the strain-displacement matrix.

Following, Eq. 7.55, the following relationship can be written

$$\begin{bmatrix} \dfrac{\partial N_1}{\partial x} & \dfrac{\partial N_2}{\partial x} & \dfrac{\partial N_3}{\partial x} & \dfrac{\partial N_4}{\partial x} & \dfrac{\partial N_5}{\partial x} & \dfrac{\partial N_6}{\partial x} & \dfrac{\partial N_7}{\partial x} & \dfrac{\partial N_8}{\partial x} \\[3mm] \dfrac{\partial N_1}{\partial y} & \dfrac{\partial N_2}{\partial y} & \dfrac{\partial N_3}{\partial y} & \dfrac{\partial N_4}{\partial y} & \dfrac{\partial N_5}{\partial y} & \dfrac{\partial N_6}{\partial y} & \dfrac{\partial N_7}{\partial y} & \dfrac{\partial N_8}{\partial y} \end{bmatrix} = [J^{-1}] \begin{bmatrix} \dfrac{\partial N_1}{\partial \xi} & \dfrac{\partial N_2}{\partial \xi} & \dfrac{\partial N_3}{\partial \xi} & \dfrac{\partial N_4}{\partial \xi} & \dfrac{\partial N_5}{\partial \xi} & \dfrac{\partial N_6}{\partial \xi} & \dfrac{\partial N_7}{\partial \xi} & \dfrac{\partial N_8}{\partial \xi} \\[3mm] \dfrac{\partial N_1}{\partial y} & \dfrac{\partial N_2}{\partial y} & \dfrac{\partial N_3}{\partial y} & \dfrac{\partial N_4}{\partial y} & \dfrac{\partial N_5}{\partial y} & \dfrac{\partial N_6}{\partial y} & \dfrac{\partial N_7}{\partial y} & \dfrac{\partial N_8}{\partial y} \end{bmatrix}$$

$$\ldots\ldots(7.79)$$

The partial derivatives obtained from Eq. 7.79 are substituted in Eq. 7.76 to obtain $[B]$ matrix.

The stiffness matrix for the eight-node quadrilateral element of thickness h can be expressed as

$$\left[K^{(e)}\right] = h\int\left[B^{T}\right][D][B]\,dA \qquad \qquad ....(7.80)$$

$$\left[K^{(e)}\right] = h\int\limits_{\xi=-1}^{1}\int\limits_{\eta=-1}^{1}\left[B(\xi,\eta)^{T}\right][D]\left[B(\xi,\eta)\right]|J|\,d\xi d\eta$$

$$\qquad \qquad .....(7.81)$$

$$= h\sum_{i=1}^{ngp}\sum_{j=1}^{ngp}W_{i}W_{j}\left[B(\xi_{i},\eta_{j})^{T}\right][D]\left[B(\xi_{i},\eta_{j})\right]|J|$$

where *ngp* represents the number of Gauss integration points.

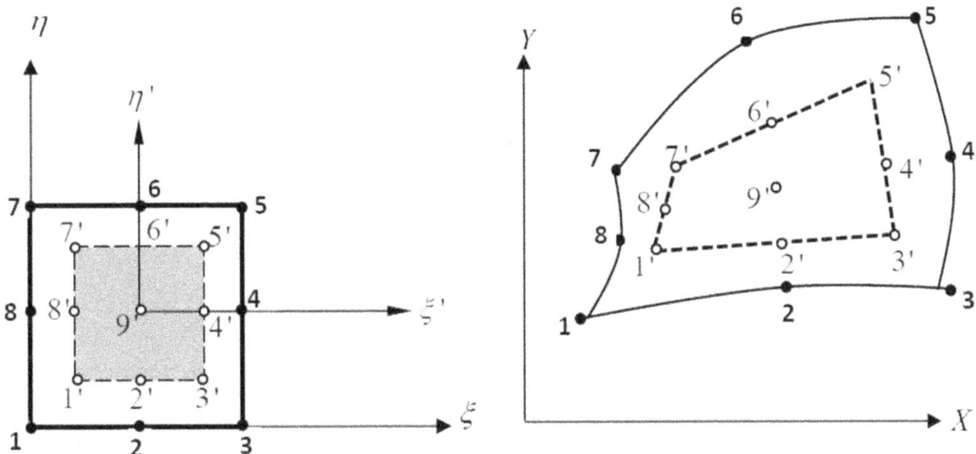

Figure 7.43 Gauss element inside eight-node quadrilateral element.

Table 7.10 Natural coordinates of eight-node quadrilateral element.

Corner node	ξ	η	ξ'	η'	Gauss node	ξ	η	ξ'	η'
1	-1	-1	$-\sqrt{3}$	$-\sqrt{3}$	1'	$\dfrac{-1}{\sqrt{3}}$	$\dfrac{-1}{\sqrt{3}}$	-1	-1
2	0	-1	0	$-\sqrt{3}$	2'	0	$\dfrac{-1}{\sqrt{3}}$	0	-1
3	1	-1	$\sqrt{3}$	$\sqrt{3}$	3'	$\dfrac{1}{\sqrt{3}}$	$\dfrac{-1}{\sqrt{3}}$	1	-1
4	1	0	$\sqrt{3}$	0	4'	0	$\dfrac{1}{\sqrt{3}}$	1	0

Table 7.10 *contd...*

Corner node	ξ	η	ξ'	η'	Gauss node	ξ	η	ξ'	η'
5	1	1	$\sqrt{3}$	$\sqrt{3}$	5'	$\dfrac{1}{\sqrt{3}}$	$\dfrac{1}{\sqrt{3}}$	1	1
6	0	1	0	$\sqrt{3}$	6'	0	$\dfrac{1}{\sqrt{3}}$	0	1
7	-1	1	$-\sqrt{3}$	$\sqrt{3}$	7'	$\dfrac{-1}{\sqrt{3}}$	$\dfrac{1}{\sqrt{3}}$	-1	1
8	-1	0	$-\sqrt{3}$	0	8'	$\dfrac{-1}{\sqrt{3}}$	0	-1	0
					9'	0	0	0	0

7.6.1   Nodal Stress Recovery

Consider a parent rectangular element with eight nodes as shown in Figure 7.43. The natural coordinates of nodes of this element are given in Table 7.10.   Natural coordinates for the Gauss integration points for this element are also given in this table. The rectangle that is obtained by joining these Gauss integration points (except the one at the centre of the element) is called Gauss element. Thus the Gauss element can be visualized as an *'nine noded rectangular element'* inside the parent rectangular element with eight nodes. Coordinate axes ξ' and η' are passed with the node 9' as origin as shown in the figure. The natural coordinates of nodes 1 through 8 with reference to the new natural coordinate system $\xi'-\eta'$ are given in columns 4 and 5 of Table 7.10. These coordinates are substituted in the shape functions of the nine noded Gauss element. These shape functions given by Eq. 4.72 are reproduced below.

$$N_1(\xi,\eta) = \frac{1}{4}\xi'\eta'(\xi'-1)(\eta'-1) \qquad N_2(\xi',\eta') = \frac{1}{2}\eta'(1-\xi'^2)(\eta'-1)$$

$$N_3(\xi',\eta') = \frac{1}{4}\xi'\eta'(\xi'+1)(\eta'-1) \qquad N_4(\xi',\eta') = \frac{1}{2}\xi'(\xi'+1)(1-\eta'^2)$$

$$N_5(\xi',\eta') = \frac{1}{4}\xi'\eta'(\xi'+1)(\eta'+1) \qquad N_6(\xi',\eta') = \frac{1}{2}\eta'(1-\xi'^2)(\eta'+1) \quad \ldots\ldots(7.82)$$

$$N_7(\xi',\eta') = \frac{1}{4}\xi'\eta'(\xi'-1)(\eta'+1) \qquad N_8(\xi',\eta') = \frac{1}{2}\xi'(\xi'-1)(1-\eta'^2)$$

$$N_9(\xi',\eta') = (1-\xi'^2)(1-\eta'^2)$$

Any scalar quantity w whose values w_i' at the corners of Gauss element are known can be interpolated through the above shape functions now expressed in terms of ξ and η as follows:

$$w(\xi',\eta') = \lfloor w_1' \quad w_2' \quad w_3' \quad w_4' \quad w_5' \quad w_6' \quad w_7' \quad w_8' \quad w_9' \rfloor \begin{Bmatrix} N_1^{(e')} \\ N_2^{(e')} \\ N_3^{(e')} \\ N_4^{(e')} \\ N_5^{(e')} \\ N_6^{(e')} \\ N_7^{(e')} \\ N_8^{(e')} \\ N_9^{(e')} \end{Bmatrix} \qquad .....(7.83)$$

The natural coordinates of the eight nodes of original element can be substituted in Eq. 7.83, to obtain the following relationship for extrapolation of nodal stresses.

$$\lfloor w_1 \quad w_2 \quad w_3 \quad w_4 \quad w_5 \quad w_6 \quad w_7 \quad w_8 \rfloor^T = [T] \begin{Bmatrix} w_1' \\ w_2' \\ w_3' \\ w_4' \\ w_5' \\ w_6' \\ w_7' \\ w_8' \\ w_9' \end{Bmatrix} \qquad .....(7.84)$$

Here $\{w'\}$ represents the each of the stress values (σ_{XX}, σ_{YY} or τ_{XY}) at the Gauss integration points.

7.6.2   Numerical Example

Analyse the skew plate shown in Figure 7.44 using eight-node quadrilateral elements under conditions of plane stress. The thickness of the plate is 10 mm, Young's modulus is $2\times10^5\ N/mm^2$ and $\mu = 0.3$. Body force is neglected in comparison with externally applied load.

Angle of skew $\alpha = 5^0$

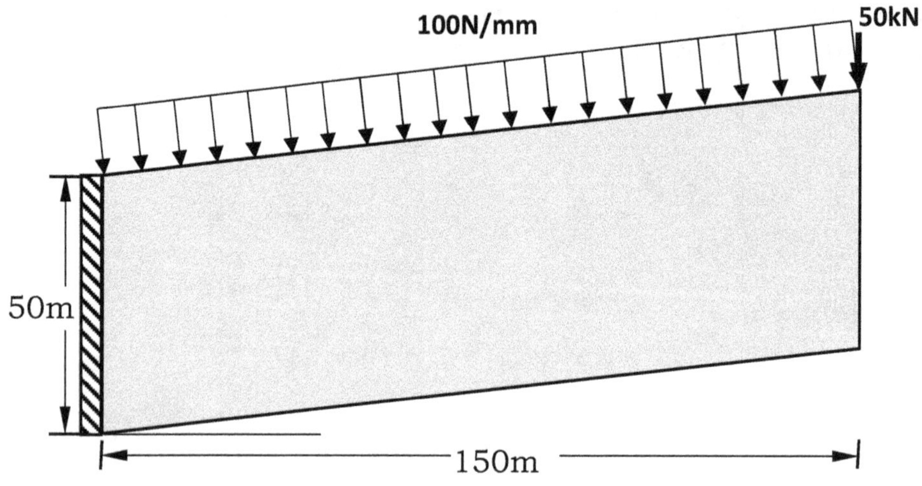

Figure 7.44 Skew plate.

The program listed in section 7.8 is utilized for obtaining the results. The following is the input file used for analysis:

```
150.0   50.0 10.0 10 4 2.0e5    77.5     0.3     0.0  1  1
1   0.0   -100.0
4   0.0   -50.0e3
%pressure on top edge followed by point load at top-right corner
```

The following is the dialog generated during execution of the program to input for other quantities. Note that the responses are marked in bold font:

```
Enter 1 for Plane Stress or  2 for Plane Strain:1
Distributed load is present on the edges
Top edge is loaded
Point loads are present at specified nodes
Enter 1 for Edge restraint OR 2 for Corner restraint:1
enter edge restraints in the order [Bottom Top Left Right]
enter 0 for restrained edge and 1 for unrestrained edge
Bottom Edge(1 or 0): 1
Top Edge(1 or 0): 1
Left Edge(1 or 0): 0
```

```
Right Edge(1 or 0): 1
Do you want plots on deformed configuration (1 for Yes 0 for No):1
```

The mesh generated is shown in Figure 7.45.

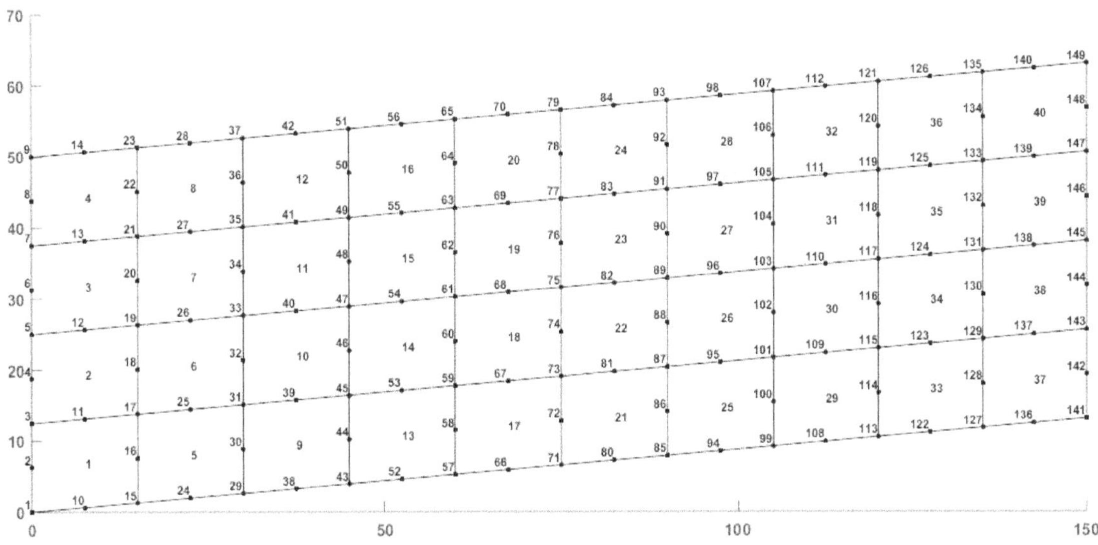

Figure 7.45 Mesh for the skew plate.

The plate is also analysed using ABAQUS software *using **CPS8** elements*. The results obtained using the present program match exactly with the results obtained using ABAQUS. Figure 7.46 and Figure 7.47 depict the variation of horizontal and vertical displacements.

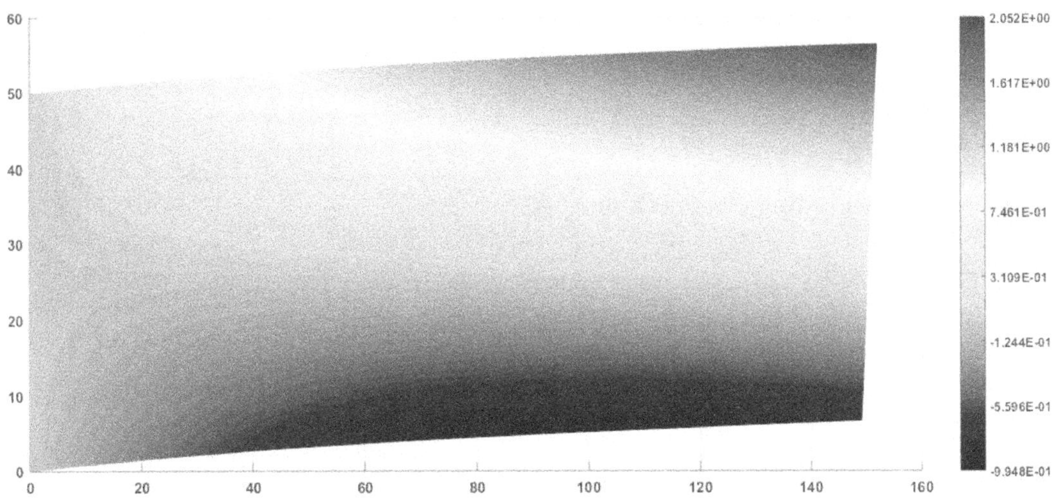

Figure 7.46 Plot of horizontal displacement.

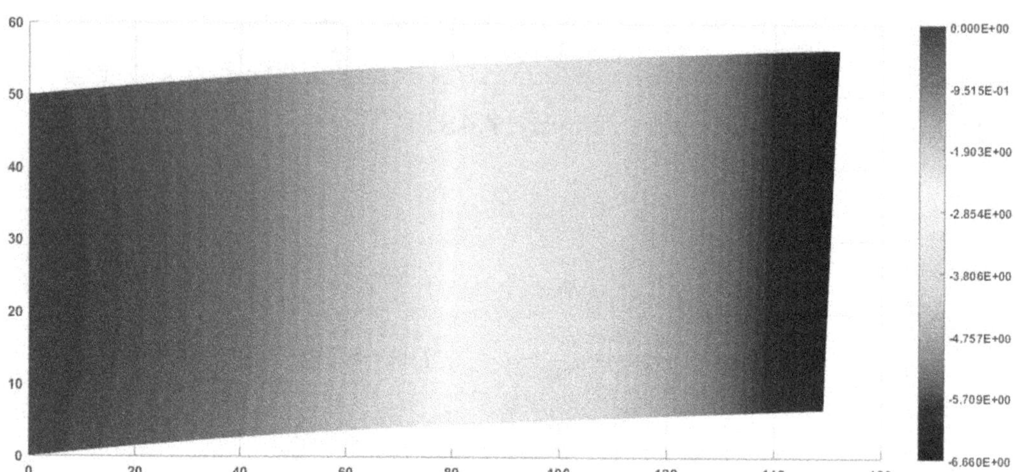

Figure 7.47 Plot of vertical displacement.

Figure 7.48, Figure 7.49, and Figure 7.50 show the plots of variation of σ_X, σ_Y and τ_{XY} across the plate. These stresses are obtained using the process of nodal stress recovery outlined earlier.

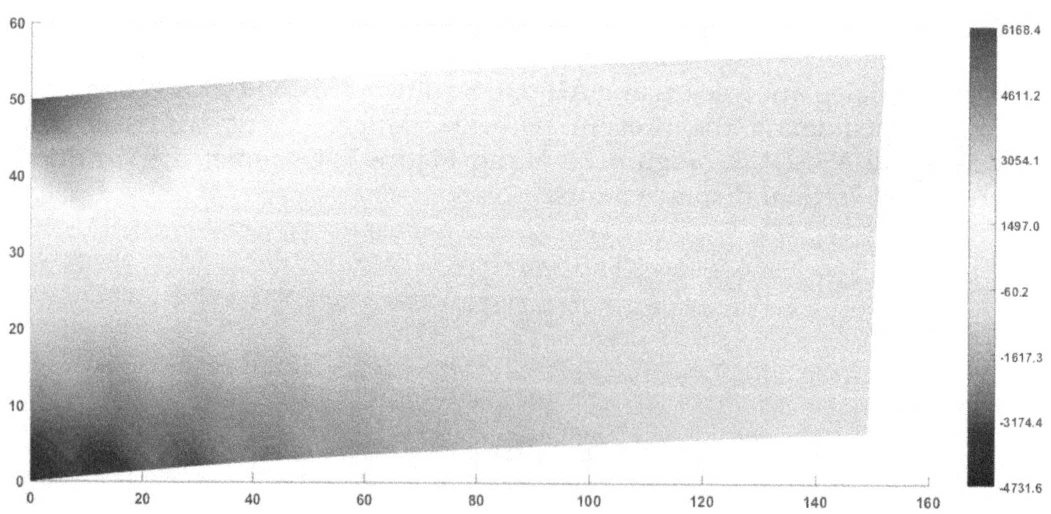

Figure 7.48 Plot of σ_X across the plate.

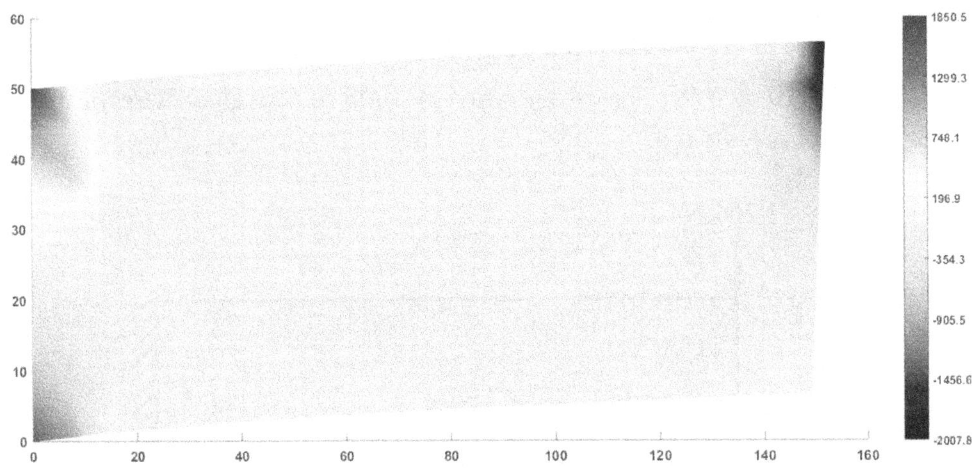

Figure 7.49 Plot of σ_Y across the plate.

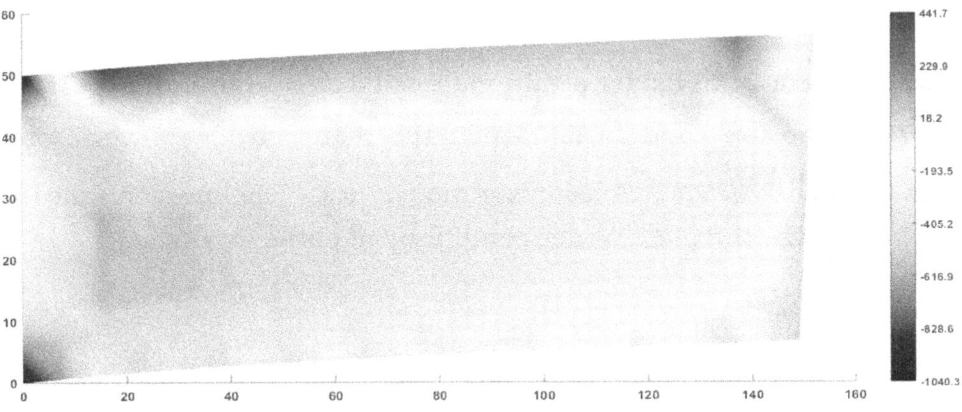

Figure 7.50 Plot of τ_{XY} across the plate.

7.7 EXERCISE PROBLEMS

1. **Exercise Problem 1:** Consider a thin plate as shown in the Figure below. Given $E=2.1\times10^8\ kN/m^2$, $\mu=0.25$ and $t=0.05\ m$. Analyse the plate under conditions of plane stress.

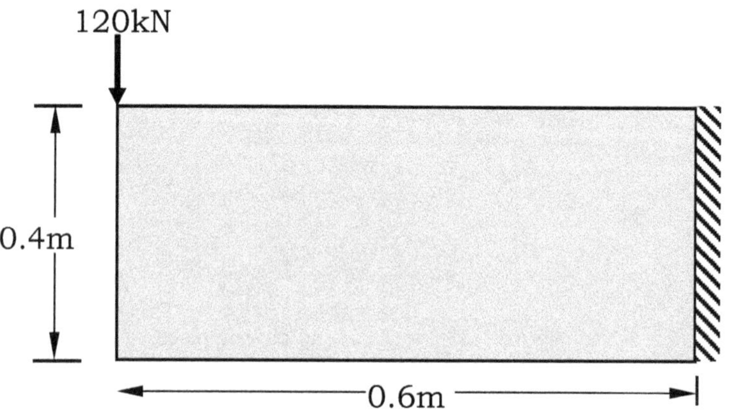

2. **Exercise Problem 2:** Consider a thin plate subjected to uniform pressure of $125\ N/mm^2$ on the top edge as shown in the Figure below. Given Young's modulus $E=2\times10^5\ N/mm^2$, Poisson's ratio $\mu=0.3$. Thickness of the plate $t=10\ mm$. Analyse the plate under conditions of plane strain.

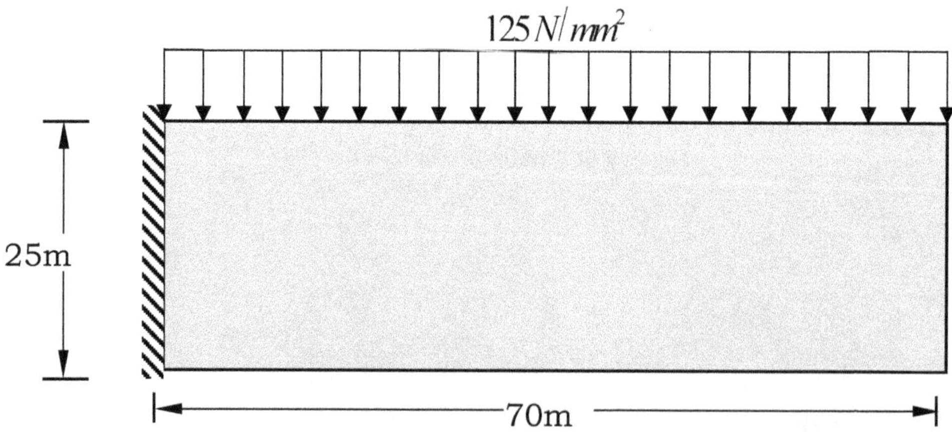

3. **Exercise problem 3:** Analyse the plate shown in Figure below under conditions of plane stress. The thickness of the plate is 15 mm, Young's modulus is $2.0\times10^5 \, N/mm^2$ and $\mu = 0.3$. Body force is neglected in comparison with externally applied load.

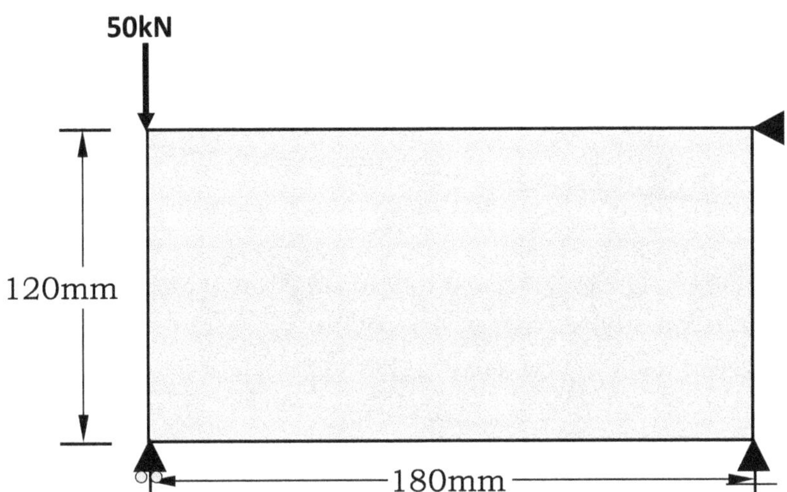

4. **Exercise Problem 4:** Consider a thin plate subjected to a point load of $1200N$ at the centre of the right edge as shown in the figure below. Adopt Young's modulus $E=2\times10^5 \, N/mm^2$, Poisson's ratio $\mu = 0.3$. Thickness of the plate $t = 15 \, mm$. Analyse the plate under conditions of plane stress using LST elements

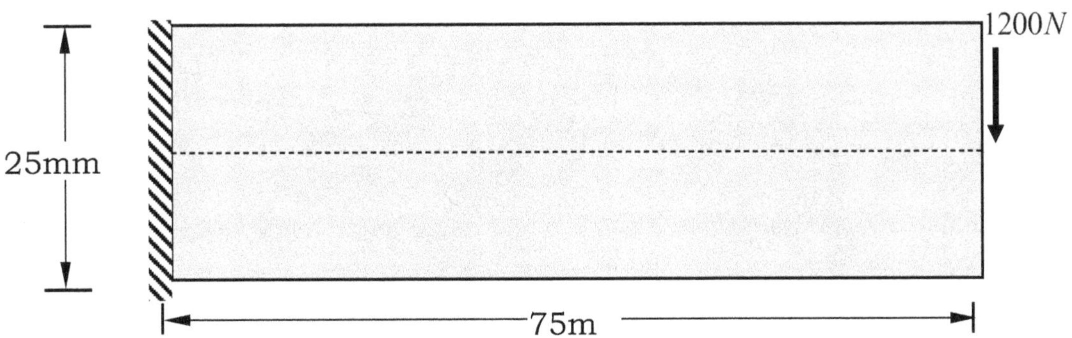

5. Exercise Problem 5: Analyse the skew plate shown in Figure below, using four-node quadrilateral elements under conditions of plane stress. The thickness of the plate is 12 mm, Young's modulus is $2\times10^5\ N/mm^2$ and $\mu = 0.3$. Body force is neglected in comparison with externally applied load. Angle of skew $\alpha = 5^0$.

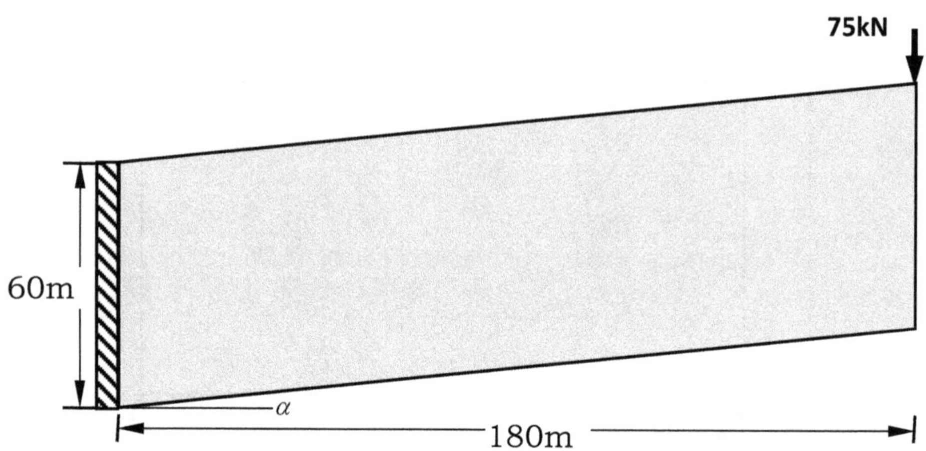

6. Exercise Problem 6: Analyse the skew plate shown in Figure below, using eight-node quadrilateral elements under conditions of plane stress. The thickness of the plate is 15 mm, Young's modulus is $2\times10^5\ N/mm^2$ and $\mu = 0.3$. Body force is neglected in comparison with externally applied load. Angle of skew is 5^0.

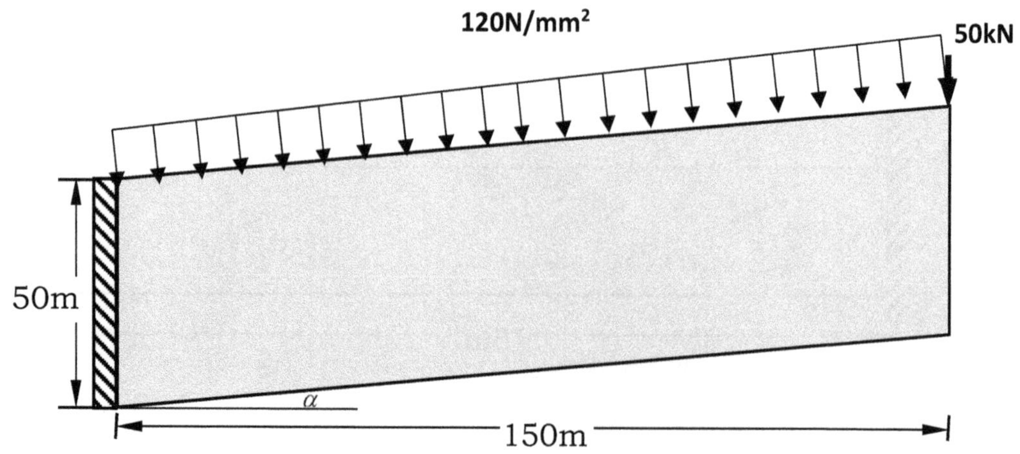

Numerical Integration

In finite element analysis, we often need to compute integrals. This is performed easily making use of numerical integration.

Consider the integral $\int_{a}^{b} f(x)dx$.

Use of analytical integration to evaluate the above integral is relatively easy in simpler problems. However, evaluation of integral by analytical means proves to be ill-suited in the implementation of finite element method on a digital computer. The alternative is to use numerical integration to evaluate the integral quite rapidly with good accuracy. The basic idea of numerical integration is to replace the continuous integral with a series of finite sums as follows:

$$\int_{a}^{b} f(x)dx = \sum_{i}^{N} W_i f(x_i) \qquad \qquad .....(8.1)$$

There exist many numerical methods for evaluating a definite integral. These methods are discussed in the subsequent sections.

8.1 NUMERICAL INTEGRATION USING POINTS WITH EQUAL SPACING

Some numerical methods make use of equally spaced intervals and a corresponding set of predetermined weights. Simpson's rule, Newton–Cotes formula, and Gauss quadrature are examples of such methods. However, the domain needs to be divided into a large number of intervals to obtain a reasonably accurate solution. This leads to an increase in computing time leading to an inefficient evaluation of integral. The alternative is to divide the domain into intervals with unequal spacing. This is described in the next section.

8.2 NUMERICAL INTEGRATION USING POINTS WITH UNEQUAL SPACING - GAUSS QUADRATURE

It is observed that all the methods of numerical integration mentioned in the previous section use equally spaced points. The accuracy of these methods can be improved only by the reduction of the interval between the sampling points. However, use of larger number of sampling points (equally spaced apart) requires large number of computations and is computationally inefficient. However, a high degree of accuracy can be obtained using the unequally spaced sampling points by making use of Legendre's polynomials. The procedure, named as Gaussian quadrature, is described below:

Consider the integral given in Eq. 8.1 once again. The integral can be changed to natural coordinates as follows:

$$x = \frac{(b-a)}{2}\xi + \frac{(b+a)}{2} \Rightarrow dx = \frac{(b-a)}{2}d\xi$$

$$if \ \xi = -1, x = \frac{(b-a)}{2} \times -1 + \frac{(b+a)}{2} = a$$

$$if \ \xi = +1, x = \frac{(b-a)}{2} \times 1 + \frac{(b+a)}{2} = b$$

$$\int_a^b f(x)dx = \frac{(b-a)}{2}\int_{-1}^{+1} f(\xi)d\xi$$

$$\qquad\qquad\qquad .....(8.2)$$

This change from the cartesian to natural coordinates is very convenient in implementation of numerical integration.

Consider the following integral represented by the sum of finite sums

$$\int_{-1}^{+1} f(\xi)d\xi = \sum_{i=1}^{r} W_i f(\xi_i) \qquad\qquad .....(8.3)$$

Expressing $f(\xi)$ as

$$f(\xi) = \alpha_1 + \alpha_2\xi + \alpha_3\xi^2 + ... + \alpha_{2r}\xi^{2r-1} \qquad\qquad .....(8.4)$$

LHS of Eq. 8.3 can be written as

$$\int_{-1}^{+1} f\left(\alpha_1 + \alpha_2\xi + \alpha_3\xi^2 + ... + \alpha_{2r}\xi^{2r-1}\right)d\xi$$

$$= \alpha_1\int_{-1}^{+1} d\xi + \alpha_2\int_{-1}^{+1} \xi d\xi + \alpha_3\int_{-1}^{+1} \xi^2 d\xi + ... + \alpha_{2r}\int_{-1}^{+1} \xi^{2r-1} d\xi \qquad .....(8.5)$$

$$= 2\alpha_1 + 0.\alpha_2 + \frac{2}{3}\alpha_3 + ... + 0.\alpha_{2r-1}$$

RHS of Eq. 8.3 can be expressed as

$$\sum_{i=1}^{r} W_i f\left(\xi_i\right) = W_1 f\left(\xi_1\right) + W_2 f\left(\xi_2\right) + W_3 f\left(\xi_3\right) + \ldots + W_r f\left(\xi_r\right)$$

$$= \left(\alpha_1 + \alpha_2 \xi_1 + \alpha_3 \xi_1^2 + \ldots + \alpha_{2r} \xi_1^{2r-1}\right) W_1 + \left(\alpha_1 + \alpha_2 \xi_2 + \alpha_3 \xi_2^2 + \ldots + \alpha_{2r} \xi_2^{2r-1}\right) W_2$$

$$+ \left(\alpha_1 + \alpha_2 \xi_3 + \alpha_3 \xi_3^2 + \ldots + \alpha_{2r} \xi_3^{2r-1}\right) W_3 + \ldots + \left(\alpha_1 + \alpha_2 \xi_r + \alpha_3 \xi_r^2 + \ldots + \alpha_{2r} \xi_r^{2r-1}\right) W_{2r}$$

$$\ldots(8.6)$$

Equating coefficients of $\alpha_1, \alpha_2, \alpha_3$ etc, we obtain

$$W_1 + W_2 + W_3 + \ldots + W_r = 2$$
$$W_1 \xi_1 + W_2 \xi_2 + W_3 \xi_3 + \ldots + W \xi_r = 0$$
$$W_1 \xi_1^2 + W_2 \xi_2^2 + W_3 \xi_3^2 + \ldots + W_r \xi_r^2 = \frac{2}{3} \tag{8.7}$$
$$\ldots$$
$$W_1 \xi_1^{2r-1} + W_2 \xi_2^{2r-1} + W_3 \xi_3^{2r-1} + \ldots + W_r \xi_r^{2r-1} = 0$$

The system of equations given by Eq. 8.7 is linear in W_i but nonlinear in ξ_i, and determines the parameters of Eq. 8.3 under the conditions

$$\left.\begin{array}{l} W_i > 0 \\ -1 \le \xi_i \le +1 \end{array}\right\} i = 1, 2, \ldots r \qquad \ldots(8.8)$$

The values of ξ_i appearing in Eq. 8.8 are the roots of Legendre Polynomials of order r. These are defined as follows:

$$P_0(\xi) = 1$$
$$P_1(\xi) = \xi$$
$$.. \qquad\qquad\qquad\qquad\qquad \ldots(8.9)$$
$$P_k(\xi) = \left(\frac{2k-1}{k}\right)\xi P_{k-1}(\xi) - \left(\frac{k-1}{k}\right)\xi P_{k-2}(\xi)$$

The weights W_i are obtained as

$$W_i = \frac{2(1-\xi_i^2)}{\left(r P_{r-1}(\xi_i)\right)^2} \qquad \ldots(8.10)$$

8.2.1 Computing Abscissa ξ_i and Weights W_i

(a) $r = 2$

Legendre polynomials of order 2 are

$$P_2(\xi) = \left(\frac{2\times 2 - 1}{2}\right)\xi \times P_1(\xi) - \left(\frac{2-1}{2}\right)\xi P_0(\xi) = \frac{3}{2}\xi \times \xi - \frac{1}{2}\xi \times 1 = \frac{3}{2}\xi^2 - \frac{1}{2}$$

Roots of $P_2(\xi)$ are $\dfrac{3}{2}\xi^2 - \dfrac{1}{2} = 0 \Rightarrow \xi = \pm\dfrac{1}{\sqrt{3}}$

Now, Eq. 8.7 can be written as

$$W_1 + W_2 = 2$$

$$W_1\xi_1 + W_2\xi_2 = 0 \Rightarrow W_1 \times -\frac{1}{\sqrt{3}} + W_2 \times \frac{1}{\sqrt{3}} = 0 \Rightarrow W_1 = W_2 = 1$$

(b) $r = 3$

Legendre polynomials of order 3 are $P_3(\xi) = \dfrac{1}{2}\xi\left(5\xi^2 - 3\right)$

Roots of $P_3(\xi)$ are $\xi = 0, \pm\sqrt{\dfrac{3}{5}}$. Now, Eq. 8.7 can be written as

$$W_1 + W_2 + W_3 = 2$$

$$W_1 \times -\sqrt{\frac{3}{5}} + W_2 \times 0 + W_3 \times \sqrt{\frac{3}{5}} = 0 \Rightarrow W_1 = W_3$$

$$W_1 \times \left(\frac{3}{5}\right) + W_2 \times 0 + W_3 \times \left(\frac{3}{5}\right) = \frac{2}{3} \Rightarrow W_1 = W_3 = \frac{10}{18} \text{ and } W_2 = \frac{16}{18}$$

Abscissa and weights for other values of r can be found out in similar manner. The roots and the corresponding weights, shown in Table 8.1 below, can be used in calculations.

Table 8.1 Abscissa and Weights for Gaussian Quadrature.

r	ξ	W
1	0.00000 00000 00000	2.00000 00000 00000
2	±0.57735 02691 89626	1.00000 00000 00000
3	±0.77459 66692 41483	0.55555 55555 55555
	0.00000 00000 00000	0.88888 88888 88888
4	±0.86113 63715 94052	0.34785 84513 74538
	±0.33998 10435 84856	0.65214 51548 62546
5	±0.90617 98459 38663	0.23692 68850 56189
	±0.53846 93101 05683	0.47862 86704 99366
	0.00000 00000 00000	0.56888 88888 88888

8.2.2   Numerical Example 8.1

Compute the integral $I = \int_{-1}^{1}\left(\dfrac{\xi^2}{2} + \sin\left(\dfrac{\xi}{2}\right)\right) d\xi$ using 3 Gauss points $r = 3$.

Taking accuracy of six decimals in both Gauss points and weights, we obtain

$$I = 0.555555\left(\frac{(-0.774596)^2}{2} + \sin\left(\frac{-0.774596}{2}\right)\right) + 0.888888\left(\frac{(0)^2}{2} + \sin\left(\frac{0}{2}\right)\right) +$$

$$0.555555\left(\frac{(0.774596)^2}{2} + \sin\left(\frac{0.774596}{2}\right)\right) = 0.666665$$

Exact solution is $I = \left|\dfrac{\xi^3}{3} - 2\cos\left(\dfrac{\xi}{3}\right)\right|_{-1}^{1} = 0.666666$

Error $= 1.5 \times 10^{-4}\%$

8.2.3   Numerical Example 8.2

Evaluate the integral $I = \int_{2}^{3} e^{-x}\cos(x)\,dx$ using 3 Gauss points $r = 3$.

$$f(x) = e^{-x}\cos(x)$$

$$a = 2 \quad b = 3 \quad \frac{(b-a)}{2} = 0.5 \quad \frac{(b+a)}{2} = 2.5$$

$$x = \frac{(b-a)}{2}\xi + \frac{(b+a)}{2} = 0.5\xi + 2.5$$

$$x_1 = 0.5 \times -0.774596 + 2.5 = 2.112702$$

$$W_1 = 0.555555 \quad f(x_1) = -0.062362$$

$$x_2 = 0.5 \times 0.000000 + 2.5 = 2.5$$

$$W_2 = 0.888888 \quad f(x_2) = -0.065761$$

$$x_3 = 0.5 \times -0.774596 + 2.5 = 2.887298$$

$$W_3 = 0.555555 \quad f(x_2) = -0.053934$$

$$I = W_1 x_1 + W_2 x_2 + W_3 x_3 = -0.061531$$

Exact solution is $I = -0.061532 \Rightarrow$ Error $= 0.0016\%$

8.3 EVALUATION OF DOUBLE AND TRIPLE INTEGRALS USING GAUSSIAN QUADRATURE

Consider the double integral $I = \int_a^b \int_c^d f(x,y)\,dxdy$.

Using

$$x = \frac{(b-a)}{2}\xi + \frac{(b+a)}{2} \Rightarrow dx = \frac{(b-a)}{2}d\xi$$

$$y = \frac{(d-c)}{2}\eta + \frac{(d+c)}{2} \Rightarrow dy = \frac{(d-c)}{2}d\eta \qquad \qquad \text{.....(8.11)}$$

$$I = \frac{(d-c)}{2} \cdot \frac{(b-a)}{2} \int_{\xi=-1}^{\xi=+1} \int_{\eta=-1}^{\eta=+1} f\left[\frac{(b-a)}{2}\xi + \frac{(b+a)}{2}, \frac{(d-c)}{2}\eta + \frac{(d+c)}{2}\right] d\xi d\eta \qquad \text{.....(8.12)}$$

$$I = \frac{(d-c)}{2} \cdot \frac{(b-a)}{2} \int_{\xi=-1}^{\xi=+1} \int_{\eta=-1}^{\eta=+1} f(\xi,\eta)\,d\xi d\eta \qquad \qquad \text{.....(8.13)}$$

$$\int_{\xi=-1}^{\xi=+1} \int_{\eta=-1}^{\eta=+1} f(\xi,\eta)\,d\xi d\eta = \sum_{i=1}^{r} \sum_{j=1}^{s} W_i W_j f(\xi_i,\eta_j) \qquad \qquad \text{.....(8.14)}$$

On similar lines, Gaussian quadrature can be used to evaluate the triple integral as follows:

$$x = \frac{(b-a)}{2}\xi + \frac{(b+a)}{2} \Rightarrow dx = \frac{(b-a)}{2}d\xi$$

$$y = \frac{(d-c)}{2}\eta + \frac{(d+c)}{2} \Rightarrow dy = \frac{(d-c)}{2}d\eta \qquad \qquad \text{.....(8.15)}$$

$$z = \frac{(f-e)}{2}\xi + \frac{(f+e)}{2} \Rightarrow dz = \frac{(f-e)}{2}d\xi$$

$$\int_a^b \int_c^d \int_e^f g(x,y,z)\,dxdydz = \frac{(b-a)}{2}\frac{(d-c)}{2}\frac{(f-e)}{2} \int_{-1}^{+1} \int_{-1}^{+1} \int_{-1}^{+1} f(\xi,\eta,\zeta)\,d\xi d\eta d\zeta \qquad \text{.....(8.16)}$$

$$\int_{-1}^{+1} \int_{-1}^{+1} \int_{-1}^{+1} g(\xi,\eta,\zeta)\,d\xi d\eta d\zeta = \sum_{i=1}^{r} \sum_{j=1}^{s} \sum_{k=1}^{t} W_i W_j W_k g(\xi_i,\eta_j,\zeta_k) \qquad \text{.....(8.17)}$$

8.3.1 Numerical Example – Double Integral

Calculate the double integral $I = \int\limits_0^\pi \int\limits_0^3 \left(x^2 - x\right)\sin y\, dxdy$

$$x = \frac{(3-0)}{2}\xi + \frac{(3+0)}{2} \Rightarrow x = \frac{3}{2}(\xi+1) \Rightarrow dx = \frac{3}{2}d\xi$$

$$y = \frac{(\pi-0)}{2}\eta + \frac{(\pi+0)}{2} \Rightarrow y = \frac{\pi}{2}(\eta+1) \Rightarrow dy = \frac{\pi}{2}d\eta$$

$$I = \frac{3\pi}{4}\int\limits_0^\pi \int\limits_0^3 f\left[\frac{3}{2}(\xi+1),\frac{\pi}{2}(\eta+1)\right]d\xi d\eta$$

$$I = \frac{3\pi}{4}\sum_{i=1}^3\sum_{j=1}^3 W_i W_j \left[\frac{9}{4}(\xi_i+1)^2 - \frac{3}{2}\xi_i\right]\sin\left[\frac{\pi}{2}(\eta_j+1)\right]$$

$$I = \frac{3\pi}{4}\begin{bmatrix}
0.55555^2\left(\frac{9}{4}(-0.774596+1)^2 - \frac{3}{2}\times(-0.774596+1)\right)\sin\left(\frac{\pi}{2}(-0.774596+1)\right)+ \\[6pt]
0.55555\times0.88888\left(\frac{9}{4}(-0.774596+1)^2 - \frac{3}{2}\times(-0.774596+1)\right)\sin\left(\frac{\pi}{2}(0+1)\right)+ \\[6pt]
0.55555^2\left(\frac{9}{4}(-0.774596+1)^2 - \frac{3}{2}\times(-0.774596+1)\right)\sin\left(\frac{\pi}{2}(0.774596+1)\right)+ \\[6pt]
0.88888\times0.55555\left(\frac{9}{4}(0+1)^2 - \frac{3}{2}\times1\right)\sin\left(\frac{\pi}{2}(-0.774596+1)\right)+ \\[6pt]
0.88888^2\left(\frac{9}{4}(0+1)^2 - \frac{3}{2}\times1\right)\sin\left(\frac{\pi}{2}(0+1)\right)+ \\[6pt]
0.88888\times0.55555\left(\frac{9}{4}(0+1)^2 - \frac{3}{2}\times1\right)\sin\left(\frac{\pi}{2}(0.774596+1)\right)+ \\[6pt]
0.55555^2\left(\frac{9}{4}(0.774596+1)^2 - \frac{3}{2}\times(0.774596+1)\right)\sin\left(\frac{\pi}{2}(-0.774596+1)\right)+ \\[6pt]
0.55555\times0.88888\left(\frac{9}{4}(0.774596+1)^2 - \frac{3}{2}\times(0.774596+1)\right)\sin\left(\frac{\pi}{2}(0+1)\right)+ \\[6pt]
0.55555^2\left(\frac{9}{4}(0.774596+1)^2 - \frac{3}{2}\times(0.774596+1)\right)\sin\left(\frac{\pi}{2}(0.774596+1)\right)
\end{bmatrix}$$

$I = 9.0061$. Exact solution is $I = \left|\frac{x^3}{3} - \frac{x^2}{2}\right|_0^3 \times \left|-\cos(x)\right|_0^\pi = 9$

Percentage error = 0.0678%

8.4 HAMMER INTEGRAL ON TRIANGLES

Consider a triangular domain as shown in Fig.8.1 below.

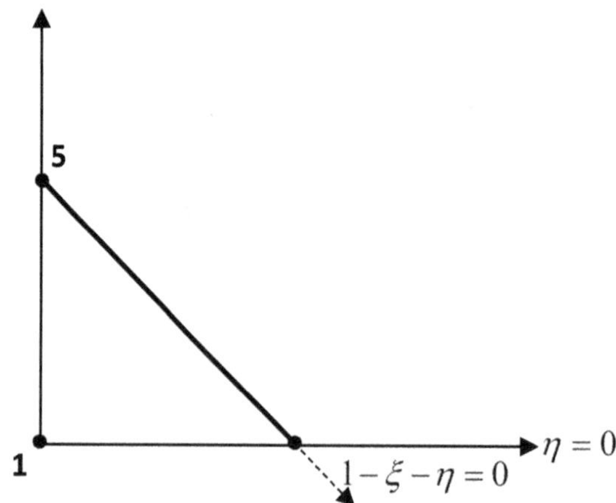

Figure 8.1 Triangular domain.

Consider the following integral represented by the sum of finite sums

$$I = \int_{\xi=0}^{1} \int_{\eta=0}^{1-\xi} f(\xi,\eta)d\xi d\eta = \sum_{i=1}^{r} W_i f(\xi_i,\eta_i) \qquad \ldots\ldots(8.18)$$

Expressing $f(\xi,\eta)$ as

$$f(\xi,\eta) = \alpha_1 + \alpha_2\xi + a_3\eta + a_4\xi^2 + \alpha_5\xi\eta + \alpha_6\eta^2 \qquad \ldots\ldots(8.19)$$

Eq. 8.19 is substituted on the LHS of Eq. 8.18 as

$$I = \alpha_1 \int_{\xi=0}^{1}\int_{\eta=0}^{1-\xi} d\xi d\eta + \alpha_2 \int_{\xi=0}^{1}\int_{\eta=0}^{1-\xi} \xi d\xi d\eta + \alpha_3 \int_{\xi=0}^{1}\int_{\eta=0}^{1-\xi} \eta d\xi d\eta$$

$$+\alpha_4 \int_{\xi=0}^{1}\int_{\eta=0}^{1-\xi} \xi^2 d\xi d\eta + \alpha_5 \int_{\xi=0}^{1}\int_{\eta=0}^{1-\xi} \xi\eta d\xi d\eta + \alpha_6 \int_{\xi=0}^{1}\int_{\eta=0}^{1-\xi} \eta^2 d\xi d\eta \qquad \ldots(8.20)$$

$$= \frac{1}{2}\alpha_1 + \frac{1}{6}.\alpha_2 + \frac{1}{6}\alpha_3 + \frac{1}{12}\alpha_4 + \frac{1}{24}\alpha_5 + \frac{1}{12}\alpha_6$$

RHS of Eq. 8.18 can be expressed as

$$\sum_{i=1}^{3} W_i f(\xi_i,\eta_i) = \sum_{i=1}^{3} W_i f\left(\alpha_1 + \alpha_2\xi_i + a_3\eta_i + a_4\xi_i^2 + \alpha_5\xi_i\eta_i + \alpha_6\eta_i^2\right)$$

$$= \alpha_1 \sum_{i=1}^{3} W_i + \alpha_2 \sum_{i=1}^{3} \xi_i W_i + \alpha_3 \sum_{i=1}^{3} \eta_i W_i + \alpha_4 \sum_{i=1}^{3} \xi_i^2 W_i + \alpha_5 \sum_{i=1}^{3} \xi_i\eta_i W_i + \alpha_4 \sum_{i=1}^{3} \eta_i^2 W_i \qquad \ldots\ldots(8.21)$$

Equating coefficients of $\alpha_1, \alpha_2, \ldots \alpha_6$, we get

$$\sum_{i=1}^{3} W_i = \frac{1}{2} \qquad \sum_{i=1}^{3} W_i \xi_i = \frac{1}{6} \qquad \sum_{i=1}^{3} W_i \eta_i = \frac{1}{6}$$

$$\sum_{i=1}^{3} W_i \xi_i^2 = \frac{1}{12} \qquad \sum_{i=1}^{3} W_i \xi_i \eta_i = \frac{1}{24} \qquad \sum_{i=1}^{3} W_i \eta_i^2 = \frac{1}{12}$$

.....(8.22)

We can verify that

$$W_1 = W_2 = W_3 = \frac{1}{3} \qquad\qquad \text{and}$$

$$(\xi_1, \eta_1) = \left(\frac{1}{6}, \frac{1}{6}\right) \quad (\xi_2, \eta_2) = \left(\frac{2}{3}, \frac{1}{6}\right) \quad (\xi_3, \eta_3) = \left(\frac{1}{6}, \frac{2}{3}\right)$$

and

$$W_1 = W_2 = W_3 = \frac{1}{3} \qquad\qquad \text{and}$$

$$(\xi_1, \eta_1) = \left(0, \frac{1}{2}\right) \quad (\xi_2, \eta_2) = \left(\frac{1}{2}, 0\right) \quad (\xi_3, \eta_3) = \left(\frac{1}{2}, \frac{1}{2}\right)$$

are two possible solutions.

Similarly, expressing $f(\xi, \eta)$ as

$$f(\xi, \eta) = \alpha_1 + \alpha_2 \xi + a_3 \eta + a_4 \xi^2 + \alpha_5 \xi \eta + \alpha_6 \eta^2 + \alpha_7 \xi^3 + \alpha_8 \xi^2 \eta + \alpha_9 \xi \eta^2 + \alpha_{10} \eta^3$$

.....(8.23)

Following the procedure outlined above, we get

$$\sum_{i=1}^{3} W_i = \frac{1}{2} \qquad \sum_{i=1}^{3} W_i \xi_i = \frac{1}{6} \qquad \sum_{i=1}^{3} W_i \eta_i = \frac{1}{6}$$

$$\sum_{i=1}^{3} W_i \xi_i^2 = \frac{1}{12} \qquad \sum_{i=1}^{3} W_i \xi_i \eta_i = \frac{1}{24} \qquad \sum_{i=1}^{3} W_i \eta_i^2 = \frac{1}{12}$$

$$\sum_{i=1}^{3} W_i \xi_i^3 = \frac{1}{20} \qquad \sum_{i=1}^{3} W_i \xi_i^2 \eta_i = \frac{1}{60} \qquad \sum_{i=1}^{3} W_i \eta_i^3 = \frac{1}{20}$$

.....(8.24)

The abscissa ξ_i, η_i and weights W_i obtained above are different from those used in Gauss quadrature and are listed (Table 8.1):

Table 8.2 Abscissa and weights for triangular element.

Order m	No. of points r		ξ	η	W
1	1		$\dfrac{1}{3}$	$\dfrac{1}{3}$	$\dfrac{1}{2}$
2	3		$\dfrac{1}{2}$	$\dfrac{1}{2}$	$\dfrac{1}{6}$
			0	$\dfrac{1}{2}$	$\dfrac{1}{6}$
			$\dfrac{1}{2}$	0	$\dfrac{1}{6}$
2	3		$\dfrac{1}{6}$	$\dfrac{1}{6}$	$\dfrac{1}{6}$
			$\dfrac{2}{3}$	$\dfrac{1}{6}$	$\dfrac{1}{6}$
			$\dfrac{1}{6}$	$\dfrac{2}{3}$	$\dfrac{1}{6}$
3	4		$\dfrac{1}{3}$	$\dfrac{1}{3}$	$-\dfrac{27}{96}$
			$\dfrac{1}{5}$	$\dfrac{1}{5}$	$\dfrac{25}{96}$
			$\dfrac{3}{5}$	$\dfrac{1}{5}$	$\dfrac{25}{96}$
			$\dfrac{1}{5}$	$\dfrac{3}{5}$	$\dfrac{25}{96}$

The locations of sample points are shown in Fig.8.2:

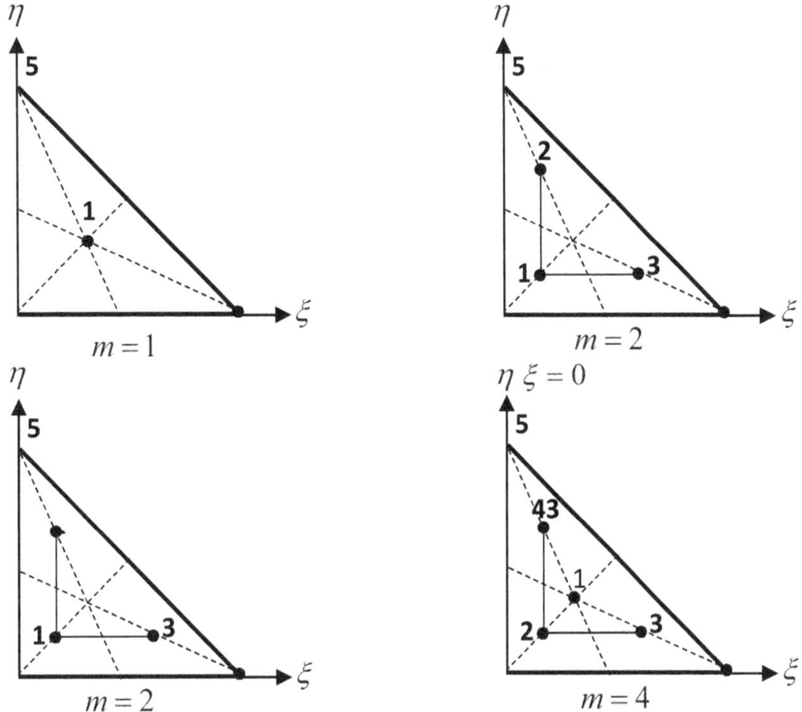

Figure 8.2 Sample points for integral on a triangle for orders m=1, 2 and 3.

8.5 EXERCISE PROBLEMS

Evaluate the following integrals:

1. $\displaystyle\int_0^3 e^{-x}\sin x\,dx$

2. $\displaystyle\int_0^\pi \cos^2 x\sin^3 x\,dx$

3. $\displaystyle\int_2^3 e^{-\tan x}\cos x\,dx$

4. $\displaystyle\int_0^\pi\int_0^2 \left(x^3 - x^2\right)\sin y\,dx\,dy$

5. $\displaystyle\int_0^\pi\int_0^3 e^{-x^2}\cos y\,dx\,dy$

Axi-Symmetric Elements

An axisymmetric solid (or a thick-walled body) of revolution is defined as a three-dimensional body that is generated by rotating a plane and is most easily described in cylindrical coordinates. Here z axis is taken as the axis of symmetry.

Problems such as soil masses subjected to circular footing loads, thick-walled pressure vessels, and a rocket nozzle subjected to thermal and pressure loading can often be analysed using axisymmetric elements. Some examples for axisymmetric structures are shown in Figure 9.1:

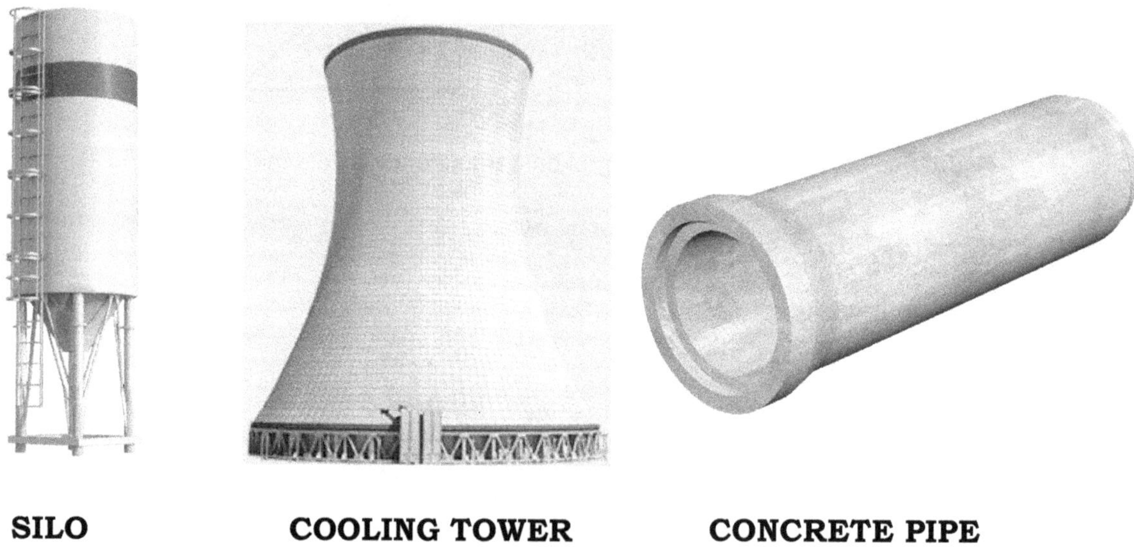

Figure 9.1 Axisymmetric Elements.

9.1 STRESS ANALYSIS OF AXISYMMETRIC ELEMENTS

To satisfy the conditions for axisymmetric stress the problem must be such that:

(a) The solid body under stress must be a solid of revolution. By convention, the axis of revolution is the z-axis, in a cylindrical polar coordinate system (r,θ,z) coordinates.

(b) The loading on the body is symmetric about the z-axis.

(c) All boundary conditions are symmetric about the z-axis.

(d) Material properties are also symmetric about the z-axis. This is automatically satisfied if the material properties are linearly elastic, homogenous, and isotropic.

If the above conditions are satisfied, the displacement field is independent of tangential (circumferential) coordinate θ. Hence the stress analysis is two dimensional though the physical problem is three dimensional.

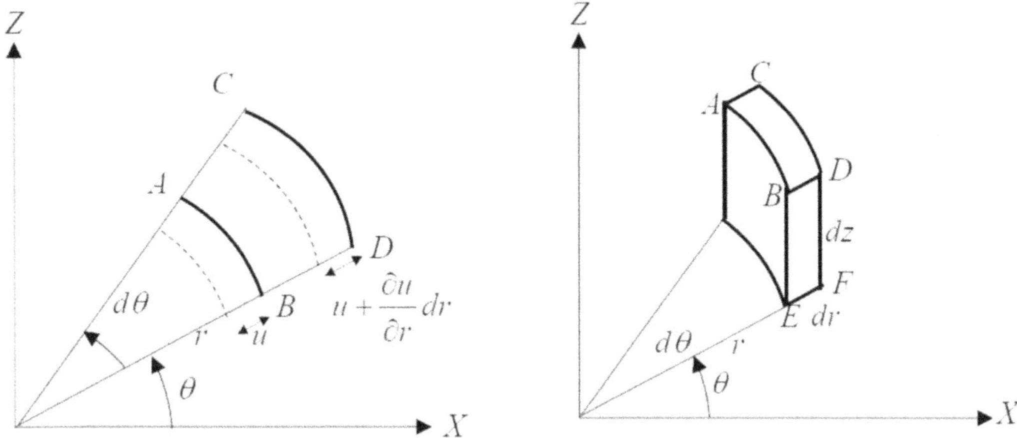

Figure 9.2 Cross Section of axisymmetric element.

9.2   STRAIN DISPLACEMENT AND STRESS-STRAIN RELATIONSHIPS

With reference to Figure 9.2 and Figure 9.3, the strains are given as follows:

Radial strain $\varepsilon_r = \dfrac{\left(u + \dfrac{\partial u}{\partial r}dr - u\right)}{dr} = \dfrac{\partial u}{\partial r}$ $\qquad .....(9.1)$

Axial strain $\varepsilon_z = \dfrac{\left(w + \dfrac{\partial w}{\partial z}dz - w\right)}{dz} = \dfrac{\partial w}{\partial z}$ $\qquad ....(9.2)$

Circumferential strain $\varepsilon_\theta = \dfrac{(r+u)d\theta - rd\theta}{r} = \dfrac{u}{r} =$ constant $\qquad$ (9.3)

As it is similar to plane strain condition, (refer to Figure 9.3)

$$\gamma_{r\theta} = \gamma_{\theta z} = 0$$

$$\gamma_{rz} = \frac{\partial w}{\partial r} + \frac{\partial u}{\partial z} \qquad\qquad .....(9.4)$$

Stress-strain relationship is given by

$$\begin{Bmatrix} \sigma_r \\ \sigma_\theta \\ \sigma_z \\ \tau_{rz} \end{Bmatrix} = \frac{E}{(1+\mu)(1-2\mu)} \begin{bmatrix} 1-\mu & \mu & \mu & 0 \\ \mu & 1-\mu & \mu & 0 \\ \mu & \mu & 1-\mu & 0 \\ 0 & 0 & 0 & \dfrac{(1-2\mu)}{2} \end{bmatrix} \begin{Bmatrix} \varepsilon_r \\ \varepsilon_\theta \\ \varepsilon_z \\ \gamma_{rz} \end{Bmatrix} \qquad .....(9.5)$$

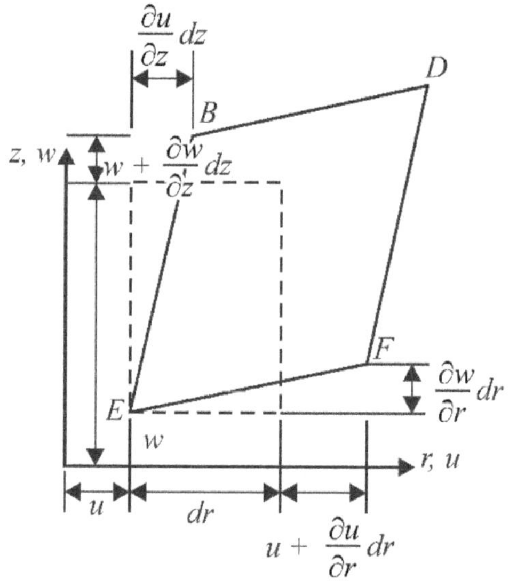

Figure 9.3 Displacements of axisymmetric element.

9.3 Finite Element Model of Axisymmetric Element

Considering an element of n nodes, the displacements u and w along r and z directions, can be represented as

$$u = u_1 N_1 + u_2 N_2 + u_3 N_3 + ... + u_n N_n$$

$$w = w_1 N_1 + w_2 N_2 + w_3 N_3 + ... + w_n N_n \qquad .....(9.6)$$

This can be represented in the matrix form as

$$\begin{Bmatrix} u \\ w \end{Bmatrix} = \begin{bmatrix} N_1 & 0 & N_2 & 0 & .. & .. & N_n & 0 \\ 0 & N_1 & 0 & N_2 & .. & .. & 0 & N_n \end{bmatrix} \lfloor u_1 \quad w_1 \quad u_2 \quad w_2 \quad .. \quad .. \quad u_n \quad w_n \rfloor^T$$

....(9.7)

Eq. 9.7 can be written as

$$\{u\} = [N]\{d\}$$

.....(9.8)

Where N is the shape function matrix and d is the vector of nodal displacements.

The vector of strains $\{\varepsilon\}$ can be represented as follows:

$$\begin{Bmatrix} \varepsilon_r \\ \varepsilon_z \\ \varepsilon_\theta \\ \gamma_{rz} \end{Bmatrix} = \begin{bmatrix} \dfrac{\partial}{\partial r} & 0 \\ 0 & \dfrac{\partial}{\partial z} \\ \dfrac{1}{r} & 0 \\ \dfrac{\partial}{\partial z} & \dfrac{\partial}{\partial r} \end{bmatrix} \begin{Bmatrix} u \\ v \end{Bmatrix}$$

.....(9.9)

Substituting Eq.9.7 on the right-hand side of Eq.9.9 leads to

$$\begin{Bmatrix} \varepsilon_r \\ \varepsilon_z \\ \varepsilon_\theta \\ \gamma_{rz} \end{Bmatrix} = \left(\begin{bmatrix} \dfrac{\partial}{\partial r} & 0 \\ 0 & \dfrac{\partial}{\partial z} \\ \dfrac{1}{r} & 0 \\ \dfrac{\partial}{\partial z} & \dfrac{\partial}{\partial r} \end{bmatrix} \begin{bmatrix} N_1 & 0 & N_2 & 0 & .. & .. & N_n & 0 \\ 0 & N_1 & 0 & N_2 & .. & .. & 0 & N_n \end{bmatrix} \right) \{d\}$$

.....(9.10)

$$\{\varepsilon\} = [B]\{d\}$$

.....(9.11)

Stiffness matrix can be written as

$$\left[K^{(e)} \right] = \iiint [B]^T [C][B] \, dV$$

.....(9.12)

Element force vector is given by

$$\left\{F^{(e)}\right\} = \iiint [N]^T \{X\}\, dV + \iint [N]^T \{T\}\, dS + \{P\} \qquad .....(9.13)$$

The first term of Eq. 9.13 is the vector of body forces and can be written as

$$\iiint [N]^T \begin{Bmatrix} R_b \\ Z_b \end{Bmatrix} dV \qquad .....(9.14)$$

The second term of Eq.9.13 represents the vector of surface tractions and can be written as

$$\iint [N]^T \begin{Bmatrix} P_r \\ P_z \end{Bmatrix} dS \qquad .....(9.15)$$

The third term of Eq.9.13 represents the vector of line loads.

The following axisymmetric elements are considered for analysis in the present chapter:

(a) 3-node triangular axisymmetric element
(b) 6-node triangular axisymmetric element
(c) 8-node quadrilateral axisymmetric element

9.4   THREE-NODE TRIANGULAR AXISYMMETRIC ELEMENT

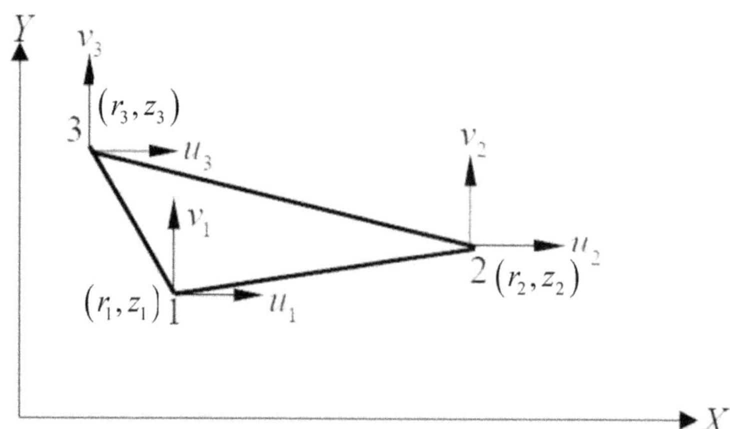

Figure 9.4 Three-node triangular axisymmetric element.

Consider the three-node axisymmetric triangular element as shown in the Figure 9.4.

The displacements u and w of any point within the element along r and z axes can be expressed in terms of nodal displacements as follows:

$$u(r,z) = a_1 + a_2 r + a_3 z$$
$$w(r,z) = a_4 + a_5 r + a_6 z \qquad \qquad .....(9.16)$$

$$\begin{Bmatrix} u \\ w \end{Bmatrix} = \begin{bmatrix} 1 & r & z & 0 & 0 & 0 \\ 0 & 0 & 0 & 1 & r & z \end{bmatrix} \{\alpha\} \qquad .....(9.17)$$

where $\{\alpha\} = \lfloor a_1 \quad a_2 \quad a_3 \quad a_4 \quad a_5 \quad a_6 \rfloor^T$ is the vector of undetermined coefficients. Substituting the nodal coordinates, we obtain

$$\begin{Bmatrix} u_1 \\ w_1 \\ u_2 \\ w_2 \\ u_3 \\ w_3 \end{Bmatrix} = \begin{bmatrix} 1 & r_1 & z_1 & 0 & 0 & 0 \\ 0 & 0 & 0 & 1 & r_1 & z_1 \\ 1 & r_2 & z_2 & 0 & 0 & 0 \\ 0 & 0 & 0 & 1 & r_2 & z_2 \\ 1 & r_3 & z_3 & 0 & 0 & 0 \\ 0 & 0 & 0 & 1 & r_3 & z_3 \end{bmatrix} \begin{Bmatrix} a_1 \\ a_2 \\ a_3 \\ a_4 \\ a_5 \\ a_6 \end{Bmatrix} \Rightarrow \{d\} = [A]\{\alpha\} \qquad .....(9.18)$$

Obtaining $\{\alpha\}$ by inverting Eq.9.18 and substituting on the right-hand side of Eq. 9.17 leads to

$$\begin{Bmatrix} u \\ w \end{Bmatrix} = \begin{bmatrix} 1 & r & z & 0 & 0 & 0 \\ 0 & 0 & 0 & 1 & r & z \end{bmatrix} [A]^{-1} \{d\} = \begin{bmatrix} N_1 & 0 & N_2 & 0 & N_3 & 0 \\ 0 & N_1 & 0 & N_2 & 0 & N_3 \end{bmatrix} \{d\} \qquad .....(9.19)$$

The interpolation functions N_i can be written as follows:

$$N_1 = \frac{1}{2A}(\alpha_1 + \beta_1 r + \gamma_1 z)$$

$$N_2 = \frac{1}{2A}(\alpha_2 + \beta_2 r + \gamma_2 z) \qquad .....(9.20)$$

$$N_3 = \frac{1}{2A}(\alpha_3 + \beta_3 r + \gamma_3 z)$$

Here,

$$A = \frac{1}{2}\begin{vmatrix} 1 & x_1 & y_1 \\ 1 & x_2 & y_2 \\ 1 & x_3 & y_3 \end{vmatrix}; \quad \begin{aligned} \alpha_i &= r_j z_k - r_k z_j \\ \beta_i &= z_j - z_k \\ \gamma_i &= r_k - r_j \\ i,j,k &= 1,2,3 \end{aligned} \qquad .....(9.21)$$

$$\begin{Bmatrix} \varepsilon_r \\ \varepsilon_z \\ \varepsilon_\theta \\ \gamma_{rz} \end{Bmatrix} = \begin{bmatrix} \dfrac{\partial}{\partial r} & 0 \\ 0 & \dfrac{\partial}{\partial z} \\ \dfrac{1}{r} & 0 \\ \dfrac{\partial}{\partial z} & \dfrac{\partial}{\partial r} \end{bmatrix} \begin{bmatrix} N_1 & 0 & N_2 & 0 & N_3 & 0 \\ 0 & N_1 & 0 & N_2 & 0 & N_3 \end{bmatrix} \{d\} \qquad \qquad .....(9.22)$$

$$\begin{Bmatrix} \varepsilon_r \\ \varepsilon_z \\ \varepsilon_\theta \\ \gamma_{rz} \end{Bmatrix} = \begin{bmatrix} \dfrac{\partial N_1}{\partial r} & 0 & \dfrac{\partial N_2}{\partial r} & 0 & \dfrac{\partial N_3}{\partial r} & 0 \\ 0 & \dfrac{\partial N_1}{\partial z} & 0 & \dfrac{\partial N_2}{\partial z} & 0 & \dfrac{\partial N_3}{\partial z} \\ \dfrac{N_1}{r} & 0 & \dfrac{N_2}{r} & 0 & \dfrac{N_3}{r} & 0 \\ \dfrac{\partial N_1}{\partial z} & \dfrac{\partial N_1}{\partial r} & \dfrac{\partial N_2}{\partial z} & \dfrac{\partial N_2}{\partial r} & \dfrac{\partial N_3}{\partial z} & \dfrac{\partial N_3}{\partial r} \end{bmatrix} \{d\} \qquad .....(9.23)$$

$$\begin{Bmatrix} \varepsilon_r \\ \varepsilon_z \\ \varepsilon_\theta \\ \gamma_{rz} \end{Bmatrix} = \dfrac{1}{2A} \begin{bmatrix} \beta_1 & 0 & \beta_2 & 0 & \beta_3 & 0 \\ 0 & \gamma_1 & 0 & \gamma_2 & 0 & \gamma_3 \\ \dfrac{1}{r}(\alpha_1 + \beta_1 r + \gamma_1 z) & 0 & \dfrac{1}{r}(\alpha_2 + \beta_2 r + \gamma_2 z) & 0 & \dfrac{1}{r}(\alpha_3 + \beta_3 r + \gamma_3 z) & 0 \\ \gamma_1 & \beta_1 & \gamma_2 & \beta_2 & \gamma_3 & \beta_3 \end{bmatrix} \{d\}$$

$$.....(9.24)$$

Eq.9.24 can be written as

$$\{\varepsilon\} = [B]\{d\} \qquad \qquad .....(9.25)$$

It is observed that the B matrix is a function of r and z.

Element stiffness matrix can be written as

$$\left[K^{(e)}\right] = 2\pi \iint_A [B]^T [C][B] \, r \, dr \, dz \qquad \qquad .....(9.26)$$

Representing the centroid of the element as $(\bar{r}, \bar{z})$, the stiffness matrix can be computed at the centroid of the element as follows:

Taking $\bar{r} = \dfrac{1}{3}(r_1 + r_2 + r_3)$; $\bar{z} = \dfrac{1}{3}(z_1 + z_2 + z_3)$ and $\bar{B} = B(\bar{r}, \bar{z})$

$$\left[K^{(e)} \right] = 2\pi \bar{r} A \left[\bar{B} \right]^{T} [D] \left[\bar{B} \right] \qquad\qquad\qquad .....(9.27)$$

$$\{ f_{bi} \} = \iiint [N]^{T} \begin{Bmatrix} R_{b} \\ Z_{b} \end{Bmatrix} r\,dr\,dz = \frac{2\pi A \bar{r}}{3} \begin{Bmatrix} \bar{R}_{b} \\ Z_{b} \end{Bmatrix} \qquad\qquad .....(9.28)$$

$$\{ f_{b} \} = \frac{2\pi A \bar{r}}{3} \begin{Bmatrix} \bar{R}_{b} \\ Z_{b} \\ \bar{R}_{b} \\ Z_{b} \\ \bar{R}_{b} \\ Z_{b} \end{Bmatrix} \qquad\qquad\qquad .....(9.29)$$

Here, $R_{b} = \omega^{2} \rho r$ and $\bar{R}_{b} = \omega^{2} \rho \bar{r}$ where ω is the angular velocity, r is the radial distance, $\bar{r}$ is the radial distance of centroid of element and ρ is the mass density. Also Z_{b} is the body force per unit volume $i.e.$, $Z_{b} = \rho g$.

9.5   EXAMPLE PROBLEM

Consider a circular footing located on a sandy soil stratum depth 8 m as shown in the Figure 9.5. The Elastic modulus and the Poisson's ratio of the soil are $E = 1 \times 10^{5} \ kN/m^{2}$ and $\mu = 0.35$ respectively. The radius of the footing is 1m and carries a load of 200kN. Eight meters beneath the footing, the soil is made up of a solid rock formation that can be considered very stiff. Assume that 7m away from the centreline of the footing the horizontal displacement of the soil is negligible.

Consider an element length of 0.5 m, analyse the footing using

(a) 3-node triangular element
(b) 6-node triangular element
(c) 8-node quadrilateral elements.

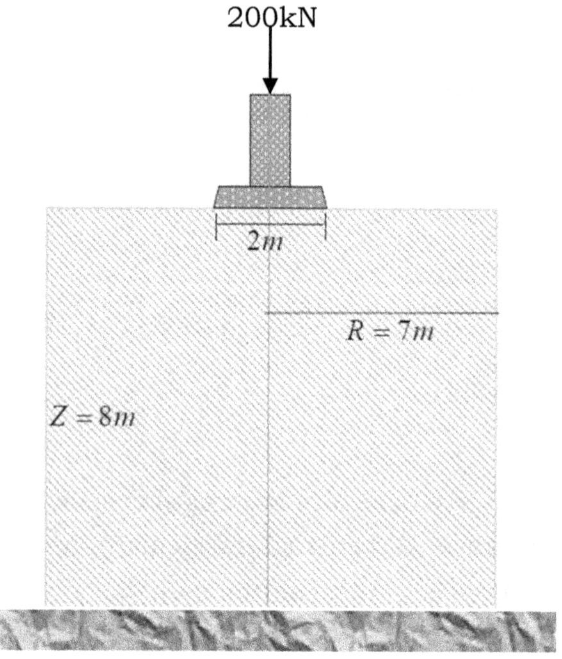

Figure 9.5 Soil underneath a circular footing.

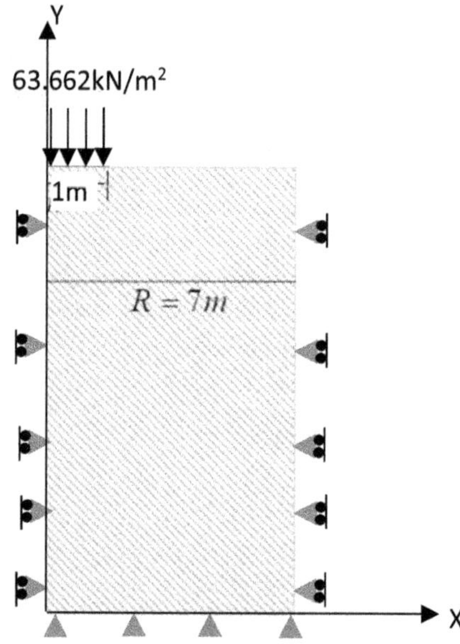

Figure 9.6 Model of the soil underneath circular footing.

Figure 9.6 shows the geometrical domain and the boundary conditions. Because of symmetry, only half the domain will be discretized. Nodes on the centreline and the nodes placed at a 7 m radius displace only in the vertical direction because the horizontal movement of the soil at this distance is assumed negligible. The nodes placed at a depth of 7 m are fixed in all directions because the rock substratum is assumed in-deformable. The 200 kN load is transformed into an equivalent uniformly distributed load of $63.662\ kN/m^2$.

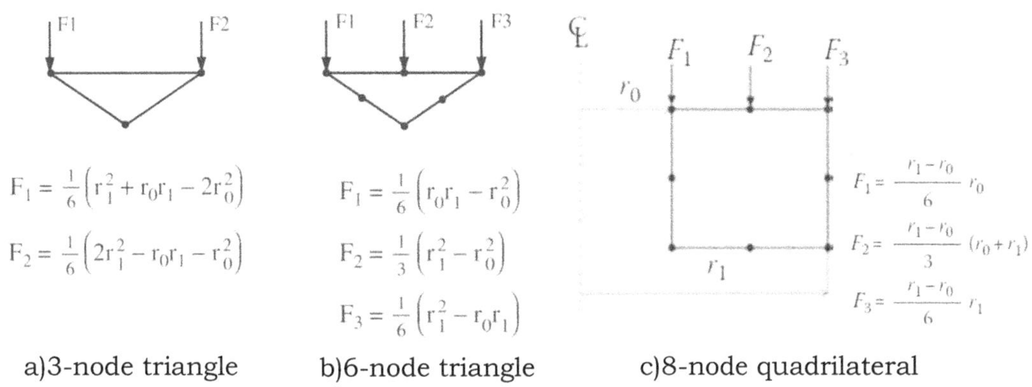

$$F_1 = \frac{1}{6}\left(r_1^2 + r_0 r_1 - 2r_0^2\right)$$

$$F_2 = \frac{1}{6}\left(2r_1^2 - r_0 r_1 - r_0^2\right)$$

$$F_1 = \frac{1}{6}\left(r_0 r_1 - r_0^2\right)$$

$$F_2 = \frac{1}{3}\left(r_1^2 - r_0^2\right)$$

$$F_3 = \frac{1}{6}\left(r_1^2 - r_0 r_1\right)$$

$$F_1 = \frac{r_1 - r_0}{6} r_0$$

$$F_2 = \frac{r_1 - r_0}{3}\left(r_0 + r_1\right)$$

$$F_3 = \frac{r_1 - r_0}{6} r_1$$

a)3-node triangle   b)6-node triangle   c)8-node quadrilateral

Figure 9.7 Equivalent nodal loads.

The equivalent nodal loads for the applied pressure in case of mesh of 3-node triangular elements, 6-node triangular elements and 8-noded quadrilateral element are shown in the Figure 9.7:

The domain is analysed using the MATLAB code provided in sections A.9, A.10 and A.11 respectively for the 3-node triangular element, 6-node triangular element and 8-node quadrilateral element. The domain is also analysed using ABAQUS using the above elements, using the procedure laid down in section 9.6.

The mesh created by the 3-node triangular elements is shown in Figure 9.8:

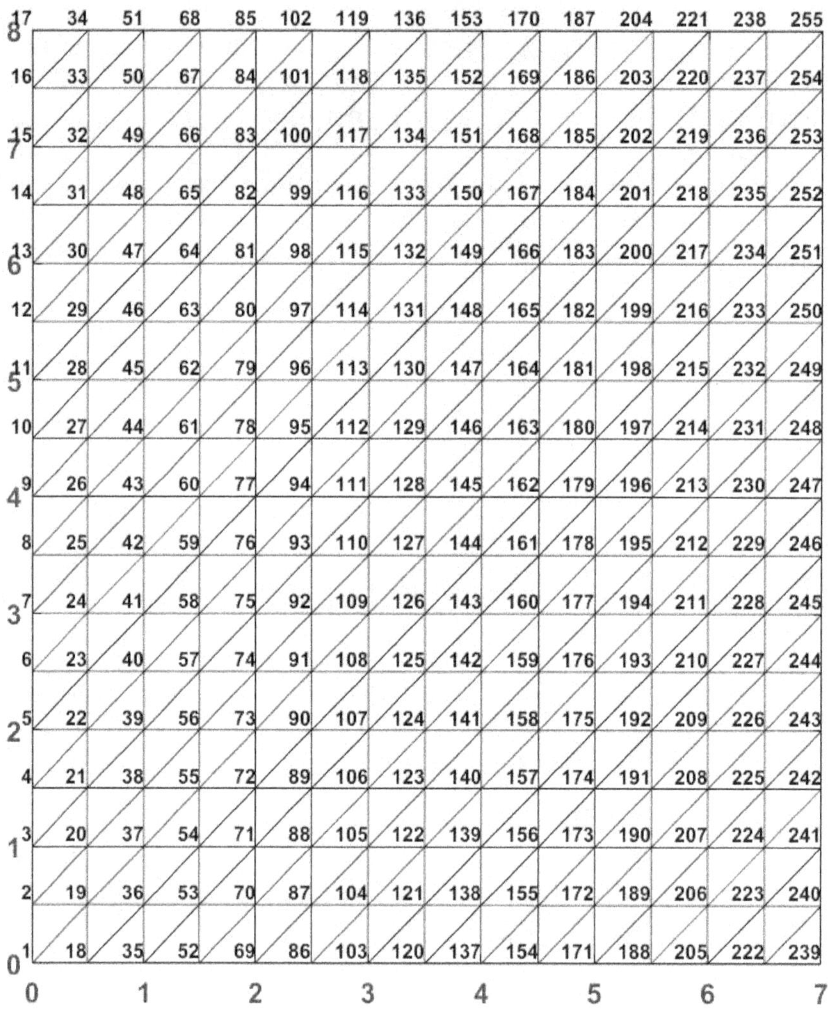

Figure 9.8 Mesh of 3-noded triangular axisymmetric elements.

Figure 9.9 and 9.10 show the radial and longitudinal displacements u_r and u_z computed using the MATLAB code as well as U_1 and U_2 using ABAQUS. It is observed that the results from both MATLAB and ABAQUS match exactly. Similarly, values of $\sigma_{rr}, \sigma_{zz}, \sigma_\theta$ and τ_{rz} obtained using MATLAB and S_{11}, S_{22}, S_{33} and S_{12} computed using ABAQUS also match exactly.

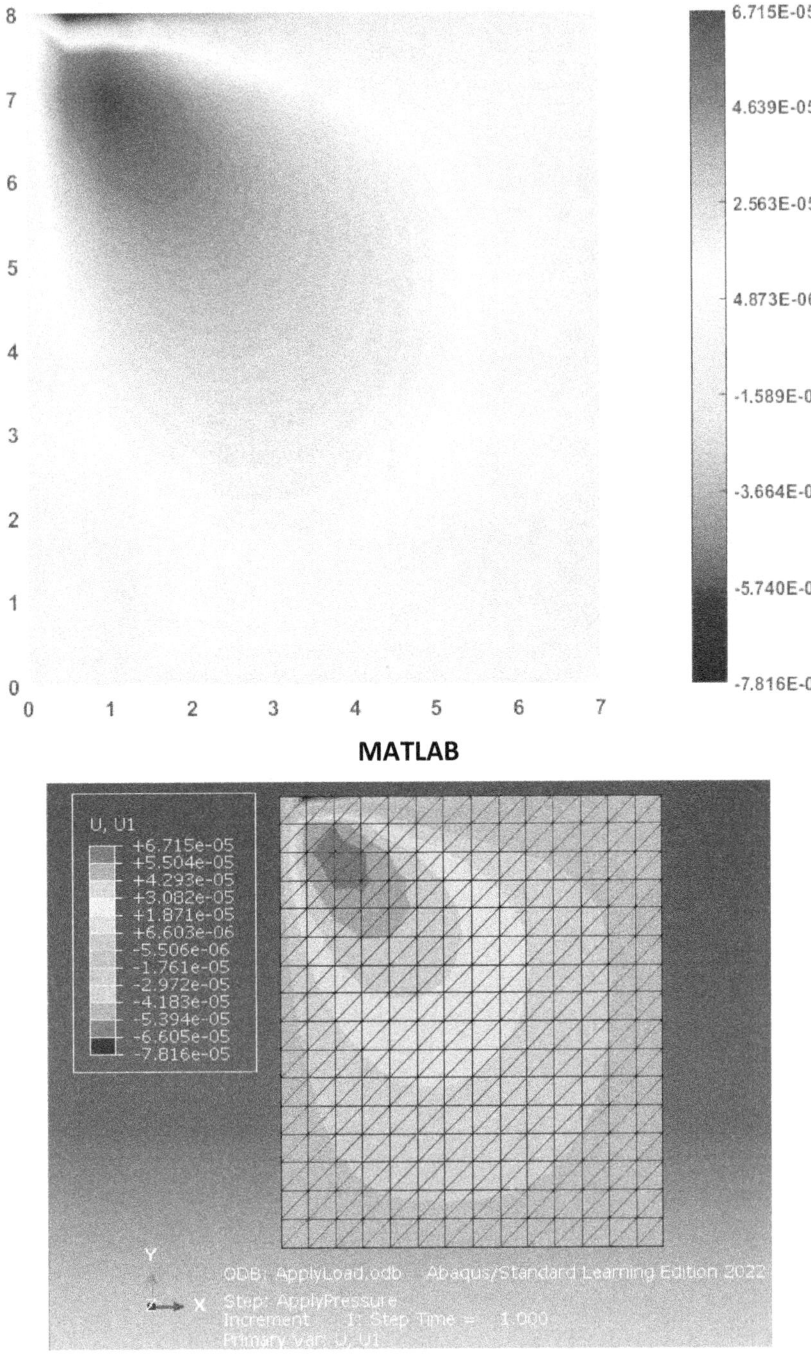

Figure 9.9 Variation of radial displacement U1 across the domain.

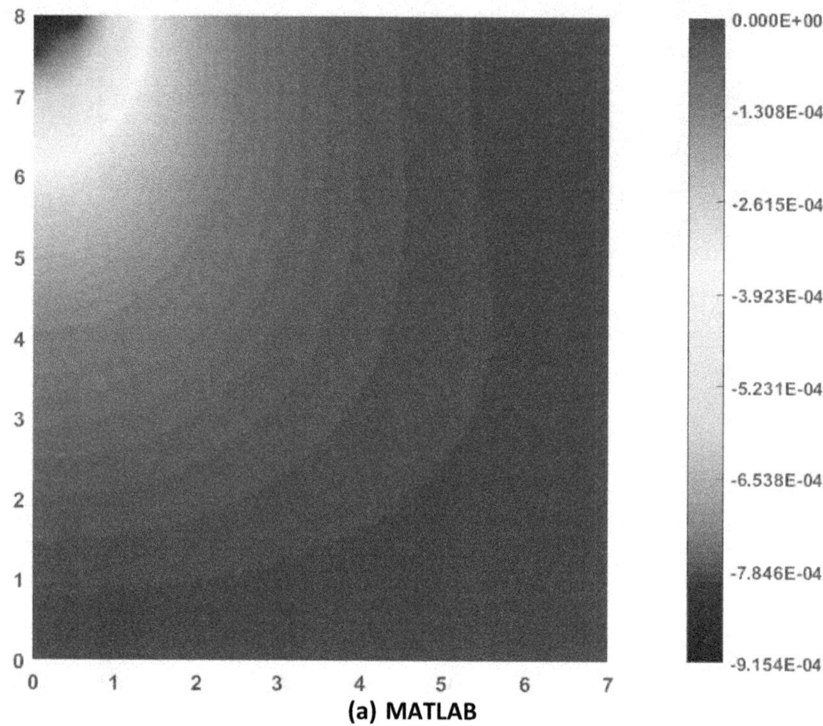

(a) MATLAB

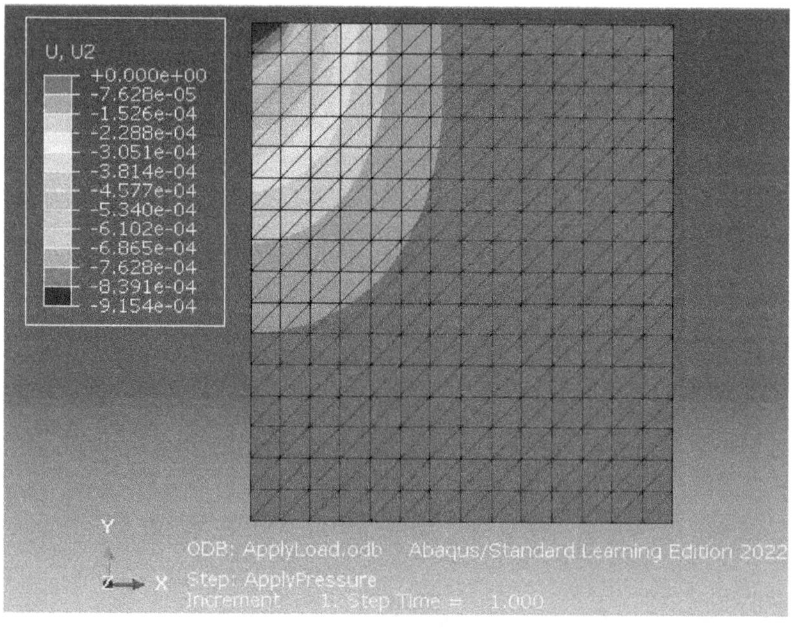

(b) ABAQUS

Figure 9.10 Variation of longitudinal displacement U2 across the domain.

The mesh created by the 6-node triangular elements is shown in Figure 9.11 below. Figures 9.12 and 9.13 show the plots of σ_{rr} and σ_{zz} (S_{11} and S_{22} in ABAQUS) computed using 6-node triangular elements. The results obtained from MATLAB match exactly with the results from ABAQUS.

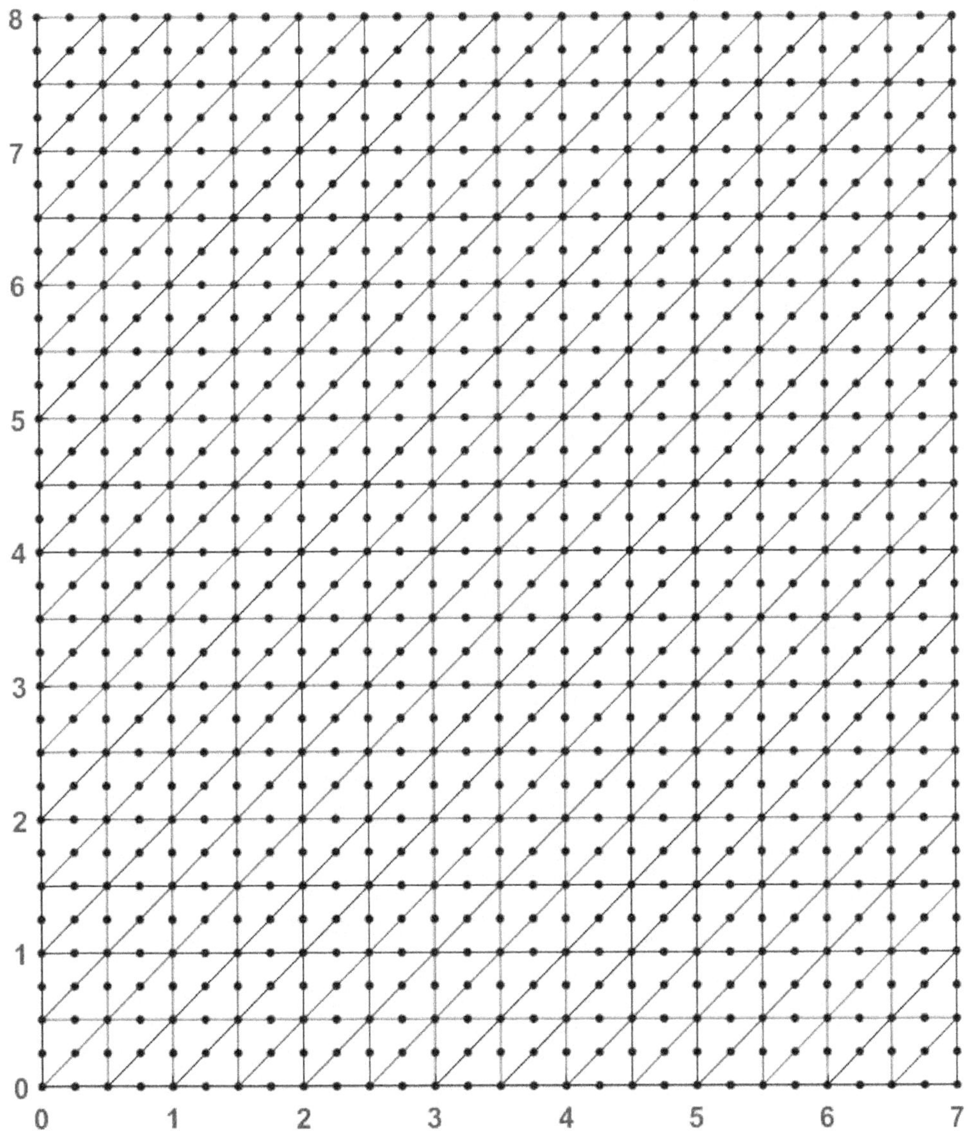

Figure 9.11 Mesh of 6-noded triangular axisymmetric elements.

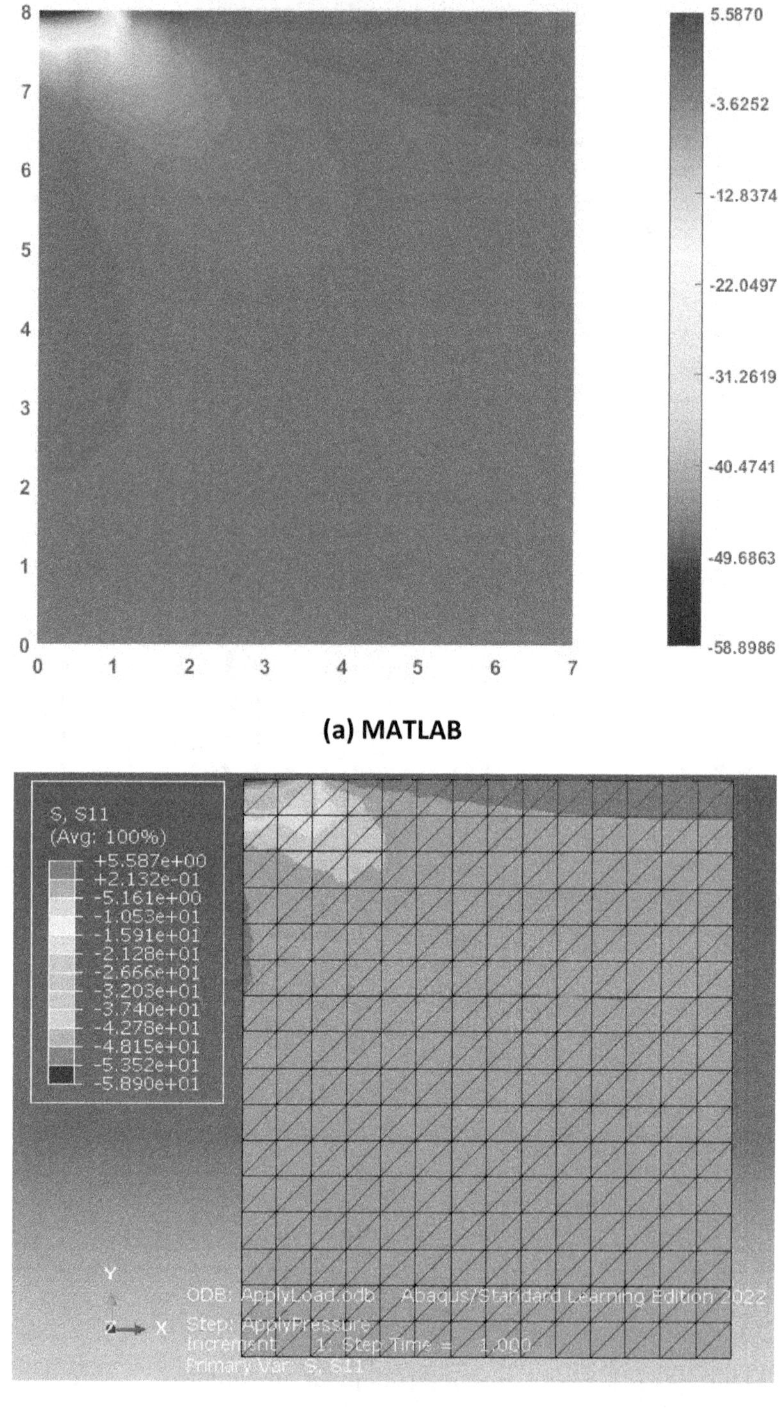

(a) MATLAB

(b) ABAQUS

Figure 9.12 Variation of S11 across the domain.

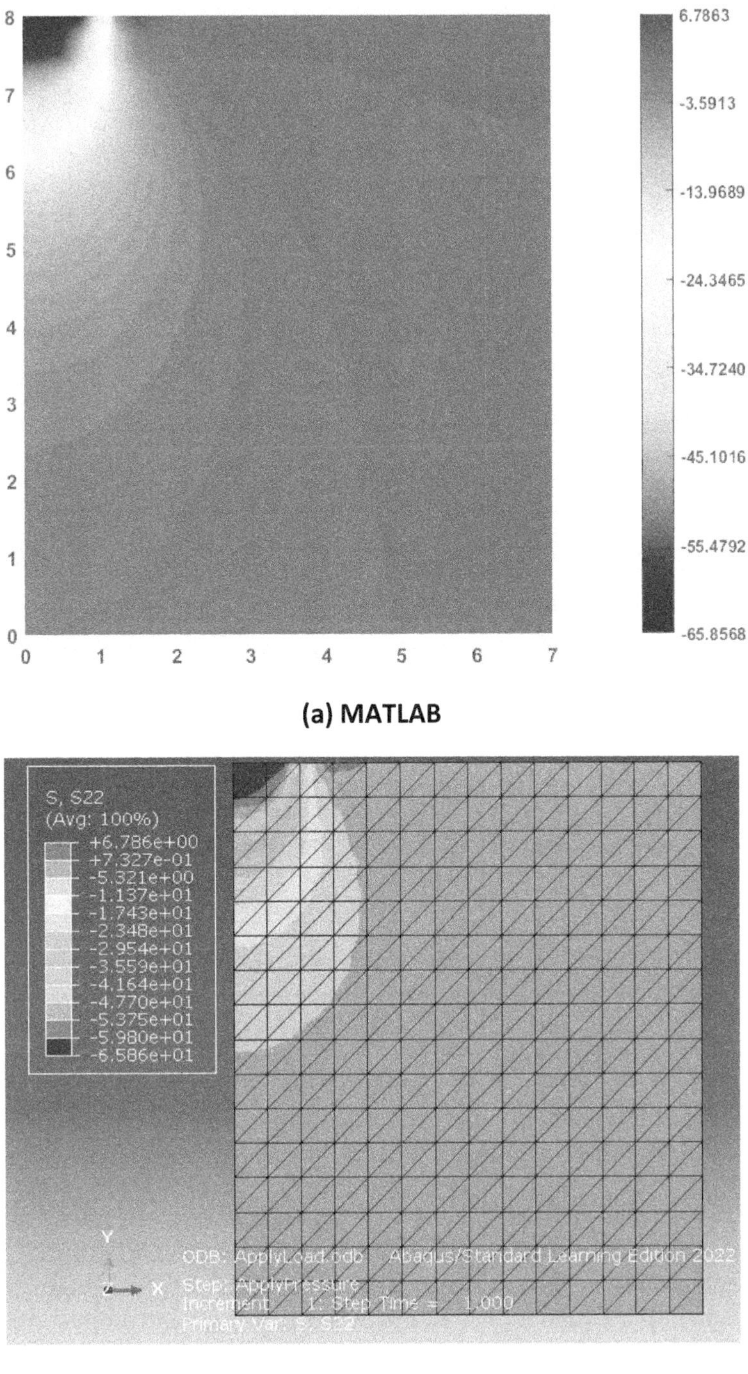

(a) MATLAB

(b) ABAQUS

Figure 9.13 Variation of S22 across the domain.

Figure 9.14 shows the mesh formed by 8 node-rectangular elements.

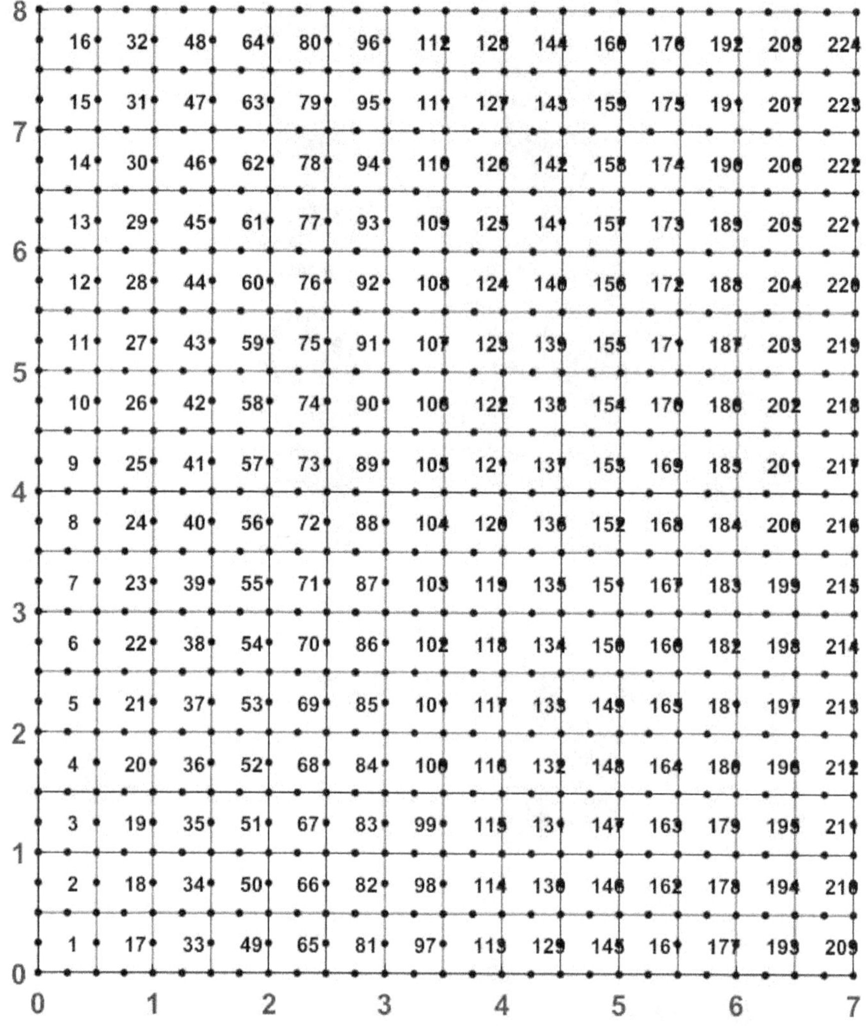

Figure 9.14 Mesh of 8 node-rectangular elements.

Values of S11, S22, S33 and S12 are obtained using Nodal stress recovery for 6-node triangular axisymmetric elements is computed using the procedure outlined in Section 7.4.2. Nodal averaging is performed to compute the average of contributions of stress received at a node from elements connected to it. Also, values of S11, S22, S33 and S12 are obtained using Nodal stress recovery outlined in Section 7.6.1 for 8-node rectangular elements. Figures 9.15, 9.16, 9.17, 9.18 represent the plots showing the variation of U1, U2, S22 and S12 across the plate. The results obtained for all displacements and forces match exactly with those obtained using ABAQUS.

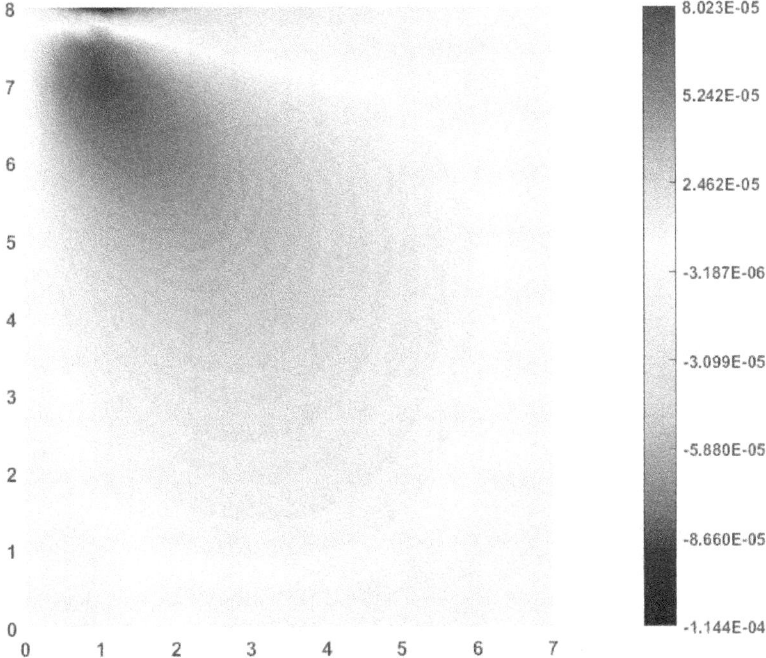

Figure 9.15 Variation of U1 across the domain.

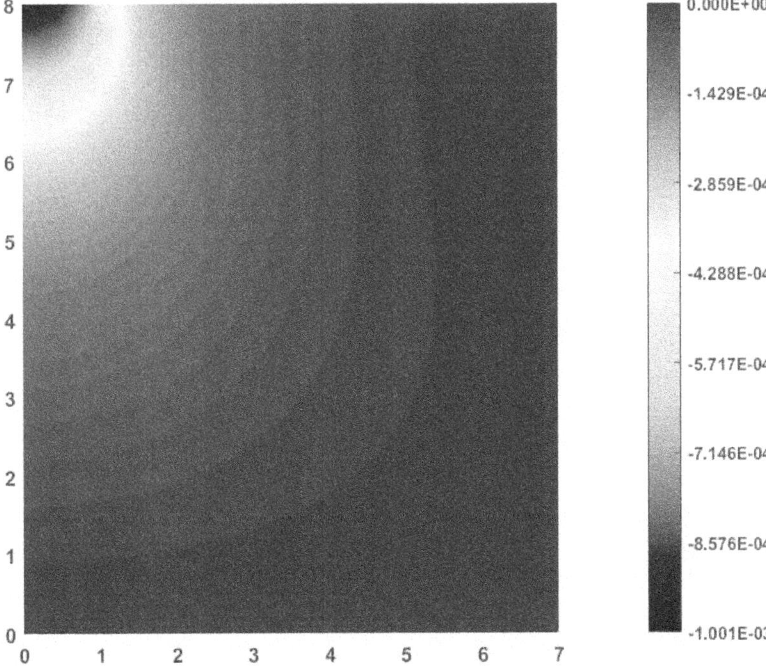

Figure 9.16 Variation of U2 across the domain.

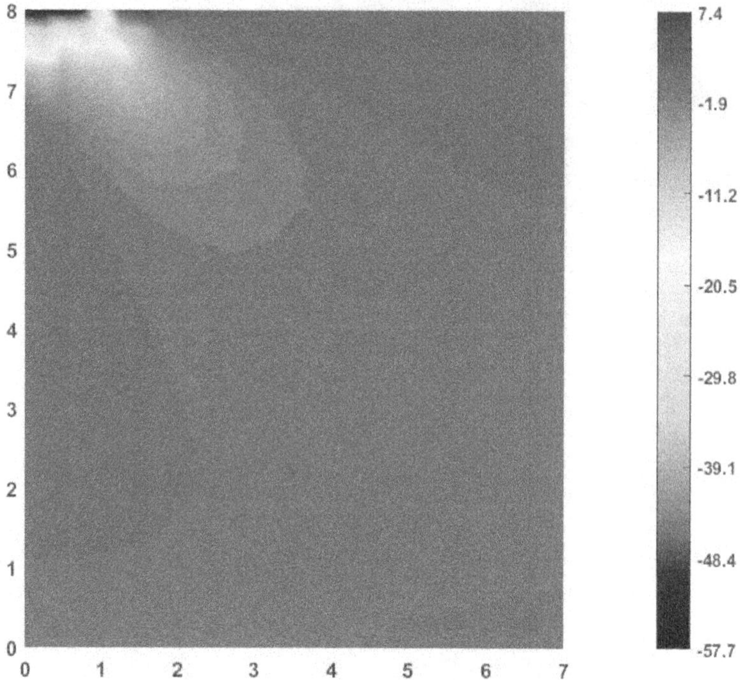

Figure 9.17 Variation of S11 across the domain.

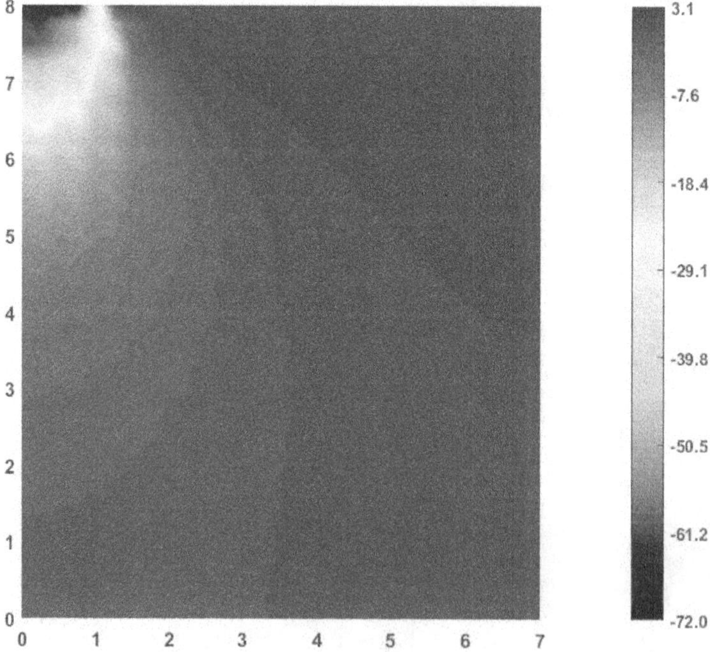

Figure 9.18 Variation of S22 across the domain.

9.6   EXERCISE PROBLEM

Consider a circular footing located on a sandy soil stratum depth 8m as shown in the Figure. The Elastic modulus and the Poisson's ratio of the soil are $E = 1 \times 10^5 \, kN/m^2$ and $\mu = 0.36$ respectively. The radius of the footing is 0.8 m and carries a load of 260 kN. Six meters beneath the footing, the soil is made up of a solid rock formation that can be considered very stiff. Assume that 8 m away from the centreline of the footing the horizontal displacement of the soil is negligible.

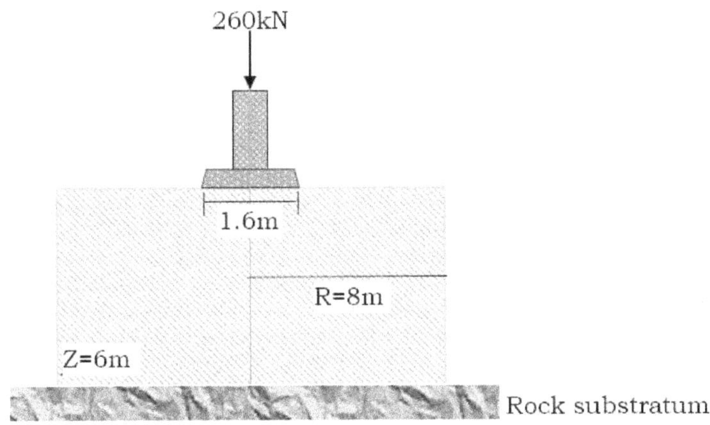

Three Dimensional Elements

Sometimes, the geometry of a structure and the applied loads are such that a general three-dimensional state of stress exists. A three-dimensional (3D) solid element is the most general of all solid finite elements because all the field variables are dependent of x, y and z. The forces acting on the 3D solid can be in any arbitrary direction in space. A 3D solid can also have any arbitrary shape, material properties and boundary conditions in space. There are altogether six possible stress components, three normal and three shear, that need to be taken into consideration.

Typically, a 3D solid element can be a tetrahedron or hexahedron in shape with either flat or curved surfaces. Each node of the element will have three translational degrees of freedom. The element can thus deform in all three directions in space. Fig. 10.1 shows a three-dimensional domain discretized into 4-node tetrahedral elements.

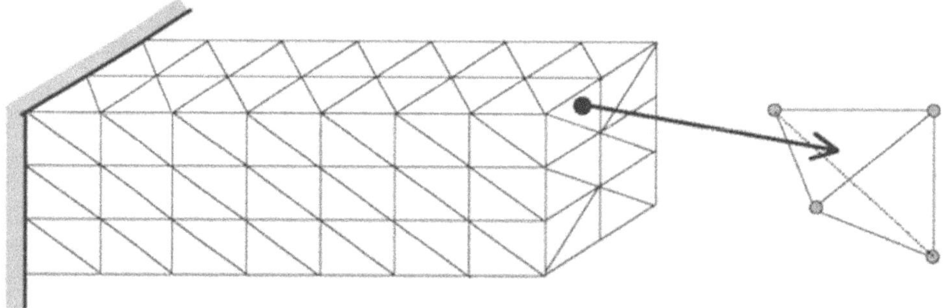

Figure 10.1 Discretization of 3D domain into 4-node tetrahedral elements.

In the general case, there are three displacement components u, v and w in the directions of the x, y, and z axes, respectively, and six strain components given by $\left(\varepsilon_x, \varepsilon_y, \varepsilon_z, \gamma_{xy}, \gamma_{yz}, \gamma_{xz}\right)$. The strain-displacement relationship in a three-dimensional continuum is defined by Eq.2.8 and Eq.2.9.

Some of the three-dimensional elements are (*Figure* 10.2):

(a)   Tetrahedral elements

(b)   Pentahedral elements

(c)   Hexahedral elements

In this chapter, the 4-node tetrahedral element and 8-node hexahedral brick element are discussed in detail.

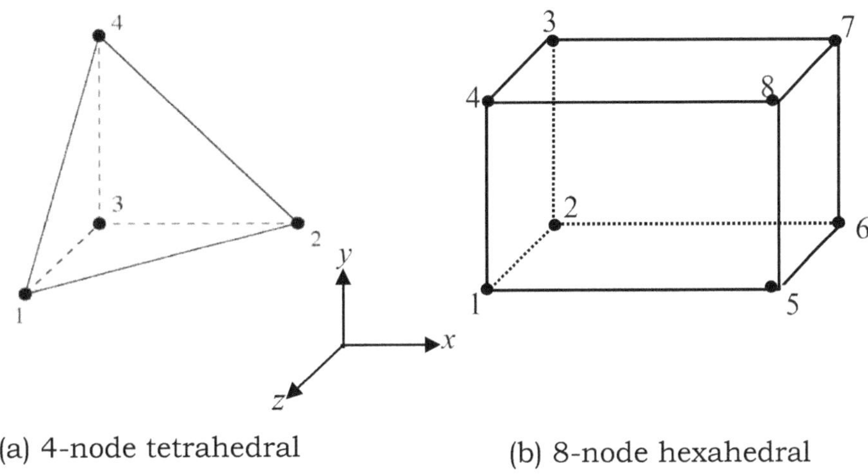

(a) 4-node tetrahedral   (b) 8-node hexahedral

Figure. 10.2 Three dimensional elements.

10.1 FOUR NODE TETRAHEDRAL ELEMENT

Figure 10.3 shows the displacements at the four nodes of a tetrahedral element. The volumetric coordinates for this element are derived in section 4.3.3. The interpolation functions for this element are given by Eq.4.62 and are reproduced below:

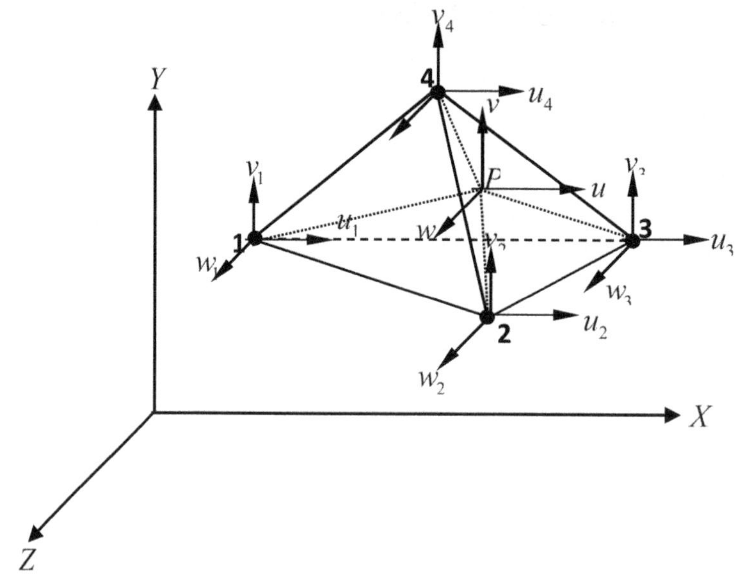

Figure 10.3 Displacements of 4-node tetrahedral element.

$$N_i = \frac{1}{6V}\left(a_i + b_i x + c_i y + d_i z\right), i = 1,4$$

$$\sum_{i=1}^{4} N_i = 1$$

$$\qquad\qquad\qquad ..... (10.1)$$

Values of a_i, b_i, c_i and d_i appearing in Eq. 10.1 are defined in Eq. 4.18.

The displacements at any point within the element are expressed in terms of nodal displacements as

$$u = u_1 N_1 + u_2 N_2 + u_3 N_3 + u_4 N_4$$
$$v = v_1 N_1 + v_2 N_2 + v_3 N_3 + v_4 N_4 \qquad .....(10.2)$$
$$w = w_1 N_1 + w_2 N_2 + w_3 N_3 + w_4 N_4$$

$$\begin{Bmatrix} u \\ v \\ w \end{Bmatrix}_{3\times 1} = \begin{bmatrix} N_1 & 0 & 0 & N_2 & 0 & 0 & N_3 & 0 & 0 & N_4 & 0 & 0 \\ 0 & N_1 & 0 & 0 & N_2 & 0 & 0 & N_3 & 0 & 0 & N_4 & 0 \\ 0 & 0 & N_1 & 0 & 0 & N_2 & 0 & 0 & N_3 & 0 & 0 & N_4 \end{bmatrix}_{3\times 12} \{d\}_{12\times 1} \qquad ....(10.3)$$

$$\{u\} = [N]\{d\}$$

where

$$\{d\} = \begin{bmatrix} u_1 & v_1 & w_1 & u_2 & v_2 & w_2 & u_3 & v_3 & w_3 & u_4 & v_4 & w_4 \end{bmatrix}^T \qquad .....(10.4)$$

The strain vector $\{\varepsilon\}$ is given by

$$\{\varepsilon\}_{6\times 1} = \begin{bmatrix} \dfrac{\partial}{\partial x} & 0 & 0 \\[2mm] 0 & \dfrac{\partial}{\partial y} & 0 \\[2mm] 0 & 0 & \dfrac{\partial}{\partial z} \\[2mm] \dfrac{\partial}{\partial y} & \dfrac{\partial}{\partial x} & 0 \\[2mm] 0 & \dfrac{\partial}{\partial z} & \dfrac{\partial}{\partial y} \\[2mm] \dfrac{\partial}{\partial z} & 0 & \dfrac{\partial}{\partial x} \end{bmatrix}_{6\times 3} \begin{Bmatrix} u \\ v \\ w \end{Bmatrix}_{3\times 1} \qquad .....(10.5)$$

Using Eq. 10.3 on the right-hand side of Eq. 10.5, we get

$$\{\varepsilon\}_{6\times 1} = [B]_{6\times 12}\{d\}_{12\times 1} \qquad (10.6)$$

The element stiffness matrix is obtained as

$$\left[K^{(e)}\right]_{12\times 12} = \iiint [B]_{12\times 6}^{T}[C]_{6\times 6}[B]_{6\times 12}\,dV \qquad (10.7)$$

In the above equation, $[C]$ matrix is given by Eq. 2.19.

10.2 EIGHT NODE BRICK ELEMENTS

As shown in Figure 10.4, the linear brick (hexahedral) element has eight nodes per element and three translational degrees of freedom per node.

The linear quadrilateral and hexahedral isoparametric elements are probably the simplest elements having constant strain within the element. However, these elements have shown some deficiency in finite element modelling where flexure is considered. The modelling accuracy can be improved by using finer mesh or by using higher order elements. The use of finer mesh resulting in large number of elements requires more computational time and computer storage. Also, the use of higher order elements having large number of unknowns, leads to large computational time.

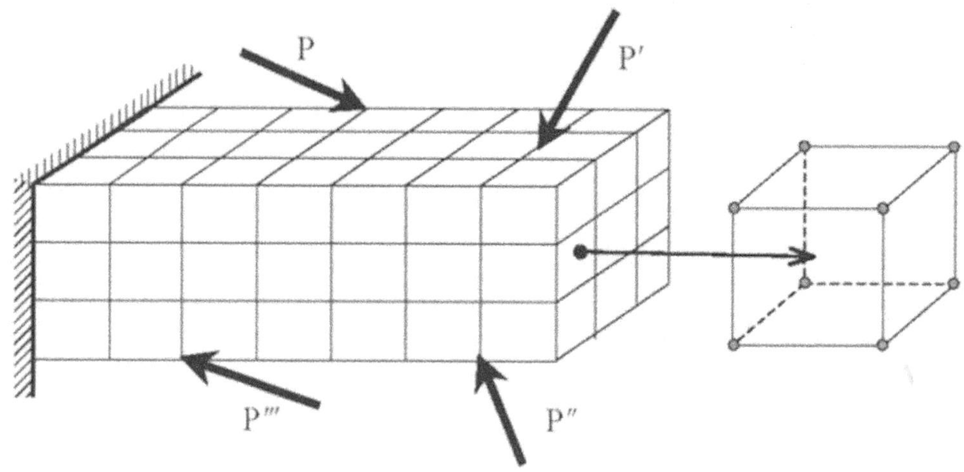

Figure 10.4 Discretization of 3D domain into 8-node hexahedral elements.

Wilson et al. (2000) improved the bending behaviour of 8 node hexahedral brick elements, by adding nine incompatible node-less displacements to the brick element. The incompatible degrees of freedom are then condensed at the element level. It was found rectangular element of this type gives better displacement under pure bending moment. This chapter describes the procedure to include incompatible modes in the analysis.

An eight-node brick element (rectangular parallelopiped) is shown in Figure 10.5 in reference to a global Cartesian coordinate system. The element is an isoparametric element in cartesian coordinate system derived from the parent element (cube) in the natural coordinates.

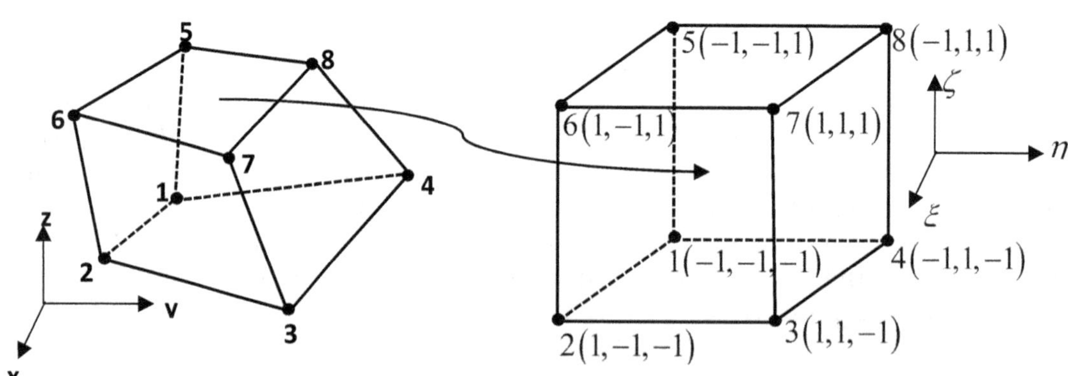

Figure 10.5 Eight node hexahedral element in cartesian coordinates.

Consider a cuboid of dimensions $2a \times 2b \times 2c$. The cartesian coordinates of the center of the cuboid can be expressed as:

$$x_C = \frac{(x_7 + x_1)}{2} \qquad y_C = \frac{(y_7 + y_1)}{2} \qquad z_C = \frac{(z_7 + z_1)}{2} \qquad \text{.....(10.8)}$$

The cartesian coordinates of the vertices of the cuboid can be normalized using the relations:

$$\xi = \frac{(x - x_c)}{a} \qquad \eta = \frac{(y - y_c)}{b} \qquad \zeta = \frac{(z - z_c)}{c} \qquad \text{.....(10.9)}$$

Using the above transformation, the vertices of the cuboid can be normalized as shown in the Figure 10.5 above.

It is observed that the cuboid represents a serendipity element as it does not have a node at its centroid. In general, the interpolation function for i[th] node with coordinates (ξ_i, η_i, ζ_i) is given by:

$$N_i = \frac{1}{8}(1 + \xi_i \xi)(1 + \eta_i \eta)(1 + \zeta_i \zeta), \ i = 1...8 \qquad \text{.....(10.10)}$$

Thus, the interpolation functions for the eight nodes of the element can be listed out as follows:

$$N_1 = \frac{1}{8}(1 - \xi)(1 - \eta)(1 - \zeta) \qquad\qquad N_2 = \frac{1}{8}(1 + \xi)(1 - \eta)(1 - \zeta)$$

$$N_3 = \frac{1}{8}(1 + \xi)(1 + \eta)(1 - \zeta) \qquad\qquad N_4 = \frac{1}{8}(1 - \xi)(1 + \eta)(1 - \zeta)$$

$$N_5 = \frac{1}{8}(1 - \xi)(1 - \eta)(1 + \zeta) \qquad\qquad N_6 = \frac{1}{8}(1 + \xi)(1 - \eta)(1 + \zeta)$$

$$N_7 = \frac{1}{8}(1 + \xi)(1 + \eta)(1 + \zeta) \qquad\qquad N_8 = \frac{1}{8}(1 - \xi)(1 + \eta)(1 + \zeta)$$

$$\text{.....(10.11)}$$

It is readily observed that these interpolation functions are linear.

10.2.1   Stiffness Matrix of Eight Node Tetrahedral Element

The cartesian coordinates (x, y, z) of a point inside the brick element can be expressed in terms of the coordinates of its vertices as

$$x = N_1x_1 + N_2x_2 + N_3x_3 + N_4x_4 + N_5x_5 + N_6x_6 + N_7x_7 + N_8x_8$$

$$y = N_1y_1 + N_2y_2 + N_3y_3 + N_4y_4 + N_5y_5 + N_6y_6 + N_7y_7 + N_8y_8 \qquad .....(10.12)$$

$$z = N_1z_1 + N_2z_2 + N_3z_3 + N_4z_4 + N_5z_5 + N_6z_6 + N_7z_7 + N_8z_8$$

This can be expressed as

$$x = \sum_{i=1}^{8} N_i x_i \qquad y = \sum_{i=1}^{8} N_i y_i \qquad z = \sum_{i=1}^{8} N_i z_i \qquad .....(10.13)$$

The derivatives of the above with reference to natural coordinates can be expressed as follows:

$$\frac{\partial x}{\partial \xi} = \sum_{i=1}^{8} \frac{\partial N_i}{\partial \xi} x_i \qquad \frac{\partial x}{\partial \eta} = \sum_{i=1}^{8} \frac{\partial N_i}{\partial \eta} x_i \qquad \frac{\partial x}{\partial \zeta} = \sum_{i=1}^{8} \frac{\partial N_i}{\partial \zeta} x_i$$

$$\frac{\partial y}{\partial \xi} = \sum_{i=1}^{8} \frac{\partial N_i}{\partial \xi} y_i \qquad \frac{\partial y}{\partial \eta} = \sum_{i=1}^{8} \frac{\partial N_i}{\partial \eta} y_i \qquad \frac{\partial y}{\partial \zeta} = \sum_{i=1}^{8} \frac{\partial N_i}{\partial \zeta} y_i \qquad .....(10.14)$$

$$\frac{\partial z}{\partial \xi} = \sum_{i=1}^{8} \frac{\partial N_i}{\partial \xi} z_i \qquad \frac{\partial z}{\partial \eta} = \sum_{i=1}^{8} \frac{\partial N_i}{\partial \eta} z_i \qquad \frac{\partial z}{\partial \zeta} = \sum_{i=1}^{8} \frac{\partial N_i}{\partial \zeta} z_i$$

Displacements can be expressed as

$$\begin{Bmatrix} u \\ v \\ w \end{Bmatrix} = \sum_{i=1}^{8} N_i \begin{Bmatrix} u_i \\ v_i \\ w_i \end{Bmatrix} \qquad .....(10.15)$$

Now the partial derivatives with respect to natural coordinates can be related to the corresponding partial derivatives with respect to cartesian coordinates as

$$\begin{Bmatrix} \dfrac{\partial}{\partial \xi} \\[2ex] \dfrac{\partial}{\partial \eta} \\[2ex] \dfrac{\partial}{\partial \zeta} \end{Bmatrix} = \begin{bmatrix} \dfrac{\partial x}{\partial \xi} & \dfrac{\partial y}{\partial \xi} & \dfrac{\partial z}{\partial \xi} \\[2ex] \dfrac{\partial x}{\partial \eta} & \dfrac{\partial y}{\partial \eta} & \dfrac{\partial z}{\partial \eta} \\[2ex] \dfrac{\partial x}{\partial \zeta} & \dfrac{\partial y}{\partial \zeta} & \dfrac{\partial z}{\partial \zeta} \end{bmatrix} \begin{Bmatrix} \dfrac{\partial}{\partial x} \\[2ex] \dfrac{\partial}{\partial y} \\[2ex] \dfrac{\partial}{\partial z} \end{Bmatrix} = [J] \begin{Bmatrix} \dfrac{\partial}{\partial x} \\[2ex] \dfrac{\partial}{\partial y} \\[2ex] \dfrac{\partial}{\partial z} \end{Bmatrix} \qquad .....(10.16)$$

$$[J] = \begin{bmatrix} \dfrac{\partial N_1}{\partial \xi} & \dfrac{\partial N_2}{\partial \xi} & \dfrac{\partial N_3}{\partial \xi} & \dfrac{\partial N_4}{\partial \xi} & \dfrac{\partial N_5}{\partial \xi} & \dfrac{\partial N_6}{\partial \xi} & \dfrac{\partial N_7}{\partial \xi} & \dfrac{\partial N_8}{\partial \xi} \\[2mm] \dfrac{\partial N_1}{\partial \eta} & \dfrac{\partial N_2}{\partial \eta} & \dfrac{\partial N_3}{\partial \eta} & \dfrac{\partial N_4}{\partial \eta} & \dfrac{\partial N_5}{\partial \eta} & \dfrac{\partial N_6}{\partial \eta} & \dfrac{\partial N_7}{\partial \eta} & \dfrac{\partial N_8}{\partial \eta} \\[2mm] \dfrac{\partial N_1}{\partial \zeta} & \dfrac{\partial N_2}{\partial \zeta} & \dfrac{\partial N_3}{\partial \zeta} & \dfrac{\partial N_4}{\partial \zeta} & \dfrac{\partial N_5}{\partial \zeta} & \dfrac{\partial N_6}{\partial \zeta} & \dfrac{\partial N_7}{\partial \zeta} & \dfrac{\partial N_8}{\partial \zeta} \end{bmatrix} \begin{bmatrix} x_1 & y_1 & z_1 \\ x_2 & y_3 & z_2 \\ x_3 & y_3 & z_3 \\ x_4 & y_4 & z_4 \\ x_5 & y_5 & z_5 \\ x_6 & y_6 & z_6 \\ x_7 & y_7 & z_7 \\ x_8 & y_8 & z_8 \end{bmatrix} \qquad .....(10.17)$$

The partial derivative with respect to cartesian coordinates can be expressed as follows:

$$\begin{Bmatrix} \dfrac{\partial}{\partial x} \\[2mm] \dfrac{\partial}{\partial y} \\[2mm] \dfrac{\partial}{\partial z} \end{Bmatrix} = [J]^{-1} \begin{Bmatrix} \dfrac{\partial}{\partial \xi} \\[2mm] \dfrac{\partial}{\partial \eta} \\[2mm] \dfrac{\partial}{\partial \zeta} \end{Bmatrix} \qquad .....(10.18)$$

$$\begin{Bmatrix} \varepsilon_x \\ \varepsilon_y \\ \varepsilon_z \\ \gamma_{xy} \\ \gamma_{yz} \\ \gamma_{xz} \end{Bmatrix} = \begin{bmatrix} \dfrac{\partial}{\partial x} & 0 & 0 \\[2mm] 0 & \dfrac{\partial}{\partial y} & 0 \\[2mm] 0 & 0 & \dfrac{\partial}{\partial z} \\[2mm] \dfrac{\partial}{\partial y} & \dfrac{\partial}{\partial x} & 0 \\[2mm] 0 & \dfrac{\partial}{\partial z} & \dfrac{\partial}{\partial y} \\[2mm] \dfrac{\partial}{\partial z} & 0 & \dfrac{\partial}{\partial x} \end{bmatrix} \begin{bmatrix} N_1 & 0 & 0 & N_2 & 0 & 0 & . & . & . & N_8 & 0 & 0 \\ 0 & N_1 & 0 & 0 & N_2 & 0 & . & . & . & 0 & N_8 & 0 \\ 0 & 0 & N_1 & 0 & 0 & N_2 & . & . & . & 0 & 0 & N_8 \end{bmatrix} \{d\}$$

$$.....(10.19)$$

$$\{\varepsilon\} = \begin{bmatrix} B_1 & B_2 & .. & B_8 \end{bmatrix} \{d\} \qquad .....(10.20)$$

The stiffness matrix can be computed as

$$\left[K^{(e)}\right] = \sum_{i=1}^{ngp}\sum_{j=1}^{ngp}\sum_{k=1}^{ngp}|J|w_i w_j w_k \left[B\right]^T \left[C\right]\left[B\right]d\xi d\eta d\zeta \qquad .....(10.21)$$

10.2.2   Incompatible Modes

To simulate adequately the flexural response, the additional displacement modes introduced to the standard eight node brick element. These additional modes are represented by functions of the type,

$$P_1 = \left(1-\xi^2\right) \quad P_2 = \left(1-\eta^2\right) \quad P_3 = \left(1-\zeta^2\right) \qquad .....(10.22)$$

These additional modes are called incompatible modes and they are not activated at the nodes of the element.

Displacement vector now includes incompatible modes as follows:

$$\begin{Bmatrix} u \\ v \\ w \end{Bmatrix} = \begin{bmatrix} N_1 & 0 & 0 & . & . & . & N_8 & 0 & 0 \\ 0 & N_1 & 0 & . & . & . & 0 & N_8 & 0 \\ 0 & 0 & N_1 & . & . & . & 0 & 0 & N_8 \end{bmatrix}\{d\} + \begin{bmatrix} P_1 & 0 & 0 & . & . & . & P_3 & 0 & 0 \\ 0 & P_1 & 0 & . & . & . & 0 & P_3 & 0 \\ 0 & 0 & P_1 & . & . & . & 0 & 0 & P_3 \end{bmatrix}\{\alpha\}$$

$$.....(10.23)$$

Stress-strain relationship gets modified as

$$\{\varepsilon\} = \begin{bmatrix} B_1 & B_2 & .. & B_8 \end{bmatrix}\{d\} + \begin{bmatrix} G_1 & G_2 & G_3 \end{bmatrix}\{\alpha\} = \begin{bmatrix} B \mid G \end{bmatrix}\begin{Bmatrix} d \\ \alpha \end{Bmatrix} \qquad .....(10.24)$$

The strain energy of the element is given by

$$U = \iiint \begin{bmatrix} \{d\}^T \mid \{\alpha\}^T \end{bmatrix}\begin{bmatrix} [B]^T \\ \hline [G]^T \end{bmatrix}[C]\begin{bmatrix} [B] \mid [G] \end{bmatrix}\begin{bmatrix} \{d\} \\ \{\alpha\} \end{bmatrix} dV \qquad .....(10.25)$$

Since there are no external forces associated with incompatible modes $\{\alpha\}$, application of principle of virtual displacements leads to

$$\begin{bmatrix} K_{uu} & K_{ua} \\ \hline K_{au} & K_{aa} \end{bmatrix}\begin{Bmatrix} \{d\} \\ \{\alpha\} \end{Bmatrix} = \begin{Bmatrix} \{F\} \\ \{0\} \end{Bmatrix} \qquad .....(10.26)$$

Displacements at the incompatible modes can be eliminated by static condensation as

$$\Big[[K_{uu}] - [K_{ua}][K_{aa}]^{-1}[K_{au}] \Big]\{d\} = \{F\} \qquad\qquad .....(10.27)$$

Displacements can be obtained by solving the above equation.

10.2.3 Numerical Example

Consider a solid cantilever beam of span $L = 3m$ and cross-sectional dimensions $b = t = 0.12\ m$ as shown in the Figure 10.6. The beam is fixed at the left face and is subjected to a point load of $W = 2.5\ kN$ in the vertically downward direction on the right face. The values of Young's modulus and Poisson's ratio are $E = 2 \times 10^8 kN / m^2$ and $\mu = 0.3$ respectively. Compute the deflection at the free end.

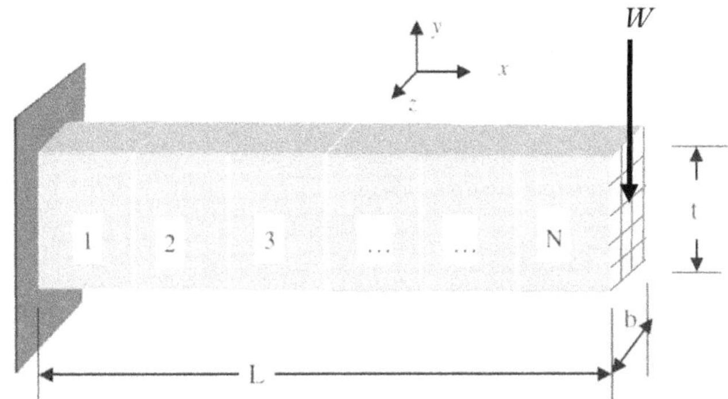

Figure 10.6 Solid cantilever beam.

The beam is discretized into N-number of 8- node hexahedral brick elements.

The beam is analysed using the MATLAB code developed and using ABAQUS. The maximum deflection on the right face using MATLAB code is 6.4017 mm while the deflection is 6.450 mm using ABAQUS.

10.3 EXERCISE PROBLEM

1. Consider a solid cantilever beam of span $L = 2m$ and cross-sectional dimensions $b = 0.2\ m$ $t = 0.3\ m$. The beam is fixed at the left face and is subjected to a point load of $W = 10.0\ kN$ in the vertically downward direction on the right face. The values of Young's modulus and Poisson's ratio are $E = 2 \times 10^8\ kN/m^2$ and $\mu = 0.3$ respectively. Compute the deflection at the free end. Use 8-node hexahedral elements.

Analysis of Plates

11.1 INTRODUCTION

Plates are structural elements that are bound by two lateral surfaces. The thickness of the plate is quite small in comparison to the lateral dimensions of the plate. One can think of a plate as being a two-dimensional equivalent of a beam in bending. Plates are also generally subject to loads normal to their plane. Plates play a major role in several important structures viz. ships, pressure vessels, building, bridge decks and other structural components. These are very important structural elements. Thus, it is important to understand their structural behaviour. The structural behaviour of thin plates in bending depends on several important factors including load, stiffness characteristics of plate and support conditions. The problem of plate bending is one of the oldest in the theory of elasticity and is discussed in several textbooks (Timoshenko and Krieger,1959, Szilard, 2004, Reddy,2006).

Based on the assumptions made with reference to the rotation of the normal to the middle plane of the plate, several theories of bending of plates have been proposed. Thin plate theory is based on the assumptions formalized by Kirchhoff in 1850 and indeed his name is often associated with this theory, though an early version was proposed by Sophie Germain in 1811. In case of the classic thin plate theory the normal remains straight and orthogonal to the middle plane after deformation.

However, a more advanced theory of thick plates was proposed by Reissner (1945) and Mindlin (1951). This theory assumes that normals remain straight, though not necessarily orthogonal to the middle plane after deformation

In this chapter, we first present the formulation of plate elements following Kirchhoff thin plate theory. Just as in the case of Euler-Bernoulli beam elements, Kirchhoff plate elements require C^1 continuity of the deflection due to the presence of second derivatives of the deflection in the virtual work expression. However, unlike beam elements, Kirchhoff plate elements have serious difficulties for satisfying the continuity requirements between elements. This leads to *non-conforming* elements, some of which can be still applied to practical situations.

Reissner-Mindlin theory of bending of thick plates is discussed in the later part of this chapter. The Reissner-Mindlin plate elements are similar to Timoshenko beam elements and thus include the effect of shear deformation. This makes them applicable for both thick and thin situations. These elements require only C^0 continuity for the deflection. Reduced integration and equivalent procedures are necessary to avoid shear locking, just as in the case of slender Timoshenko beam elements.

However, an important question that arises in the minds of the readers is to know which plate theory would be appropriate to a given situation. In fact, it is very important to learn both Kirchhoff and Reissner-Mindlin theories to gain a complete understanding of the behaviour plates in bending. Many commercial finite element analysis codes provide for Kirchhoff's plate elements. Also there has been a continued research to enhance the performance of these elements. On the other hand, Reissner-Mindlin plate elements are attractive, as they are simple to formulate and are versatile in being applicable for the analysis of both thick and thin plates. However, one needs to exercise special care and caution to utilize Reissner-Mindlin plate elements to avoid problems like shear locking or spurious mechanisms.

11.2 BENDING OF THIN PLATES – KIRCHHOFF'S PLATE THEORY

A plate is defined as a flat solid whose thickness is much smaller than its other dimensions. We assume that the middle plane is equidistant from the upper and lower faces. This plane is taken as the reference plane (z = 0) for deriving the plane kinematic equations. A plate with homogeneous isotropic material carries lateral loads by bending, like a straight beam. Hence the axial straining is zero, the middle plane coincides with the neutral plane and the displacement field can be expressed in terms of the lateral deflection and the rotations of the normal.

11.2.1   Assumptions made in Thin Plate Theory

(a) The x–y plane coincides with the middle plane of the plate in the undeformed geometry. *For the points on the x-y plane, $u = v = 0$, this means that these points only move vertically.*

(b) The lateral dimension of the plate is at least 10 times its thickness.

(c) The vertical displacement of any point of the plate can be taken equal to that of the point (below or above it) in the middle plane.

(d) A straight line normal to the undeformed middle plane remains straight and normal to the deformed middle plane (normal orthogonality condition).

(e) Strains are small: deflections are less than the order of (1/100) of the span length.

(f) The strain of the middle surface is zero or negligible.

(g) The normal stress σ_z is negligible (*plane stress assumption*).

11.2.2   Differential Equations in Thin Plate Theory

Consider the plate element as shown in Figure.11.1. The plate has larger dimensions along x and y directions. The thickness of the plate is along the z direction and is very small with reference to the lateral dimensions of the plate.

Figure 11.2 shows the deformation of the plate. As is observed from the figure, the normal passing through the middle plane is assumed to remain a straight line even after bending. But this assumption is valid only in case of small thickness of plate. As the thickness of the plate increases, the actual deformation of the normal is not linear.

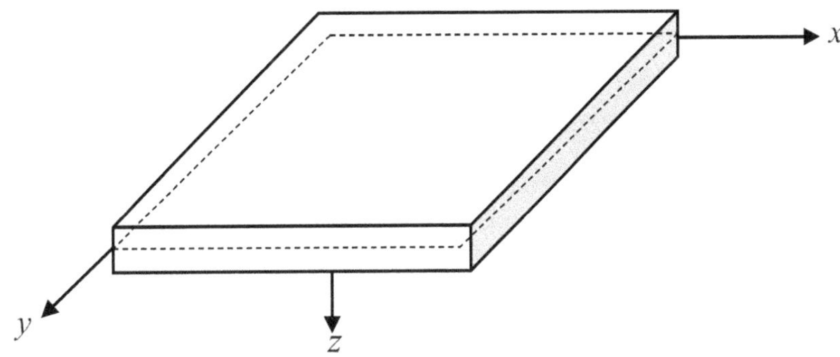

Figure 11.1 Deformed configuration of a thin plate in bending.

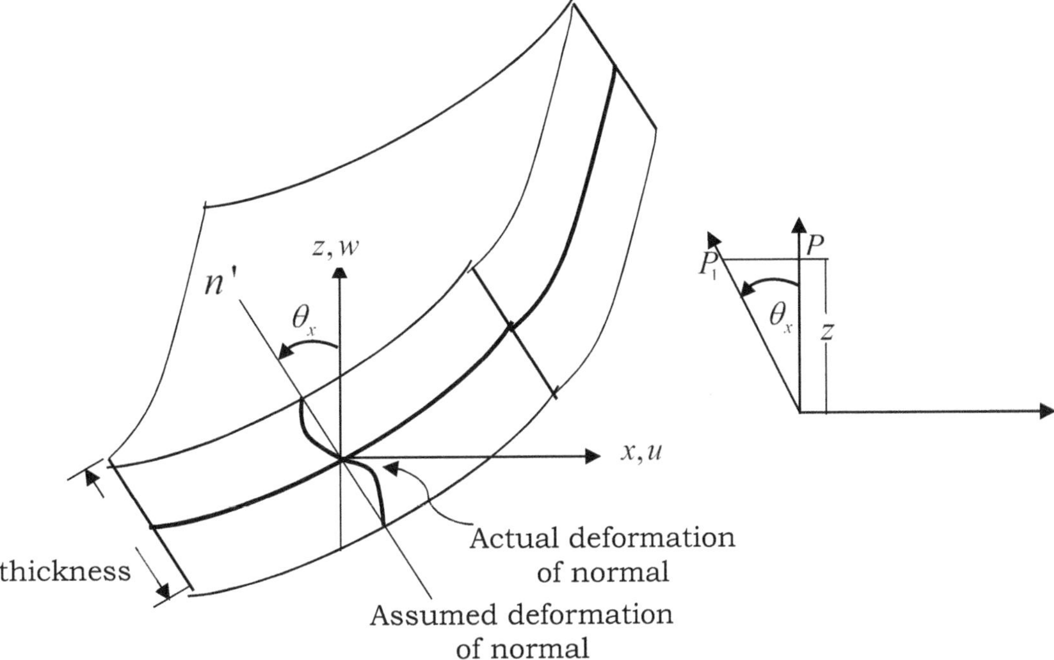

Figure 11.2 Deformation of the normal vector and in-plane deformation field in a thin plate.

$$\theta_x = \frac{\partial w}{\partial x}$$

$$PP_1 = u = -z\theta_x$$

$$u = -z\frac{\partial w}{\partial x}$$

Similarly,

$$\theta_y = \frac{\partial w}{\partial y}$$

$$v = -z\frac{\partial w}{\partial y}$$

As observed from the Figure 11.2, the displacement field can be expressed as follows:

$$\theta_x = \frac{\partial w}{\partial x}$$

$$\theta_y = \frac{\partial w}{\partial y}$$

.....(11.1a)

$$u(x, y, z) = -z\frac{\partial w(x, y)}{\partial x}$$

$$v(x, y, z) = -z\frac{\partial w(x, y)}{\partial y}$$

.....(11.1b)

$$w(x, y, z) = w(x, y)$$

It is observed from Eq.11.1 that w is the out-of-plane displacement at the mid-surface of the plate in the z-direction. Further, the in-plane displacements u and v (along x and y axes respectively) can be expressed as functions of the out-of-plane displacement w (along z-axis). Thus, the vector of displacements at any node has three components and is given as $\lfloor w \quad \theta_x \quad \theta_y \rfloor^T$ or $\left\lfloor w \quad \dfrac{\partial w}{\partial x} \quad \dfrac{\partial w}{\partial y} \right\rfloor^T$.

Owing to the assumption that "*A straight line normal to the undeformed middle plane remains straight and normal to the deformed middle plane*", the transverse shear deformation is negligible. Thus, using Eq. 11.1b, the in-plane stresses can be expressed as functions of displacements w as follows:

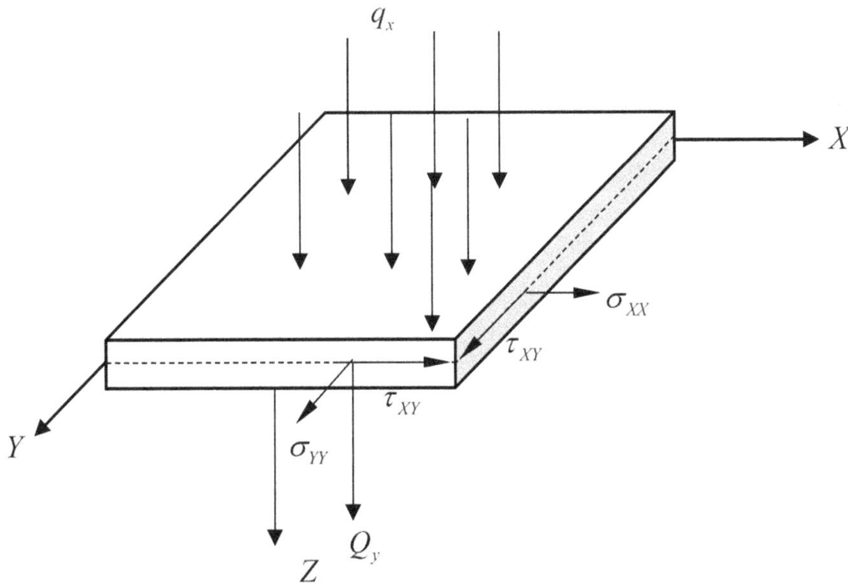

Figure 11.3 Internal stresses in a thin plate.

$$\{\varepsilon\} = \left\{ \begin{array}{c} \varepsilon_x \\ \varepsilon_y \\ \gamma_{xy} \end{array} \right\} = \left\{ \begin{array}{c} \dfrac{\partial u}{\partial x} \\ \dfrac{\partial v}{\partial y} \\ \dfrac{\partial u}{\partial y} + \dfrac{\partial v}{\partial x} \end{array} \right\} = \left\{ \begin{array}{c} -z\dfrac{\partial^2 w}{\partial x^2} \\ -z\dfrac{\partial^2 w}{\partial y^2} \\ -2\dfrac{\partial^2 w}{\partial x \partial y} \end{array} \right\} = -z \left\{ \begin{array}{c} \chi_x \\ \chi_y \\ \chi_z \end{array} \right\} \qquad (11.2)$$

where $\{\chi\} = \lfloor \chi_x \quad \chi_y \quad \chi_z \rfloor^T$ is the vector of curvatures.

Internal stresses in the plate are shown in Figure 11.3. Moments and shear forces produced are given in the Figure 11.4. Taking the thickness of the plate as t, these moments and shears can be expressed in terms of internal stresses as follows:

$$M_{xx} = \int_{-\frac{t}{2}}^{\frac{t}{2}} \sigma_{xx} z\, dz \; ; \; M_{yy} = \int_{-\frac{t}{2}}^{\frac{t}{2}} \sigma_{yy} z\, dz \; ; \; M_{xy} = \int_{-\frac{t}{2}}^{\frac{t}{2}} \tau_{xy} z\, dz \; ; \; Q_x = \int_{-\frac{t}{2}}^{\frac{t}{2}} \sigma_{xx}\, dz \; ; \; Q_y = \int_{-\frac{t}{2}}^{\frac{t}{2}} \sigma_{yy}\, dz$$

$$M_{xx} = \int_{-\frac{t}{2}}^{\frac{t}{2}} \sigma_{xx} z\, dz \; ; \; M_{yy} = \int_{-\frac{t}{2}}^{\frac{t}{2}} \sigma_{yy} z\, dz \; ; \; M_{xy} = \int_{-\frac{t}{2}}^{\frac{t}{2}} \tau_{xy} z\, dz \qquad \dots(11.3)$$

$$Q_x = \int_{-\frac{t}{2}}^{\frac{t}{2}} \sigma_{xx}\,dz \;;\; Q_y = \int_{-\frac{t}{2}}^{\frac{t}{2}} \sigma_{yy}\,dz \qquad \qquad \dots(11.4)$$

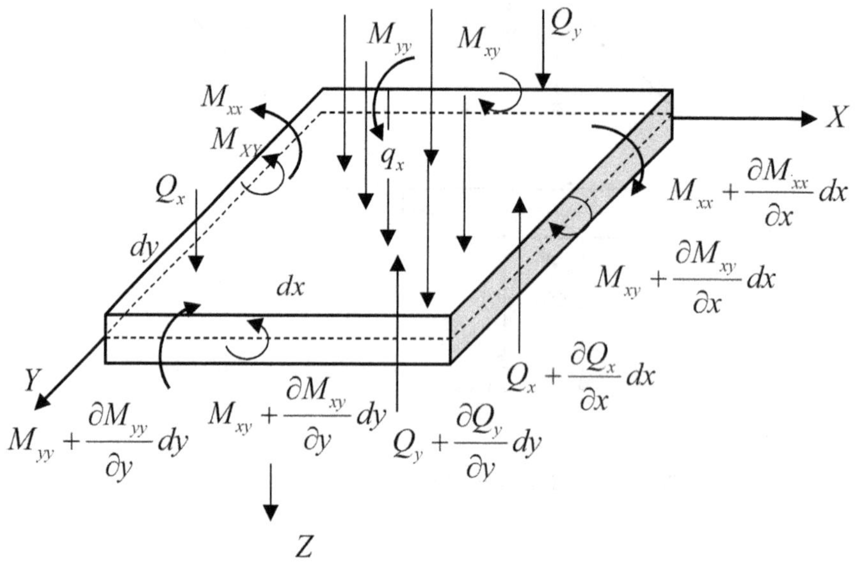

Figure 11.4 Forces and moments in a thin plate.

Assuming a state of plane stress, the stress vector $\{\sigma\}$ is given by

$$\{\sigma\} = [C]\{\varepsilon\} \qquad \qquad \dots(11.5)$$

where
$$[C] = \frac{E}{\left(1-\mu^2\right)} \begin{bmatrix} 1 & \mu & 0 \\ \mu & 1 & 0 \\ 0 & 0 & \dfrac{(1-\mu)}{2} \end{bmatrix} \qquad \qquad \dots(11.6)$$

is the constitutive matrix for plane stress condition. Substituting Eq. 11.2 and Eq. 11.6 in Eq. 11.5, we obtain the following relationship.

$$\{\sigma\} = -z[C]\{\chi\} \qquad \qquad \dots(11.7)$$

Substituting Eq. 11.6 in Eq.11.3 we obtain

$$\{M\} = \begin{bmatrix} M_{xx} \\ M_{yy} \\ M_{xy} \end{bmatrix} = \frac{t^3}{12}[C]\{\chi\} \qquad \qquad \dots(11.8)$$

Considering the equilibrium of the plate shown in the Figure 11.4

$$-Q_x dy - Q_y dx + \left(Q_x + \frac{\partial Q_x}{\partial x}dx\right)dy + \left(Q_y + \frac{\partial Q_y}{\partial y}dy\right)dx + q(x,y)dxdy = 0 \quad .....(11.9a)$$

$$\frac{\partial Q_x}{\partial x} + \frac{\partial Q_y}{\partial y} + q(x,y) = 0 \qquad .....(11.9b)$$

Considering the moment equilibrium about x and y axes leads to

$$\frac{\partial M_{xy}}{\partial x} + \frac{\partial M_{yy}}{\partial y} = Q_y$$

$$\frac{\partial M_{xy}}{\partial y} + \frac{\partial M_{xx}}{\partial x} = Q_x$$
$$(11.10)$$

Substituting Eq. 11.10 in Eq. 11.9 yields

$$\frac{\partial^2 M_{xx}}{\partial x^2} + \frac{\partial^2 M_{xy}}{\partial x \partial y} + \frac{\partial^2 M_{yy}}{\partial y^2} + q(x,y) = 0 \qquad .....(11.11)$$

11.2.3   Governing Equations for Bending of Thin Plates

Substituting Eq.11.7 into Eq.11.11 leads to

$$\frac{\partial^4 w}{\partial x^4} + \frac{\partial^4 w}{\partial x^2 \partial y^2} + \frac{\partial^4 w}{\partial y^4} = \frac{q(x,y)}{D_r}$$

$$\nabla^4 w = \frac{q}{D_r}$$
$$.....(11.12)$$

where $D_r = \dfrac{Et^3}{12(1-\mu^2)}$ is the bending stiffness of the plate.

Solution of Eq. 11.12 is dependent of on finding suitable w that can satisfy Eq. 11.12 and also on boundary conditions of the plate.

11.3 FINITE ELEMENT FORMULATION OF THIN PLATES

For thin plates based on Kirchhoff's theory, plate elements require C[1] continuity. This means that w and its derivatives up to the second order shall be continuous and

$$\frac{\partial^2 w}{\partial y \partial x} = \frac{\partial^2 w}{\partial x \partial y} \qquad .....(11.13)$$

In case of a conventional plate element, there are three degrees of freedom at each node viz. one translation w along the z direction and two rotations θ_x and θ_y (about x and y axes respectively). In case of small deformations,

$$\theta_x = \frac{\partial w}{\partial x}$$

$$\theta_y = \frac{\partial w}{\partial y}$$

.....(11.14)

where rotation components θ_x and θ_y are continuous functions of x and y, otherwise kinks will develop owing to discontinuity of slopes. Also, it is to be noted that θ_x and M_x represent the rotation and moment about y axis. Similarly, θ_y and M_y represent the rotation and moment about x axis.

11.3.1 Three Noded Triangular Plate Elements

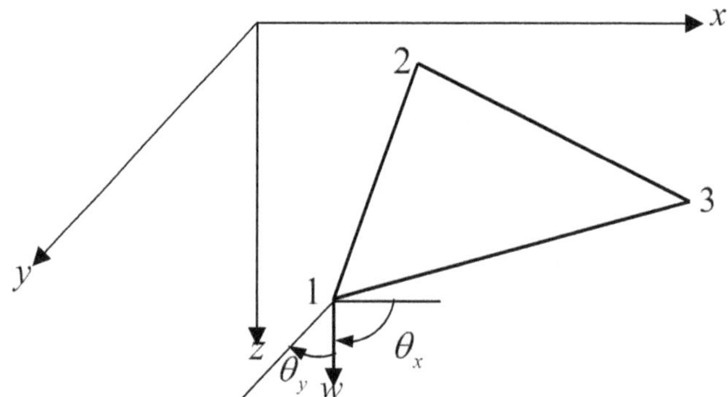

Figure 11.5 Three noded triangular plate element.

Consider a three node triangular plate element shown in Figure 11.5. Each node has three degrees of freedom viz. one z-translation, x-rotation and y-rotation. Thus, there are nine degrees of freedom for the element. Accordingly, the interpolation function for the vertical displacement w shall consist of nine terms and can be written as

$$w(x, y) = \alpha_0 + \alpha_1 x + \alpha_2 y + \alpha_3 xy + \alpha_4 x^2 + \alpha_5 xy + \alpha_6 y^2 + \alpha_7 x^3 + \alpha_8 (x^2 y + xy^2) + \alpha_9 y^3$$

$$w = \lfloor \phi \rfloor \{\alpha\}$$

.....(11.15)

However, a complete polynomial in x and y contains 10 terms. It can be observed from Eq. 11.15 that $x^2 y$ and xy^2 terms are combined to get 9 terms in the expression.

$$\frac{\partial w}{\partial x} = \alpha_2 + 2\alpha_4 x + \alpha_5 y + 3\alpha_7 x^2 + \alpha_8 \left(2xy + y^2\right)$$

$$\frac{\partial w}{\partial y} = \alpha_3 + \alpha_5 x + 2\alpha_6 y + \alpha_8 \left(x^2 + 2xy\right) + 3\alpha_9 x^2$$

$$\quad\quad\quad .....(11.16)$$

$$\begin{Bmatrix} w \\ \theta_x \\ \theta_y \end{Bmatrix} = \begin{bmatrix} 1 & x & y & x^2 & xy & y^2 & x^3 & x^2 y + xy^2 & y^3 \\ 0 & 1 & 0 & 2x & y & 0 & 3x^2 & 2xy + y^2 & 0 \\ 0 & 0 & 1 & 0 & x & 2y & 0 & x^2 + 2xy & 3y^2 \end{bmatrix} \{\alpha\}$$

$$\quad\quad\quad .....(11.17)$$

We now substitute the following nodal values at each node in Eq. 11.15 and Eq. 11.16 as follows:

Node 1: $w = w_1, \theta_x = \theta_{x1}, \theta_y = \theta_{y1}$

Node 2: $w = w_2, \theta_x = \theta_{x2}, \theta_y = \theta_{y2}$

Node 3: $w = w_3, \theta_x = \theta_{x3}, \theta_y = \theta_{y3}$

Substitution will lead to the following system of equations:

$$\begin{Bmatrix} w_1 \\ \theta_{x1} \\ \theta_{y1} \\ w_2 \\ \theta_{x2} \\ \theta_{y2} \\ w_3 \\ \theta_{x3} \\ \theta_{y2} \end{Bmatrix} = \begin{bmatrix} 1 & x_1 & y_1 & x_1^2 & x_1 y_1 & y_1^2 & x_1^3 & x_1^2 y_1 + x_1 y_1^2 & y_1^3 \\ 0 & 1 & 0 & 2x_1 & y_1 & 0 & 3x_1^2 & 2x_1 y_1 + y_1^2 & 0 \\ 0 & 0 & 1 & 0 & x_1 & 2y_1 & 0 & x_1^2 + 2x_1 y_1 & 3y_1^2 \\ 1 & x_2 & y_2 & x_2^2 & x_2 y_2 & y_2^2 & x_2^3 & x_2^2 y_2 + x_2 y_2^2 & y_2^3 \\ 0 & 1 & 0 & 2x_2 & y_2 & 0 & 3x_2^2 & 2x_2 y_2 + y_2^2 & 0 \\ 0 & 0 & 1 & 0 & x_2 & 2y_2 & 0 & x_2^2 + 2x_2 y_2 & 3y_2^2 \\ 1 & x_3 & y_3 & x_3^2 & x_3 y_3 & y_3^2 & x_3^3 & x_3^2 y_3 + x_3 y_3^2 & y_3^3 \\ 0 & 1 & 0 & 2x_3 & y_3 & 0 & 3x_3^2 & 2x_3 y_3 + y_3^2 & 0 \\ 0 & 0 & 1 & 0 & x_3 & 2y_3 & 0 & x_3^2 + 2x_3 y_3 & 3y_2^2 \end{bmatrix} \begin{Bmatrix} \alpha_1 \\ \alpha_2 \\ \alpha_3 \\ \alpha_4 \\ \alpha_5 \\ \alpha_6 \\ \alpha_7 \\ \alpha_8 \\ \alpha_9 \end{Bmatrix}$$

$$\quad\quad\quad .....(11.18)$$

$$\{d\} = [A]\{\alpha\} \quad\quad\quad ......(11.19)$$

Inverting the relationship, we obtain,

$$\{\alpha\} = [A]^{-1}\{d\} \qu\quad\quad .....(11.20)$$

Substituting Eq. 11.20 in Eq. 11.

$$w = \lfloor \phi \rfloor [A]^{-1}\{d\} = [N]\{d\} \quad\quad\quad .....(11.21)$$

where $\lfloor N \rfloor = \lfloor \phi \rfloor [A]^{-1}$. Substituting Eq. 11.21 in Eq.11.2, we obtain

$$\{\varepsilon\} = \begin{Bmatrix} \varepsilon_x \\ \varepsilon_y \\ \gamma_{xy} \end{Bmatrix} = -z \begin{Bmatrix} \dfrac{\partial^2 w}{\partial x^2} \\ \dfrac{\partial^2 w}{\partial y^2} \\ 2\dfrac{\partial^2 w}{\partial x \partial y} \end{Bmatrix} = -z \begin{bmatrix} 0 & 0 & 0 & 2 & 0 & 0 & 6x & 2y & 0 \\ 0 & 0 & 0 & 0 & 0 & 2 & 0 & 2x & 6y \\ 0 & 0 & 0 & 0 & 2 & 0 & 0 & 4(x+y) & 0 \end{bmatrix} [A]^{-1}\{d\}$$

$$\ldots\ldots(11.22)$$

$$\{\varepsilon\} = [B]\{d\} \qquad\qquad \ldots\ldots(11.23)$$

where

$$[B] = -z \begin{bmatrix} 0 & 0 & 0 & 2 & 0 & 0 & 6x & 2y & 0 \\ 0 & 0 & 0 & 0 & 0 & 2 & 0 & 2x & 6y \\ 0 & 0 & 0 & 0 & 2 & 0 & 0 & 4(x+y) & 0 \end{bmatrix} [A]^{-1} \qquad \ldots\ldots(11.24)$$

Stiffness matrix is obtained as follows:

$$\left[K^{(e)} \right] = \int_{A_e} \int_{z=-\frac{t}{2}}^{z=+\frac{t}{2}} [B]^T [D][B] \, dz \, dA \qquad\qquad \ldots\ldots(11.25)$$

11.3.2 Four Noded Rectangular Plate Elements

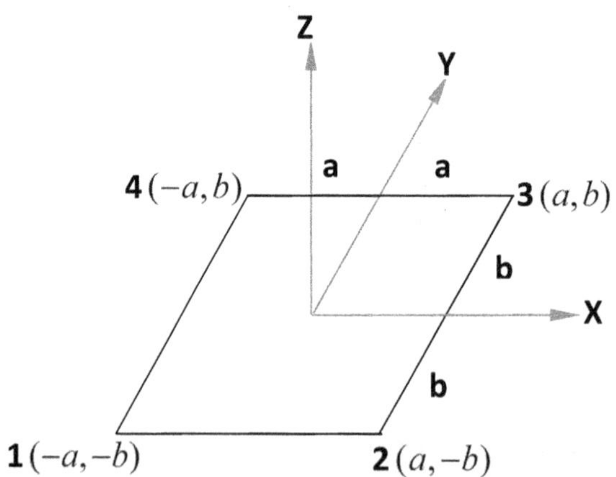

Figure 11.6 Rectangular element with 12 degrees of freedom.

Consider a four noded rectangular plate element of dimensions $2a \times 2b$ shown in Figure 11.6. Each node has three degrees of freedom viz. one z-translation,

x-rotation and y-rotation. Thus, there are twelve degrees of freedom for the element. Accordingly, the interpolation function for the vertical displacement w shall consist of nine terms and can be written as

$$w(x, y) = \alpha_1 + \alpha_2 x + \alpha_3 y + \alpha_4 x^2 + \alpha_5 xy + \alpha_6 y^2 + \alpha_7 x^3 + \alpha_8 x^2 y + \alpha_9 xy^2$$
$$+ \alpha_{10} y^3 + \alpha_{11} x^3 y + \alpha_{12} xy^3$$

.....(11.26)

$$w = \lfloor \phi \rfloor \{\alpha\}$$

.....(11.27)

$$\begin{Bmatrix} w \\ \theta_x \\ \theta_y \end{Bmatrix} = \begin{bmatrix} 1 & x & y & x^2 & xy & y^2 & x^3 & x^2 y & xy^2 & y^3 & x^3 y & xy^3 \\ 0 & 1 & 0 & 2x & y & 0 & 3x^2 & 2xy & y^2 & 0 & 3x^2 y & y^3 \\ 0 & 0 & 1 & 0 & x & 2y & 0 & x^2 & 2xy & 3y^2 & x^3 & 3xy^2 \end{bmatrix} \{\alpha\}$$

...(11.28)

It is observed from Eq.11.28 that w and its derivatives are defined as cubic polynomials in x and y. It is to be noted that a cubic polynomial is defined uniquely by 4 constants. Thus, displacements along any boundary can be defined uniquely by the two end values of displacements and slopes at each node. However, in case of slopes, only two end values of slopes exist and thus the cubic polynomial is not defined uniquely. Thus, leads to a discontinuity of normal slopes in general. Thus, the interpolation function is called *non-conforming* and the rectangular plate element is called a *non-conforming* element.

The nodal displacements $\lfloor w \quad \theta_x \quad \theta_y \rfloor$ take the values $\lfloor w_1 \quad \theta_{x1} \quad \theta_{y1} \rfloor$, $\lfloor w_2 \quad \theta_{x2} \quad \theta_{y2} \rfloor$, $\lfloor w_3 \quad \theta_{x3} \quad \theta_{y3} \rfloor$ and $\lfloor w_4 \quad \theta_{x4} \quad \theta_{y4} \rfloor$ at nodes 1, 2, 3 and 4 respectively. We now substitute the following nodal coordinates at each node in Eq. 11.28 to obtain the following system of equations:

$$\begin{Bmatrix} w_1 \\ \theta_{x1} \\ \theta_{y1} \\ w_2 \\ \theta_{x2} \\ \theta_{y2} \\ w_3 \\ \theta_{x3} \\ \theta_{y3} \\ w_4 \\ \theta_{x4} \\ \theta_{y4} \end{Bmatrix} = \begin{bmatrix} 1 & -a & -b & a^2 & ab & b^2 & -a^3 & -a^2b & -ab^2 & -b^3 & a^3b & ab^3 \\ 0 & 1 & 0 & -2a & -b & 0 & 3a^2 & 2ab & b^2 & 0 & -3a^2b & -b^3 \\ 0 & 0 & 1 & 0 & -a & -2b & 0 & a^2 & 2ab & 3b^2 & -a^3 & -3ab^2 \\ 1 & a & -b & a^2 & -ab & b^2 & a^3 & -a^2b & ab^2 & -b^3 & -a^3b & -ab^3 \\ 0 & 1 & 0 & 2a & -b & 0 & 3a^2 & -2ab & b^2 & 0 & 3a^2b & -b^3 \\ 0 & 0 & 1 & 0 & a & -2b & 0 & a^2 & -2ab & 3b^2 & a^3 & 3ab^2 \\ 1 & a & b & a^2 & ab & b^2 & a^3 & a^2b & ab^2 & b^3 & a^3b & ab^3 \\ 0 & 1 & 0 & 2a & b & 0 & 3a^2 & 2ab & b^2 & 0 & 3a^2b & b^3 \\ 0 & 0 & 1 & 0 & a & 2b & 0 & a^2 & 2ab & 3b^2 & a^3 & 3ab^2 \\ 1 & -a & b & a^2 & -ab & b^2 & -a^3 & a^2b & ab^2 & -b^3 & -a^3b & -ab^3 \\ 0 & 1 & 0 & -2a & b & 0 & 3a^2 & -2ab & b^2 & 0 & 3a^2b & -b^3 \\ 0 & 0 & 1 & 0 & -a & 2b & 0 & a^2 & -2ab & 3b^2 & -a^3 & -3ab^2 \end{bmatrix} \{\alpha\}$$

$$\text{.....(11.29)}$$

$$\{d\} = [A]\{\alpha\} \qquad \text{.....(11.30)}$$

Inverting the relationship, we obtain,

$$\{\alpha\} = [A]^{-1}\{d\} \qquad \text{.....(11.31)}$$

Substituting Eq. 11.31 in Eq. 11.27

$$w = \lfloor \phi \rfloor [A]^{-1}\{d\} = [N]\{d\} \qquad \text{......(11.32)}$$

where $\lfloor N \rfloor = \lfloor \phi \rfloor [A]^{-1}$ is the shape function matrix. Substituting Eq. 11.32 in Eq.11.2 , we obtain

$$\{\varepsilon\} = [B]\{d\} \qquad \text{.....(11.33)}$$

where

$$[B] = -z \begin{bmatrix} 0 & 0 & 0 & -2 & 0 & 0 & -6x & -2y & 0 & 0 & -6xy & 0 \\ 0 & 0 & 0 & 0 & 0 & -2 & 0 & 0 & -2x & -6y & 0 & -6xy \\ 0 & 0 & 0 & 0 & -2 & 0 & 0 & -4x & -4y & 0 & -6x^2 & -6y^2 \end{bmatrix}[A]^{-1}$$

$$\text{.....(11.34)}$$

Stiffness matrix is obtained as follows:

$$\left[K^{(e)} \right] = \iiint_V [B]^T [C][B] \, dV \qquad (11.35)$$

For a plate of uniform thickness h, Eq. 11.35 can be written as

$$\left[K^{(e)}\right] = h \iint [B]^T [C][B] dA \tag{11.36}$$

The element stiffness matrix $\left[K^{(e)}\right]$ can be evaluated numerically using Gaussian quadrature as follows:

$$\left[K^{(e)}\right] = h \int_{\xi=-1}^{1} \int_{\eta=-1}^{1} \left[B(\xi,\eta)^T\right][D]\left[B(\xi,\eta)\right]|J| d\xi d\eta$$

$$= h \sum_{i=1}^{ngp} \sum_{j=1}^{ngp} W_i W_j \left[B(\xi_i,\eta_j)^T\right][D]\left[B(\xi_i,\eta_j)\right]|J| \tag{11.37}$$

where ngp represents the number of Gauss integration points.

The curvatures of the plate element viz. κ_{xx}, κ_{yy} and κ_{xy} are obtained from the second order partial differentiation of $w(x,y)$ w.r.t x and y as follows:

$$\begin{Bmatrix} \kappa_x \\ \kappa_y \\ \kappa_{xy} \end{Bmatrix} = - \begin{Bmatrix} \dfrac{\partial^2 w}{\partial x^2} \\ \dfrac{\partial^2 w}{\partial y^2} \\ \dfrac{\partial^2 w}{\partial x \partial x} \end{Bmatrix} = \begin{bmatrix} 0 & 0 & 0 & -2 & 0 & 0 & -6x & -2y & 0 & 0 & -6xy & 0 \\ 0 & 0 & 0 & 0 & 0 & -2 & 0 & 0 & -2x & -6y & 0 & -6xy \\ 0 & 0 & 0 & 0 & -2 & 0 & 0 & -4x & -4y & 0 & -6x^2 & -6y^2 \end{bmatrix}$$

$$\dots\dots(11.38)$$

Thus, the vector of curvatures $\left\{\kappa^{(e)}\right\}$ for the plate element can be expressed as

$$\left\{\kappa^{(e)}\right\} = \begin{Bmatrix} \kappa_x \\ \kappa_y \\ \kappa_{xy} \end{Bmatrix} = [B][\Phi]^{-1}\{d\} \tag{11.39}$$

From the moment-curvature relationship we obtain

$$\begin{Bmatrix} M_x \\ M_y \\ M_{xy} \end{Bmatrix} = [D] \begin{Bmatrix} \kappa_x \\ \kappa_y \\ \kappa_{xy} \end{Bmatrix} = [D][B][\Phi]^{-1}\{d\} \tag{11.40}$$

The force vector is given as

$$\left\{Q^{(e)}\right\} = \iiint \{N\}^T \{X\} dV + \iint_{S_1} \{N_S\}^T \{f\} dA + \{Q\} \tag{11.41}$$

The first, second and third terms in the above equation correspond to the vector of body forces, vector of surface tractions and point loads, respectively. Considering the vector of body forces,

$$[X] = [X_b \quad Y_b \quad Z_b]^T = [0 \quad -\rho g \quad 0]^T \qquad \dots\dots(11.42)$$

$$\iiint \{N\}^T \{X\} dV = -4\rho ghab \left[\frac{1}{4} \quad \frac{a}{12} \quad \frac{b}{12} \quad \frac{1}{4} \quad -\frac{a}{12} \quad \frac{b}{12} \quad \frac{1}{4} \quad -\frac{a}{12} \quad -\frac{b}{12} \quad \frac{1}{4} \quad \frac{a}{12} \quad -\frac{b}{12} \right]^T \qquad \dots\dots(11.43)$$

Considering the vector of surface traction, if q is the normal pressure acting on the plate, $\{f\} = \lfloor 0 \quad 0 \quad q \rfloor^T$,

$$\iint_{S_1} \{N_S\}^T \{f\} dA = 4qab \left[\frac{1}{4} \quad \frac{a}{12} \quad \frac{b}{12} \quad \frac{1}{4} \quad -\frac{a}{12} \quad \frac{b}{12} \quad \frac{1}{4} \quad -\frac{a}{12} \quad -\frac{b}{12} \quad \frac{1}{4} \quad \frac{a}{12} \quad -\frac{b}{12} \right]^T \qquad \dots\dots(11.44)$$

Any point loads as given in the third term of Eq.11.41 can be applied normal to the plate at specified points.

11.3.3 Boundary Conditions

Free edge: No boundary conditions need to specified.

Clamped edge: $w = 0$

Note that as θ_x and θ_y are functions of w, setting w to zero automatically sets θ_x and θ_y to zero.

Simply supported edge:

On *edge* $x = a$, $w = 0$ *(essential boundary condition) and*

$$\frac{\partial^2 w}{\partial x^2} = 0 \text{ (natural boundary condition)}.$$

However, in FEM only essential boundary condition needs to be specified.

11.3.4 Description of the Program ACM Plate.m

A MATLAB program titled **ACMplate.m** is given in section A.12 for the analysis of a rectangular plate using the 12 degrees of freedom rectangular element outlined above. The following are the important steps involved in the program:

(a) *The following data is input:* Length, width and thickness of plate, number of divisions along x and y axes, Young's modulus, Poisson's ratio, self-weight of the plate, normal pressure on the plate and number of loaded nodes.

(b) A MATLAB script titled *Q4mesh*.m is called to create a rectangular mesh of given length and width and given number of divisions along x and y axes. Node numbers and element numbers are generated and the mesh is drawn.

(c) Symbolic logic toolbox is utilized to define the displacement function $w(x, y)$ as per Eq.11.26.

(d) $[A]$ matrix as given in Eq. 11.29 and Eq.11.30 is generated.

(e) Shape function matrix $[N]$ is generated using Eq.11.32 and is utilized to compute the $[B]$ matrix as per Eq.11.34

(f) Element stiffness matrix $\left[K^{(e)}\right]$ is computed using Gaussian quadrature using Eq.11.37.

(g) Element force vector $\left\{F^{(e)}\right\}$ is computed as per Eq.11.41, Eq.11.43 and Eq.11.44

(h) Assembly is performed to obtain the global stiffness matrix $[K]$ and force vector $\{F\}$.

(i) Boundary conditions are applied and the resulting system of equations are solved to obtain the vector of displacements and rotations.

(j) Moments are computed using Eq. 11.40

(k) Finally surface plots are made to depict the variation of vertical displacement, rotations and moments.

11.3.5   Numerical Examples

Numerical Example - 1

Analyse the square plate of side 2 m as shown in Figure 11.7, using four noded rectangular plate elements under conditions of plane stress. The thickness of the plate is 0.05m, Young's modulus is $2.1\times10^8\ kN/mm^2$ and $\mu=0.3$. The plate is subjected to a pressure of 100.0 kPa normal to its surface.

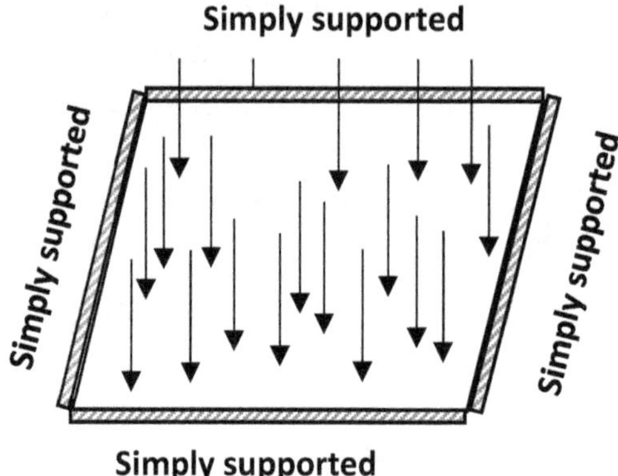

Figure 11.7 Simply supported Square plate subjected to normal pressure.

The program listed in section A.12 is utilized for obtaining the results. The following is the input file used for analysis:

```
2.0  2.0  0.05  10  10  2.1e8  0.0  0.3  -100.0  0
%rectangular plate subjected to pressure
%E and q are in kPa dimensions in metres
```

The following is the dialog generated during execution of the program to input for other quantities. Note that the responses are marked in bold font:

```
enter edge restraints in the order [Bottom Top Left Right]
enter 0 for clamped edge
enter 1 for simply-supported edge
enter 2 for symmetric edge condition
Bottom Edge(0/1/2): 1
Top Edge(0/1/2): 1
Left Edge(0/1/2): 1
Right Edge(0/1/2): 1
maximum deflection is within limits
```

The mesh generated is shown in Figure 11.8. The plate is also analysed using ABAQUS software *using **S4R5** elements*. The results obtained using the present program match very well those obtained using ABAQUS.

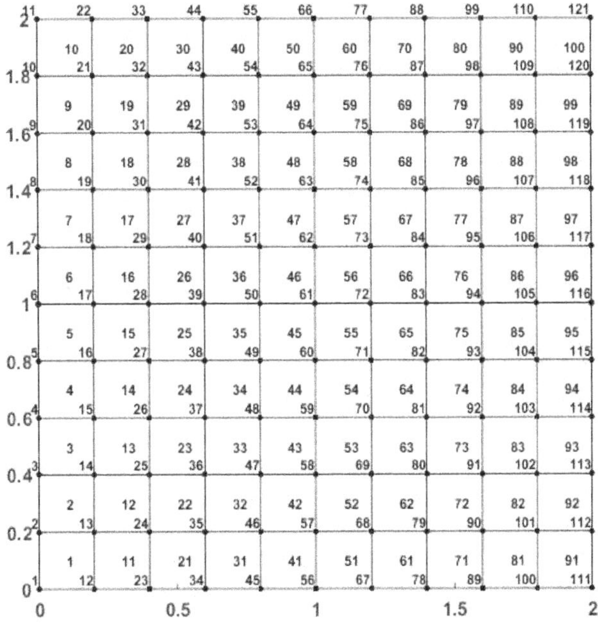

Figure 11.8 Mesh for the square plate.

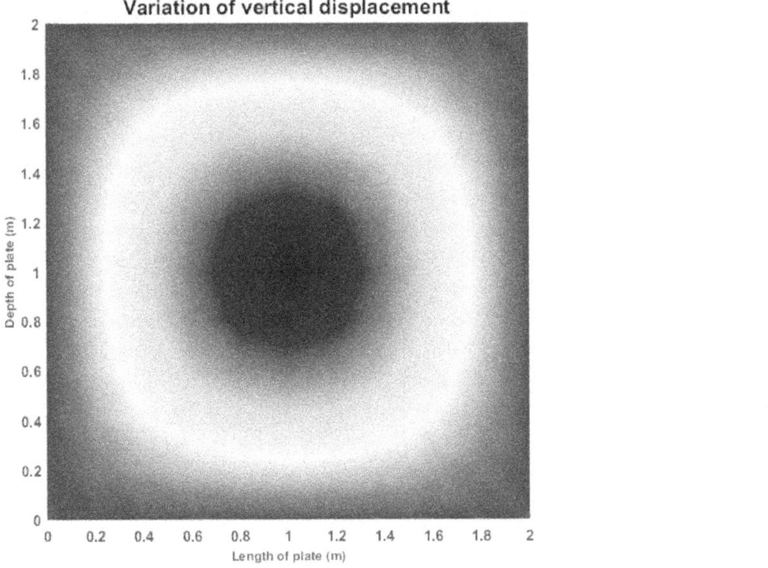

Figure 11.9 Variation of vertical displacement across square plate.

Figure 11.9 depicts the variation of vertical displacement across the square plate obtained using the MATLAB code given in Section A.12. The variation of vertical displacement obtained using ABAQUS is shown in Figure 11.10.

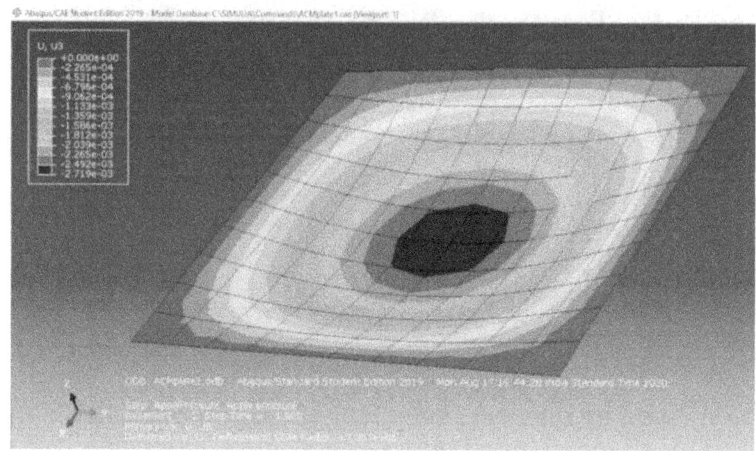

Figure 11.10 ABAQUS-Variation of vertical displacement across square plate.

Figure 11.11 and Figure 11.12 show the variation of M_{XX} across the plate computed using the MATLAB code and ABAQUS, respectively. Owing to symmetry, values of M_{YY} are identical to the values of M_{XX} and thus are not plotted here. Figure 11.13 and Figure 11.14 show the variation of M_{XY} across the plate computed using the MATLAB code and ABAQUS, respectively. The results obtained match closely with the results obtained using the present MATLAB code.

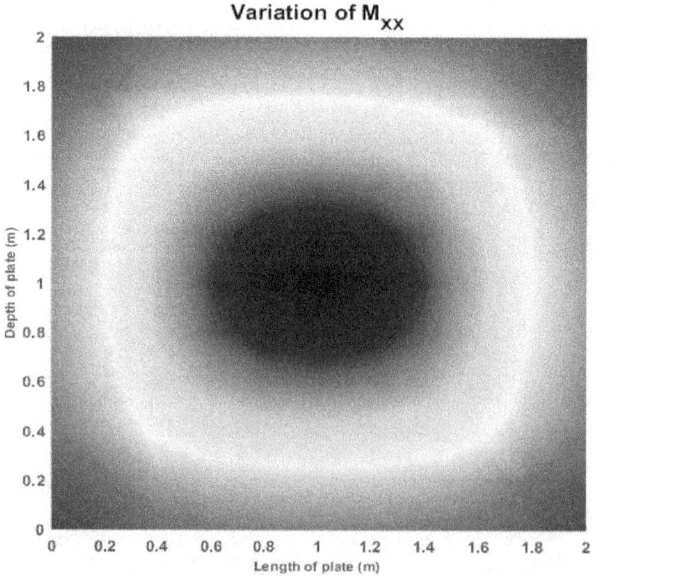

Figure 11.11 Variation of M_{XX} across the square plate.

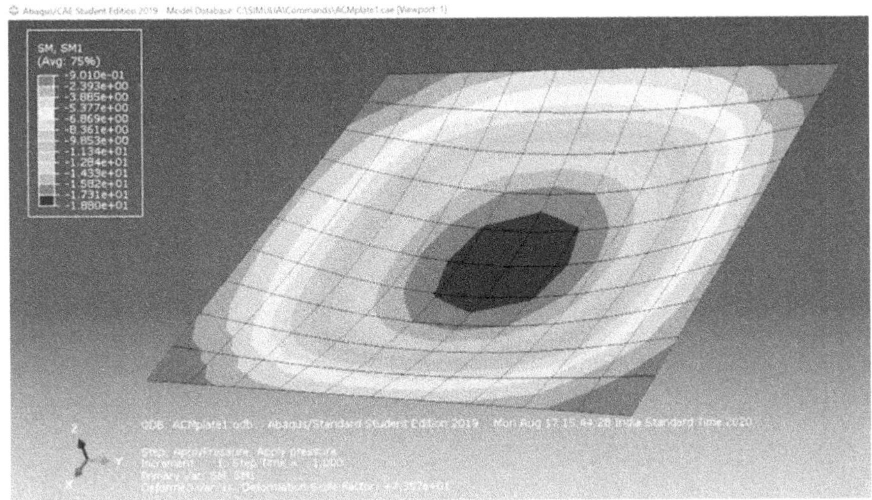

Figure 11.12 *ABAQUS*-Variation of M_{XX} across the square plate.

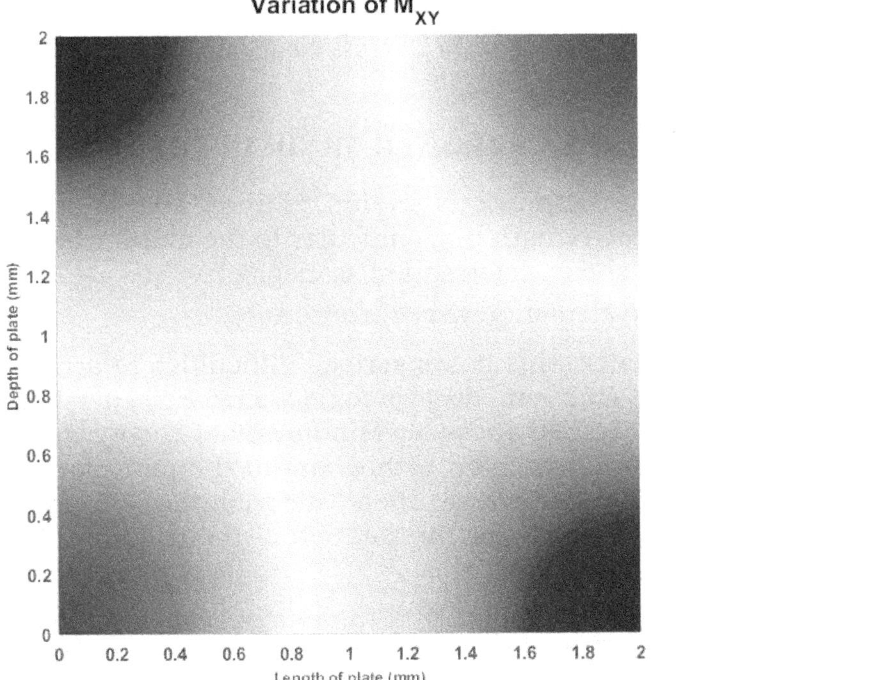

Figure 11.13 Variation of M_{XY} across the square plate.

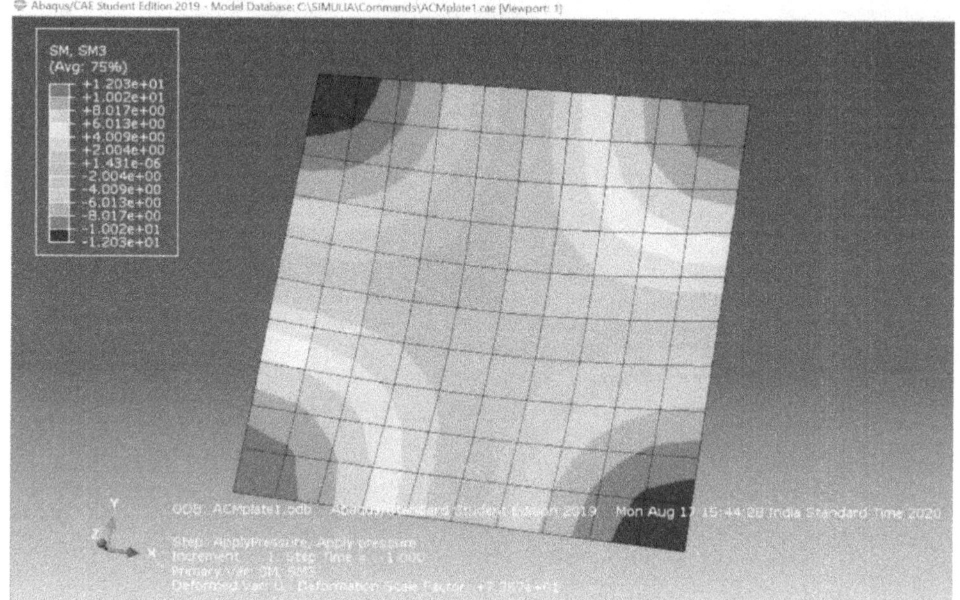

Figure 11.14 ABAQUS- Variation of M_{XY} across the square plate.

11.4 THEORY OF THICK PLATES – REISSNER-MINDLIN THEORY

Kirchhoff's theory, described in the previous section, is applicable to only thin plates where normals to the plate remain perpendicular to the mid-surface after deformation. This is because shear deformation is negligible in case of thin plates. Also, thin plates require elements with C^1 continuity.

However, in case of thick plates, this poses serious difficulties in arriving at a conforming deflection field. This can be overcome using Reissner-Mindlin theory for thick plates. The so called Reissner-Mindlin plate theory assumes that the normal to the plate do not remain orthogonal to the mid–plane after deformation, thus allowing for transverse shear deformation effects. This assumption is analogous to that made for the rotation of the transverse cross section in Timoshenko beam theory.

In thick plates, we relax the assumption that "*a vertical element of the plate before bending remains perpendicular to the middle surface of the plate after bending*". Accordingly, transverse normals may rotate without remaining normal to the mid-plane. This facilitates the use of C^0 elements for thick plates. However, in cases when the Reissner-Mindlin's theory is applied to thin plates, "*shear locking effect*" due to the excessive influence of the transverse deformation terms. One may note that a similar situation arises when Timoshenko plate theory is applied to the analysis of slender beams. *Shear locking* can be avoided using reduced integration, as would be described later in this section.

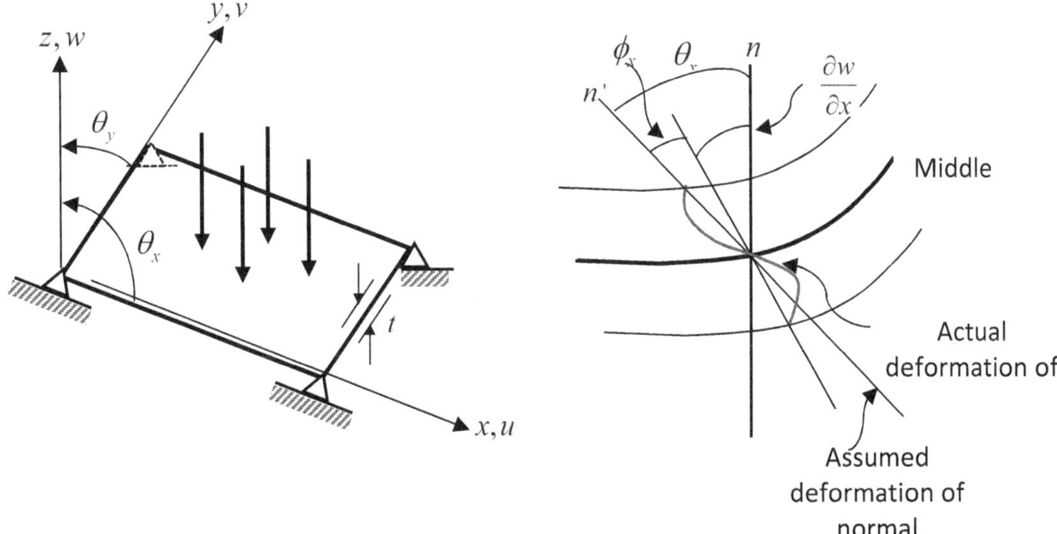

Figure 11.15 Middle Plane of thick plate.

Referring to the Figure 11.15, it is observed that

$$\theta_x = \frac{\partial w}{\partial x} + \phi_x$$

$$\theta_y = \frac{\partial w}{\partial y} + \phi_y$$

.....(11.45)

in xz and yz planes respectively.

Thus, the displacement field becomes

$$u = z\theta_x$$

$$v = z\theta_y$$

$$w = w(x, y)$$

.....(11.46)

The strain vector $\{\varepsilon\}$ is given by

$$\{\varepsilon\} = \left\{\begin{array}{c} \varepsilon_x \\ \varepsilon_y \\ \gamma_{xy} \\ \gamma_{yz} \\ \gamma_{xz} \end{array}\right\} = \left\{\begin{array}{c} z\dfrac{\partial\theta_x}{\partial x} \\[2mm] z\dfrac{\partial\theta_y}{\partial y} \\[2mm] z\left(\dfrac{\partial\theta_x}{\partial y} + \dfrac{\partial\theta_y}{\partial x}\right) \\[2mm] \left(\theta_y - \dfrac{\partial w}{\partial y}\right) \\[2mm] \left(\theta_z - \dfrac{\partial w}{\partial x}\right) \end{array}\right\} \qquad .....(11.47)$$

Eq. 11.47 constitutes the main equation of Reissner-Mindlin plate theory. It is very important to observe here that transverse displacement w and slopes θ_x and θ_y are independent. In case of thin plates, these slopes themselves are functions of the transverse displacement viz. $\theta_x = -\dfrac{\partial w}{\partial x}$ and $\theta_y = -\dfrac{\partial w}{\partial y}$. Thus, in such a case, the theory of thick plates reduces to the theory of thin plates.

Assuming that the material is homogeneous and isotropic, the in-plane stresses σ_x, σ_y and τ_{xy} are related to the corresponding strains ε_x, ε_y and γ_{xy} as follows:

$$\left\{\begin{array}{c} \sigma_x \\ \sigma_y \\ \tau_{xy} \end{array}\right\} = \frac{E}{\left(1-\mu^2\right)}\left[\begin{array}{ccc} 1 & \mu & 0 \\ \mu & 1 & 0 \\ 0 & 0 & \dfrac{(1-\mu)}{2} \end{array}\right]\left\{\begin{array}{c} \varepsilon_x \\ \varepsilon_y \\ \gamma_{xy} \end{array}\right\} \qquad .....(11.48)$$

The out-of-plane stresses τ_{xz} and τ_{yz} are related to the corresponding out-of-plane strains γ_{xz} and γ_{yz} as follows:

$$\left\{\begin{array}{c} \tau_{xz} \\ \tau_{yz} \end{array}\right\} = \left[\begin{array}{cc} G & 0 \\ 0 & G \end{array}\right]\left\{\begin{array}{c} \gamma_{xz} \\ \gamma_{yz} \end{array}\right\} \qquad .....(11.49)$$

where G (shear modulus) is related to E (Young's modulus) as follows:

$$G = \frac{E}{2(1+\mu)} \qquad\qquad \dots(11.50)$$

Substituting the first three terms of Eq. 11.47 on the right-hand side of Eq.11.48, we obtain

$$\begin{Bmatrix} \sigma_{xx} \\ \sigma_{yy} \\ \tau_{xy} \end{Bmatrix} = \frac{E}{(1-\mu^2)} \begin{bmatrix} 1 & \mu & 0 \\ \mu & 1 & 0 \\ 0 & 0 & \dfrac{(1-\mu)}{2} \end{bmatrix} \begin{Bmatrix} z\dfrac{\partial\theta_x}{\partial x} \\ z\dfrac{\partial\theta_y}{\partial y} \\ z\left(\dfrac{\partial\theta_x}{\partial y}+\dfrac{\partial\theta_y}{\partial x}\right) \end{Bmatrix} \qquad \dots(11.51)$$

$$M_{xx} = \int_{-\frac{t}{2}}^{\frac{t}{2}} \sigma_{xx} z\,dz = \frac{E}{(1-\mu^2)} \int_{-\frac{t}{2}}^{\frac{t}{2}} z^2 \left(\frac{\partial\theta_x}{\partial x}+\mu\frac{\partial\theta_y}{\partial y}\right) dz = \frac{Et^3}{12(1-\mu^2)}\left(\frac{\partial\theta_x}{\partial x}+\mu\frac{\partial\theta_y}{\partial y}\right)$$

$$M_{yy} = \int_{-\frac{t}{2}}^{\frac{t}{2}} \sigma_{yy} z\,dz = \frac{E}{(1-\mu^2)} \int_{-\frac{t}{2}}^{\frac{t}{2}} z^2 \left(\mu\frac{\partial\theta_x}{\partial x}+\frac{\partial\theta_y}{\partial y}\right) dz = \frac{Et^3}{12(1-\mu^2)}\left(\mu\frac{\partial\theta_x}{\partial x}+\frac{\partial\theta_y}{\partial y}\right)$$

$$M_{xy} = \int_{-\frac{t}{2}}^{\frac{t}{2}} \tau_{yy} z\,dz = \frac{E}{(1-\mu^2)} \int_{-\frac{t}{2}}^{\frac{t}{2}} z^2 \left(\frac{\partial\theta_x}{\partial y}+\frac{\partial\theta_y}{\partial x}\right) dz = \frac{Et^3}{12(1-\mu^2)}\left(\frac{\partial\theta_x}{\partial y}+\frac{\partial\theta_y}{\partial x}\right)$$

$$Q_y = \int_{-\frac{t}{2}}^{\frac{t}{2}} \tau_{yz}\,dz = G\int_{-\frac{t}{2}}^{\frac{t}{2}} \gamma_{yz}\,dz = Gt\gamma_{yz} = Gt\left(\theta_y - \frac{\partial w}{\partial y}\right)$$

$$Q_x = \int_{-\frac{t}{2}}^{\frac{t}{2}} \tau_{xz}\,dz = G\int_{-\frac{t}{2}}^{\frac{t}{2}} \gamma_{xz}\,dz = Gt\gamma_{xz} = Gt\left(\theta_x - \frac{\partial w}{\partial x}\right)$$

$$\dots(11.52)$$

The term $\dfrac{Et^3}{12\left(1-\mu^2\right)}$ represents the bending stiffness of the plate and is denoted by D_r. Writing Eq. 11.52 in matrix form we obtain

$$\begin{Bmatrix} M_{xx} \\ M_{yy} \\ M_{xy} \\ Q_y \\ Q_x \end{Bmatrix} = \begin{bmatrix} D_r & \mu D_r & 0 & 0 & 0 \\ \mu D_r & D_r & 0 & 0 & 0 \\ 0 & 0 & \dfrac{D_r\left(1-\mu\right)}{2} & 0 & 0 \\ 0 & 0 & 0 & Gt & 0 \\ 0 & 0 & 0 & 0 & Gt \end{bmatrix} \begin{Bmatrix} \dfrac{\partial \theta_x}{\partial x} \\ \dfrac{\partial \theta_y}{\partial y} \\ \dfrac{\partial \theta_x}{\partial y}+\dfrac{\partial \theta_y}{\partial x} \\ \theta_y - \dfrac{\partial w}{\partial y} \\ \theta_z - \dfrac{\partial w}{\partial x} \end{Bmatrix} \qquad \dots\dots(11.53)$$

Eq. 11.53 can be written concisely as

$$M = \left[D_M\right]\{\chi\} \qquad \dots\dots(11.54)$$

The total strain energy of the plate is given as

$$U = \int_A \{\chi\}^T \left[D_M\right]\{\chi\}\, dA \qquad \dots\dots(11.55)$$

Eq. 11.55 can be written as

$$U = U_M + U_S = \frac{1}{2}\int_A \{\chi_M\}^T \left[D_M\right]\{\chi_M\}\, dA + \frac{k}{2}\int_A \{\chi_S\}^T \left[D_S\right]\{\chi_S\}\, dA \qquad \dots\dots(11.56)$$

Here U_B and U_S are the components of strain energy due to bending and shear respectively. κ is the shear energy correction factor and is usually assigned a value of $\dfrac{5}{6}$.

$$\{\chi_B\} = \begin{Bmatrix} \dfrac{\partial \theta_x}{\partial x} \\[2mm] \dfrac{\partial \theta_y}{\partial y} \\[2mm] \dfrac{\partial \theta_x}{\partial y} + \dfrac{\partial \theta_y}{\partial x} \end{Bmatrix} \text{ and } \{\chi_S\} = \begin{Bmatrix} \theta_y - \dfrac{\partial w}{\partial y} \\[2mm] \theta_z - \dfrac{\partial w}{\partial x} \end{Bmatrix} \qquad \text{.....(11.57)}$$

$$[D_M] = \dfrac{Et^3}{12(1-\mu^2)} \begin{bmatrix} 1 & \mu & 0 \\ \mu & 1 & 0 \\ 0 & 0 & \dfrac{(1-\mu)}{2} \end{bmatrix} \text{ and } [D_S] = \begin{bmatrix} Gt & 0 \\ 0 & Gt \end{bmatrix} \qquad \text{.....(11.58)}$$

11.4.1 Finite Element Formulation of Thick Plates

From the previous section we observe that the slopes θ_x and θ_y are independent of the transverse displacement w. This implies that the finite element formulation of thick plates does not require C^1 continuity. The transverse displacement and the slopes can be interpolated independently using elements with C^0 continuity as follows:

$$w(x, y) = \sum_{i=1}^{n} N_i w_i$$

$$\theta_x(x, y) = \sum_{i=1}^{n} N_i \theta_{xi} \qquad \text{.....(11.59)}$$

$$\theta_y(x, y) = \sum_{i=1}^{n} N_i \theta_{yi}$$

Eq. 11.59 can be represented in the matrix form as

$$\begin{Bmatrix} w \\ \theta_x \\ \theta_y \end{Bmatrix} = \begin{bmatrix} N_1 & 0 & 0 & N_2 & 0 & 0 & .. & .. & .. & N_n & 0 & 0 \\ 0 & N_1 & 0 & 0 & N_2 & 0 & .. & .. & .. & 0 & N_n & 0 \\ 0 & 0 & N_1 & 0 & 0 & N_2 & .. & .. & .. & 0 & 0 & N_n \end{bmatrix} \{d\} \qquad \text{.....(11.60)}$$

Note that n is the number of nodes. is the vector of nodal displacements and rotations $\{d\}$ is given as follows:

$$\{d\} = \begin{bmatrix} w_1 & \theta_{x1} & \theta_{y1} & w_2 & \theta_{x2} & \theta_{y2} & .. & .. & .. & w_n & \theta_{xn} & \theta_{yn} \end{bmatrix}^T \qquad \text{.....(11.61)}$$

The curvature $\{\chi_M\}$ in Eq. 11.57 can be combined with the second and third rows of Eq. 11.60 as follows:

$$\{\chi_B\} = \begin{bmatrix} \dfrac{\partial}{\partial x} & 0 \\ 0 & \dfrac{\partial}{\partial y} \\ \dfrac{\partial}{\partial y} & \dfrac{\partial}{\partial x} \end{bmatrix} \begin{Bmatrix} \theta_x \\ \theta_y \end{Bmatrix} = \begin{bmatrix} \dfrac{\partial}{\partial x} & 0 \\ 0 & \dfrac{\partial}{\partial y} \\ \dfrac{\partial}{\partial y} & \dfrac{\partial}{\partial x} \end{bmatrix} \begin{bmatrix} 0 & N_1 & 0 & 0 & N_2 & 0 & .. & .. & .. & 0 & N_n & 0 \\ 0 & 0 & N_1 & 0 & 0 & N_2 & .. & .. & .. & 0 & 0 & N_n \end{bmatrix}$$

$$.....(11.62)$$

$$\{\chi_M\} = \begin{bmatrix} 0 & \dfrac{\partial N_1}{\partial x} & 0 & 0 & \dfrac{\partial N_2}{\partial x} & 0 & .. & .. & .. & 0 & \dfrac{\partial N_n}{\partial x} & 0 \\ 0 & 0 & \dfrac{\partial N_1}{\partial y} & 0 & 0 & \dfrac{\partial N_2}{\partial y} & .. & .. & .. & 0 & 0 & \dfrac{\partial N_n}{\partial y} \\ 0 & \dfrac{\partial N_1}{\partial y} & \dfrac{\partial N_1}{\partial x} & 0 & \dfrac{\partial N_2}{\partial y} & \dfrac{\partial N_2}{\partial x} & .. & .. & .. & 0 & \dfrac{\partial N_n}{\partial y} & \dfrac{\partial N_n}{\partial x} \end{bmatrix} \{d\} = [B_M]\{d\}$$

$$.....(11.63)$$

Here, $[B_M]$ is the strain-displacement matrix for bending terms.

Curvature $\{\chi_S\}$ from Eq. 11.57 can expressed as follows:

$$\{\chi_S\} = \begin{Bmatrix} \theta_y - \dfrac{\partial w}{\partial y} \\ \theta_x - \dfrac{\partial w}{\partial x} \end{Bmatrix} = \begin{bmatrix} -\dfrac{\partial}{\partial y} & 0 & 1 \\ -\dfrac{\partial}{\partial x} & 1 & 0 \end{bmatrix} \begin{Bmatrix} w \\ \theta_x \\ \theta_y \end{Bmatrix}$$

$$(11.64)$$

Substituting Eq.11.60 in the right hand side of Eq.11.64 leads to

$$\{\chi_S\} = \begin{bmatrix} -\dfrac{\partial}{\partial y} & 0 & 1 \\ -\dfrac{\partial}{\partial x} & 1 & 0 \end{bmatrix} \begin{bmatrix} N_1 & 0 & 0 & N_2 & 0 & 0 & .. & .. & .. & N_n & 0 & 0 \\ 0 & N_1 & 0 & 0 & N_2 & 0 & .. & .. & .. & 0 & N_n & 0 \\ 0 & 0 & N_1 & 0 & 0 & N_2 & .. & .. & .. & 0 & 0 & N_n \end{bmatrix} \{d\}$$

$$.....(11.65)$$

$$\{\chi_S\} = \begin{bmatrix} -\dfrac{\partial N_1}{\partial y} & 0 & N_1 & -\dfrac{\partial N_2}{\partial y} & 0 & N_2 & .. & .. & .. & -\dfrac{\partial N_n}{\partial y} & 0 & N_n \\[2ex] -\dfrac{\partial N_1}{\partial x} & N_1 & 0 & -\dfrac{\partial N_2}{\partial x} & N_2 & 0 & .. & .. & .. & -\dfrac{\partial N_n}{\partial x} & N_n & 0 \end{bmatrix} \{d\}$$

$$.....(11.66)$$

$$\{\chi_S\} = [B_S]\{d\} \qquad\qquad .....(11.67)$$

where $[B_S]$ is the strain-displacement matrix for shear terms.

Thus, the overall stiffness matrix for the thick plate is split into two components as follows:

$$\left[K^{(e)}\right] = \left[K_M^{(e)}\right] + \left[K_S^{(e)}\right] = \int_A [B_M]^T [D_M][B_M] dA + \kappa \int_A [B_S]^T [D_S][B_S] dA \quad .....(11.68)$$

It is to be observed from the above equation that $\left[K_M^{(e)}\right]$ matrix has h^3 term as it appears in $[D_M]$ matrix. However, $\left[K_S^{(e)}\right]$ has h term as it appears in $[D_S]$ matrix. Owing to this, as the thickness of the plate h decreases, the numerical value of $\left[K_M^{(e)}\right]$ matrix decreases much faster than the numerical value of $\left[K_S^{(e)}\right]$ matrix. Thus, shear terms dominate the bending terms as the thickness of the plate decreases leading to a phenomenon of shear locking. To avoid this, the shear terms are integrated to a lower order compared to the bending terms. For example, if 3×3 Gauss points are used for numerical integration of the bending terms, 2×2 Gauss points are used for numerical integration of the shear terms. This will ensure a proper solution.

11.4.2   Boundary Conditions

For a clamped edge: $w = \theta_x = \theta_y = 0$

For a simply supported edge $x = a$: $w = \theta_x = 0$

For a simply supported edge $y = a$: $w = \theta_y = 0$

For a free edge: No boundary conditions are required.

11.5 DESCRIPTION OF MATLAB PROGRAM S4PLATE.M (APPENDIX A.13)

(a) Input the values of length, width and thickness of plate, Young's modulus and Poisson's ratio, number of divisions along x and y axes, normal pressure, number of loaded nodes, angle of skew.

(b) Call the MATLAB script Q4skewplate.m to generate the node and element numbers and connectivity of elements. The 4 nodes of the element are numbered anti-clockwise starting from the lower left-hand corner and ending at the upper left-hand corner. The mesh for the discretised plate is plotted along with node and element numbering. The nodes on the four edges of the plate (top, bottom, left and right) are identified.

(c) Global stiffness matrix and force vector are initialised. Gauss points chosen are 3×3 for bending and 2×2 for shear. D matrices are formulated for bending and shear.

(d) A for-loop is initialised for index i ranging from 1 to number of elements. Inside this loop, for each iteration, the following computations are performed.

   (i) Stiffness matrices $K_B^{(e)}$ and $K_S^{(e)}$ are initialized.

   (ii) Two nested loops are used to perform numerical integration using Gaussian quadrature for bending. Inside these loops, MATLAB script *smlin.m* is called to compute the function values and partial derivatives at various integration points. The MATLAB script *smlin.m* makes use of symbolic logic toolbox to compute the functions and their derivatives at each integration point.

   (iii) These values are used to compute Jacobian J and also strain displacement matrix for bending B. These values are utilized to compute the element stiffness matrix for bending $K_b^{(e)}$.

   (iv) Element force vector $F^{(e)}$ is computed in the same loop. Similar computations are performed to obtain the element stiffness matrix for shear $K_s^{(e)}$ using reduced integration.

   (v) The global stiffness matrix K and global force vector F are computed in this section.

   (vi) Point loads, if any, are added to the corresponding degrees of freedom of the global force vector.

(e) Boundary conditions are set based on the user dialog and are applied on the matrix system of equations which is then solved to obtain the overall displacement vector.

(f) Numerical integration is performed using a single Gauss point for both bending and shear. A loop is run over the number of elements. Inside the loop, a pair of nested loops are run. In these loops, the function values and derivatives are computed at each integration point to compute the B and B_s matrices. These matrices are used to compute the values of M_{XX}, M_{YY} and M_{XY} are computed at the centroid of the element.

(g) The obtained values of vertical displacements and in-plane moments are plotted using MATLAB graphing tools.

11.5.1 Example Problems

Example Problem 1: A rectangular plate of dimensions $3\,m \times 2\,m \times 0.025\,m$ is simply supported on all four edges and is subjected to a uniform pressure of 140kPa, as shown in Figure 11.16. The Young's modulus of the plate is $2.1 \times 10^8\,kPa$ and Poisson's ratio is 0.3. Analyse the plate by discretising it, using 4 node quadrilateral elements.

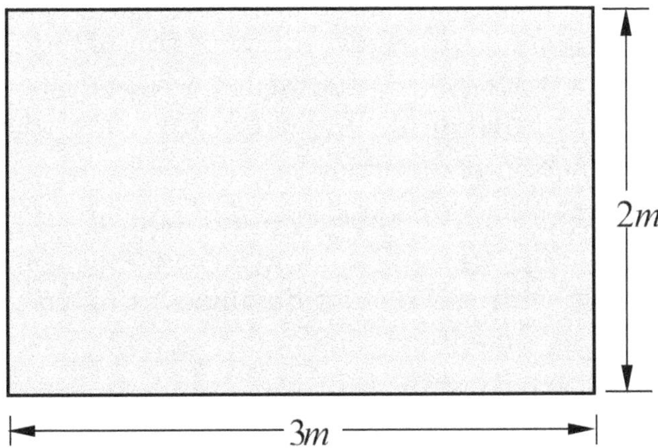

Figure 11.16 Simply supported rectangular plate.

Owing to the symmetry of plate, only a quarter of the plate is analysed. The program listed in *Appendix A.13* is utilized for obtaining the results.

The following is the input file used for analysis:

```
1.5    1.0   0.025   10 7    2.1e8   0.0      0.3     -140.0   0 0.0
%uniform pressure(kPa) over QUARTER rectangular plate
%E in kPa, dimensions in metres (NUMERICAL EXAMPLE-1)
```

The following is the dialog generated during execution of the program to input for other quantities. Note that the responses are marked in bold font:

```
enter edge restraints in the order [Bottom Top Left Right]
enter 0 for clamped edge
enter 1 for simply-supported edge
```

```
enter 2 for symmetric edge condition
Bottom Edge(0/1/2): 1
Top Edge(0/1/2): 2
Left Edge(0/1/2): 1
Right Edge(0/1/2): 2
```

The mesh generated is shown in Figure 11.17. The plate is also analysed using ABAQUS software *using S4R5 elements.* The results obtained using the present program match very well those obtained using ABAQUS. Figure 11.18 depicts the variation of vertical displacement across the square plate obtained using the MATLAB code given in *Appendix A.13.* The variation of vertical displacement obtained using ABAQUS is shown in Figure 11.19.

Figure 11.20 and Figure 11.21 show the variation of M_{XX} across the plate computed using the MATLAB code and ABAQUS, respectively. Figure 11.22 and Figure 11.23 show the variation of M_{YY} across the plate computed using the MATLAB code and ABAQUS, respectively.

Figure 11.24 and Figure 11.25 show the variation of M_{XY} across the plate computed using the MATLAB code and ABAQUS, respectively. The results obtained match closely with the results obtained using the present MATLAB code.

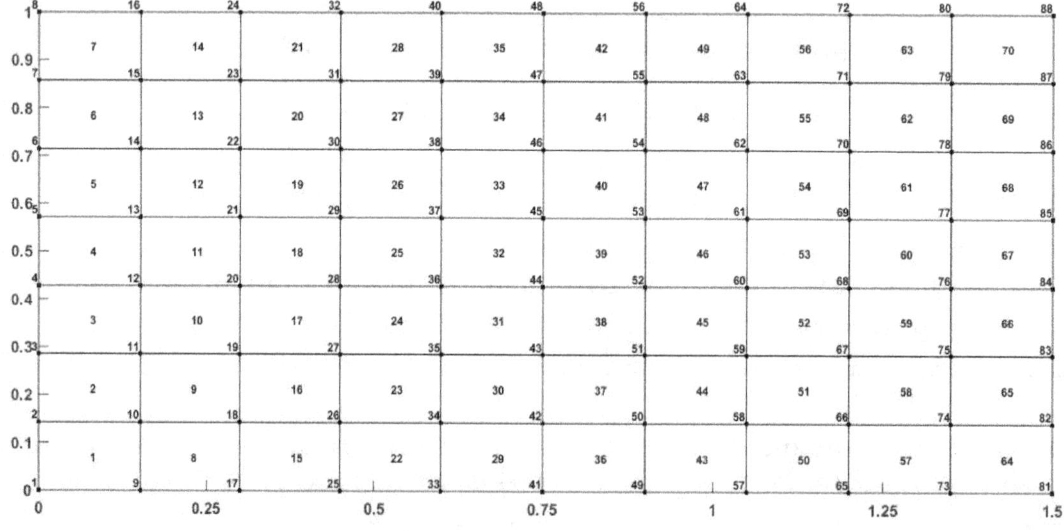

Figure 11.17 Mesh for the quarter-rectangular plate.

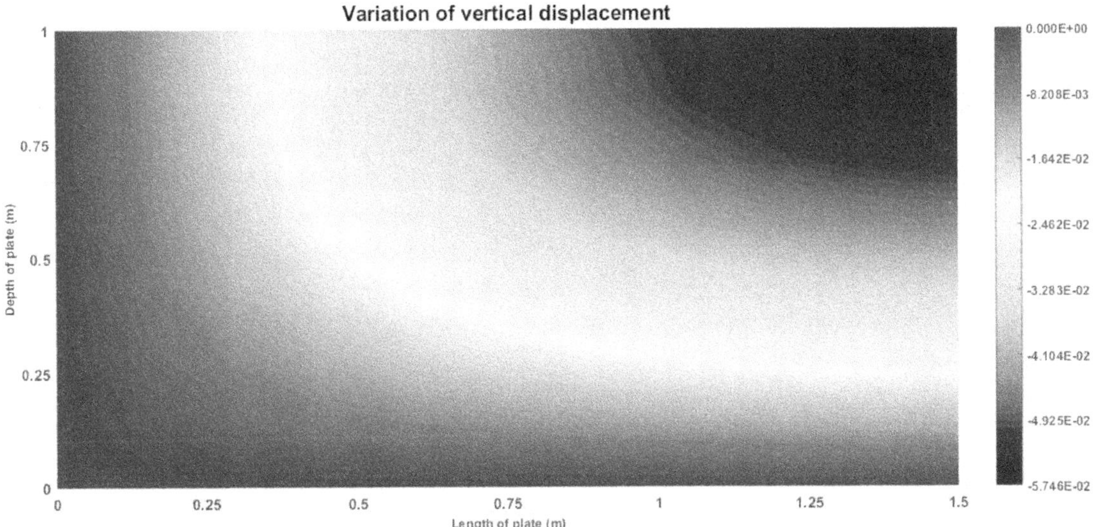

Figure 11.18 Variation of vertical displacement across quarter rec tangular plate.

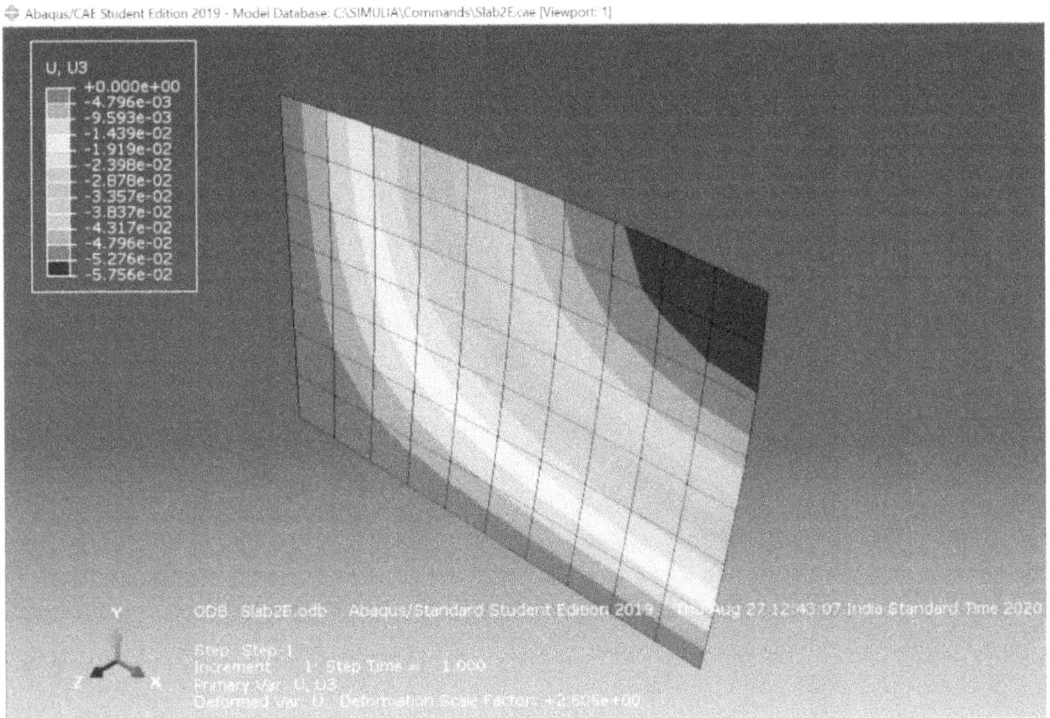

Figure 11.19 ABAQUS-Variation of vertical displacement across quarter rectangular plate.

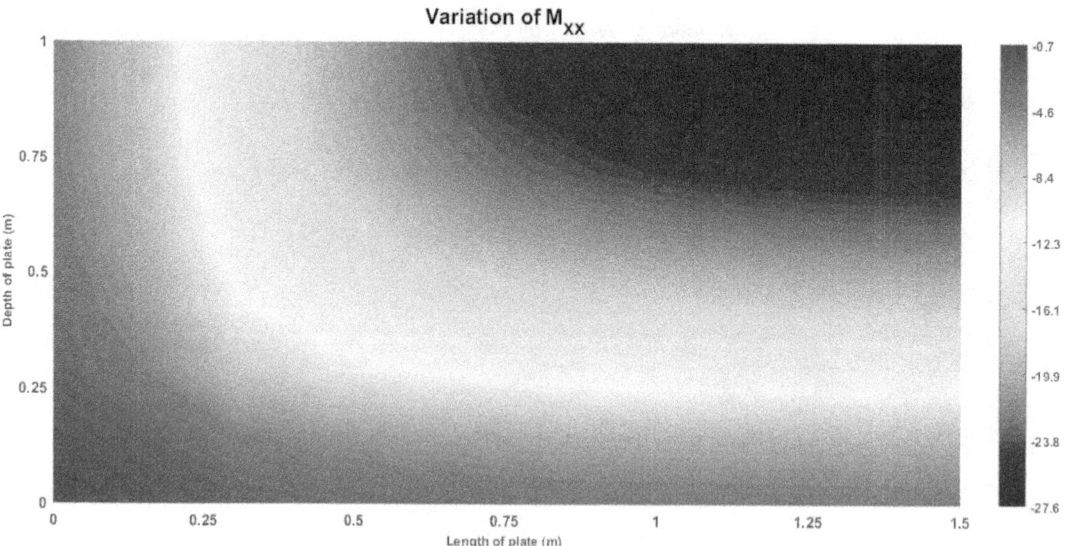

Figure 11.20 Variation of M_{XX} across quarter rectangular plate.

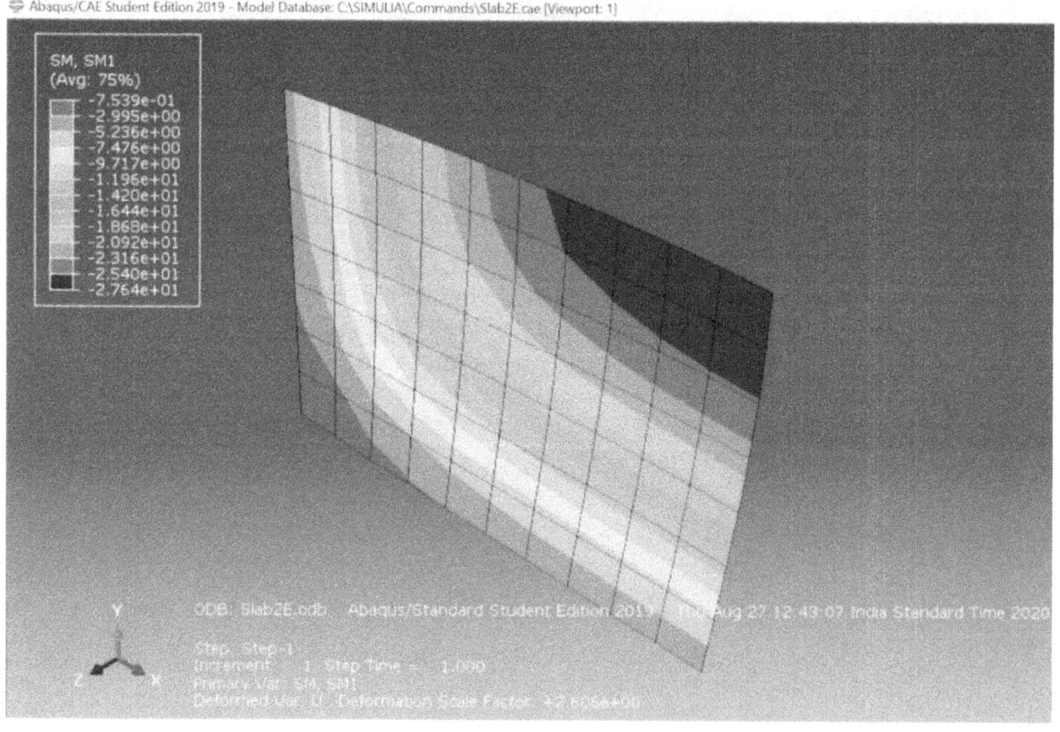

Figure 11.21 ABAQUS-Variation of M_{XX} across quarter rectangular plate.

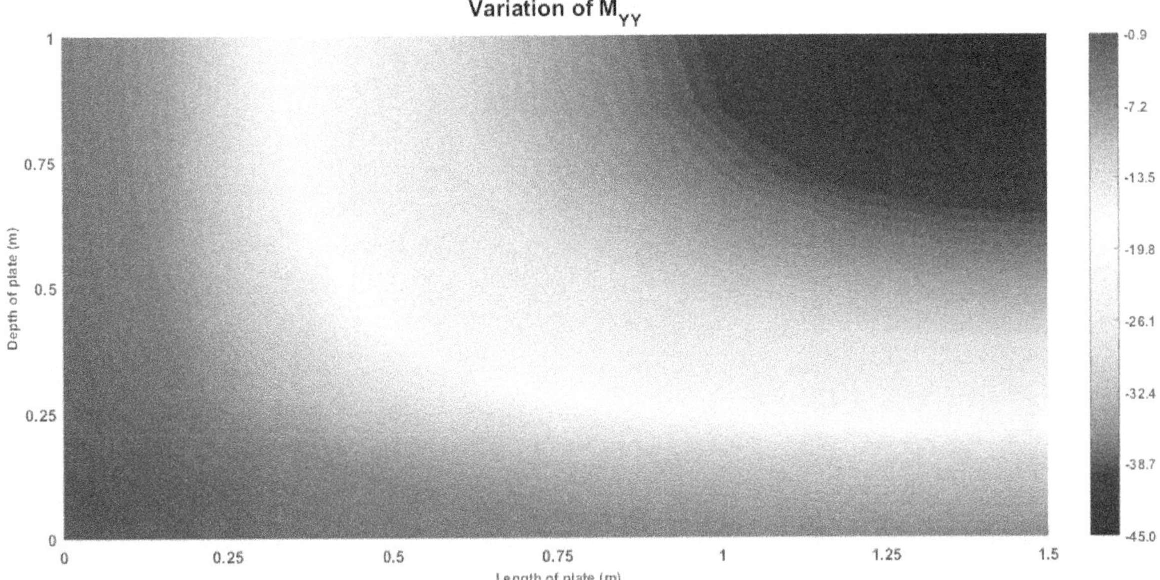

Figure 11.22 Variation of M_{YY} across quarter rectangular plate.

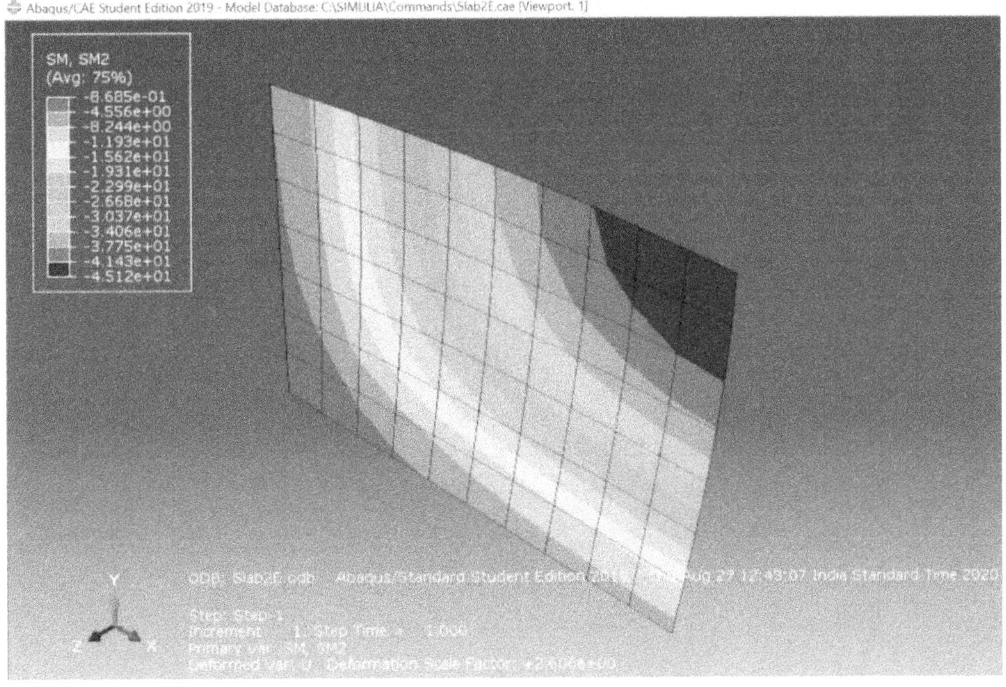

Figure 11.23 ABAQUS- Variation of M_{YY} across quarter rectangular plate.

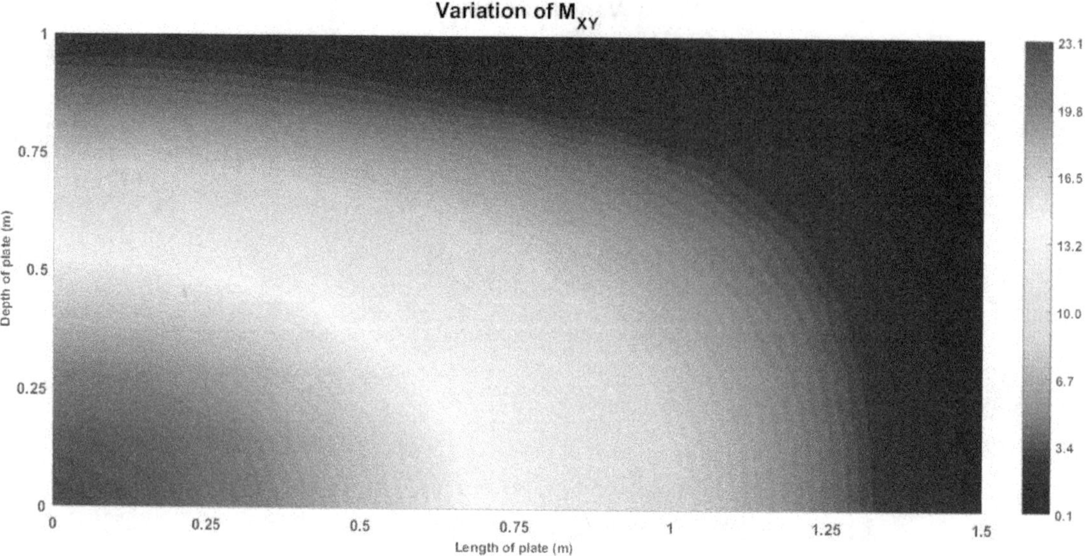

Figure 11.24 Variation of M_{XY} across quarter rectangular plate.

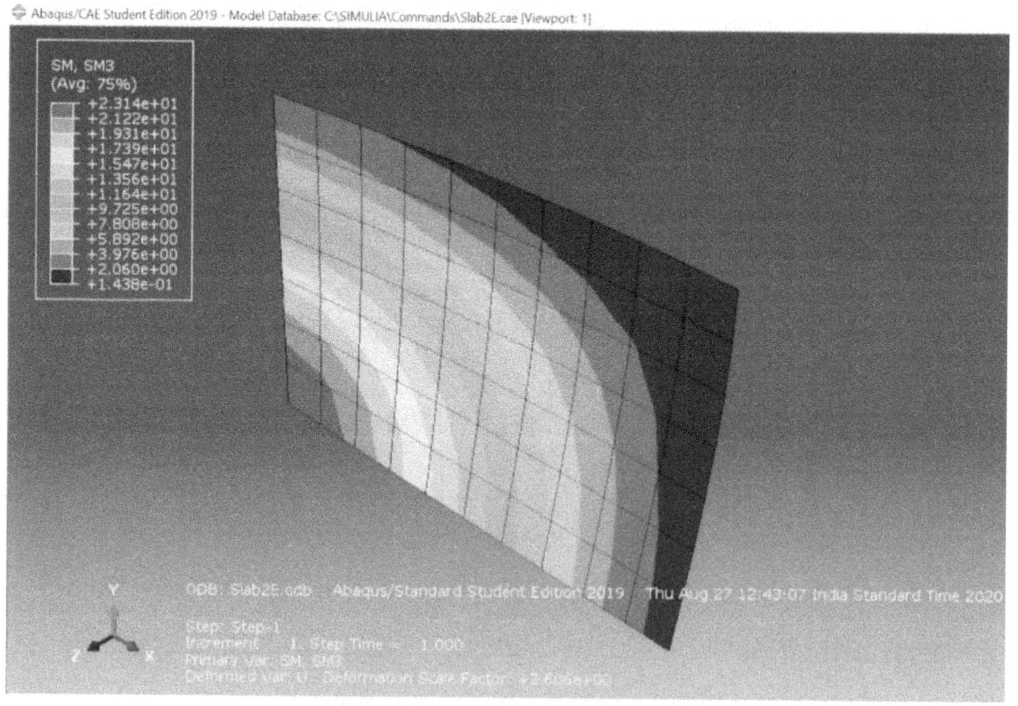

Figure 11.25 ABAQUS-Variation of M_{XY} across quarter rectangular plate.

Example Problem-2: Rectangular plate clamped on opposite edges

Analyse a rectangular plate of dimensions $1.5\,m \times 1\,m \times 0.025\,m$ shown in the Figure 11.26. The plate is clamped on four edges and is subjected to a point load of 50 kN at the center. The Young's modulus of the material of the plate is $2.1 \times 10^8\,kPa$ and Poisson's ratio is 0.3. Discretise the plate using four node quadrilateral elements.

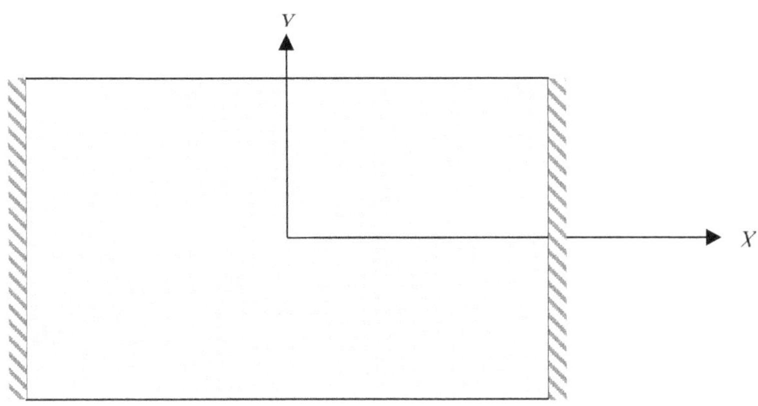

Figure 11.26 Rectangular plate subjected to point load at center.

The program listed in section A.12 is utilized for obtaining the results.

The following is the input file used for analysis:

```
1.5 1.0   0.025   10   8    2.1e8   0.0       0.3     0.0   1 0.0
7   -50.0
%load(kN)at bottom-left quarter rectangular plate
%E in kPa , dimensions in metres (NUMERICAL EXAMPLE-2)
```

The following is the dialog generated during execution of the program to input for other quantities. Note that the responses are marked in bold font:

```
enter edge restraints in the order [Bottom Top Left Right]
enter 0 for clamped edge
enter 1 for simply-supported edge
enter 2 for symmetric edge condition
enter 3 for free edge
Bottom Edge(0/1/2): 3
Top Edge(0/1/2): 3
Left Edge(0/1/2): 0
Right Edge(0/1/2): 0
```

The mesh generated using 4 noded quadrilateral elements is shown in Figure 11.27 below. The plate is also analysed using ABAQUS software *using **S4R5** elements*. The results obtained using the present program match very well those obtained using ABAQUS.

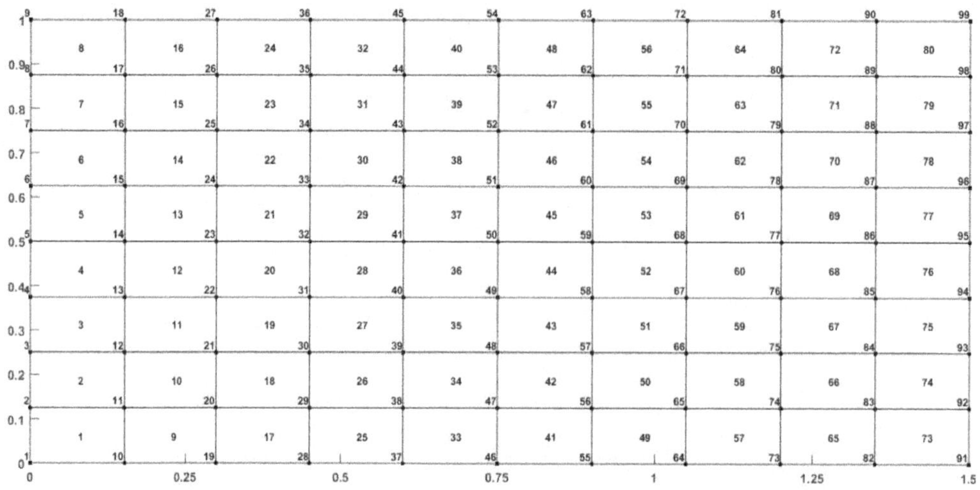

Figure 11.27 Mesh for the Rectangular plate.

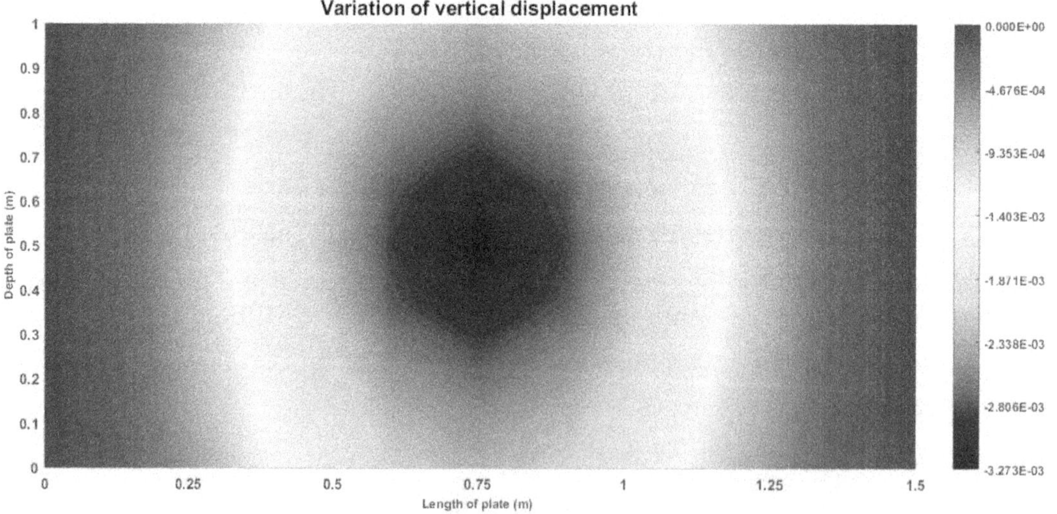

Figure 11.28 Variation of vertical displacement across the Rectangular plate.

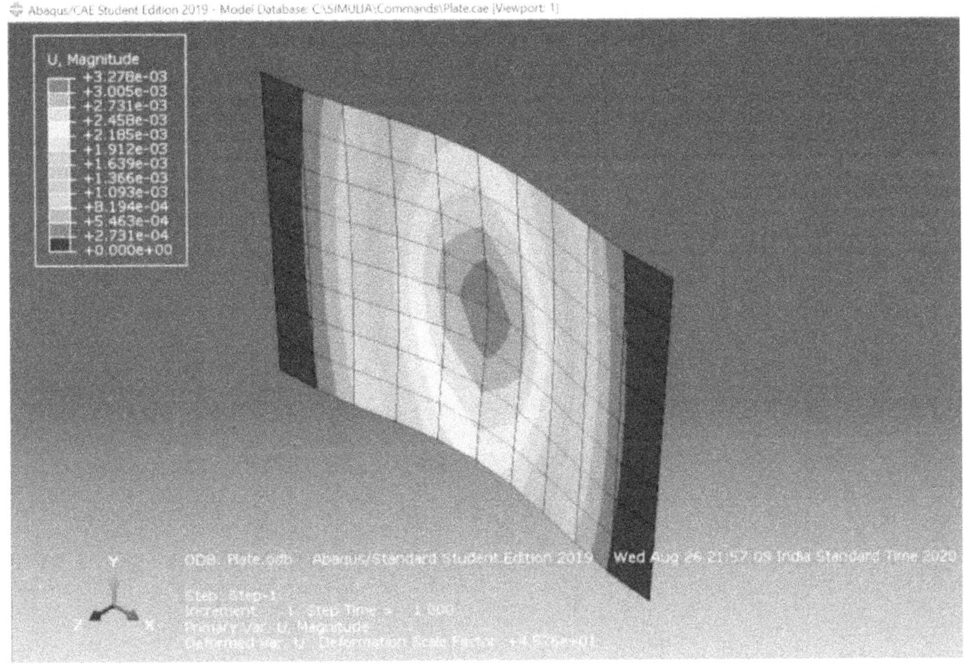

Figure 11.29 ABAQUS -Variation of vertical displacement across the Rectangular plate.

Figure 11.28 depicts the variation of vertical displacement across the square plate obtained using the MATLAB code given in *Appendix A.13*. The variation of vertical displacement obtained using ABAQUS is shown in Figure 11.29.

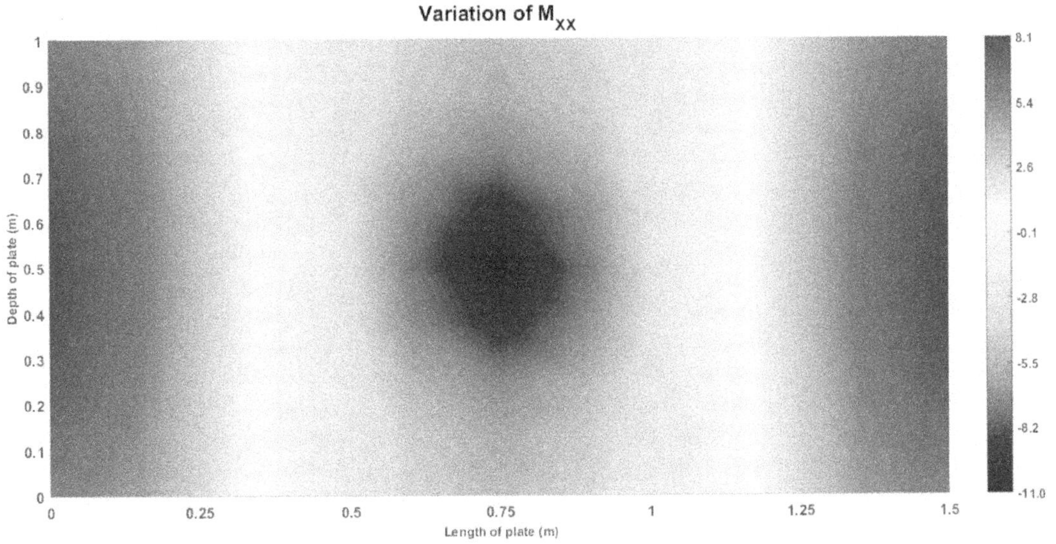

Figure 11.30 Variation of M_{xx} across the rectangular plate.

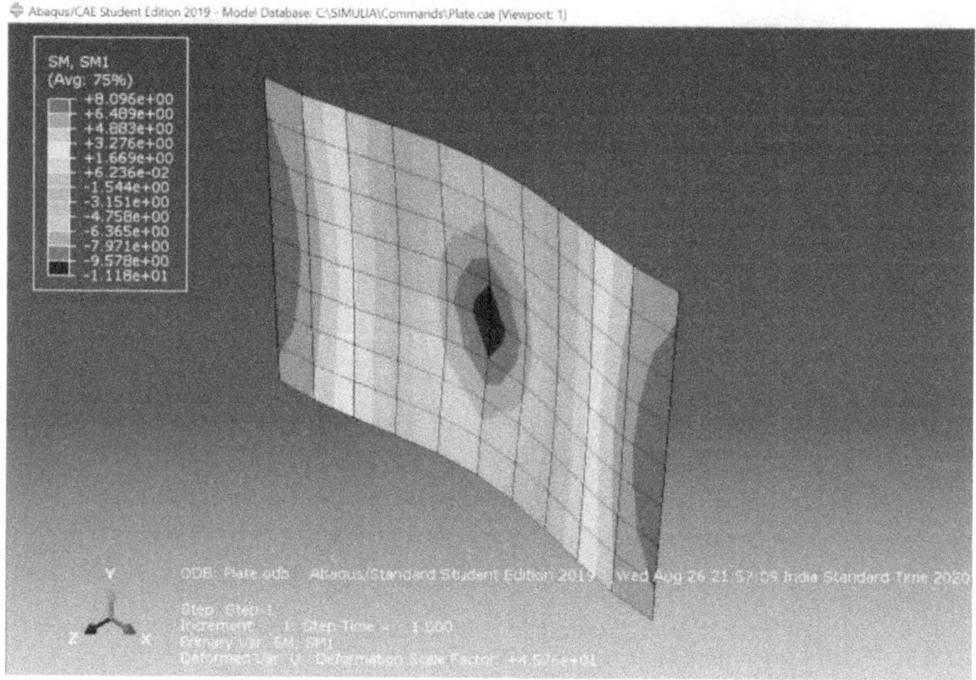

Figure 11.31 ABAQUS-Variation of M_{XX} across the rectangular plate.

Figure 11.30 and Figure 11.31 show the variation of M_{XX} across the plate computed using the MATLAB code and ABAQUS, respectively.

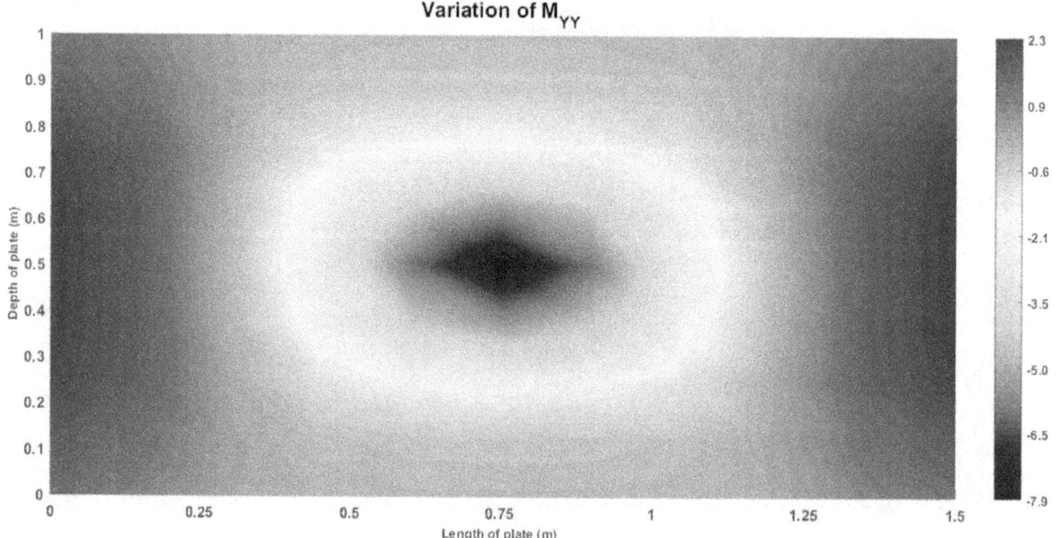

Figure 11.32 Variation of M_{YY} across the rectangular plate.

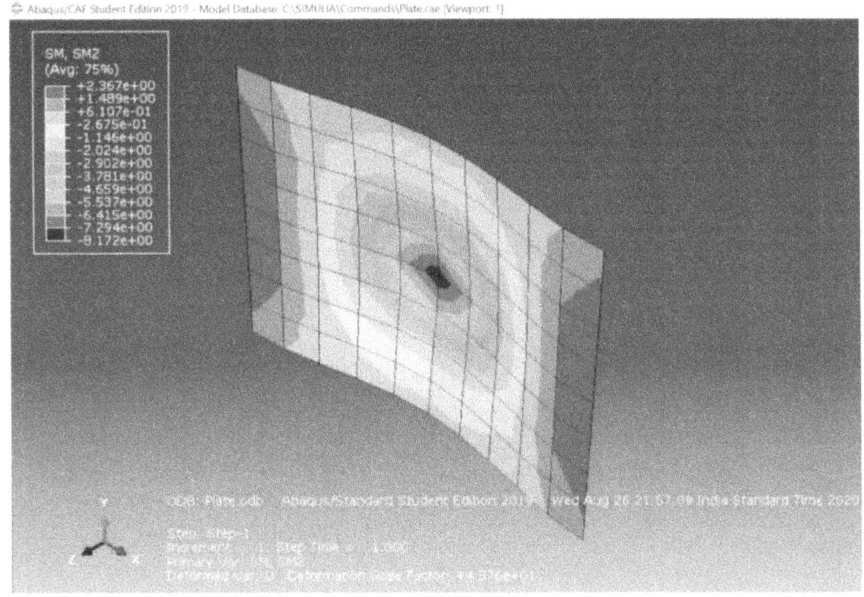

Figure 11.33 ABAQUS-Variation of M_{YY} across the rectangular plate.

Figure 11.32 and Figure 11.33 show the variation of M_{YY} across the plate computed using the MATLAB code and ABAQUS, respectively. Figure 11.34 and Figure 11.35 show the variation of M_{XY} across the plate computed using the MATLAB code and ABAQUS, respectively.

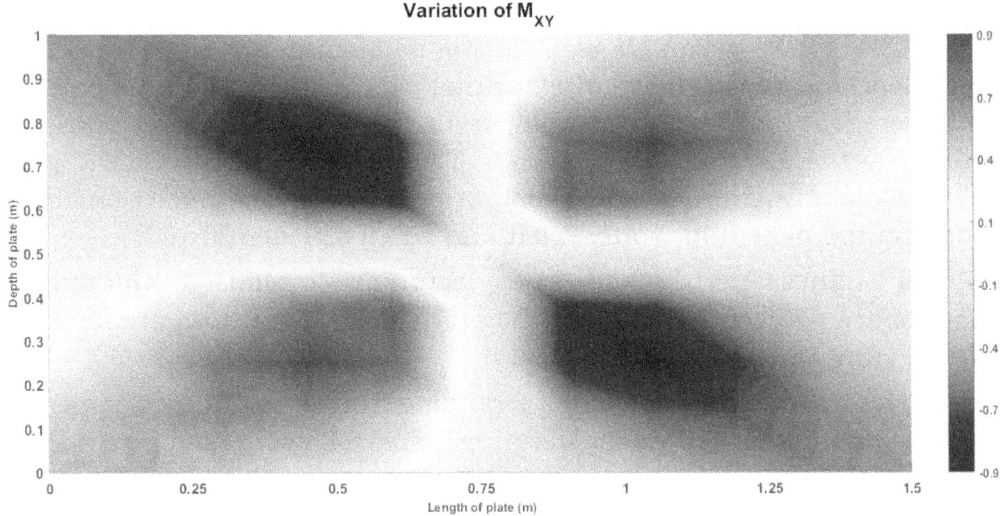

Figure 11.34 Variation of M_{XY} across the rectangular plate.

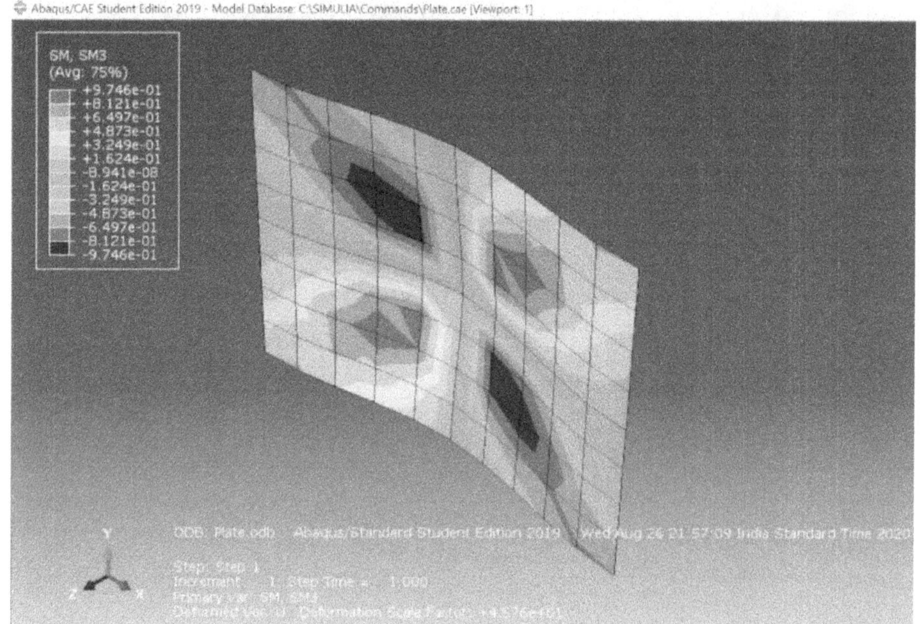

Figure 11.35 ABAQUS-Variation of M_{XY} across the rectangular plate.

11.6 DESCRIPTION OF MATLAB CODE S8 PLATEC.M (APPENDIX A.14)

(a) Input the values of length, width and thickness of plate, Young's modulus and Poisson's ratio, number of divisions along x and y axes, normal pressure, number of loaded nodes, angle of skew.

(b) Call the MATLAB script *Q8skewplate.m* to generate the node and element numbers and connectivity of elements. The 8 nodes of the element are numbered anti-clockwise starting from the lower left-hand corner and ending at the upper left-hand corner. The mesh for the discretised plate is plotted along with node and element numbering. The nodes on the four edges of the plate (top, bottom, left and right) are identified.

(c) Global stiffness matrix and force vector are initialised. Gauss points chosen are 3×3 for bending and 2×2 for shear. D matrices are formulated for bending and shear.

(d) A for-loop is initialised for index i ranging from 1 to number of elements. Inside this loop, for each iteration, the following computations are performed.

(e) Stiffness matrices $K_B^{(e)}$ and $K_S^{(e)}$ are initialized.

(f) Two nested loops are used to perform numerical integration using Gaussian quadrature for bending. Inside these loops, MATLAB script *fmquad.m* is called to compute the function values and partial derivatives at various integration points. The MATLAB script *fmquad.m* makes use of symbolic logic toolbox to compute the functions and their derivatives at each integration point.

(g) These values are used to compute Jacobian J and also strain displacement matrix for bending B. These values are utilized to compute the element stiffness matrix for bending $K_b^{(e)}$.

(h) Element force vector $F^{(e)}$ is computed in the same loop. Similar computations are performed to obtain the element stiffness matrix for shear $K_s^{(e)}$ using reduced integration.

(i) The global stiffness matrix K and global force vector F are computed in this section.

(j) Point loads, if any, are added to the corresponding degrees of freedom of the global force vector.

(k) Boundary conditions are set based on the user dialog and are applied on the matrix system of equations which is then solved to obtain the overall displacement vector.

(l) Numerical integration is performed using 2×2 Gauss points for both bending and shear. A loop is run over the number of elements. Inside the loop, a pair of nested loops are run. In these loops, the function values and derivatives are computed at each integration point to compute to compute the B and B_s matrices. These matrices are used to compute the values of

M_{XX}, M_{YY} and M_{XY} are computed at the integration points of the element. Stresses are then extrapolated to the eight nodes of the element by a process of stress recovery, similar to that.

Eq. 7.84 is modified to extrapolate stress from the four nodes of a Gauss element to the eight nodes of the element as follows:

$$\lfloor w_1 \quad w_2 \quad w_3 \quad w_4 \quad w_5 \quad w_6 \quad w_7 \quad w_8 \rfloor^T = [T] \begin{Bmatrix} w_1' \\ w_2' \\ w_3' \\ w_4' \end{Bmatrix} \text{ where } [T] \text{ is given by}$$

$$[T] = \begin{bmatrix} 1.866 & -0.5 & 0.134 & -0.5 \\ 0.683 & -0.183 & -0.183 & 0.683 \\ -0.5 & 1.866 & -0.5 & 0.134 \\ -0.183 & 0.683 & 0.683 & -0.183 \\ 0.134 & -0.5 & 1.866 & -0.5 \\ -0.183 & -0.183 & 0.683 & 0.683 \\ -0.5 & 0.134 & -0.5 & 1.866 \\ 0.683 & -0.183 & -0.183 & 0.683 \end{bmatrix}$$

The obtained values of vertical displacement and in-plane moments are plotted using MATLAB graphing tools.

11.6.1 Example Problems

Numerical Example 1-Simply supported square plate

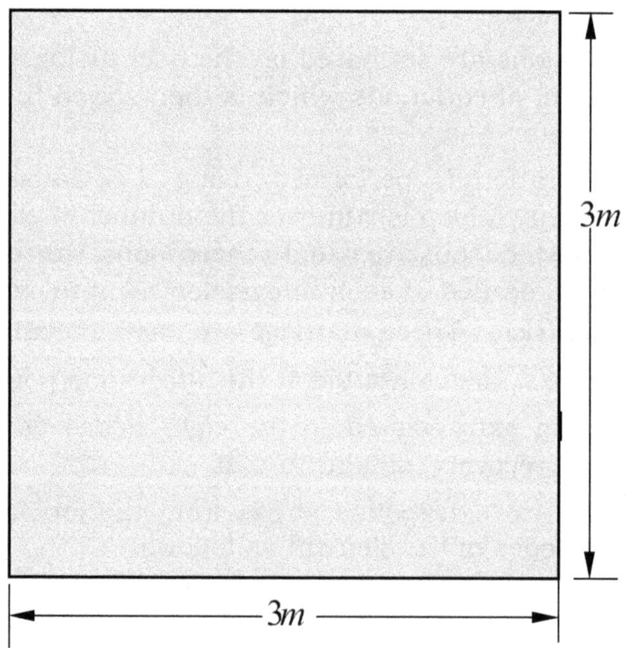

Figure 11.36 Simply supported square plate.

Analyse a square plate of dimensions $3\ m \times 3\ m \times 0.1\ m$ simply supported on all edges as shown in the Figure 11.36. Young's modulus of the plate is $2.0 \times 10^8\ kPa$

and Poisson's ratio is 0.3. The plate is subjected to a uniform pressure of 6850 kPa normal to its surface.

Owing to symmetry, only a quarter of the plate is analysed. The MATLAB code is given in *Appendix A.14.* Results are also computed using ABAQUS.

The program listed in section A.14 is utilized for obtaining the results.

The following is the input file used for analysis:

```
1.5 1.5  0.1 10 10 2.0e8  0.0  0.3  -6895.0 0  0.0
%E in kPa , dimensions in metres (EXAMPLE-1)
%pressure in kPa
```

The following is the dialog generated during execution of the program to input for other quantities. Note that the responses are marked in bold font:

```
enter edge restraints in the order [Bottom Top Left Right]
enter 0 for restrained edge and 1 for simply-supported edge
Bottom Edge(1 or 0): 1
Top Edge(1 or 0): 2
Left Edge(1 or 0): 1
Right Edge(1 or 0): 2
```

Figure 11.37 shows the mesh generated using the MATLAB program.

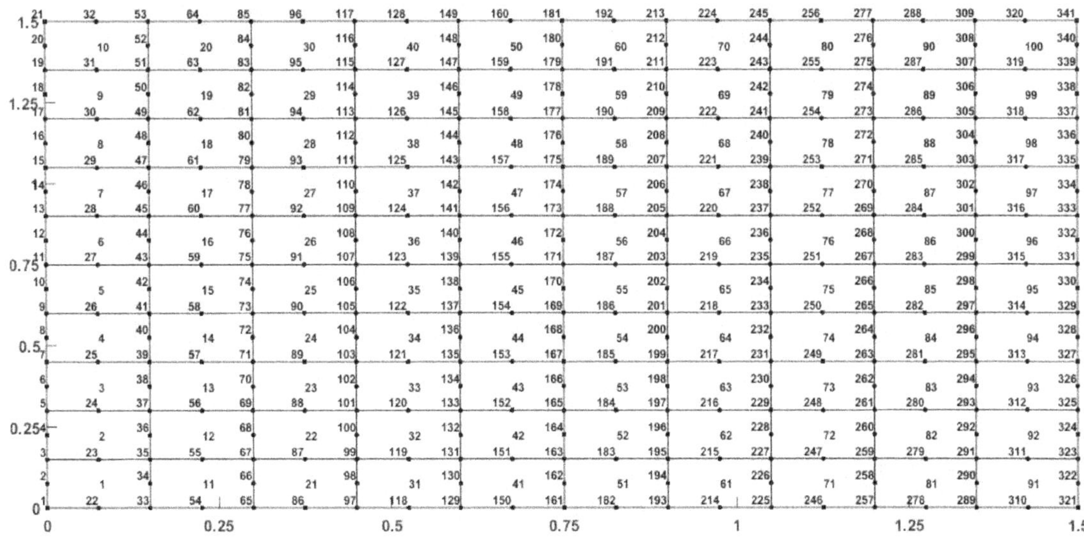

Figure 11.37 Mesh generated for quarter square plate.

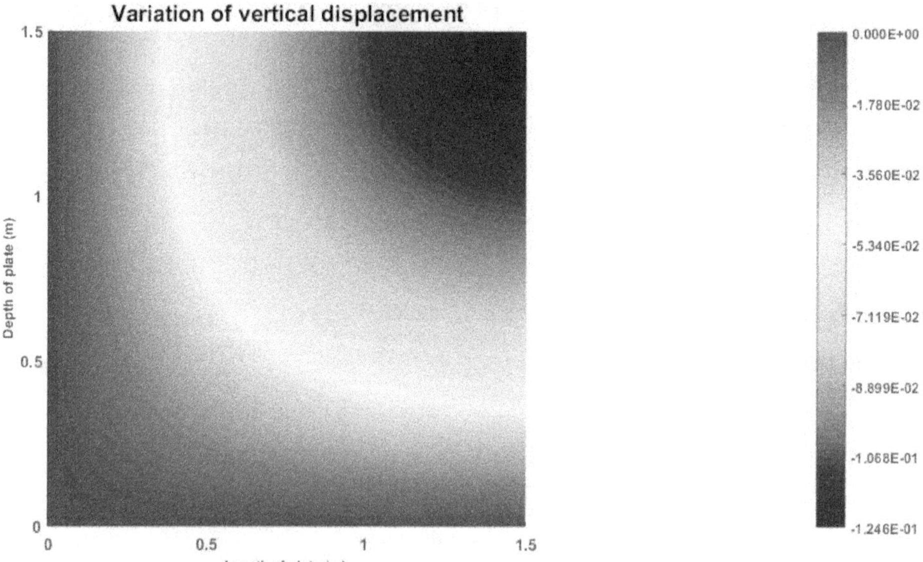

Figure 11.38 Variation of vertical *displacement* across the quarter square plate.

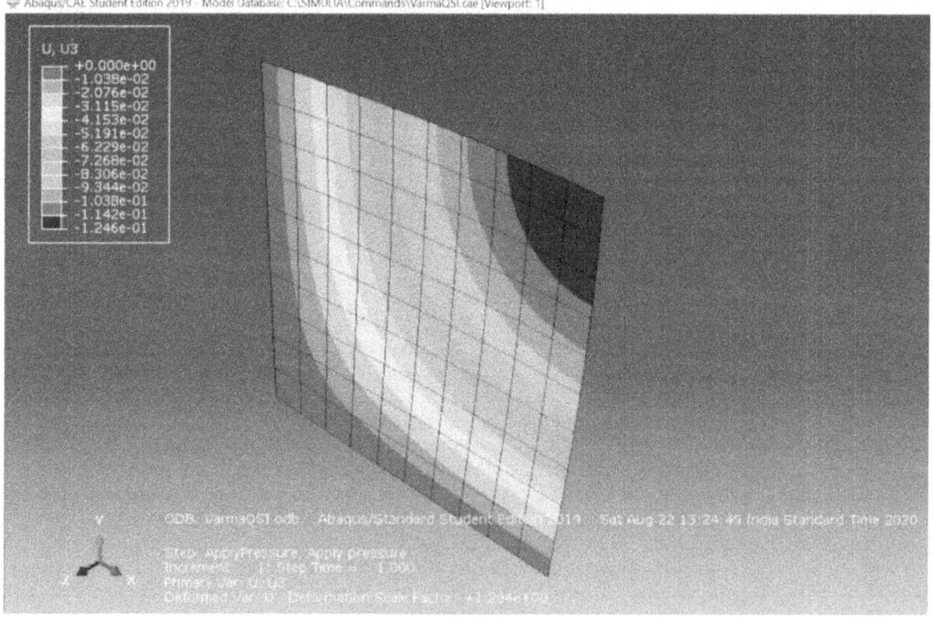

Figure 11.39 Variation of vertical displacement across the quarter square plate.

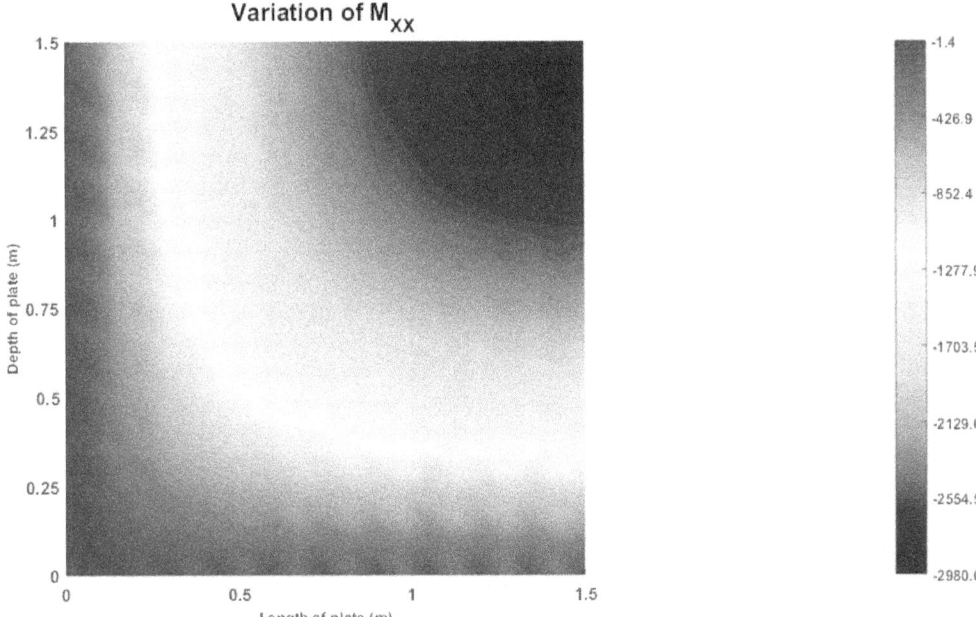

Figure 11.40 Variation of M_{XX} across the quarter square plate.

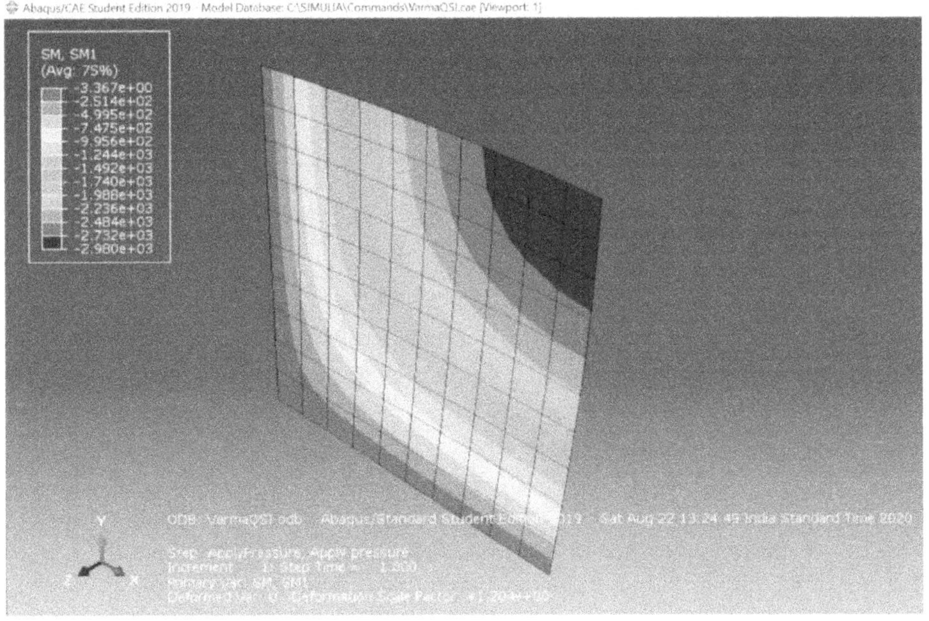

Figure 11.41 ABAQUS -Variation of M_{XX} across the quarter square plate.

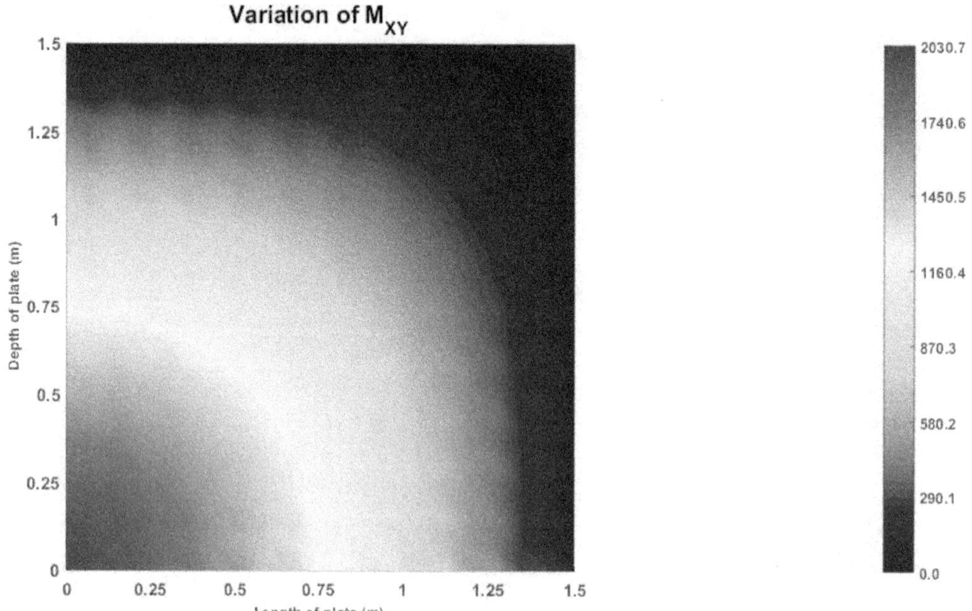

Figure 11.42 Variation of M_{XY} across the quarter square plate.

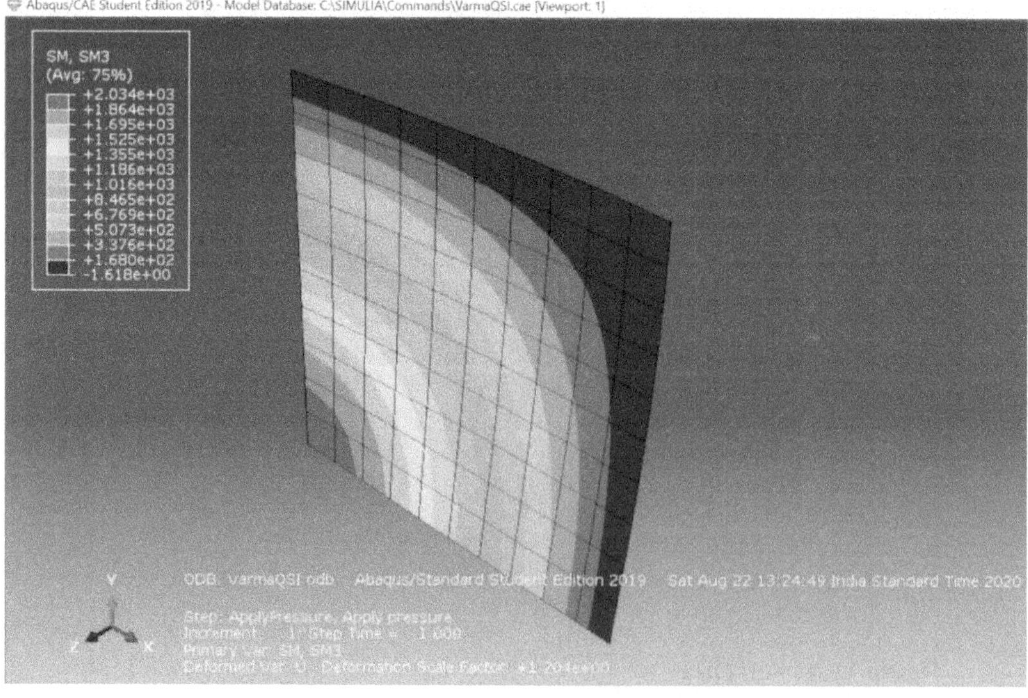

Figure 11.43 ABAQUS-Variation of M_{XY} across the quarter square plate.

Figure 11.38 and Figure 11.39 depict the variation of vertical displacement across the quarter plate obtained using MATLAB code and ABAQUS respectively. Figure 11.40 and Figure 11.41 depict the variation of M_{XX} across the quarter plate obtained using MATLAB code and ABAQUS respectively. Figure 11.42 and Figure 11.43 depict the variation of M_{XY} across the quarter plate obtained using MATLAB code and ABAQUS respectively.

Numerical Example 2- Rectangular plate

Analyse a rectangular plate of dimensions $1.5\ m \times 1\ m \times 0.05\ m$ shown in the figure 11.44. The plate is clamped on four edges and is subjected to a point load of 500kN at the center. The Young's modulus of the material of the plate is $2 \times 10^8\ kPa$ and Poisson's ratio is 0.3. Discretise the plate using eight-node quadrilateral elements.

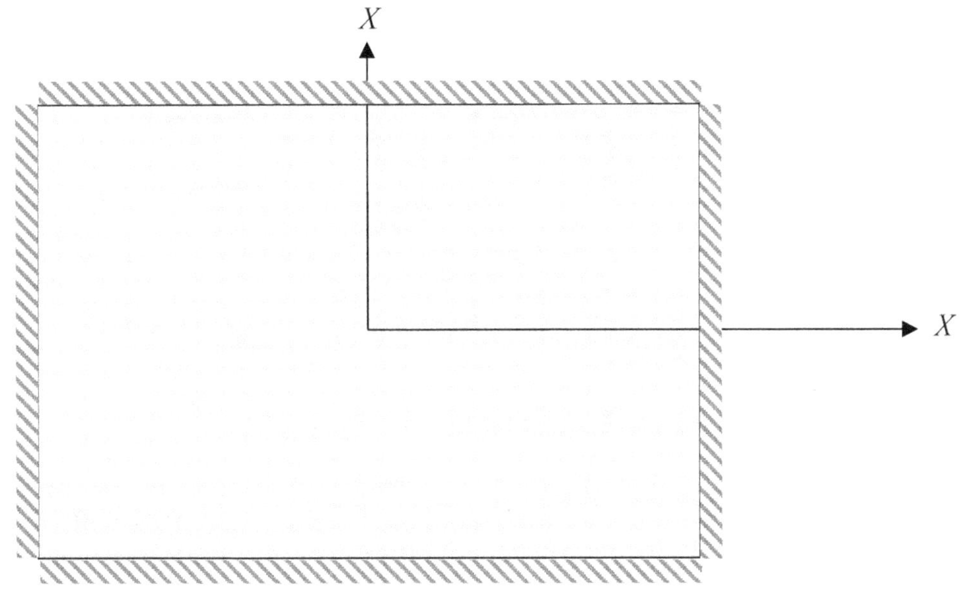

Figure 11.44 Clamped rectangular plate subjected to point load at center.

The program listed in *Appendix A.14* is utilized for obtaining the results.

The following is the input file used for analysis:

```
1.5 1.0  0.025  10  8   2.1e8  0.0   0.3   0.0  1 0.0
7   -50.0
%load(kN)at bottom-left quarter rectangular plate
%E in kPa , dimensions in metres (EXAMPLE-2)
```

The following is the dialog generated during execution of the program to input for other quantities. Note that the responses are marked in bold font:

```
enter edge restraints in the order [Bottom Top Left Right]
enter 0 for restrained edge and 1 for simply-supported edge
Bottom Edge(1 or 0): 0
Top Edge(1 or 0): 0
Left Edge(1 or 0): 0
Right Edge(1 or 0): 0
```

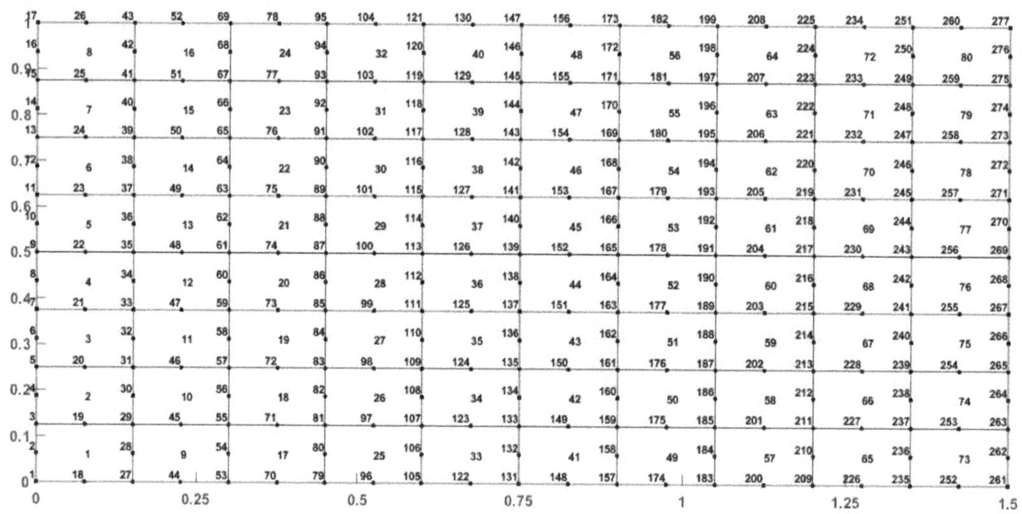

Figure 11.45 Mesh for the rectangular plate.

The mesh generated using 8 node quadrilateral elements is shown in Figure 11.45 above. The plate is also analysed using ABAQUS software **S8R** *elements*. The results obtained using the present program match very well those obtained using ABAQUS.

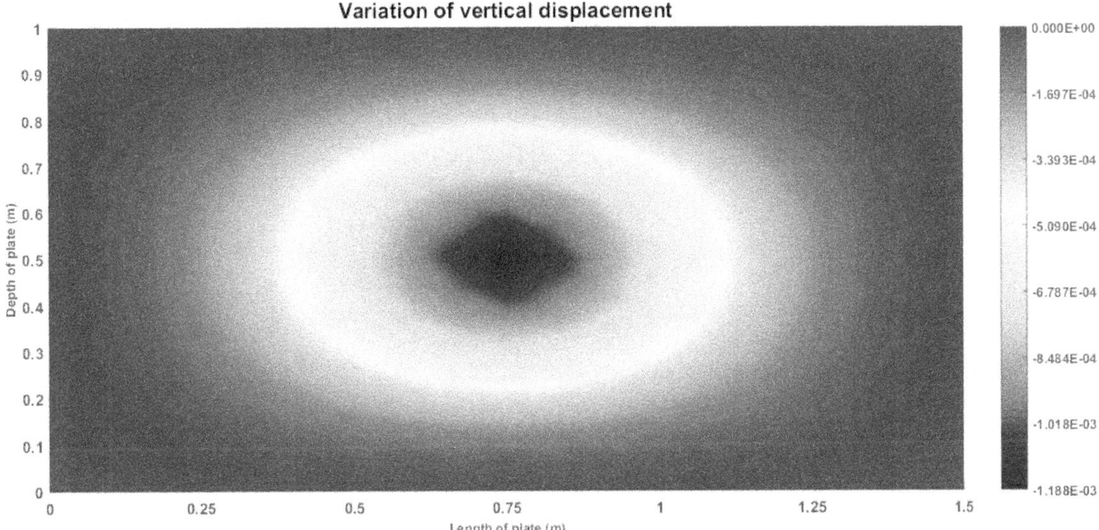

Figure 11.46 Variation of Vertical displacement across the rectangular plate.

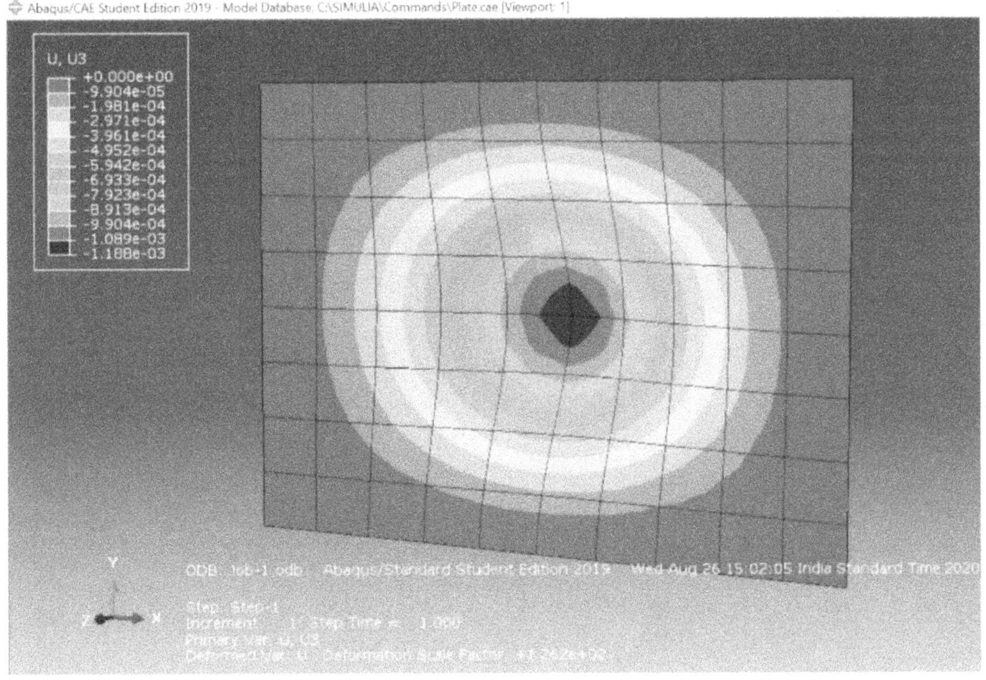

Figure 11.47 ABAQUS- Variation of vertical displacement across the rectangular plate.

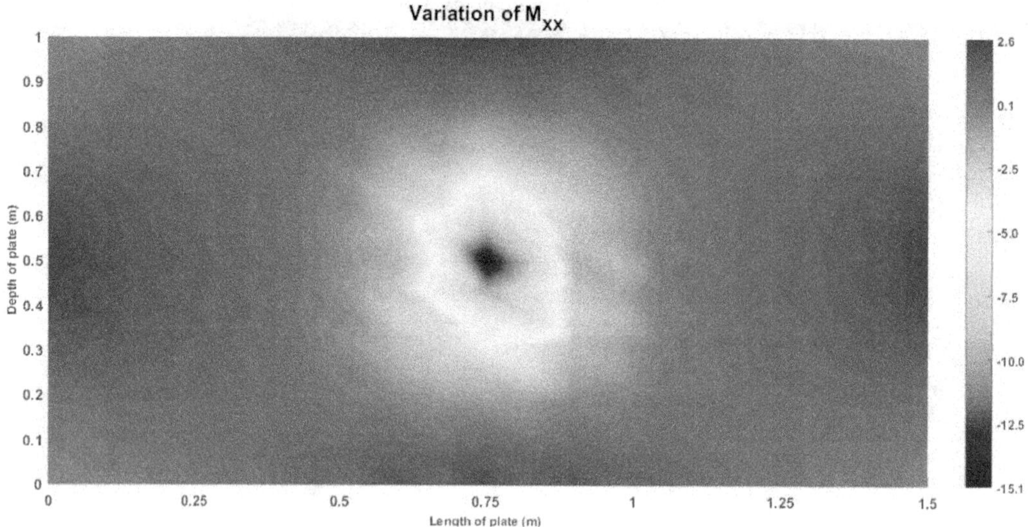

Figure 11.48 Variation of M_{XX} across the rectangular plate.

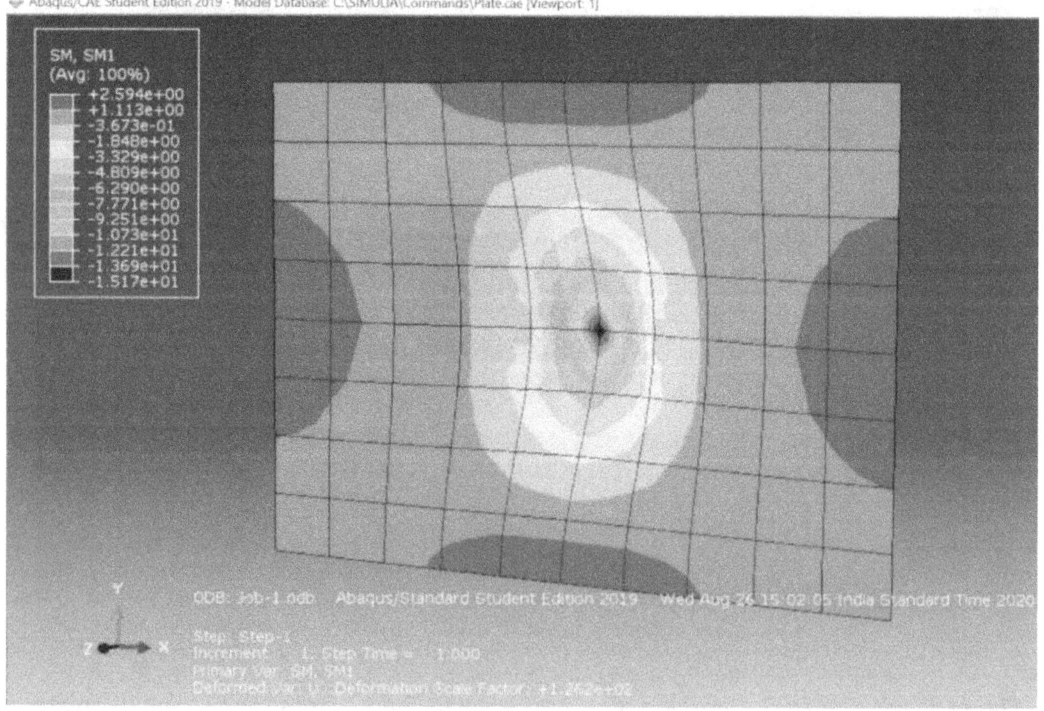

Figure 11.49 ABAQUS-Variation of M_{XX} across the rectangular plate.

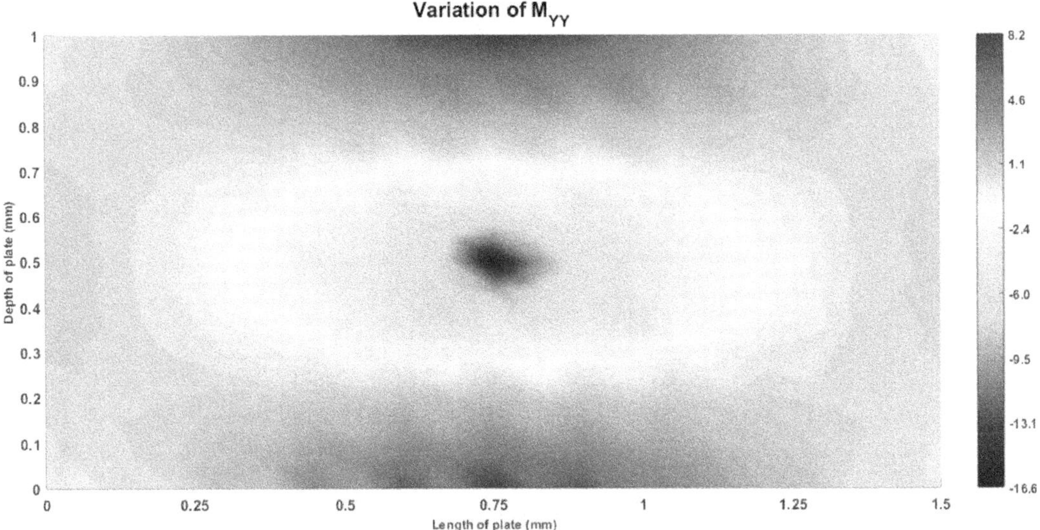

Figure 11.50 Variation of M_{YY} across the rectangular plate.

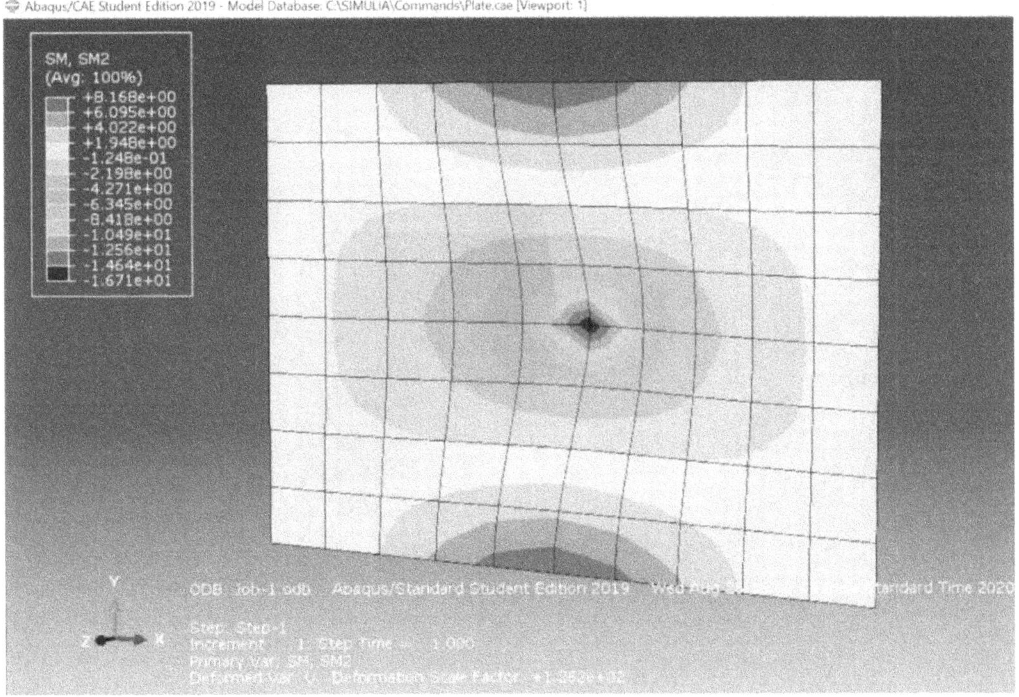

Figure 11.51 ABAQUS-Variation of M_{YY} across the rectangular plate.

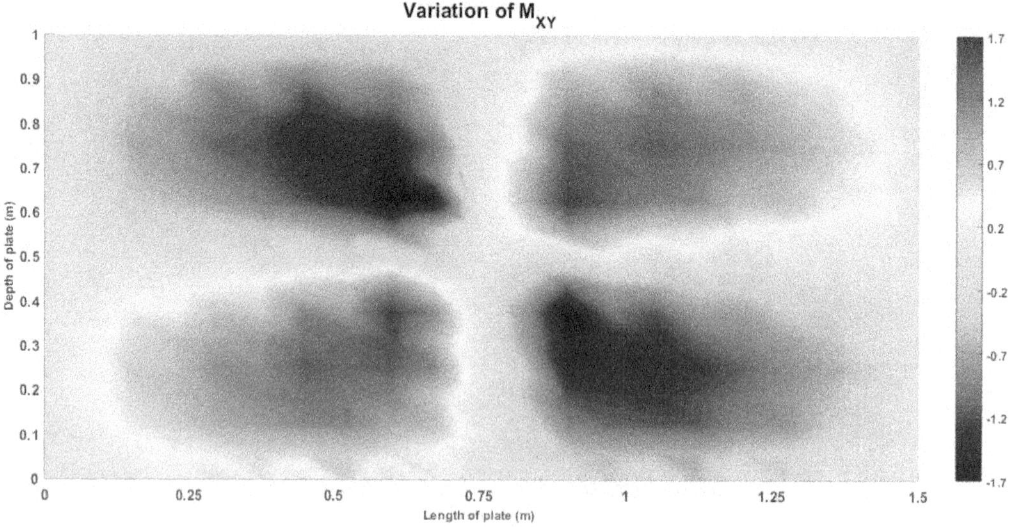

Figure 11.52 Variation of M_{XY} across the rectangular plate.

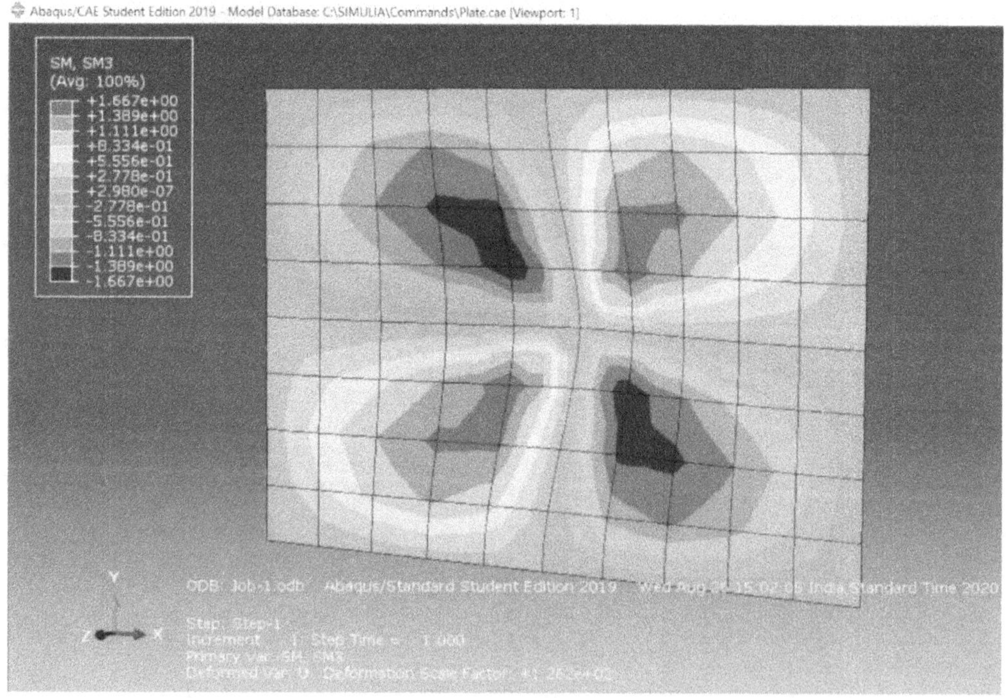

Figure 11.53 ABAQUS-Variation of M_{XY} across the rectangular plate.

Figure 11.46 depicts the variation of vertical displacement across the square plate obtained using the MATLAB code given in *Appendix A.14*. The variation of vertical displacement obtained using ABAQUS is shown in Figure 11.47. Figure 11.48 and Figure 11.49 show the variation of M_{XX} across the plate computed using the MATLAB code and ABAQUS, respectively. Figure 11.50 and Figure 11.51 show the variation of M_{YY} across the plate computed using the MATLAB code and ABAQUS, respectively. Figure 11.52 and Figure 11.53 show the variation of M_{XY} across the plate computed using the MATLAB code and ABAQUS, respectively.

11.7 EXERCISE PROBLEMS

Solve the following problems using

   (a) *4-node rectangular plate elements*

   (b) *4-node quadrilateral shell elements*

   (c) *8- node quadrilateral shell elements*

1. Analyse the square plate of side 2 m, using four node rectangular plate elements under conditions of plane stress. The thickness of the plate is 0.05m, Young's modulus is $2.1 \times 10^8 \, kN/m^2$ and $\mu = 0.3$. The plate is subjected to a pressure of 80.0 kPa normal to its surface. The plate is clamped on all edges.

2. A rectangular plate of dimensions $2\,m \times 1.5\,m \times 0.025\,m$ is subjected to a point load of 50 kN. The Young's modulus of the plate is $2.1 \times 10^8 \, kN/m^2$ and Poisson's ratio is 0.3. Analyse the plate for each of the following support conditions:

   (a) all edges clamped
   (b) all edges simply supported
   (c) opposite edges clamped

3. Analyse a rectangular plate of dimensions $1.5\,m \times 1\,m \times 0.02\,m$. The plate is clamped on short edges and is subjected to a uniform normal pressure of $80\,kPa$. The Young's modulus of the material of the plate is $2.1 \times 10^8 \, kN/m^2$ and Poisson's ratio is 0.3.

Free Vibration Analysis of Structures

Static analysis of structures with one-dimensional, two-dimensional, axisymmetric and three-dimensional elements was taken up in the preceding chapters. In each structure, element stiffness matrix and force vectors were computed and are used in the assembly of structure stiffness matrix and force vector. Mass matrix of the structure can be computed on similar lines and the dynamic analysis of structures can be performed. The present chapter deals with the dynamic analysis of structures.

12.1 FREE VIBRATION ANALYSIS OF STRUCTURES

Consider a multi-degree of freedom system with $[m], [c]$ and $[k]$ being mass, damping and stiffness matrices respectively. Here, $\{F(t)\}$ is the vector of externally applied forces. Following the D'Alembert's principle, *Inertial force+ Viscous force + Elastic force = Externally applied force.*

Here,

(a) inertial force is the product of mass matrix $[m]$ and vector of accelerations $\{\ddot{v}\}$

(b) damping force is the product of damping matrix $[c]$ and the vector of velocities $\{\dot{v}\}$

(c) elastic force is the product of stiffness matrix $[k]$ and the vector of displacements $\{v\}$

Thus, the d'Alembert's principle can be stated as

$$[m]\{\ddot{v}\}+[c]\{\dot{v}\}+[k]\{v\}=\{F(t)\} \qquad .....(12.1)$$

Considering free vibration, the above equation can be written as

$$[m]\{\ddot{v}(t)\}+[k]\{v(t)\}=\{0\} \qquad .....(12.2)$$

Substituting $\{v\} = \{A\} e^{i\omega t}$

$$\qquad\qquad\qquad\qquad\qquad\qquad\qquad\qquad\qquad\qquad .....(12.3)$$

$$\{\ddot{v}(t)\} = -\omega^2 \{v(t)\} \qquad\qquad\qquad\qquad\qquad\qquad .....(12.4)$$

We obtain,

$$\left| k - \omega^2 m \right| = 0 \qquad\qquad\qquad\qquad\qquad\qquad .....(12.5)$$

We need to solve the eigen value problem given by Eq.12.5 to obtain natural frequencies and mode shapes of the structure.

The free vibration analysis of various structures is described in the succeeding sections.

12.2 FREE VIBRATION ANALYSIS OF BEAMS

Consider a beam as shown in the Figure 12.1. The beam is divided into 4 elements. Nodes 1 and 5 are fixed. The objective now is to perform a dynamic analysis of the beam.

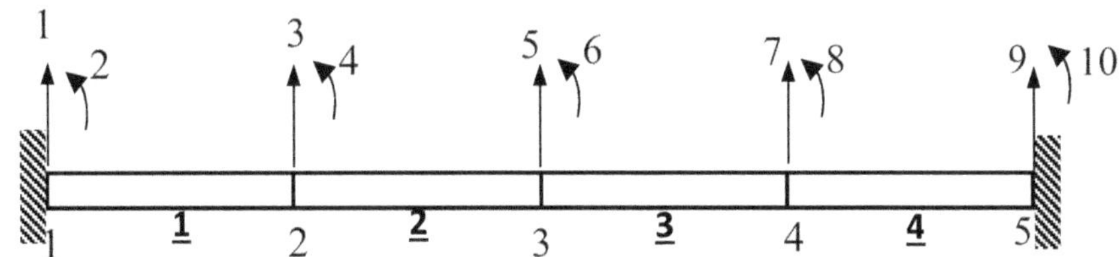

Figure 12.1 Fixed beam.

Stiffness matrix for each element can be computed and the stiffness matrix for the entire structure can be developed using the procedure described in section 63. The next sections explain about Lumped mass matrix, consistent mass, and damping matrices.

12.2.1 Lumped Mass Matrix

Lumped mass approach is the simplest method to define the inertial properties of a dynamic system. In this approach, the mass of an element of the beam is obtained by the product of linear mass (mass per unit length of beam) and length of element. Half of this mass is lumped at each of the nodes of the element, where translational displacements are defined. This distribution of the mass is determined by statics. To assemble the mass matrix, one just needs to add contributions of lumped masses at the translational degrees of nodes.

Figure 12.2 shows the distribution of lumped mass for a) uniform distribution of mass, b) triangular distribution of mass and c) general mass distribution respectively.

In this method, the rotational degrees of freedom usually are assigned zero values. However, a finite value of mass moment of inertia may be assigned to the rotational degrees of freedom.

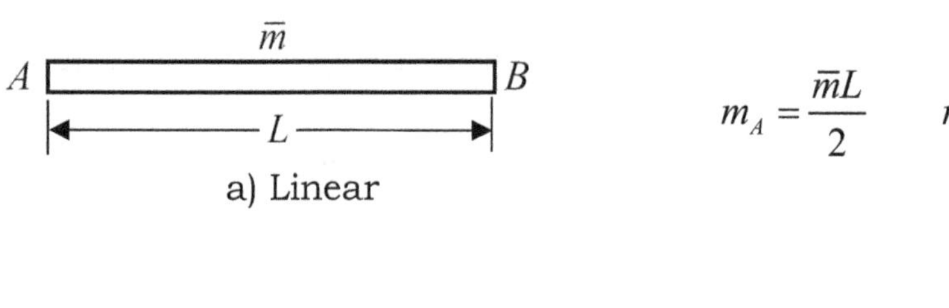

$$m_A = \frac{\overline{m}L}{2} \qquad m_B = \frac{\overline{m}L}{2}$$

a) Linear

$$m_A = \frac{\overline{m}L}{6} \qquad m_B = \frac{\overline{m}L}{3}$$

b) Triangular

$$m_A = \frac{\int_0^L (L-x)\overline{m}(x)\,dx}{L} \qquad m_B = \frac{\int_0^L x\overline{m}(x)\,dx}{L}$$

c) General

Figure 12.2 Distribution of Lumped mass.

12.2.2 Consistent Mass Matrix

Mass coefficients corresponding to the degrees of freedom of a beam element are evaluated using a procedure similar to that of obtaining stiffness matrix. The Hermite cubic polynomial interpolation functions N_i for the beam element are given by Eq. 4.32. These functions are used to evaluate the mass matrix as follows:

The linear mass (*defined as mass per unit length*) of the beam is denoted by can be expressed as $\overline{m} = \dfrac{m}{L}$, where m is the total mass of the beam and L is the corresponding length of the beam.

The mass matrix of an element can now be evaluated as

$$m_{ij} = \int_0^L \overline{m}(x) N_i(x) N_j(x)\, dx, \quad i, j = 1, 4 \qquad \dots\dots(12.6)$$

$$\left[m^{(e)}\right] = \frac{\overline{m}L}{420}
\begin{bmatrix}
156 & 22L & 54 & -13L \\
22L & 4L^2 & 13L & -3L^2 \\
54 & 13L & 156 & -22L \\
-13L & -3L^2 & -22L & 4L^2
\end{bmatrix} \qquad \dots\dots(12.7)$$

The mass matrix for the entire system can be obtained by a process of assembly similar to that of stiffness matrix.

A few points need to be understood here. Compared to consistent mass matrix, the lumped mass matrix formulation can be made with relative ease and lesser computation effort compared to the consistent mass matrix formulation. However, the lumped mass matrix contains zeroes (that correspond to the rotational degrees of freedom) in the diagonal matrix. The corresponding entries in a consistent mass matrix are not zero owing to *mass coupling.* The zeroes in the diagonal of lumped mass matrix are eliminated by a process of static condensation of rotational degrees of freedom, thus reducing the size of the problem. Nevertheless, consistent mass matrix gives results that are closer to the exact solution compared to the lumped mass matrix.

12.2.3   Numerical Example 1

Compute the lumped mass matrix and consistent mass matrix for the cantilever

beam of 3m span as shown in the Figure 12.3. Given $\overline{m} = 420\dfrac{kg}{m}$:

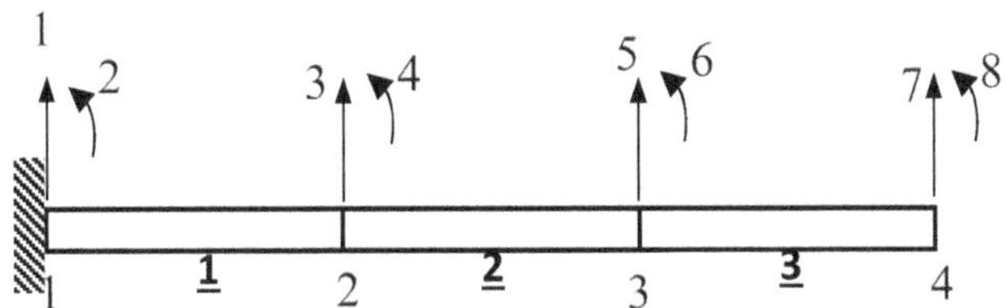

Figure 12.3 Cantilever beam.

The above cantilever beam is divided into 3 elements, each of 1 m length. The beam has 4 nodes, with each node having a translational degree of freedom and a rotational degree of freedom as shown in the Figure 12.3. Thus, the entire beam has 4 translational degrees of freedom and 4 rotational degrees of freedom.

So, in case of lumped mass matrix, mass lumped at each node at the translational degree of freedom $= \dfrac{\overline{m}L}{2} = \dfrac{420 \times 1}{2} = 210 \ kg$. It is to be noted that the node 1 receives contribution from element 1 alone, node 2 receives contribution from the elements 1 and 2, node 3 receives contribution from the elements 2 and 3 and node 4 receives contribution from element 3 alone. The lumped mass matrix for each element is as follows:

$$\left[m^{(e)} \right] = \frac{\overline{m}L}{2} \begin{bmatrix} 1 & 0 & 0 & 0 \\ 0 & 0 & 0 & 0 \\ 0 & 0 & 1 & 0 \\ 0 & 0 & 0 & 0 \end{bmatrix} = \begin{bmatrix} 210 & 0 & 0 & 0 \\ 0 & 0 & 0 & 0 \\ 0 & 0 & 210 & 0 \\ 0 & 0 & 0 & 0 \end{bmatrix} \qquad .....(12.8)$$

The consistent mass matrix, following the formulation of Eq. 12.7, is written as

$$\left[m^{(e)} \right] = \frac{\overline{m}L}{420} \begin{bmatrix} 156 & 22L & 54 & -13L \\ 22L & 4L^2 & 13L & -3L^2 \\ 54 & 13L & 156 & -22L \\ -13L & -3L^2 & -22L & 4L^2 \end{bmatrix} = \begin{bmatrix} 156 & 22 & 54 & -13 \\ 22 & 4 & 13L & -3 \\ 54 & 13 & 156 & -22 \\ -13 & -3 & -22 & 4 \end{bmatrix} \qquad .....(12.9)$$

The assembled mass matrix for the beam element for lumped mass is given as $diag \lfloor 210 \ 0 \ 420 \ 0 \ 420 \ 0 \ 210 \ 0 \rfloor$. The degrees of freedom 1 and 2 are restrained for the cantilever beam. Thus, rows 1 and 2 and corresponding columns 1,2 are eliminated. Thus, the lumped mass matrix for the beam can be expressed as $diag \lfloor 420 \ 0 \ 420 \ 0 \ 210 \ 0 \rfloor$.

The assembled consistent mass matrix is given by

$$[M] = \begin{bmatrix} 156 & 22 & 54 & -13 & 0 & 0 & 0 & 0 \\ 22 & 4 & 13 & -3 & 0 & 0 & 0 & 0 \\ 54 & 13 & 312 & 0 & 54 & -13 & 0 & 0 \\ -13 & -3 & 0 & 8 & 13 & -3 & 0 & 0 \\ 0 & 0 & 54 & 13 & 312 & 0 & 54 & -13 \\ 0 & 0 & -13 & -3 & 0 & 8 & 13 & -3 \\ 0 & 0 & 0 & 0 & 54 & 13 & 156 & -22 \\ 0 & 0 & 0 & 0 & -13 & -3 & -22 & 4 \end{bmatrix} \qquad .....(12.10)$$

After striking off rows and columns 1, 2 from the above mass matrix we obtain

$$[M] = \begin{bmatrix} 312 & 0 & 54 & -13 & 0 & 0 \\ 0 & 8 & 13 & -3 & 0 & 0 \\ 54 & 13 & 312 & 0 & 54 & -13 \\ -13 & -3 & 0 & 8 & 13 & -3 \\ 0 & 0 & 54 & 13 & 156 & -22 \\ 0 & 0 & -13 & -3 & -22 & 4 \end{bmatrix} \qquad \dots\dots(12.11)$$

12.2.4 Numerical Example 2

Compute the natural frequencies for the fixed beam given in Figure 12.4. The span of the beam is 4m and the cross-sectional dimensions of the beam are $0.02m \times 0.03\ m$ .Adopt $E = 2 \times 10^{11}\ N/m^2$ and $\rho = 7850\ kg/m^3$.

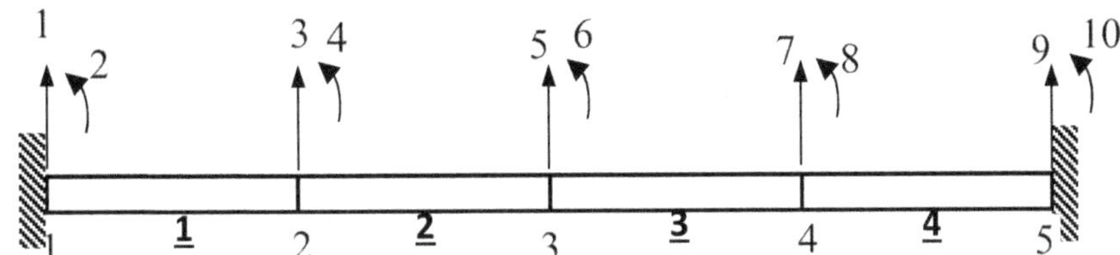

Figure 12.4 Fixed beam.

The beam is divided into 4 elements, each of length $L_e = 1\ m$. Linear mass $\bar{m}$ is calculated by computing the product of mass density ρ with length of each element $viz., \bar{m} = \rho A$. Results are also obtained using the MATLAB code given in *Section A.16.*

$$I = \frac{0.02 \times 0.03^3}{12} = 4.5 \times 10^{-8}\ m^4 \Rightarrow EI = 9 \times 10^3\ Nm^2$$

$$A = 0.02 \times 0.03 = 6 \times 10^{-4}\ m^2$$

$\bar{m} = 7850 \times 0.02 \times 0.03 = 4.71\ kg/m$ is the linear mass

The element stiffness matrix for elements 1, 2, 3 and 4 is

$$[K^{(e)}] = 10^5 \begin{bmatrix} 1.08 & 0.54 & -1.08 & 0.54 \\ 0.54 & 0.36 & -0.54 & 0.18 \\ -1.08 & -0.54 & 1.08 & -0.54 \\ 0.54 & 0.18 & -0.54 & 0.36 \end{bmatrix} \qquad \dots\dots(12.12)$$

The structure stiffness matrix is assembled and the rows and columns corresponding to 1, 2, 9 and 10 degrees of freedom (*restrained degrees of freedom*) are struck off, resulting in

$$[K] = 10^9 \begin{bmatrix} 2.16 & 0 & -1.08 & 0.54 & 0 & 0 \\ 0 & 0.72 & -0.54 & 0.18 & 0 & 0 \\ -1.08 & -0.54 & 2.16 & 0 & -1.08 & 0.54 \\ 0.54 & 0.18 & 0 & 0.72 & -0.54 & 0.18 \\ 0 & 0 & -1.08 & -0.54 & 1.08 & -0.54 \\ 0 & 0 & 0.54 & 0.18 & -0.54 & 0.36 \end{bmatrix} \qquad ......(12.13)$$

Consistent mass matrix for the element is obtained by substituting $L_e = 1\,m$ in Eq. 12.9 as

$$\left[m^{(e)}\right] = \begin{bmatrix} 1.7494 & 0.2467 & 0.6056 & -0.1458 \\ 0.2467 & 0.0449 & 0.1458 & -0.0336 \\ 0.6056 & 0.1458 & 1.7494 & -0.2467 \\ -0.1458 & -0.0336 & -0.2467 & 0.0449 \end{bmatrix} \qquad .....(12.14)$$

The consistent mass matrix for the entire structure is assembled in the same way as the stiffness matrix. Then rows and columns corresponding to restrained degrees of freedom are struck off, resulting in the following consistent matrix:

$$[M] = \begin{bmatrix} 3.4989 & 0 & 0.6056 & -0.1458 & 0 & 0 \\ 0 & 0.6056 & 0.1458 & -0.0336 & 0 & 0 \\ 0.6056 & 0.1458 & 3.4989 & 0 & 0.6056 & -0.1458 \\ -0.1458 & -0.0336 & 0 & 0.0897 & 0.1458 & -0.0336 \\ 0 & 0 & 0.6056 & 0.1458 & 1.7494 & -0.2467 \\ 0 & 0 & -0.1458 & -0.0336 & -0.2467 & 0.0449 \end{bmatrix} \qquad .....(12.15)$$

The stiffness and mass matrices $[K]$ and $[M]$ are substituted in Eq. 12.10 and the eigen value problem is solved. The resulting natural frequencies are shown in the second column of Table 12.1.

Lumped mass matrix for the element is obtained by

$$\left[m^{(e)}\right] = \frac{\bar{m}L_e}{2}\begin{bmatrix} 1 & 0 & 0 & 0 \\ 0 & 0 & 0 & 0 \\ 0 & 0 & 1 & 0 \\ 0 & 0 & 0 & 0 \end{bmatrix} = \begin{bmatrix} 2.355 & 0 & 0 & 0 \\ 0 & 0 & 0 & 0 \\ 0 & 0 & 2.355 & 0 \\ 0 & 0 & 0 & 0 \end{bmatrix} \quad .....(12.16)$$

The lumped mass matrix for the entire structure after the process of assembly and striking of rows corresponding to restrained degrees of freedom can be written as $diag\begin{bmatrix} 4.71 & 0 & 4.71 & 0 & 2.335 & 0 \end{bmatrix}$. The degrees of freedom 2, 4 and 6 contain zero entries corresponding to rotational inertia that is neglected in the lumped mass formulation. These rotational degrees of freedom are eliminated both in the lumped mass matrix and assembled stiffness matrix using a process of static condensation, resulting in the following reduced mass and stiffness matrices respectively.

$$M_R = \begin{bmatrix} 4.71 & 0 & 0 \\ 0 & 4.71 & 0 \\ 0 & 0 & 2.355 \end{bmatrix} \text{ and } [K_R] = 10^5\begin{bmatrix} 1.6615 & -0.9554 & 0.2492 \\ -0.9554 & 0.9138 & -0.3323 \\ 0.2492 & -0.3323 & 0.1454 \end{bmatrix} \quad ...(12.17)$$

Substituting $[K]$ and $[M]$ in Eq.12.5 and solving eigen-value problem, we obtain the natural frequencies as shown in the third column of Table 12.1 as follows:

Table 12.1 Natural frequencies of beam $\omega\left(rad/\sec\right)$.

Mode	Consistent mass matrix	Lumped mass matrix
1	17.079	16.250
2	107.373	91.729
3	303.397	228.417
4	683.240	
5	1285.859	
6	2563.507	

Natural Frequencies of beams and plane frames can also be computed using the MATLAB code given section A.16.

12.3 FREE VIBRATION ANALYSIS OF BARS AND TRUSSES

12.3.1 Numerical Example 3

Consider a bar element shown in the Figure 12.5

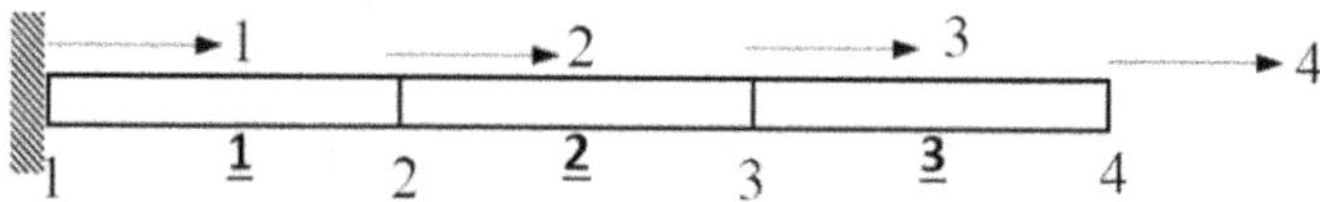

Figure 12.5 Bar with fixed-free end conditions.

The length of the bar is $1.5m$ and cross-sectional dimensions of the bar are $0.02\ m \times 0.02\ m$. Adopt $E = 2 \times 10^{11}\ N/m^2$ and $\rho = 7850\ kg/m^3$.

Free vibration analysis is performed using MATLAB code provided in Section A.17. The total degrees of freedom of the beam are 4 and degree of freedom 1 is restrained. Length of each element = $L_e = 0.5\ m$

Linear mass $\bar{m} = \rho A = 7850 \times 0.02 \times 0.02 = 3.14\ kg/m$

The stiffness matrix of elements $1, 2$ and 3 is given as

$$\left[k^{(e)}\right] = \frac{AE}{L_e}\begin{bmatrix} 1 & -1 \\ -1 & 1 \end{bmatrix} = \frac{0.02 \times 0.02 \times 2 \times 10^{11}}{0.5}\begin{bmatrix} 1 & -1 \\ -1 & 1 \end{bmatrix} = 10^8 \begin{bmatrix} 1 & -1 \\ -1 & 1 \end{bmatrix} \qquad .....(12.18)$$

The corresponding consistent mass matrix is given as

$$\left[m^{(e)}\right] = \frac{\bar{m}L_e}{6}\begin{bmatrix} 2 & 1 \\ 1 & 2 \end{bmatrix} = \frac{3.14 \times 0.5}{6}\begin{bmatrix} 2 & 1 \\ 1 & 2 \end{bmatrix} = \begin{bmatrix} 0.5233 & 0.2617 \\ 0.2617 & 0.5233 \end{bmatrix} \qquad .....(12.19)$$

The structure stiffness matrix and mass matrix are assembled and rows and columns corresponding to the restrained degrees of freedom (1st row and 1st column) are struck off, resulting in the following matrices

$$[K] = 10^9 \begin{bmatrix} 4 & -2 & 0 \\ -2 & 4 & -2 \\ 0 & -2 & 2 \end{bmatrix} \text{ and } [M] = \begin{bmatrix} 52.333 & 13.0833 & 0 \\ 13.0833 & 52.333 & 13.0833 \\ 0 & 13.0833 & 26.167 \end{bmatrix} \qquad .....(12.20)$$

Substituting $[K]$ and $[M]$ in Eq.12.5 and solving eigen-value problem, we obtain the natural frequencies as shown in the first column of Table 12.2. The exact values of frequency for this beam are given by

$$\omega_r = \frac{(2r-1)\pi}{2L}\sqrt{\frac{EA}{\bar{m}}}, \quad r = 1, 2, 3, ... N \qquad .....(12.21)$$

Table 12.2 Natural frequencies of bar $\omega \times 10^4 \left(rad \, / \sec \right)$.

Mode	Using MATLAB code	Exact values
1	0.2673	0.2649
2	0.8742	0.7929
3	1.5860	1.3214

As the number of elements increase, the system becomes more flexible and consequently, the natural frequencies will reduce and approach the exact values.

12.3.2 Numerical Example 4

Consider a Warren truss consisting of two bays, each of span 6 meters, as shown in Figure 12.6. The Young's modulus of material of element is $E = 2 \times 10^{11} \, N/m^2$. The cross-sectional area of each element of truss is $0.01 \, m^2$ and the density is $\rho = 7850 \, kg/m^3$.

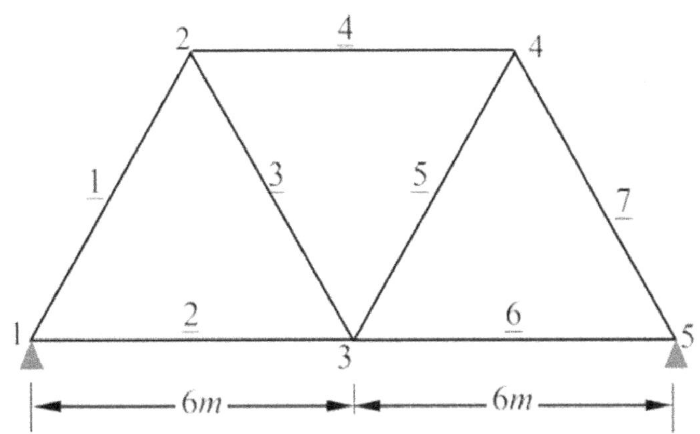

Figure 12.6 Two-bay Warren truss.

The problem is solved using the MATLAB code provided in Section A.17.

The element stiffness matrix and mass matrix w.r.t member axes are

$$\left[k^{(e)} \right] = 10^8 \begin{bmatrix} 3.3333 & 0 & -3.3333 & 0 \\ 0 & 0 & 0 & 0 \\ -3.3333 & 0 & 3.3333 & 0 \\ 0 & 0 & 0 & 0 \end{bmatrix} \quad \text{and} \quad \begin{bmatrix} 157 & 0 & 78.5 & 0 \\ 0 & 157 & 0 & 78.5 \\ 78.5 & 0 & 157 & 0 \\ 0 & 78.5 & 0 & 157 \end{bmatrix}$$

The stiffness and mass matrices are assembled and the rows and columns corresponding to restrained degrees of freedom are struck off. The resulting stiffness and mass matrices are given as follows:

$$[K] = 10^8 \begin{bmatrix} 5.0 & 0 & -0.833 & 1.4434 & -3.333 & 0 & 0 \\ 0 & 5.0 & 1.4434 & -2.5 & 0 & 0 & 0 \\ -0.8333 & 1.4434 & 8.3333 & 0 & -0.833 & -1.4434 & -3.333 \\ 1.4434 & -2.5 & 0 & 5.0 & -1.4434 & -2.5 & 0 \\ -3.333 & 0 & -0.8333 & -1.4434 & 5.0 & 0 & -0.833 \\ 0 & 0 & -1.4434 & -2.5 & 0 & 5.0 & 1.4434 \\ 0 & 0 & -3.3333 & 0 & -0.833 & 1.4434 & 4.1667 \end{bmatrix}$$

$$[M] = \begin{bmatrix} 471 & 0 & 78.5 & 0 & 78.5 & 0 & 0 \\ 0 & 471 & 0 & 78.5 & 0 & 78.5 & 0 \\ 78.5 & 0 & 628 & 0 & 78.5 & 0 & 78.5 \\ 0 & 78.5 & 0 & 628 & 0 & 78.5 & 0 \\ 78.5 & 0 & 78.5 & 0 & 471 & 0 & 78.5 \\ 0 & 78.5 & 0 & 78.5 & 0 & 471 & 0 \\ 0 & 0 & 78.5 & 0 & 78.5 & 0 & 314 \end{bmatrix}$$

Substituting $[K]$ and $[M]$ in Eq. 12.5 and solving eigen-value problem, we obtain the natural frequencies as shown in the third column of Table 12.3 as follows:

Table 12.3 Natural frequencies of truss.

Mode	$\omega \times 10^3 \ (rad / \sec)$
1	0.3088
2	0.4277
3	0.8658
4	1.0251
5	1.2856
6	1.5045
7	1.7405

Free vibration analysis of bars and plane trusses can also be performed using MATLAB code given in section A.17.

12.4 FREE VIBRATION ANALYSIS OF PLANE FRAMES

12.4.1   Numerical Example

Consider a plane frame consisting of four elements, as shown in Figure 12.7. The Young's modulus of material of element is $E = 2 \times 10^{11} \, N/m^2$. The cross-sectional dimensions of each element are $0.1 \, m \times 0.1 \, m$ and the density is $\rho = 7850 \, kg/m^3$.

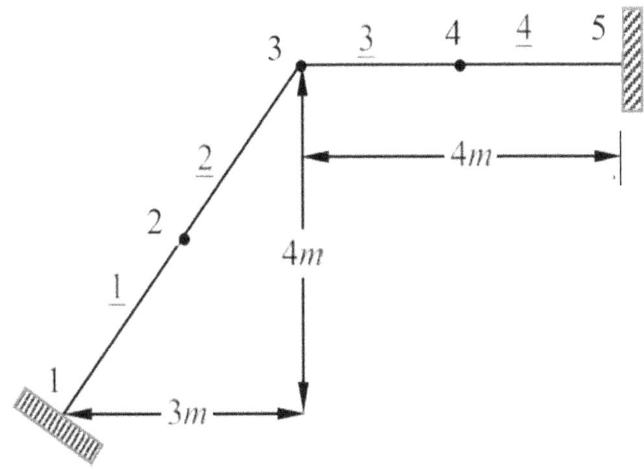

Figure 12.7 Plane Frame.

The element stiffness and mass matrices of the elements 1 and 2 with reference to the local axes are

$$\left[k_{1M}^{(e)}\right] = \left[k_{2M}^{(e)}\right] = 10^8 \begin{bmatrix} 8 & 0 & 0 & -8 & 0 & 0 \\ 0 & 0.0128 & 0.016 & 0 & -0.0128 & 0.016 \\ 0 & 0.016 & 0.0267 & 0 & -0.016 & 0.0133 \\ -8 & 0 & 0 & 8 & 0 & 0 \\ 0 & -0.0128 & -0.016 & 0 & 0.0128 & -0.016 \\ 0 & 0.016 & 0.0133 & 0 & -0.016 & 0.0267 \end{bmatrix}$$

$$
\left[m_{1M}^{(e)} \right] = \left[m_{2M}^{(e)} \right] = \begin{bmatrix}
65.4167 & 0 & 0 & 32.7083 & 0 & 0 \\
0 & 72.8929 & 25.6994 & 0 & 25.2321 & -15.1860 \\
0 & 25.6994 & 11.6815 & 0 & 15.1860 & -8.7612 \\
32.7083 & 0 & 0 & 65.4167 & 0 & 0 \\
0 & 25.2321 & 15.1860 & 0 & 72.8929 & -25.6994 \\
0 & -15.1860 & -8.7612 & 0 & -25.6994 & 11.6815
\end{bmatrix}
$$

The rotation transformation matrix is given by

$$
[T] = \left[\begin{array}{c|c} [T_1] & [0] \\ \hline [0] & [T_1] \end{array} \right] \text{ where } [T_1] = \begin{bmatrix}
0.6 & 0.8 & 0 \\
-0.8 & 0.6 & 0 \\
0 & 0 & 1
\end{bmatrix}
$$

The corresponding element and mass stiffness matrices of elements 1 and 2 with reference to the structure axes are given by

$$
\left[k_{1S}^{(e)} \right] = \left[k_{2S}^{(e)} \right] = 10^8 \begin{bmatrix}
2.8882 & 3.8339 & -0.0128 & -2.8882 & -3.8339 & -0.0128 \\
3.8339 & 5.1246 & 0.0096 & -3.8339 & -5.1246 & 0.0096 \\
-0.0128 & 0.0096 & 0.0267 & 0.0128 & -0.0096 & 0.0133 \\
-2.8882 & -3.8339 & 0.0128 & 2.8882 & 3.8339 & 0.0128 \\
-3.8339 & -5.1246 & -0.0096 & 3.8339 & 5.1246 & -0.0096 \\
-0.0128 & 0.0096 & 0.0133 & 0.0128 & -0.0096 & 0.0267
\end{bmatrix}
$$

$$
\left[m_{1S}^{(e)} \right] = \left[m_{2S}^{(e)} \right] = \begin{bmatrix}
70.2014 & -3.5886 & -20.5595 & 27.9236 & 3.5886 & 12.1488 \\
-3.5886 & 68.1081 & 15.4196 & 3.5886 & 30.0169 & -9.1116 \\
-20.5595 & 15.4196 & 11.6815 & -12.1488 & 9.1116 & -8.7612 \\
27.9236 & 3.5886 & -12.1488 & 70.2014 & -3.5886 & 20.5595 \\
3.5886 & 30.0169 & 9.1116 & -3.5886 & 68.1081 & -15.4196 \\
12.1488 & -9.1116 & -8.7612 & 20.5595 & -15.4196 & 11.6815
\end{bmatrix}
$$

The element stiffness and mass matrices of the elements 3 and 4 with reference to the local axes are

$$\left[k_{3M}^{(e)}\right]=\left[k_{4M}^{(e)}\right]=10^9\begin{bmatrix} 1 & 0 & 0 & -1 & 0 & 0 \\ 0 & 0.0025 & 0.0025 & 0 & -0.0025 & 0.0025 \\ 0 & 0.0025 & 0.0033 & 0 & -0.0025 & 0.0017 \\ -1 & 0 & 0 & 1 & 0 & 0 \\ 0 & -0.0025 & -0.0025 & 0 & 0.0025 & -0.0025 \\ 0 & 0.0025 & 0.0017 & & -0.0025 & 0.0033 \end{bmatrix}$$

$$\left[m_{3M}^{(e)}\right]=\left[m_{4M}^{(e)}\right]=\begin{bmatrix} 52.3333 & 0 & 0 & 26.1667 & 0 & 0 \\ 0 & 58.3143 & 16.4476 & 0 & 20.1857 & -9.7190 \\ 0 & 16.4476 & 5.9810 & 0 & 9.7190 & -4.4857 \\ 26.1667 & 0 & 0 & 52.3333 & 0 & 0 \\ 0 & 20.1857 & 9.7190 & 0 & 58.3143 & -16.4476 \\ 0 & -9.7190 & -4.4857 & 0 & -16.4476 & 5.9810 \end{bmatrix}$$

As the elements 3 and 4 are horizontal, the rotation transformation matrix equals to identity matrix. Thus the stiffness and mass matrices of elements 3 and 4 are equal to their corresponding matrices along the member axes.

$$\left[k_{3S}^{(e)}\right]=\left[k_{4S}^{(e)}\right]=\left[k_{3M}^{(e)}\right]=\left[k_{4M}^{(e)}\right] \text{ and } \left[m_{3S}^{(e)}\right]=\left[m_{4S}^{(e)}\right]=\left[m_{3M}^{(e)}\right]=\left[m_{4M}^{(e)}\right]$$

As the structure has 5 nodes with each node having 3 degrees of freedom, the total number of degrees of freedom for the entire structure are 15.

Thus, the assembled stiffness and mass matrices have a size of 15×15. The restrained degrees of freedom correspond to fixed supports at nodes 1 and 5. Thus the restrained degrees of freedom are 1, 2, 3, 13, 14, 15 respectively. Eliminating these restrained degrees of freedom, the resulting stiffness and mass matrices with reference to the structure axes have the size 9×9 and are given by

$$[K]=10^9\begin{bmatrix} 0.5776 & 0.7668 & 0 & -0.2888 & -0.3834 & -0.0013 & 0 & 0 & 0 \\ 0.7668 & 1.0249 & 0 & -0.3834 & -0.5125 & 0.0010 & 0 & 0 & 0 \\ 0 & 0 & 0.0053 & 0.0013 & -0.0010 & 0.0013 & 0 & 0 & 0 \\ -0.2888 & -0.3834 & 0.0013 & 1.2888 & 0.3834 & 0.0013 & -1 & 0 & 0 \\ -0.3834 & -0.5125 & -0.0010 & 0.3834 & 0.5150 & 0.0015 & 0 & -0.0025 & 0.0025 \\ -0.0013 & 0.0010 & 0.0013 & 0.0013 & 0.0015 & 0.0060 & 0 & -0.0025 & 0.0017 \\ 0 & 0 & 0 & -1 & 0 & 0 & 2 & 0 & 0 \\ 0 & 0 & 0 & 0 & -0.0025 & -0.0025 & 0 & 0.0050 & 0 \\ 0 & 0 & 0 & 0 & 0.0025 & 0.0017 & 0 & 0 & 0.0067 \end{bmatrix}$$

$$[M]=\begin{bmatrix} 140.4029 & -7.1771 & 0 & 27.9236 & 3.5886 & 12.1488 & 0 & 0 & 0 \\ -7.1771 & 136.2162 & 0 & 3.5886 & 30.0169 & -9.1116 & 0 & 0 & 0 \\ 0 & 0 & 23.3631 & -12.1488 & 9.1116 & -8.7612 & 0 & 0 & 0 \\ 27.9236 & 3.5886 & -12.1488 & 122.5348 & -3.5886 & 20.5595 & 26.1667 & 0 & 0 \\ -3.5886 & 30.0169 & 9.1116 & -3.5886 & 126.4224 & 1.0280 & 0 & 20.1857 & -9.7190 \\ 12.1488 & -9.1116 & -8.7612 & 20.5595 & 1.0280 & 17.6625 & 0 & 9.7190 & -4.4857 \\ 0 & 0 & 0 & 26.1667 & 0 & 0 & 104.6667 & 0 & 0 \\ 0 & 0 & 0 & 0 & 20.1857 & 9.7190 & 0 & 116.6286 & 0 \\ 0 & 0 & 0 & 0 & -9.7190 & -4.4857 & 0 & 0 & 11.9619 \end{bmatrix}$$

Substituting $[K]$ and $[M]$ in Eq.12.5 and solving eigen-value problem using MATLAB code given in section A.16, we obtain the natural frequencies as shown in the second column of Table 12.4 as follows:

Table 12.4 Natural frequencies of plane frame.

Mode	$\omega \times 10^3 \, (rad\,/\sec)$
1	103.711
2	178.434
3	389.416
4	620.490
5	1047.701

12.5 EXERCISE PROBLEMS

1. Compute the natural frequencies of the stepped bar as shown in the Figure 12.8 below. The bar consists of three elements with areas of cross section $0.02m^2$, $0.01\ m^2$ and $0.03\ m^2$ respectively. Use lumped mass matrix as well as consistent mass matrix and compare the results.

   Adopt $E = 2.1 \times 10^8 \ kN/m^2$.

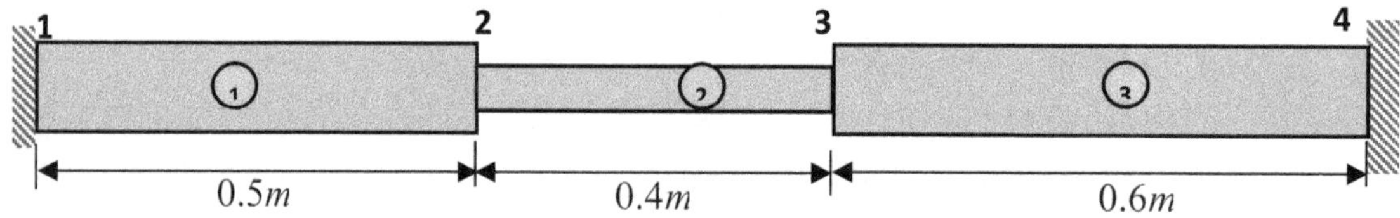

Figure 12.8 Stepped bar.

2. Compute the natural frequencies of the triangular truss shown in Figure 12.9. Each element has a cross sectional area of $0.01\, m^2$ and $E = 2\times10^8\, kN/m^2$

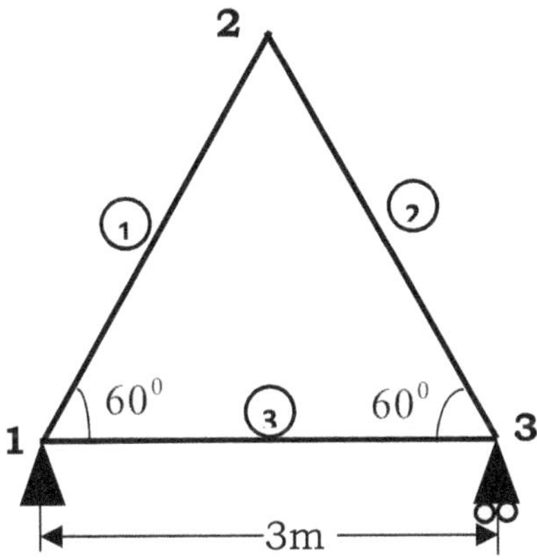

Figure 12.9 Triangular Truss.

3. Compute the natural frequencies for a cantilever beam shown in the Figure 12.10. The beam has a span of 3m and is divided into 3 elements. The cross-sectional dimensions of the beam are $0.01\, m \times 0.0\,2m$.Adopt $E = 2\times10^{11}\, N/m^2$ and $\rho = 7850\, kg/m^3$. Use lumped mass matrix as well as consistent mass matrix and compare the results.

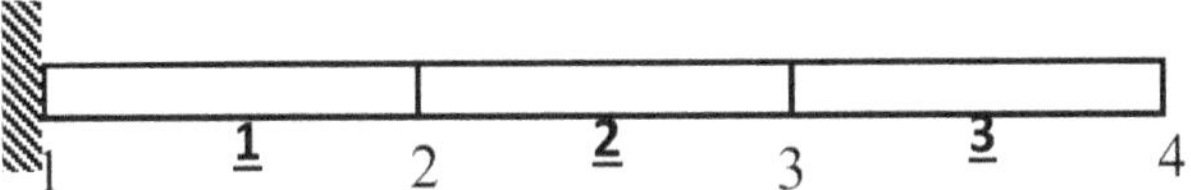

Figure 12.10 Cantilever beam.

4. Compute the natural frequencies for a continuous beam shown in the Figure 12.11. The cross-sectional dimensions of the beam are $0.01m \times 0.02m$.Adopt $E = 2\times10^{11}\, N/m^2$ and $\rho = 7850\, kg/m^3$. Use lumped mass matrix as well as consistent mass matrix and compare the results.

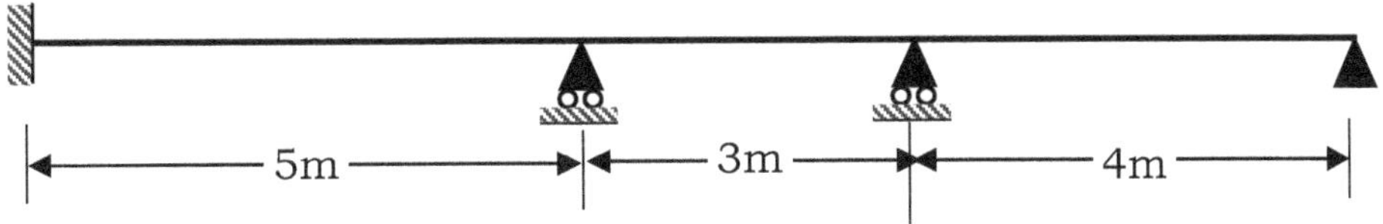

Figure 12.11 Three span continuous beam.

5. Compute the natural frequencies of the plane frame as shown in the Figure 12.12. Adopt $E = 2 \times 10^8 \, kN / m^2$ for all elements.

$A_1 = 0.06 m^2$, $I_1 = 5 \times 10^{-4} m^4$, $\quad A_2 = 0.04 \, m^2$, $\quad I_2 = 4.5 \times 10^{-4} m^4$, $\quad A_2 = 0.05 \, m^2$,

$I_3 = 5 \times 10^{-4} m^4$, $A_2 = 0.03 \, m^2$

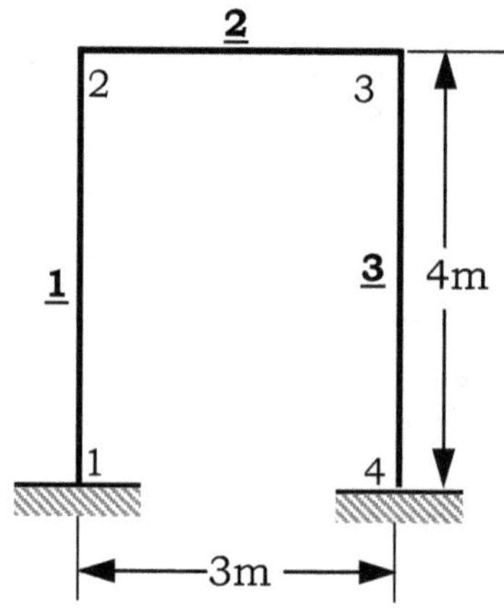

Figure 12.12 Portal Frame.

Matlab Code, Sample Input and Output for each Element

A.1. ANALYSIS OF BARS AND PLANE TRUSSES

```
% Finite Element Analysis of bars and Plane Trusses
clear
clc
format long
%-----------------------------------------------------------
%First, the input and output files are defined
filename = 'PlaneTruss';
infile = strcat(filename,'.inp');
inp= fopen(infile,'r');
outfile =strcat(filename,'.out');
out = fopen(outfile,'w');
%___________________________________________________________
%P R E P R O C E S S O R
%___________________________________________________________
%R E A D I N G   N O D A L   I N F O R M A T I O N
[row] = fscanf(inp,'%d %d %f',3);
% this line reads number of nodes and number of elements
%in the truss
nNode = row(1);   %number of nodes
nElem = row(2);   %number of elements
gamma = row(3);   %unit weight of material;
ndof=2*nNode;
% 'ndof' is the total number of degrees of freedom
%for the Plane Truss
%Read nodal information from the input file node by node
[mat] = fscanf(inp,'%d %f %f %d %d %e %e',[7,nNode]);
mat = mat';
%reading x and y coordinates of nodes
%(columns 2 and 3 of input as above)
Xval = mat(:,2);
Yval = mat(:,3);
%X and Y restraints of a node
resx = mat(:,4); %reading restraint along x-direction from column 4
resy = mat(:,5); %reading restraint along y-direction from column 5
```

```
%-----------------------------------------------------------------
%Generation of nodal force vector (F) (of size ndof) along global axes
%-----------------------------------------------------------------
fx = mat(:,6); %reading fx from column 6
fy = mat(:,7); %reading fy from column 7
F= zeros(ndof,1);
%Creating global nodal force vector of size ndof and initializing %all
elements to %zero
for i=1:nNode
  F(2*i-1) = fx(i);
  F(2*i)   = fy(i);
end
%bn is a counter for the total number of restraints in the Plane Truss
bn = 0; %initialized to zero
%-----------------------------------------------------------------
%Notation used to indicate restraint associated with a given degree of
%freedom
     % 0 indicates RESTRAINED degree of freedom
     % 1 indicates UNRESTRAINED degree of freedom
%-----------------------------------------------------------------
for i=1:nNode
     if(resx(i)==0)
         %if this degree of freedom along x-direction is restrained
         bn = bn+1;
         ifix(bn) = 2*i-1;
         %add to the list of restrained degrees of freedom
     end
     if(resy(i)==0)
         %if this degree of freedom along x-direction is restrained
         bn = bn+1;
         ifix(bn) = 2*i;
         %add to the list of restrained degrees of freedom
     end
end
%NOTE: bn is incremented by 2  for a pin support
% bn is incremented by 1  for a roller support in the above for loop
% at the end of the above loop, bn = total number of restrained
% degrees of freedom
%ifix is the list of restrained degrees of freedom for the Truss
fprintf(out,'A N A L Y S I S    O F    A    P L A N E    T R U S S \n\n');
fprintf(out,'N O D A L    I N F O R M A T I O N \n');
fprintf(out,'________________________________________________\n');
```

```matlab
fprintf(out,'Node   Coordinates     Restraints*            Forces        Type
of support\n');
fprintf(out,'    X       Y       X       Y          Fx        Fy \n');
fprintf(out,'------------------------------------------------------------\n');
for i=1:nNode
    fprintf(out,'%2d    %6.3f    %6.3f     %d         %d %9.1f
         %9.1f',i,Xval(i),Yval(i),resx(i),resy(i),F(2*i-1),F(2*i));
    if(resy(i)==0)
        if(resx(i)==0)
           fprintf(out,'       Pin support\n');
        else
           fprintf(out,'     Roller support\n');
        end
    else
        fprintf(out,'\n');
    end
end
fprintf(out,'* NOTE: 1 indicates an unrestrained degree of freedom\n');
fprintf(out,'        0 indicates a  restrained  degree of freedom\n');
%------------------------------------------------------------
%READING ELEMENT INFORMATION
[mat] = fscanf(inp,'%d %d %d %f %e',[5,nElem]);
%nElem is the total number of elements in the Truss
mat = mat';
FirstNode = zeros(nElem,1); %Initializing vector of first nodes of
elements
SecondNode = zeros(nElem,1);
%Initializing vector of second nodes of elements
A = zeros(nElem,1) ;
%initializing vector of cross section areas of elements
E = zeros(nElem,1);
%initializing vector of Modulus of elasticity of elements
Cx= zeros(nElem,1);
%initializing vectors of direction cosines of elements Cx and Cy
Cy= zeros(nElem,1);
Le= zeros(nElem,1); %Initializing Length of each element
%reading first and second nodes of each element from columns 2 and 3
FirstNode  = mat(:,2);
SecondNode = mat(:,3);
A = mat(:,4); %reading cross sectional areas of elements from column 4
E= mat(:,5); %reading Modulus of Elasticity of elements from column 5
%------------------------------------------------------------
fprintf(out,'________________________________________\n');
```

```matlab
fprintf(out,'E L E M E N T    I N F O R M A T I O N\n');
fprintf(out,'Element first    second      Area of c/s Young''modulus \n');
fprintf(out,'         node    node        (A)                 (E)       \n');
fprintf(out,'----------------------------------------------------------------\n');
for e=1:nElem
    fprintf(out,'%2d           %2d         %2d            %f
                 %.3e\n',e,FirstNode(e),SecondNode(e),A(e),E(e));
end
K=zeros(ndof,ndof); %initialize the global stiffness matrix of size ndof
for e=1:nElem
    %x and y coordinates of first node
    x1=Xval(FirstNode(e));
    y1=Yval(FirstNode(e));
    %x and y coordinates of second node
    x2=Xval(SecondNode(e));
    y2=Yval(SecondNode(e));
    % (x1,y1) &  (x2,y2) are coordinates of first and second nodes  of
    %element e respectively
    L(e)=sqrt((x2-x1)^2+(y2-y1)^2); %length of element e
    %direction cosines of element e
    c =(x2-x1)/L(e);
    s =(y2-y1)/L(e);
    [x1 y1 x2 y2 c s]
    %Storing values of direction cosines c and s in
    %vectors {Cx} and {Cy} for subsequent use
    Cx(e)=c;
    Cy(e)=s;
    %Degrees of freedom for element e with nodes m and n
    %are 2m-1, 2m , 2n-1 and 2n respectively.
    %These degrees of freedom are stored in the vector {dof} as shown
    %below
    dof=[2*FirstNode(e)-1 2*FirstNode(e)  2*SecondNode(e)-1
         2*SecondNode(e)];
    %Rotation transformation matrix for element e
    T=[c     s    0     0;
      -s     c    0     0;
       0     0    c     s;
       0     0   -s     c];
    %Element stiffness matrix with reference to local axes
    estif = A(e)*E(e)/L(e)* [ 1    0    -1    0
                              0    0     0    0
                             -1    0     1    0
                              0    0     0    0];
```

```matlab
    f = 0.5*gamma*A(e)*L(e)*[0  1  0  1]';
    % element force vector for self weight of element
    ke = T'*estif*T; %element stiffness matrix with reference global axes
    K(dof,dof) = K(dof,dof)+ke; %assembly
    F(dof) = F(dof)+f;
end
%Assembly of global stiffness matirx
%store the global stiffness matrix for future use
%to compute support reactions
K1 = K;
%applying boundary conditions (specified by restraints along X and Y
%axes)
%on global stiffness matrix and global force vector
K(ifix,:)=0;
%Elements in rows with d.o.f corresponding to ifix are made 0
K(:,ifix)=0   ;
%Elements in columns with d.o.f corresponding to ifix are made 0
K(ifix,ifix)=eye(length(ifix));
%Diagonal elements with d.o.f corresponding to ifix are made 1
F(ifix) = 0;
%corresponding columns of Global force vector are also made zero
%obtaining vector of nodal displacements along global axes
%by solving [K]{u} = {F}
u=inv(K)*F;
fprintf(out,'_____________________________________________________\n');
fprintf(out,'N O D A L    D I S P L A C E M E N T S\n');
fprintf(out,'Node        X- displacement              Y- displacement\n');
fprintf(out,'------------------------------------------------------------\n');
for i=1:nNode
    fprintf(out,'%2d      %22.10e      %22.10e\n',i,u(2*i-1),u(2*i));
end
%------------------------------------------------------------
%COMPUTATION OF INTERNAL FORCES IN TRUSS ELEMENTS
mat = [  1    0    -1    0
         0    0     0    0
        -1    0     1    0
         0    0     0    0];
for e=1:nElem
    %degrees of freedom corresponding to element e
    dof=[2*FirstNode(e)-1 2*FirstNode(e)  2*SecondNode(e)-1
        2*SecondNode(e)];
    %rotation transformation matrix
    c = Cx(e);
```

```
    s = Cy(e); %direction cosines
    T=[c       s     0        0;
      -s       c     0        0;
       0       0     c        s;
       0       0    -s        c];
    ke =(A(e)*E(e)/L(e))*mat;
    %Element stiffness matrix w.r.t. local axes
    g1 =ke*T*u(dof);
    %Vector of internal forces of element w.r.t. local axes
    lforce(e) = g1(3);   %axial force in element e
end
fprintf(out,'________________________________________________________\n');
fprintf(out,'A X I A L   F O R C E    I N    M E M B E R S * \n');
fprintf(out,'ELEMENT                    Force                Strain\n');
fprintf(out,'-----------------------------------------------------------\n');
for e=1:nElem
    strain(e) = lforce(e)/(A(e)*E(e));
    fprintf(out,'%3d      %20.7f        %20.7e\n',e,lforce(e),strain(e));
end
fprintf(out,'* Note: positive sign indicates TENSION and negative sign
                  indicates COMPRESSION\n');
K = K1;
codevec = ones(ndof,1); %code vector to indicate restrained dof
ifix = ifix';
for row=1:size(ifix,1)
   codevec(ifix(row)) = 0;
end
%picking up rows of K corresponding to restrained dof and columns
%corresponding to unrestrained dof
nrow=0;
for row=1:ndof
    ncol=0;
    if(codevec(row)==0) %pickup rows corresponding to restrained dof
        nrow = nrow+1;
        for col=1:ndof
            if(codevec(col)==1)
                %pickup columns corresponding to unrestrained dof
                ncol = ncol+1;
                uR(ncol) = u(col);
                KRF(nrow,ncol) = K(row,col); %reduced stiffness matrix
            end
        end
    end
end
```

```matlab
end
supp = KRF*uR';%vector of support reactions
fprintf(out,'________________________________________________\n');
fprintf(out,'S U P P O R T     R E A C T I O N S\n');
fprintf(out,'Support node        Rx                  Ry\n');
fprintf(out,'--------------------------------------------------------\n');
count= 0;
for j=1:nNode
    j1 = 2*j-1;
    j2 = 2*j;
    if((codevec(j1)==0)||(codevec(j2)==0))
        if(codevec(j1)==0)
            count = count+1;
            Rx = supp(count);
        else
            Rx = 0;
        end
        if(codevec(j2)==0)
            count = count+1;
            Ry = supp(count);
        else
            Ry = 0;
        end
        fprintf(out,'    %2d           %10.3f     %10.3f\n',j,Rx,Ry);
    end
end
fprintf(out,'________________________________________________\n');
%close the file identifier for output file
fclose(out);
type PlaneTruss.out
```

A.1.1. Input and output for numerical examples

a) Numerical example 6.1 (Stepped bar-1)

<u>Input:</u>
```
3   2   0
1       0.0     0.0     0   0       0.0     0.0
2       0.5     0.0     1   0     250.0     0.0
3       0.9     0.0     1   0     400.0     0.0
1   1     2         0.0026        2.0e8
2   2     3         0.0013        2.0e8
%Example 6.1
%Length in m Area in m^2  Load in kN Young's modulus kN/m^2
```

Output:

```
A N A L Y S I S   O F   A   P L A N E   T R U S S
N O D A L   I N F O R M A T I O N
```

Node	Coordinates		Restraints*		Forces		Type of support
	X	Y	X	Y	Fx	Fy	
1	0.000	0.000	0	0	0.0	0.0	Pin support
2	0.500	0.000	1	0	250.0	0.0	Roller support
3	0.900	0.000	1	0	400.0	0.0	Roller support

```
* NOTE: 1 indicates an unrestrained degree of freedom
        0 indicates a restrained   degree of freedom
```

```
E L E M E N T   I N F O R M A T I O N
```

Element	first node	second node	Area of c/s (A)	Young'modulus (E)
1	1	2	0.002600	2.000e+008
2	2	3	0.001300	2.000e+008

```
N O D A L   D I S P L A C E M E N T S
```

Node	X- displacement	Y- displacement
1	0.0000000000e+000	0.0000000000e+000
2	6.2500000000e-004	0.0000000000e+000
3	1.2403846154e-003	0.0000000000e+000

```
A X I A L   F O R C E   I N   M E M B E R S *
```

ELEMENT	Force	Strain
1	650.0000000	1.2500000e-003
2	400.0000000	1.5384615e-003

```
* Note: positive sign indicates TENSION and negative sign indicates
COMPRESSION
S U P P O R T   R E A C T I O N S
```

Support node	Rx	Ry
1	-650.000	0.000
2	0.000	0.000
3	0.000	0.000

---

b)   *Input and output for numerical example 6.2*

Input:

```
4  3  0
1      0.0    0.0    0  0   0.0      0.0
2      0.5    0.0    1  0   200.0    0.0
```

```
3      0.9     0.0      1   0  500.0      0.0
4      1.5     0.0      0   0   0.0       0.0
1   1     2  0.0020  2.0e8
2   2     3  0.0010  0.7e8
3   3     4  0.0030  1.0e8
%Example 6.2
%Length in m     Area in m^2 Load in kN  Young's modulus kN/m^2
```

Output:

A N A L Y S I S   O F   A   P L A N E   T R U S S

N O D A L   I N F O R M A T I O N

Node	Coordinates		Restraints*		Forces		Type of support
	X	Y	X	Y	Fx	Fy	
1	0.000	0.000	0	0	0.0	0.0	Pin support
2	0.500	0.000	1	0	200.0	0.0	Roller support
3	0.900	0.000	1	0	500.0	0.0	Roller support
4	1.500	0.000	0	0	0.0	0.0	Pin support

```
* NOTE: 1 indicates an unrestrained degree of freedom
        0 indicates a  restrained   degree of freedom
```

E L E M E N T   I N F O R M A T I O N

Element	first node	second node	Area of c/s (A)	Young'modulus (E)
1	1	2	0.002000	2.000e+008
2	2	3	0.001000	7.000e+007
3	3	4	0.003000	1.000e+008

N O D A L   D I S P L A C E M E N T S

Node	X- displacement	Y- displacement
1	0.0000000000e+000	0.0000000000e+000
2	3.5458167331e-004	0.0000000000e+000
3	8.3266932271e-004	0.0000000000e+000
4	0.0000000000e+000	0.0000000000e+000

A X I A L   F O R C E   I N   M E M B E R S *

ELEMENT	Force	Strain
1	283.6653386	7.0916335e-004
2	83.6653386	1.1952191e-003
3	-416.3346614	-1.3877822e-003

```
* Note: positive sign indicates  TENSION  and negative  sign  indicates
COMPRESSION
```

S U P P O R T   R E A C T I O N S

```
Support node        Rx               Ry
-------------------------------------------------------------------------
      1          -283.665           0.000
      2             0.000           0.000
      3             0.000           0.000
      4          -416.335           0.000
```

c) Input and output for numerical example 6.3

Input:

```
3  2 0.785e-4
1    0.0      0.00        0   0   0.0        0.0
2    0.0      250.0       0   1   0.0        0.0
3    0.0      500.0       0   1   0.0      500.0
1    1      2   2500.0   2.0e5
2    2      3   2000.0   2.0e5
%Example 6.3
%Length in mm    Area in mm^2 Load in N  Young's modulus N/mm^2
```

Output:

```
A N A L Y S I S    O F   A    P L A N E    T R U S S

N O D A L    I N F O R M A T I O N
```

Node	Coordinates		Restraints*		Forces		Type of support
	X	Y	X	Y	Fx	Fy	
1	0.000	0.000	0	0	0.0	0.0	Pin support
2	0.000	250.000	0	1	0.0	0.0	
3	0.000	500.000	0	1	0.0	500.0	

```
* NOTE: 1 indicates an unrestrained degree of freedom
        0 indicates a  restrained  degree of freedom
```

```
E L E M E N T    I N F O R M A T I O N
```

Element	first node	second node	Area of c/s (A)	Young'modulus (E)
1	1	2	2500.000000	2.000e+005
2	2	3	2000.000000	2.000e+005

```
N O D A L    D I S P L A C E M E N T S
```

Node	X- displacement	Y- displacement
1	0.0000000000e+000	0.0000000000e+000
2	0.0000000000e+000	2.8189062500e-004
3	0.0000000000e+000	6.0665625000e-004

```
A X I A L   F O R C E   I N   M E M B E R S *
ELEMENT                 Force                    Strain
-----------------------------------------------------------------------
   1                563.7812500              1.1275625e-006
   2                519.6250000              1.2990625e-006
* Note: positive sign indicates TENSION and negative sign indicates
COMPRESSION
```

```
S U P P O R T     R E A C T I O N S
Support node      Rx                  Ry
-----------------------------------------------------------------------
   1             0.000            -563.781
   2             0.000               0.000
   3             0.000               0.000
```

d)   *Input and output for numerical example 6.4*

Input:

```
3  3  0
1  0.0  0.0                     0   0       0.0        0.0
2  1.0  1.73205080756888        1   1     100.0     -150.0
3  2.0  0.0                     1   0       0.0        0.0
1   1     2  0.01   2.0e8
2   2     3  0.01   2.0e8
3   1     3  0.01   2.0e8
%Example 6.4 Triangular truss
```

Output:

```
A N A L Y S I S   O F   A   P L A N E   T R U S S
N O D A L   I N F O R M A T I O N
```

Node	Coordinates		Restraints*		Forces		Type of support
	X	Y	X	Y	Fx	Fy	
1	0.000	0.000	0	0	0.0	0.0	Pin support
2	1.000	1.732	1	1	100.0	-150.0	
3	2.000	0.000	1	0	0.0	0.0	Roller support

```
* NOTE: 1 indicates an unrestrained degree of freedom
        0 indicates a  restrained   degree of freedom
```

```
E L E M E N T   I N F O R M A T I O N
```

Element	first node	second node	Area of c/s (A)	Young'modulus (E)
1	1	2	0.010000	2.000e+08
2	2	3	0.010000	2.000e+08
3	1	3	0.010000	2.000e+08

```
N O D A L   D I S P L A C E M E N T S
Node          X- displacement           Y- displacement
------------------------------------------------------------------
 1            0.0000000000e+00           0.0000000000e+00
 2            2.4664774834e-04          -1.2691875673e-04
 3            9.3295496687e-05           0.0000000000e+00
```

```
A X I A L   F O R C E   I N   M E M B E R S *
ELEMENT               Force                  Strain
------------------------------------------------------------------
 1                 13.4090066            6.7045033e-06
 2               -186.5909934           -9.3295497e-05
 3                 93.2954967            4.6647748e-05
* Note: positive sign indicates TENSION and negative sign indicates
COMPRESSION
```

```
S U P P O R T   R E A C T I O N S
Support node      Rx                Ry
------------------------------------------------------------------
 1            -100.000          -11.613
 3               0.000          161.593
```

e) Input and output for numerical example 6.5

Input:

```
4   5   0
1   0.0   0.0     0    0      0.0      0.0
2   2.0   0.0     1    0      0.0      0.0
3   0.0   2.0     1    1     50.0    -100.0
4   2.0   2.0     1    1      0.0    -125.0
1   1       3   0.01   2.0e8
2   3       4   0.01   2.0e8
3   2       4   0.01   2.0e8
4   1       4   0.01   2.0e8
5   2       3   0.01   2.0e8
% Example 6.5  Triangular truss
```

Output:

```
A N A L Y S I S   O F   A   P L A N E   T R U S S
N O D A L   I N F O R M A T I O N
```

Node	Coordinates		Restraints*		Forces		Type of support
	X	Y	X	Y	Fx	Fy	
1	0.000	0.000	0	0	0.0	0.0	Pin support
2	2.000	0.000	1	0	0.0	0.0	Roller support
3	0.000	2.000	1	1	50.0	-100.0	
4	2.000	2.000	1	1	0.0	-125.0	

```
* NOTE: 1 indicates an unrestrained degree of freedom
        0 indicates a restrained   degree of freedom
```

```
E L E M E N T    I N F O R M A T I O N
Element       first    second    Area of c/s      Young'modulus
              node     node        (A)                (E)
-----------------------------------------------------------------------
    1           1        3        0.010000          2.000e+08
    2           3        4        0.010000          2.000e+08
    3           2        4        0.010000          2.000e+08
    4           1        4        0.010000          2.000e+08
    5           2        3        0.010000          2.000e+08


N O D A L    D I S P L A C E M E N T S
Node          X- displacement          Y- displacement
-----------------------------------------------------------------------
    1         0.0000000000e+00         0.0000000000e+00
    2         4.6635307197e-04         0.0000000000e+00
    3         3.6638721410e-04        -9.9965857864e-05
    4         3.1638721410e-04        -1.7496585786e-04


A X I A L   F O R C E   I N   M E M B E R S *
ELEMENT              Force                Strain
-----------------------------------------------------------------------
    1             -99.9658579          -4.9982929e-05
    2             -50.0000000          -2.5000000e-05
    3            -174.9658579          -8.7482929e-05
    4              70.7106781           3.5355339e-05
    5              -0.0000000          -7.1054274e-21
*Note:  positive  sign  indicates  TENSION;negative  sign  indicates
COMPRESSION
-----------------------------------------------------------------------
S U P P O R T    R E A C T I O N S
Support node      Rx              Ry
-----------------------------------------------------------------------
    1            -50.000          49.966
    2              0.000         174.966
```

f)   Input and output for numerical example 6.6
Input:
```
5  7
1  0.0  0.0                         0   0      0.0       0.0
2  1.0  1.73205080756888           1   1      0.0      -50.0
3  2.0  0.0                        1   1      0.0     -150.0
4  3.0  1.73205080756888           1   1      0.0     -100.0
5  4.0  0.0                        0   0      0.0       0.0
1    1      2   0.01   2.0e8
2    1      3   0.01   2.0e8
3    2      3   0.01   2.0e8
4    2      4   0.01   2.0e8
5    3      4   0.01   2.0e8
```

```
6    3     5   0.01   2.0e8
7    4     5   0.01   2.0e8
% Example 6.6  Warren truss
```

Output:

```
A N A L Y S I S   O F   A   P L A N E   T R U S S
N O D A L   I N F O R M A T I O N
```

Node	Coordinates		Restraints*		Forces		Type of support
	X	Y	X	Y	Fx	Fy	
1	0.000	0.000	0	0	0.0	0.0	Pin support
2	1.000	1.732	1	1	0.0	-50.0	
3	2.000	0.000	1	1	0.0	-150.0	
4	3.000	1.732	1	1	0.0	-100.0	
5	4.000	0.000	0	0	0.0	0.0	Pin support

```
* NOTE: 1 indicates an unrestrained degree of freedom
        0 indicates a  restrained   degree of freedom
```

```
E L E M E N T   I N F O R M A T I O N
```

Element	first node	second node	Area of c/s (A)	Young'modulus (E)
1	1	2	0.010000	2.000e+08
2	1	3	0.010000	2.000e+08
3	2	3	0.010000	2.000e+08
4	2	4	0.010000	2.000e+08
5	3	4	0.010000	2.000e+08
6	3	5	0.010000	2.000e+08
7	4	5	0.010000	2.000e+08

```
N O D A L   D I S P L A C E M E N T S
```

Node	X- displacement	Y- displacement
1	0.0000000000e+00	0.0000000000e+00
2	6.1323258842e-05	-2.1867166667e-04
3	-7.2168783649e-06	-3.7488333333e-04
4	-6.8540137207e-05	-2.5617166667e-04
5	0.0000000000e+00	0.0000000000e+00

```
A X I A L   F O R C E   I N   M E M B E R S *
```

ELEMENT	Force	Strain
1	-158.7135890	-7.9356795e-05
2	-7.2168784	-3.6084392e-06
3	101.0132031	5.0506602e-05
4	-129.8633960	-6.4931698e-05
5	72.1456896	3.6072845e-05
6	7.2168784	3.6084392e-06
7	-187.5811025	-9.3790551e-05

** Note: positive sign indicates TENSION and negative sign indicates COMPRESSION*

```
S U P P O R T     R E A C T I O N S
Support node      Rx              Ry
---------------------------------------------------------------------
     1           86.574         137.450
     5          -86.574         162.450
```

A.2. ANALYSIS OF SPACE TRUSSES

```matlab
% Finite Element Analysis of Space Truss
clc
clear
%------------------------------------------------------------------
%First, the input and output files are defined
filename = 'SpaceTruss';
%Here we provide the name of the input/output files to be used
infile = strcat(filename,'.inp');
%Input file name  specified by concatenating '.inp' to filename
inp= fopen(infile,'r');
% Input file is opened and file identifier 'inp' is associated with it
outfile =strcat(filename,'.out');
%Output file name specified by concatenating '.out' to filename
out = fopen(outfile,'w');
% Output file is opened and file identifier 'out' is associated with it
%________________________________________________________________
%P R E P R O C E S S O R
%________________________________________________________________
%R E A D I N G    N O D A L    I N F O R M A T I O N
[row] = fscanf(inp,'%d',2);
% this line reads number of nodes and number of elements in the truss
nNode = row(1);   %number of nodes
nElem = row(2);   %number of elements
ndof=3*nNode;     % 'ndof' is the total number of degrees of freedom for
the Space Truss
%Read nodal information from the input file node by node
[mat] = fscanf(inp,'%d %f %f %f  %d %d %d %e %e %e',[10,nNode]);
mat = mat'; %Transposed for future use
Xval = mat(:,2);
%reading x coordinates of nodes (column 2 of input as above)
Yval = mat(:,3);
%reading y coordinates of nodes (column 3 of input as above)
Zval = mat(:,4);
%reading y coordinates of nodes (column 4 of input as above)
%X,Y and Z restraints of a node
resx = mat(:,5); %reading restraint along x-direction from column 5
resy = mat(:,6); %reading restraint along y-direction from column 6
resz = mat(:,7); %reading restraint along z-direction from column 7
%-----------------------------------------------------------------
%Generation of  nodal force vector (F) (of size ndof) along global axes
%-----------------------------------------------------------------
```

```matlab
fx = mat(:,8); %reading fx from column 8
fy = mat(:,9); %reading fy from column 9
fz = mat(:,10); %reading fy from column 10
F= zeros(ndof,1);
%Creating global nodal force vector of size ndof and initializing all
%elements to zero
for i=1:nNode
  F(3*i-2)    = fx(i);
  F(3*i-1)    = fy(i);
  F(3*i)      = fz(i);
end
%bn is a counter for the total number of restraints in the Space Truss
bn = 0; %initialized to zero
%---------------------------------------------------------------------
%Notation used to indicate restraint associated with a given degree of
freedom
    % 0 indicates RESTRAINED degree of freedom
    % 1 indicates UNRESTRAINED degree of freedom
%---------------------------------------------------------------------
for i=1:nNode
    if(resx(i)==0)
        %if this degree of freedom along x-direction is restrained
        bn = bn+1;
        ifix(bn) = 3*i-2;
        %add to the list of restrained degrees of freedom
    end
    if(resy(i)==0)
        %if this degree of freedom along y-direction is restrained
        bn = bn+1;
        ifix(bn) = 3*i-1;
        %add to the list of restrained degrees of freedom
    end
    if(resz(i)==0)
        %if this degree of freedom along z-direction is restrained
        bn = bn+1;
        ifix(bn) = 3*i;
        %add to the list of restrained degrees of freedom
    end
end
fprintf(out,'A N A L Y S I S   O F   A   S P A C E   T R U S S \n\n');
fprintf(out,'N O D A L   I N F O R M A T I O N \n');
fprintf(out,'________________________________________________________\n');
fprintf(out,'Node      Coordinates      Restraints*        Forces \n');
```

```matlab
fprintf(out,'         X      Y      Z    X  Y  Z Fx      Fy    Fz\n');
fprintf(out,'-----------------------------------------------------------\n');
for i=1:nNode
    fprintf(out,'%2d %4.1f %4.1f  %4.1f  %d %d %d %9.1f %9.1f %9.1f\n',...
            i,Xval(i),Yval(i),Zval(i),resx(i),resy(i),resz(i),...
            F(3*i-2),F(3*i-1),F(3*i));
end
fprintf(out,'* NOTE: 1 indicates an unrestrained degree of freedom\n');
fprintf(out,'        0 indicates a restrained  degree of freedom\n');
%-------------------------------------------------------------------
%READING ELEMENT INFORMATION
[mat] = fscanf(inp,'%d %d %d %f %e',[5,nElem]);
mat = mat'; %transposed for future sure
FirstNode = zeros(nElem,1); %Initializing vector of first nodes of
elements
SecondNode = zeros(nElem,1);
%Initializing vector of second nodes of elements
A= zeros(nElem,1) ;
%initializing vector of cross section areas of elements
E= zeros(nElem,1);
%initializing vector of Modulus of elasticity of elements
Cx = zeros(nElem,1);
%initializing vectors of direction cosines of elements Cx,Cy and Cz
Cy= zeros(nElem,1);
Cz= zeros(nElem,1);
Le= zeros(nElem,1); %Initializing Length of each element
%reading first and second nodes of each element from columns 2 and 3
FirstNode  = mat(:,2);
SecondNode = mat(:,3);
A= mat(:,4); %reading cross sectional areas of elements from column 4
E= mat(:,5); %reading Modulus of Elasticity of elements from column 5
%-------------------------------------------------------------------
fprintf(out,'_____________________________________________\n');
fprintf(out,'E L E M E N T   I N F O R M A T I O N\n');
fprintf(out,'Element   first   second  Area of c/s Young''modulus \n');
fprintf(out,'          node    node        (A)         (E)      \n');
fprintf(out,'-----------------------------------------------------\n');
for e=1:nElem
    fprintf(out,'%2d        %2d        %2d          %f %.3e\n',...
            e,FirstNode(e),SecondNode(e),A(e),E(e));
end
K=zeros(ndof,ndof); %initialize the global stiffness matrix of size ndof
T=zeros(6,6);
```

```matlab
for e=1:nElem
    %x and y coordinates of first node
    x1=Xval(FirstNode(e));
    y1=Yval(FirstNode(e));
    z1=Zval(FirstNode(e));
    %x and y coordinates of second node
    x2=Xval(SecondNode(e));
    y2=Yval(SecondNode(e));
    z2=Zval(SecondNode(e));
    % (x1,y1) &  (x2,y2) are coordinates of first and second nodes  of
    %  element e respectively
    L(e)=sqrt((x2-x1)^2+(y2-y1)^2+(z2-z1)^2); %length of element e
    %direction cosines of element e
    cx =(x2-x1)/L(e);
    cy =(y2-y1)/L(e);
    cz =(z2-z1)/L(e);
    cxz = sqrt(cx^2+cz^2);
    %Storing values of direction cosines c and s in vectors {Cx} and
    %{Cy}for subsequent use
    Cx(e)=cx;
    Cy(e)=cy;
    Cz(e)=cz;
    %Degrees of freedom for element e with nodes m and n are
    % 2m-1, 2m , 2n-1 and 2n
    %These degrees of freedom are stored in the vector {dof}
    %as shown below
    dof=[3*FirstNode(e)-2  3*FirstNode(e)-1 3*FirstNode(e)
          3*SecondNode(e)-2 3*SecondNode(e)-1 3*SecondNode(e)];
    %Element stiffness matrix with reference to local axes
    ke = (A(e)*E(e)/L(e))*[cx^2 cx*cy     cx*cz   -cx^2     -cx*cy    -cx*cz
                           cx*cy    cy^2     cy*cz   -cx*cy    -cy^2    -cy*cz
                           cx*cz    cy*cz    cz^2    -cx*cz    -cy*cz    -cz^2
                           -cx^2    -cx*cy   -cx*cz   cx^2     cx*cy     cx*cz
                           -cx*cy   -cy^2    -cy*cz   cx*cy    cy^2      cy*cz
                           -cx*cz   -cy*cz   -cz^2    cx*cz    cy*cz     cz^2];
    K(dof,dof) = K(dof,dof)+ke; %assembly
end
%Assembly of global stiffness matirx
%store the global stiffness matrix for future use to compute support
%reactions
K1 = K;
%applying boundary conditions(specified by restraints along X and Y axes)
%on global stiffness matrix and global force vector
```

```matlab
K(ifix,:)=0    ;
%Elements in rows with d.o.f corresponding to ifix are made 0
K(:,ifix)=0    ;
%Elements in columns with d.o.f corresponding to ifix are made 0
K(ifix,ifix)=eye(length(ifix));
%Diagonal elements with d.o.f corresponding to ifix are made 1
F(ifix) = 0;
%corresponding columns of Global force vector are also made zero
%obtaining  vector of nodal displacements along global axes by solving
[K]{u} = {F}
u=inv(K)*F;
fprintf(out,'____________________________________________________\n');
fprintf(out,'N O D A L   D I S P L A C E M E N T S\n');
fprintf(out,'Node          X- displacement              Y- displacement
Z- displacement\n');
fprintf(out,'----------------------------------------------------------\n');
for i=1:nNode
    fprintf(out,'%2d     %22.10e     %22.10e      %22.10e\n',i,
    u(3*i-2),u(3*i-1),u(3*i));
end
%-------------------------------------------------------------------------
%COMPUTATION OF INTERNAL FORCES IN TRUSS ELEMENTS
for e=1:nElem
    cx = Cx(e); cy = Cy(e); cz=Cz(e)
    %degrees of freedom corresponding to element e
    dof=[3*FirstNode(e)-2  3*FirstNode(e)-1   3*FirstNode(e)
        3*SecondNode(e)-2 3*SecondNode(e)-1  3*SecondNode(e)];
    ke = (A(e)*E(e)/L(e)); %Element stiffness matrix w.r.t. local axes
    lforce(e=ke*[-cx -cy -cz cx cy cz]*u(dof);%axial force in element e
end
fprintf(out,'____________________________________________________\n');
fprintf(out,'A X I A L   F O R C E   I N   M E M B E R S * \n');
fprintf(out,'ELEMENT               Force \n');
fprintf(out,'----------------------------------------------------------\n');
for i=1:nElem
    fprintf(out,'%3d       %20.7f\n',i,lforce(i));
end
fprintf(out,'* Note: positive sign indicates TENSION and negative sign
indicates COMPRESSION\n');
K = K1;
ifix=ifix';
codevec = ones(ndof,1); %code vector to indicate restrained dof
```

```matlab
for row=1:size(ifix,1)
    codevec(ifix(row)) = 0;
end
fdof = [1:ndof]'; %list of all degrees of freedom
pdof = ifix(:,1); %list of restrained degrees of freedom
fdof(pdof)=[]; %list of unrestrained degrees of freedom
KRF = K(pdof,fdof);
%reduced stiffness matrix with rows corresponding to restrained dof
%and columns corresponding to unrestrained dof
[nrow ncol] = size(KRF);
uF = u(fdof); %extracting {uF}, the vector of displacements at free dof
supp = KRF*uF; %vector of support reactions
fprintf(out,'____________________________________________________\n');
fprintf(out,'S U P P O R T    R E A C T I O N S\n');
fprintf(out,'Support node        Rx              Ry              Rz\n');
fprintf(out,'----------------------------------------------------\n');
count= 0;
for j=1:nNode
    j1 = 3*j-2;
    j2 = 3*j-1;
    j3 = 3*j;
    if((codevec(j1)==0)||(codevec(j2)==0)||(codevec(j3)==0))
        if(codevec(j1)==0)
            count = count+1;
            Rx = supp(count);
        else
            Rx = 0;
        end
        if(codevec(j2)==0)
            count = count+1;
            Ry = supp(count);
        else
            Ry = 0;
        end
        if(codevec(j3)==0)
            count = count+1;
            Rz = supp(count);
        else
            Rz = 0;
        end
        fprintf(out,'  %2d        %10.3f    %10.3f   %10.3f\n',j,Rx,Ry,Rz);
    end
end
```

```
fprintf(out,'______________________________________________________\n');
%close the file identifier for output file
fclose(out);
type SpaceTruss.out
```

A.2.1. Input and output for numerical examples

Numerical example 6.7

Input:

```
4  3
1    6.0    0.0      0.0    1      0      1    0.0      0.0   -45.0
2    0.0    3.0      0.0    0      0      0    0.0      0.0    0.0
3    0.0    3.0      6.0    0      0      0    0.0      0.0    0.0
4    0.0    0.0     -4.0    0      0      0    0.0      0.0    0.0
1    1    2   2.0e-4   2.0e8
2    1    3   5.0e-4   2.0e8
3    1    4   1.0e-4   2.0e8
```

Output:

```
A N A L Y S I S   O F   A   S P A C E   T R U S S
N O D A L   I N F O R M A T I O N
```

Node	Coordinates			Restraints*			Forces		
	X	Y	Z	X	Y	Z	Fx	Fy	Fz
1	6.0	0.0	0.0	1	0	1	0.0	0.0	-45.0
2	0.0	3.0	0.0	0	0	0	0.0	0.0	0.0
3	0.0	3.0	6.0	0	0	0	0.0	0.0	0.0
4	0.0	0.0	-4.0	0	0	0	0.0	0.0	0.0

```
* NOTE: 1 indicates an unrestrained degree of freedom
        0 indicates a  restrained  degree of freedom
```

```
E L E M E N T   I N F O R M A T I O N
```

Element	first node	second node	Area of c/s (A)	Young'modulus (E)
1	1	2	0.000200	2.000e+08
2	1	3	0.000500	2.000e+08
3	1	4	0.000100	2.000e+08

```
N O D A L   D I S P L A C E M E N T S
```

Node	X- displacement	Y- displacement	Z- displacement
1	-3.0503675298e-03	0.0000000000e+00	-9.6965068074e-03
2	0.0000000000e+00	0.0000000000e+00	0.0000000000e+00
3	0.0000000000e+00	0.0000000000e+00	0.0000000000e+00

```
4              0.0000000000e+00      0.0000000000e+00           0.0000000000e+00
```

```
A X I A L   F O R C E   I N   M E M B E R S *
ELEMENT                Force
-------------------------------------------------------------------
   1                -16.2686268
   2                 49.2306613
   3                -21.9570125
* Note: positive sign indicates TENSION and negative sign indicates
COMPRESSION
```

```
S U P P O R T    R E A C T I O N S
Support node      Rx              Ry         Rz
-------------------------------------------------------------------
   1            0.000          -9.135        0.000
   2           14.551          -7.276        0.000
   3          -32.820          16.410       32.820
   4           18.269           0.000       12.180
```

Numerical example 6.8

Input

```
24 80
 1   -0.4   -0.4    0.0    0    0    0    0.0    0.0    0.00
 2    0.4   -0.4    0.0    0    0    0    0.0    0.0    0.00
 3    0.4    0.4    0.0    0    0    0    0.0    0.0    0.00
 4   -0.4    0.4    0.0    0    0    0    0.0    0.0    0.00
 5   -0.4   -0.4    3.0    1    1    1    5.0    0.0    0.00
 6    0.4   -0.4    3.0    1    1    1    0.0    0.0    0.00
 7    0.4    0.4    3.0    1    1    1    0.0    0.0    0.00
 8   -0.4    0.4    3.0    1    1    1    5.0    0.0    0.00
 9   -0.4   -0.4    6.0    1    1    1    5.0    0.0    0.00
10    0.4   -0.4    6.0    1    1    1    0.0    0.0    0.00
11    0.4    0.4    6.0    1    1    1    0.0    0.0    0.00
12   -0.4    0.4    6.0    1    1    1    5.0    0.0    0.00
13   -0.4   -0.4    9.0    1    1    1    5.0    0.0    0.00
14    0.4   -0.4    9.0    1    1    1    0.0    0.0    0.00
15    0.4    0.4    9.0    1    1    1    0.0    0.0    0.00
16   -0.4    0.4    9.0    1    1    1    5.0    0.0    0.00
17   -0.4   -0.4   12.0    1    1    1    5.0    0.0    0.00
18    0.4   -0.4   12.0    1    1    1    0.0    0.0    0.00
19    0.4    0.4   12.0    1    1    1    0.0    0.0    0.00
20   -0.4    0.4   12.0    1    1    1    5.0    0.0    0.00
21   -0.4   -0.4   15.0    1    1    1    5.0    0.0    0.00
22    0.4   -0.4   15.0    1    1    1    0.0    0.0    0.00
23    0.4    0.4   15.0    1    1    1    0.0    0.0    0.00
24   -0.4    0.4   15.0    1    1    1    5.0    0.0    0.00
 1    1    5    21.06e-4    2.0e8
 2    2    6    21.06e-4    2.0e8
```

3	3	7	21.06e-4	2.0e8
4	4	8	21.06e-4	2.0e8
5	1	6	21.06e-4	2.0e8
6	2	5	21.06e-4	2.0e8
7	2	7	21.06e-4	2.0e8
8	3	6	21.06e-4	2.0e8
9	3	8	21.06e-4	2.0e8
10	4	7	21.06e-4	2.0e8
11	4	5	21.06e-4	2.0e8
12	1	8	21.06e-4	2.0e8
13	5	6	21.06e-4	2.0e8
14	6	7	21.06e-4	2.0e8
15	7	8	21.06e-4	2.0e8
16	8	5	21.06e-4	2.0e8
17	5	9	21.06e-4	2.0e8
18	6	10	21.06e-4	2.0e8
19	7	11	21.06e-4	2.0e8
20	8	12	21.06e-4	2.0e8
21	5	10	21.06e-4	2.0e8
22	6	9	21.06e-4	2.0e8
23	6	11	21.06e-4	2.0e8
24	7	10	21.06e-4	2.0e8
25	7	12	21.06e-4	2.0e8
26	8	11	21.06e-4	2.0e8
27	8	9	21.06e-4	2.0e8
28	5	12	21.06e-4	2.0e8
29	9	10	21.06e-4	2.0e8
30	10	11	21.06e-4	2.0e8
31	11	12	21.06e-4	2.0e8
32	12	9	21.06e-4	2.0e8
33	9	13	21.06e-4	2.0e8
34	10	14	21.06e-4	2.0e8
35	11	15	21.06e-4	2.0e8
36	12	16	21.06e-4	2.0e8
37	9	14	21.06e-4	2.0e8
38	10	13	21.06e-4	2.0e8
39	10	15	21.06e-4	2.0e8
40	11	14	21.06e-4	2.0e8
41	11	16	21.06e-4	2.0e8
42	12	15	21.06e-4	2.0e8
43	12	13	21.06e-4	2.0e8
44	9	16	21.06e-4	2.0e8
45	13	14	21.06e-4	2.0e8
46	14	15	21.06e-4	2.0e8
47	15	16	21.06e-4	2.0e8
48	16	13	21.06e-4	2.0e8
49	13	17	21.06e-4	2.0e8
50	14	18	21.06e-4	2.0e8
51	15	19	21.06e-4	2.0e8
52	16	20	21.06e-4	2.0e8

```
53   13   18   21.06e-4      2.0e8
54   14   17   21.06e-4      2.0e8
55   14   19   21.06e-4      2.0e8
56   15   18   21.06e-4      2.0e8
57   15   20   21.06e-4      2.0e8
58   16   19   21.06e-4      2.0e8
59   16   17   21.06e-4      2.0e8
60   13   20   21.06e-4      2.0e8
61   17   18   21.06e-4      2.0e8
62   18   19   21.06e-4      2.0e8
63   19   20   21.06e-4      2.0e8
64   20   17   21.06e-4      2.0e8
65   17   21   21.06e-4      2.0e8
66   18   22   21.06e-4      2.0e8
67   19   23   21.06e-4      2.0e8
68   20   24   21.06e-4      2.0e8
69   17   22   21.06e-4      2.0e8
70   18   21   21.06e-4      2.0e8
71   18   23   21.06e-4      2.0e8
72   19   22   21.06e-4      2.0e8
73   19   24   21.06e-4      2.0e8
74   20   23   21.06e-4      2.0e8
75   20   21   21.06e-4      2.0e8
76   17   24   21.06e-4      2.0e8
77   21   22   21.06e-4      2.0e8
78   22   23   21.06e-4      2.0e8
79   23   24   21.06e-4      2.0e8
80   24   21   21.06e-4      2.0e8
```

Output

```
A N A L Y S I S   O F   A   S P A C E   T R U S S

N O D A L   I N F O R M A T I O N
```

Node	Coordinates			Restraints*			Forces		
	X	Y	Z	X	Y	Z	Fx	Fy	Fz
1	-0.4	-0.4	0.0	0	0	0	0.0	0.0	0.0
2	0.4	-0.4	0.0	0	0	0	0.0	0.0	0.0
3	0.4	0.4	0.0	0	0	0	0.0	0.0	0.0
4	-0.4	0.4	0.0	0	0	0	0.0	0.0	0.0
5	-0.4	-0.4	3.0	1	1	1	5.0	0.0	0.0
6	0.4	-0.4	3.0	1	1	1	0.0	0.0	0.0
7	0.4	0.4	3.0	1	1	1	0.0	0.0	0.0
8	-0.4	0.4	3.0	1	1	1	5.0	0.0	0.0
9	-0.4	-0.4	6.0	1	1	1	5.0	0.0	0.0
10	0.4	-0.4	6.0	1	1	1	0.0	0.0	0.0
11	0.4	0.4	6.0	1	1	1	0.0	0.0	0.0
12	-0.4	0.4	6.0	1	1	1	5.0	0.0	0.0

13	-0.4	-0.4	9.0	1	1	1	5.0	0.0	0.0
14	0.4	-0.4	9.0	1	1	1	0.0	0.0	0.0
15	0.4	0.4	9.0	1	1	1	0.0	0.0	0.0
16	-0.4	0.4	9.0	1	1	1	5.0	0.0	0.0
17	-0.4	-0.4	12.0	1	1	1	5.0	0.0	0.0
18	0.4	-0.4	12.0	1	1	1	0.0	0.0	0.0
19	0.4	0.4	12.0	1	1	1	0.0	0.0	0.0
20	-0.4	0.4	12.0	1	1	1	5.0	0.0	0.0
21	-0.4	-0.4	15.0	1	1	1	5.0	0.0	0.0
22	0.4	-0.4	15.0	1	1	1	0.0	0.0	0.0
23	0.4	0.4	15.0	1	1	1	0.0	0.0	0.0
24	-0.4	0.4	15.0	1	1	1	5.0	0.0	0.0

```
* NOTE: 1 indicates an unrestrained degree of freedom
        0 indicates a  restrained  degree of freedom
```

E L E M E N T   I N F O R M A T I O N

Element	first node	second node	Area of c/s (A)	Young'modulus (E)
1	1	5	0.002106	2.000e+08
2	2	6	0.002106	2.000e+08
3	3	7	0.002106	2.000e+08
4	4	8	0.002106	2.000e+08
5	1	6	0.002106	2.000e+08
6	2	5	0.002106	2.000e+08
7	2	7	0.002106	2.000e+08
8	3	6	0.002106	2.000e+08
9	3	8	0.002106	2.000e+08
10	4	7	0.002106	2.000e+08
11	4	5	0.002106	2.000e+08
12	1	8	0.002106	2.000e+08
13	5	6	0.002106	2.000e+08
14	6	7	0.002106	2.000e+08
15	7	8	0.002106	2.000e+08
16	8	5	0.002106	2.000e+08
17	5	9	0.002106	2.000e+08
18	6	10	0.002106	2.000e+08
19	7	11	0.002106	2.000e+08
20	8	12	0.002106	2.000e+08
21	5	10	0.002106	2.000e+08
22	6	9	0.002106	2.000e+08
23	6	11	0.002106	2.000e+08
24	7	10	0.002106	2.000e+08
25	7	12	0.002106	2.000e+08
26	8	11	0.002106	2.000e+08
27	8	9	0.002106	2.000e+08
28	5	12	0.002106	2.000e+08
29	9	10	0.002106	2.000e+08
30	10	11	0.002106	2.000e+08
31	11	12	0.002106	2.000e+08

32	12	9	0.002106	2.000e+08
33	9	13	0.002106	2.000e+08
34	10	14	0.002106	2.000e+08
35	11	15	0.002106	2.000e+08
36	12	16	0.002106	2.000e+08
37	9	14	0.002106	2.000e+08
38	10	13	0.002106	2.000e+08
39	10	15	0.002106	2.000e+08
40	11	14	0.002106	2.000e+08
41	11	16	0.002106	2.000e+08
42	12	15	0.002106	2.000e+08
43	12	13	0.002106	2.000e+08
44	9	16	0.002106	2.000e+08
45	13	14	0.002106	2.000e+08
46	14	15	0.002106	2.000e+08
47	15	16	0.002106	2.000e+08
48	16	13	0.002106	2.000e+08
49	13	17	0.002106	2.000e+08
50	14	18	0.002106	2.000e+08
51	15	19	0.002106	2.000e+08
52	16	20	0.002106	2.000e+08
53	13	18	0.002106	2.000e+08
54	14	17	0.002106	2.000e+08
55	14	19	0.002106	2.000e+08
56	15	18	0.002106	2.000e+08
57	15	20	0.002106	2.000e+08
58	16	19	0.002106	2.000e+08
59	16	17	0.002106	2.000e+08
60	13	20	0.002106	2.000e+08
61	17	18	0.002106	2.000e+08
62	18	19	0.002106	2.000e+08
63	19	20	0.002106	2.000e+08
64	20	17	0.002106	2.000e+08
65	17	21	0.002106	2.000e+08
66	18	22	0.002106	2.000e+08
67	19	23	0.002106	2.000e+08
68	20	24	0.002106	2.000e+08
69	17	22	0.002106	2.000e+08
70	18	21	0.002106	2.000e+08
71	18	23	0.002106	2.000e+08
72	19	22	0.002106	2.000e+08
73	19	24	0.002106	2.000e+08
74	20	23	0.002106	2.000e+08
75	20	21	0.002106	2.000e+08
76	17	24	0.002106	2.000e+08
77	21	22	0.002106	2.000e+08
78	22	23	0.002106	2.000e+08
79	23	24	0.002106	2.000e+08
80	24	21	0.002106	2.000e+08

```
N O D A L   D I S P L A C E M E N T S
Node          X- displacement          Y- displacement              Z-
displacement
-------------------------------------------------------------------------
    1      0.0000000000e+00   0.0000000000e+00       0.0000000000e+00
    2      0.0000000000e+00   0.0000000000e+00       0.0000000000e+00
    3      0.0000000000e+00   0.0000000000e+00       0.0000000000e+00
    4      0.0000000000e+00   0.0000000000e+00       0.0000000000e+00
    5      4.7030025210e-03   4.5649304152e-05       8.8360762421e-04
    6      4.6983349222e-03  -4.5611577006e-05      -8.8320397711e-04
    7      4.6983349222e-03   4.5611576995e-05      -8.8320397711e-04
    8      4.7030025210e-03  -4.5649304141e-05       8.8360762420e-04
    9      1.4579919301e-02   2.7656204635e-05       1.4549623929e-03
   10      1.4575278303e-02  -2.7606141682e-05      -1.4537526753e-03
   11      1.4575278303e-02   2.7606141638e-05      -1.4537526753e-03
   12      1.4579919301e-02  -2.7656204591e-05       1.4549623929e-03
   13      2.7525060816e-02   1.4384491191e-05       1.7766239470e-03
   14      2.7520419630e-02  -1.4334572987e-05      -1.7746093790e-03
   15      2.7520419630e-02   1.4334572896e-05      -1.7746093790e-03
   16      2.7525060816e-02  -1.4384491101e-05       1.7766239470e-03
   17      4.1933349094e-02   5.5359277548e-06       1.9199609583e-03
   18      4.1928708282e-02  -5.4857213900e-06      -1.9171415435e-03
   19      4.1928708282e-02   5.4857212420e-06      -1.9171415435e-03
   20      4.1933349093e-02  -5.5359276093e-06       1.9199609583e-03
   21      5.6734868386e-02   1.1088910869e-06       1.9563055998e-03
   22      5.6730174000e-02  -1.0836431535e-06      -1.9526788909e-03
   23      5.6730173999e-02   1.0836429427e-06      -1.9526788909e-03
   24      5.6734868386e-02  -1.1088908816e-06       1.9563055998e-03
```

```
A X I A L   F O R C E   I N   M E M B E R S *
ELEMENT            Force
-------------------------------------------------------------------------
     1          124.0585104
     2         -124.0018384
     3         -124.0018384
     4          124.0585104
     5           48.4579243
     6          -48.5681676
     7         -114.1750279
     8         -114.1750279
     9          -48.5681676
    10           48.4579243
    11          114.2266187
    12          114.2266187
    13           -2.4574907
    14           48.0289906
    15           -2.4574907
    16          -48.0687173
    17           80.2182095
    18          -80.1050372
```

19	-80.1050372
20	80.2182095
21	38.7005781
22	-38.9202954
23	-72.2275822
24	-72.2275822
25	-38.9202954
26	38.7005781
27	72.3301724
28	72.3301724
29	-2.4434854
30	29.0692672
31	-2.4434854
32	-29.1219835
33	45.1612822
34	-45.0482812
35	-45.0482812
36	45.1612822
37	28.9983506
38	-29.2173045
39	-40.5915132
40	-40.5915132
41	-29.2173045
42	28.9983506
43	40.6935173
44	40.6935173
45	-2.4435845
46	15.0943053
47	-2.4435845
48	-15.1468692
49	20.1245164
50	-20.0115159
51	-20.0115159
52	20.1245164
53	19.2957444
54	-19.5146923
55	-17.9901449
56	-17.9901449
57	-19.5146923
58	19.2957444
59	18.0921435
60	18.0921435
61	-2.4433871
62	5.7764645
63	-2.4433871
64	-5.8293318
65	5.1027877
66	-4.9894436
67	-4.9894436
68	5.1027877

```
69                    9.5923659
70                   -9.8128525
71                   -4.4285663
72                   -4.4285663
73                   -9.8128525
74                    9.5923659
75                    4.5317480
76                    4.5317480
77                   -2.4715944
78                    1.1410761
79                   -2.4715944
80                   -1.1676622
```

* Note: positive sign indicates TENSION and negative sign indicates COMPRESSION

```
S U P P O R T     R E A C T I O N S
Support node      Rx              Ry            Rz
-----------------------------------------------------------------
       1         -12.486         -29.432       -281.250
       2         -12.514          29.419        281.250
       3         -12.514         -29.419        281.250
       4         -12.486          29.432       -281.250
```

A.3. ANALYSIS OF BEAMS AND PLANE FRAMES

```matlab
% Finite Element Analysis of Beams and Plane Frames
%----------------------------------------------------------------
clear
clc
format long
%----------------------------------------------------------------
filename = 'PFrame';
infile = strcat(filename,'.inp');
outfile =strcat(filename,'.out');
inp= fopen(infile,'r');
out = fopen(outfile,'w');
%----------------------------------------------------------------
%P R E P R O C E S S O R
%----------------------------------------------------------------
%R E A D I N G   N O D A L   I N F O R M A T I O N
[mat] = fscanf(inp,'%d',2);
% this line reads number of nodes and number of elements in the truss
nNode = mat(1);   %number of nodes
nElem = mat(2);   %number of elements
ndof=3*nNode;
% 'ndof' is the total number of degrees of freedom for the Plane Frame
%Read nodal information from the input file node-by-node
[mat] = fscanf(inp,'%d %f %f %d %d %d  %e %e  %e',[9,nNode]);
mat = mat';
Xval = mat(:,2); %reading x coordinates of nodes (column 2 of input)
Yval = mat(:,3); %reading y coordinates of nodes (column 3 of input)
%X and Y restraints of a node
resx = mat(:,4); %reading restraint along x-direction from column 4
resy = mat(:,5); %reading restraint along y-direction from column 5
resz = mat(:,6); %reading restraint about z-direction from column 6
%NOTE : 0 indicates restrained degree of freedom
%         1 indicates unrestrained degree of freedom
%----------------------------------------------------------------
%Generation of  nodal force vector (F) (of size ndof) along global axes
%----------------------------------------------------------------
%Degrees of freedom corresponding to i'th node are 3i-2, 3i-1 and 3i
%fx , fy  and Mz are components of nodal forces along global X and Y
directions and Moment about z-direction respectively
%At node i, fx acts along degree of freedom  '3i-2'
%             fy acts along degree of freedom  '3i-1'
%             Mz acts about degree of freedom  '3i'
```

```matlab
fx = mat(:,7); %reading fx from column 7
fy = mat(:,8); %reading fy from column 8
Mz = mat(:,9); %reading Mz from column 9
F= zeros(ndof,1); %initializing global nodal force vector
for i=1:nNode
  F(3*i-2) = fx(i);
  F(3*i-1) = fy(i);
  F(3*i)   = Mz(i);
end
%bn is a counter for the total number of restraints in the Plane Truss
bn = 0;%initialized to zero
%-------------------------------------------------------------------
%Notation used to indicate restraint associated with a given degree of
freedom
    % 0 indicates RESTRAINED degree of freedom
    % 1 indicates UNRESTRAINED degree of freedom
%-------------------------------------------------------------------
for i=1:nNode
  if(resx(i)==0)
      %if this degree of freedom along x-direction is restrained
      bn = bn+1;
      ifix(bn) = 3*i-2;
      %add to the list of restrained degrees of freedom
  end
  if(resy(i)==0)
      %if this degree of freedom along x-direction is restrained
      bn = bn+1;
      ifix(bn) = 3*i-1;
      %add to the list of restrained degrees of freedom
  end
  if(resz(i)==0)
      %if this degree of freedom ABOUT z-direction is restrained
      bn = bn+1;
      ifix(bn) = 3*i; %add to the list of restrained degrees of freedom
  end
end
end
fprintf(1,'A N A L Y S I S  OF  B E A M S  AND  P L A N E  F R A M E S \n');
fprintf(1,'N O D A L    I N F O R M A T I O N \n');
fprintf(1,'_____________________________________________________\n');
fprintf(1,'Node   Coordinates        *Nodal          Nodal Loads
          In-plane    Type of support\n');
fprintf(1,'                            Restraints                      moment\n');
fprintf(1,'-----------------------------------------------------------\n');
```

```matlab
fprintf(1,'        X        Y      x   y theta(z)    Fx       Fy       Mz\n');
fprintf(1,'-------------------------------------------------------------\n');
for i=1:nNode
    fprintf(1,'%2d    %4.1f    %4.1f    %d   %d   %d    %7.1f  %7.1f
%7.1f',i,Xval(i),Yval(i),resx(i),resy(i),resz(i),F(3*i-2),
F(3*i-1),F(3*i));
    if(resx(i)==0 &&resy(i)==0&&resz(i)==0)
        fprintf(1,'         Fixed end support');
    end
    if(resx(i)==0&&resy(i)==0&&resz(i)==1)
        fprintf(1,'         Pin support');
    end
    if(resx(i)==1&&resy(i)==0&&resz(i)==1)
        fprintf(1,'         Roller support');
    end
    fprintf(1,'\n');
end
fprintf(1,'* NOTE: 1 indicates an unrestrained degree of freedom\n');
fprintf(1,'        0 indicates a  restrained  degree of freedom\n');
fprintf(1,'\nThe  vector  of  restrained  degrees  of  freedom  for  the
structure is\n');
for i=1:bn
    fprintf(1,'%d ',ifix(i));
end
fprintf(1,'\n');
%READING ELEMENT INFORMATION
[mat] = fscanf(inp,'%d %d %d %f %e %e %d %f %f %f',[10,nElem]);
mat = mat';
FirstNode = zeros(nElem,1);
SecondNode = zeros(nElem,1);
A        = zeros(nElem,1) ;
Iz       = zeros(nElem,1) ;
E        = zeros(nElem,1);
Cx       = zeros(nElem,1);
Cy       = zeros(nElem,1);
Le       = zeros(nElem,1);
eforce = zeros(6,nElem);
FirstNode  = mat(:,2);SecondNode = mat(:,3);
A        = mat(:,4);
Iz       = mat(:,5);
E        = mat(:,6);
Ltype    = mat(:,7);
w1 = mat(:,8);
```

```matlab
d1 = mat(:,9);
d2 = mat(:,10);
loadflag = zeros(nElem);%flag used to mark loaded elements
                        %flag =0 indicates element NOT carrying any load
                        %flag =1 indicates element carrying load
if(Ltype==1)
    fprintf(1,'Distributed load\n');
elseif(Ltype==2)
    fprintf(1,'Point load\n');
elseif(Ltype==3)
    fprintf(1,'No load\n');
end
fprintf(1,'_________________________________________________\n');
fprintf(1,'E L E M E N T    I N F O R M A T I O N\n');
fprintf(1,'Element first second Area    Moment of    Young''s    Element
          Description\n');
fprintf(1,'          node   node    (A)     Intertia(Iz) modulus(E)   load*
          w1        x1       w2        x2\n');
fprintf(1,'---------------------------------------------------------\n');
for e=1:nElem
   if(Ltype(e)==1)
      fprintf(1,'%2d %2d   %2d    %.3f %.3e %.2e   DL   %7.2f %7.2f    %7.2f %7.2f\n',...
      e,FirstNode(e),SecondNode(e),A(e),Iz(e),E(e),w1(e),d1(e),w1(e),d2(e));
   elseif(Ltype(e)==2)
       fprintf(1,'%2d   %2d   %2d %.3f %.3e %.2e   PL   %7.2f %7.2f    %7.2f 7.2f\n',...
e,FirstNode(e),SecondNode(e),A(e),Iz(e),E(e),w1(e),d1(e),w1(e),d2(e));
   else
       fprintf(1,'%2d %2d %2d %.3f    %.3e    %.2e    %7.2f %7.2f    %7.2f %7.2f\n',...
e,FirstNode(e),SecondNode(e),A(e),Iz(e),E(e),w1(e),d1(e),w1(e),d2(e));
   end
end
fprintf(1,'*NOTE: DL = Distributed load w on the element\n');
fprintf(1,'*NOTE: PL = Point load W on the element\n');
fprintf(1,'from distance of d1 to a distance of d2 from left end\n');
fprintf(1,'_________________________________________________\n');
fprintf(1,'DETAILED  COMPUTATIONS  FOR  EACH  ELEMENT  ARE  PRESENTED
BELOW\n');
fprintf(1,'{eforce} the vector of equivalent nodal forces for each
element along local axes\n');
fprintf(1,'Element       Vector of Element Forces  {eforce}\n');
```

```matlab
%Computation of equivalent nodal loads owing to uniformly distributed
load %over some portion of the span
for e=1:nElem
    x1=Xval(FirstNode(e));      y1=Yval(FirstNode(e));
    %x and y coordinates of first node
    x2=Xval(SecondNode(e));     y2=Yval(SecondNode(e));
    %x and y coordinates of second node
    le =sqrt((x2-x1)^2+(y2-y1)^2); %length of element e
    %direction cosines of element e
    c=(x2-x1)/le;
    s=(y2-y1)/le;
    %Storing values of direction cosines c and s in vectors {Cx} and
    %{Cy}for subsequent use
    Cx(e)=c;
    Cy(e)=s;
    L(e)=le;
    if(Ltype(e)==1) %for distributed loading
        %computation of equivalent nodal loads owing to uniformly
        %distributed load over %some portion of the span
        c1 = d2(e)-d1(e);
        a = d1(e)+0.5*c1;
        b = 0.5*c1+le-d2(e);
        p1 = (-w1(e)*c1/(12*le^2))*(12*a*b^2+c1^2*(le-3*b));
        p2 = (w1(e)*c1/(12*le^2))* (12*a^2*b+c1^2*(le-3*a));
        p3 = (w1(e)*c1*b/le)-(p1+p2)/le;
        p4 = (w1(e)*c1*a/le)+(p1+p2)/le;
        eforce(:,e) = eforce(:,e)+ [ 0.0 p3 -p1 0.0 p4 -p2]';
        %Negative of vector of fixed end forces in added to the element
        %force vector
    elseif(Ltype(e)==2) %for point loads
        a = d1(e); b = d2(e);
        if((a+b)~=le)
            fprintf(1,'Wrong input for point load\n');
            exit();
        end
        p2 = w1(e)*b/le - w1(e)*a*b*(a-b)/le^3 ;
        p3 = w1(e)*a*b^2/le^2;
        p5 = w1(e)*a/le+w1(e)*a*b*(a-b)/le^3;
        p6 = -w1(e)*a^2*b/le^2;
        eforce(:,e) = eforce(:,e)+[0.0 p2 p3 0.0 p5 p6]';
    end
    fprintf(1,'  %d        ',e);
    for row=1:6
```

```matlab
        fprintf(1,'%8.3f    ',eforce(row,e));
    end
    fprintf(1,'\n');
end
%-------------------------------------------------------------------
% P R O C E S S O R
%-------------------------------------------------------------------
%in this loop the following computations are performed for each element
% a) identification of degrees of freedom associated with element
% b) computation of direction cosines of each element
% c) computation of Rotation transformation matrix of each element
% d) computation of element stiffness matrix along element axes
% e) computation of element stiffness matrix along global axes using
%    rotation transformation matrix
% f) assembly of global stiffness matrix and global force vector
%processing element load information
aa=zeros(nElem,ndof);
K=zeros(ndof,ndof); %initialize the global stiffness matrix of size ndof
Feq=zeros(ndof,1); %nodal force vector due to element loads
for e=1:nElem
    c =Cx(e);
    s =Cy(e);
    le = L(e);
    ae = A(e)*E(e); %axial rigidity
    ei = E(e)*Iz(e); %flexural rigidity
    %Degrees of freedom for element e with nodes m and n are
    % 3m-2, 3m-1 ,3m,
    % and 3n-2, 3n-1 ,3n
    %These degrees of freedom are stored in the vector {dof}
    %as shown below
    dof=[3*FirstNode(e)-2  3*FirstNode(e)-1 3*FirstNode(e)
         3*SecondNode(e)-2 3*SecondNode(e)-1 3*SecondNode(e)];
    %rotation transformation matrix for element e
    T=[c    s    0    0    0    0;
      -s    c    0    0    0    0;
       0    0    1    0    0    0;
       0    0    0    c    s    0;
       0    0    0   -s    c    0;
       0    0    0    0    0    1;];
    %element stiffness matrix w.r.t local axes
    [estif] = hermite(le,ae,ei);
    ke=(T'*estif*T);
    %element stiffness matrix with reference to global axes
```

```
    K(dof,dof)=K(dof,dof)+ke;   %assembly of global stiffness matrix K
    %{Feq} is vector of equivalent nodal forces due to element loads
    Feq(dof) = Feq(dof) + T'*eforce(:,e);
end
Fc = F+Feq;
%combined force vector {Fc} is obtained by adding vector of nodal loads
{F} to {Feq}
%store the global stiffness matrix for future use to compute support
reactions
K1 = K;
%----------------------------------------------------------------
%applying boundary conditions (specified by restraints along X and Y
axes)
%on global stiffness matrix and global force vector
K(ifix,:)=0;
%Elements in rows with d.o.f corresponding to ifix are made 0
K(:,ifix)=0;
%Elements in columns with d.o.f corresponding to ifix are made 0
K(ifix,ifix)=eye(length(ifix));
%Diagonal elements with d.o.f corresponding to ifix are made 1
Fc(ifix) = 0;
 %corresponding columns of Global force vector are also made zero
%obtaining vector of nodal displacements along global axes
u=inv(K)*Fc;
fprintf(1,'_____________________________________________\n');
fprintf(1,'N O D A L    D I S P L A C E M E N T S\n');
fprintf(1,'Node       X- displacement           Y- displacement        Z-
rotation\n');
fprintf(1,'-------------------------------------------------------\n');
for i=1:nNode
    fprintf(1,'%2d    %23.14e    %23.14e    %23.14e\n',i,u(3*i-2),
    u(3*i-1),u(3*i));
end
%----------------------------------------------------------------
%P O S T P R O C E S S I N G
%----------------------------------------------------------------
%Computation of internal forces and moments in elements of Plane Frame
syms eta MM
MM = zeros(nElem,1);
MM = sym(MM);
xval = 0:0.01:3;
for e=1:nElem
    %degrees of freedom corresponding to element e
```

```matlab
dof=[3*FirstNode(e)-2  3*FirstNode(e)-1 3*FirstNode(e)
     3*SecondNode(e)-2 3*SecondNode(e)-1 3*SecondNode(e)];
c =Cx(e);  s =Cy(e);  %direction cosines
le = L(e);
ae = A(e)*E(e); %axial rigidity
ei = E(e)*Iz(e); %flexural rigidity
%rotation transformation matrix for element e
T=[c    s    0    0    0    0;
   -s   c    0    0    0    0;
    0   0    1    0    0    0;
    0   0    0    c    s    0;
    0   0    0   -s    c    0;
    0   0    0    0    0    1;];
%element stiffness matrix w.r.t local axes
[estif] = hermite(le,ae,ei);
lforce(:,e) = estif*T*u(dof)-eforce(:,e);
%Vector of internal forces of element w.r.t. local axes
uu = T*u(dof);
Ax = (ae/le)*[-1 1]*[uu(1) uu(4)]';
Mx = ei*[0 (-6+12*eta)/le^2 (-4+6*eta)/le 0 (6-12*eta)/le^2
         (-2+6*eta)/le]*T*u(dof);
Vx = E(e)*Iz(e)*[0 12/le^3 6/le^2 0 -12/le^3  6/le^2]*T*u(dof);
if(Ltype(e)==1)
    Mx = Mx+(w1(e)*le^2/12)*(1-6*eta+6*eta^2);
    Mx1 = double(subs(Mx,eta,0));
    Mx2 = double(subs(Mx,eta,1));
    Vx = Vx-(w1(e)*le/2)*(1-2*eta);
    Vx1 = double(subs(Vx,eta,0));
    Vx2 = double(subs(Vx,eta,1));
elseif(Ltype(e)==2)
    W = w1(e); a= d1(e); b= d2(e);
    Mxx = (W*a*b^2/le^2)- (W*b^2*(3*a+b)/le^2)*eta+W*le*
          (eta-a/le)*heaviside(eta-a/le);
    Vxx = -(W*b^2/le^3)*(3*a+b)+W*heaviside(eta-a/le);
    Mx = Mx+Mxx;
    Vx = Vx+Vxx;
end
Mx1 = double(subs(Mx,eta,0));
Mx2 = double(subs(Mx,eta,1));
Vx1 = double(subs(Vx,eta,0));
Vx2 = double(subs(Vx,eta,1));
lforce(:,e)= [-Ax Vx1 Mx1 Ax Vx2 Mx2]';
end
```

```matlab
fprintf(1,'______________________________________________\n');
fprintf(1,'F O R C E S AND  M O M E N T S    I N    E L E M E N T S \n');
fprintf(1,'----------------------------------------------------------\n');
fprintf(1,'ELEMENT  Axial Shear Bending   Axial    Shear    Bending\n');
fprintf(1,'         Force Force  Moment    Force    Force    Moment\n');
fprintf(1,'--------------------    --------------------------\n');
fprintf(1,'          at FIRST NODE          at SECOND NODE\n');
fprintf(1,'----------------------------------------------------------\n');
%using appropriate sign convention to obtain axial forces, shear
%forces and bending moment from element internal forces --
%Axial force    : Tension positive
%Shear force    : Clockwise shear positive
%Bending moment: Sagging moment positive
for e=1:nElem
    f1= double(lforce(1,e));
    f2= double(lforce(2,e));
    f3= double(lforce(3,e));
    f4= double(lforce(4,e));
    f5= double(lforce(5,e));
    f6= double(lforce(6,e));
     fprintf(1,'%3d %10.3f %10.3f %10.3f %10.3f %10.3f
              %10.3f\n',e,f1,f2,f3,f4,f5,f6);
end
K = K1;
codevec = ones(ndof,1); %code vector to indicate restrained dof
ifix = ifix';
for row=1:size(ifix,1)
   codevec(ifix(row)) = 0;
end
%picking up rows of K corresponding to restrained dof and columns
%corresponding to unrestrained dof
nrow=0;
for row=1:ndof
    ncol=0;
    if(codevec(row)==0) %pickup rows corresponding to restrained dof
        nrow = nrow+1;
        for col=1:ndof
            if(codevec(col)==1)
                %pickup columns corresponding to unrestrained dof
                ncol = ncol+1;
                uF(ncol) = u(col);
                KRF(nrow,ncol) = K(row,col); %reduced stiffness matrix
            end
```

```matlab
        end
    end
end
supp = KRF*uF';%vector of support reactions
fprintf(1,'_______________________________________________\n');
fprintf(1,'S U P P O R T     R E A C T I O N S\n');
fprintf(1,'Support node      Rx              Ry          Mz\n');
fprintf(1,'-----------------------------------------------------------\n');
count= 0;
for j=1:nNode
    j1 = 3*j-2;
    j2 = 3*j-1;
    j3 = 3*j;
    if((codevec(j1)==0)||(codevec(j2)==0)||(codevec(j3)==0))
        if(codevec(j1)==0)
            count = count+1;
            Rx = supp(count)-Feq(j1);
        else
            Rx = 0;
        end
        if(codevec(j2)==0)
            count = count+1;
            Ry = supp(count)-Feq(j2);
        else
            Ry = 0;
        end
        if(codevec(j3)==0)
            count = count+1;
            Mz = supp(count)-Feq(j3);
        else
            Mz = 0;
        end
        fprintf(1,'   %2d        %10.3f    %10.3f   %10.3f\n',j,Rx,Ry,Mz);
    end
end
fprintf(1,'_______________________________________________\n');
fclose(out);
%-----------------------------------------------------------
function [K1]= hermite(le,ae,ei);
syms x L eta AE EI p alpha a W
phi = [1 x x^2 x^3];
Dphi = diff(phi);
A =[subs(phi,0) ; subs(Dphi,0); subs(phi,L);subs(Dphi,L)];
```

```
phi = [1 x x^2 x^3];
N1 = phi*inv(A);
N1 =subs(N1,x,eta*L);
DN1 = diff(diff(N1));
K1 = zeros(4,4);
K1=sym(K1);
for i=1:4
    for j=1:4
        K1(i,j)=(EI/L^3)*int(DN1(i)*DN1(j),0,1);
    end
end
K1;
psi = [1 x];
Dpsi = diff(psi);
A = [subs(psi,0);subs(psi,1)];
N2 = psi*inv(A);
N2 =subs(N2,x,eta);
DN2 = diff(N2);
K2 = zeros(2,2);
K2=sym(K2);
for i=1:2
    for j=1:2
        K2(i,j)=(AE/L)*int(DN2(i)*DN2(j),0,1);
    end
end
K2;
K = zeros(6,6);
K = sym(K);
K(1,1) = K2(1,1); K(1,4)=K2(1,2);
K(4,1) = K2(2,1); K(4,4)=K2(2,2);
K(2,:)=[0 K1(1,1) K1(1,2) 0 K1(1,3) K1(1,4)];
K(3,:)=[0 K1(2,1) K1(2,2) 0 K1(2,3) K1(2,4)];
K(5,:)=[0 K1(3,1) K1(3,2) 0 K1(3,3) K1(3,4)];
K(6,:)=[0 K1(4,1) K1(4,2) 0 K1(4,3) K1(4,4)];
K;
K1 = subs(K,{L AE EI},[le ae ei]);
```

A.3.1. Input and output for numerical examples

Numerical Example 6.9

Input Data:

```
4    3
1    0.0      0.0      0    0    0    0.0      0.0      0.0
```

```
2   5.0      0.0       1   0   1   0.0      0.0      0.0
3   8.0      0.0       1   0   1   0.0      0.0      0.0
4  12.0      0.0       0   0   1   0.0      0.0      0.0
1   1   2   0.06      4.5e-4    2.0e8   1   -25.0  0.0    5.0
2   2   3   0.06      4.5e-4    2.0e8   2  -100.0  2.0    1.0
3   3   4   0.06      4.5e-4    2.0e8   1   -50.0  0.0    4.0
```

Output:

```
A N A L Y S I S   O F   B E A M S   A N D       P L A N E   F R A M E S
N O D A L   I N F O R M A T I O N
```

Node	Coordinates		*Nodal Restraints			Nodal Loads		In-plane moment	Type of support
	X	Y	x	y	theta(z)	Fx	Fy	Mz	
1	0.0	0.0	0	0	0	0.0	0.0	0.0	Fixed support
2	5.0	0.0	1	0	1	0.0	0.0	0.0	Roller support
3	8.0	0.0	1	0	1	0.0	0.0	0.0	Roller support
4	12.0	0.0	0	0	1	0.0	0.0	0.0	Pin support

```
* NOTE: 1 indicates an unrestrained degree of freedom
        0 indicates a  restrained  degree of freedom
The vector of restrained degrees of freedom for the structure is
1 2 3 5 8 10 11
```

```
E L E M E N T   I N F O R M A T I O N
```

Element	first node	second node	Area (A)	Moment of Inertia(Iz)	Young's modulus(E)	Element load*	w1	x1	w2	x2
1	1	2	0.060	4.500e-04	2.00e+08	DL	-25.00	0.00	-25.00	5.00
2	2	3	0.060	4.500e-04	2.00e+08	DL	-100.00	2.00	-100.00	1.00
3	3	4	0.060	4.500e-04	2.00e+08	DL	-50.00	0.00	-50.00	4.00

```
*NOTE: DL = Distributed load w on the element
       PL = Point load W on the element
            from a distance of d1 to a distance of d2 from left end
```

```
DETAILED COMPUTATIONS FOR EACH ELEMENT ARE PRESENTED BELOW
the vector of equivalent nodal forces for each element along local axes
```

Element	Vector of Element Forces					
1	0.000	-62.500	-52.083	0.000	-62.500	52.083
2	0.000	-25.926	-22.222	0.000	-74.074	44.444
3	0.000	-100.000	-66.667	0.000	-100.000	66.667

```
N O D A L   D I S P L A C E M E N T S
```

Node	X- displacement	Y- displacement	Z-rotation
1	0.00000000000000e+00	0.00000000000000e+00	0.00000000000000e+00
2	0.00000000000000e+00	0.00000000000000e+00	2.75688014403292e-04
3	0.00000000000000e+00	0.00000000000000e+00	-3.84516460905350e-04
4	0.00000000000000e+00	0.00000000000000e+00	9.32998971193416e-04

```
F O R C E S    A N D    M O M E N T S    I N    E L E M E N T S
-------------------------------------------------------------------
ELEMENT  Axial        Shear        Bending      Axial       Shear       Bending
         Force        Force        Moment       Force       Force       Moment
         --------------------------------       ------------------------------
              at FIRST NODE                           at SECOND NODE
-------------------------------------------------------------------
   1     0.000        68.455       -62.008      0.000      -56.545      -32.234
   2     0.000        19.396       -32.234      0.000      -80.604      -74.045
   3     0.000       118.511       -74.045      0.000      -81.489        0.000

S U P P O R T    R E A C T I O N S
Support node     Rx              Ry           Mz
------------------------------------------------------
    1          0.000          68.455       62.008
    2          0.000          75.941        0.000
    3          0.000         199.115        0.000
    4          0.000          81.489        0.000
```

Numerical Example 6.10

Input Data:

```
3   2
1   0.0      0.0      0   0   0   0.0      0.0      0.0
2   4.0      0.0      1   0   1   0.0      0.0      0.0
3   7.0      0.0      1   0   1   0.0      0.0      0.0
1   1   2   0.09     6.0e-4   2.0e8   2   -120.0  2.0    2.0
2   2   3   0.09     6.0e-4   2.0e8   1    -25.0  0.0    3.0
%Numerical Example 6.10
```

Output:

```
A N A L Y S I S    O F    B E A M S    A N D    P L A N E    F R A M E S
N O D A L    I N F O R M A T I O N
```

Node of Restraints	Coordinates		*Nodal			Nodal Loads			In-plane support	Type
	X	Y	x	y	theta(z)	Fx	Fy	Mz		
1	0.0	0.0	0	0	0	0.0	0.0	0.0	Fixed support	
2	4.0	0.0	1	0	1	0.0	0.0	0.0	Roller support	
3	7.0	0.0	1	0	1	0.0	0.0	0.0	Roller support	

```
* NOTE: 1 indicates an unrestrained degree of freedom
        0 indicates a  restrained  degree of freedom
```

```
The vector of restrained degrees of freedom for the structure is
1 2 3 5 8
```

```
E L E M E N T     I N F O R M A T I O N
Element first second Area    Moment of  Young's Element    Description
        node  node  (A)    Intertia(Iz) modulus(E) load* w1 x1  w2   x2
-----------------------------------------------------------------------
   1     1    2 0.090  6.000e-04   2.00e+08 DL -120.00 2.00 -120.00 2.0
   2     2    3 0.090  6.000e-04   2.00e+08 DL  -25.00 0.00  -25.00 3.0
*NOTE: DL = Distributed load w on the element
       PL = Point load W on the element
            from a distance of d1 to a distance of d2 from left end
```

```
DETAILED COMPUTATIONS FOR EACH ELEMENT ARE PRESENTED BELOW
{eforce} the vector of equivalent nodal forces for each element along
local axes
Element      Vector of Element Forces   {eforce}
   1         0.000    -60.000    -60.000     0.000    -60.000    60.000
   2         0.000    -37.500    -18.750     0.000    -37.500    18.750
```

```
N O D A L    D I S P L A C E M E N T S
Node        X- displacement          Y- displacement        Z-rotation
----------------------------------------------------------------------
   1      0.00000000000000e+00 0.00000000000000e+00 0.00000000000000e+00
   2      0.00000000000000e+00 0.00000000000000e+00 1.32812500000000e-04
   3      0.00000000000000e+00 0.00000000000000e+00 5.07812500000000e-05
```

```
F O R C E S    A N D    M O M E N T S    I N    E L E M E N T S
----------------------------------------------------------------------
ELEMENT  Axial      Shear      Bending     Axial      Shear      Bending
         Force      Force      Moment      Force      Force      Moment
         ----------------------------       ----------------------------
             at FIRST NODE                      at SECOND NODE
----------------------------------------------------------------------
   1     0.000     65.977     -67.969      0.000     -54.023     -44.063
   2     0.000     52.188     -44.063      0.000     -22.813       0.000
```

```
S U P P O R T    R E A C T I O N S
Support node     Rx              Ry          Mz
----------------------------------------------------------------
   1           0.000          65.977      67.969
   2           0.000         106.211       0.000
   3           0.000          22.813       0.000
```

Numerical Example 6.11

Input Data:

```
4   3
1   0.0     0.0    0  0  0   0.0    0.0    0.0
2   0.0     4.0    1  1  1   0.0    0.0    0.0
3   3.0     4.0    1  1  1   0.0    0.0    0.0
```

```
4   3.0      0.0       0    0    0     0.0      0.0       0.0
1   1   2    0.06      4.5e-4    2.0e8   2       -100.0  2.0   2.0
2   2   3    0.06      4.5e-4    2.0e8   1        -20.0  0.0   3.0
3   3   4    0.06      4.5e-4    2.0e8   3          0.0  0.0   0.0
%Numerical Example 6.11
```

Output:

```
A N A L Y S I S   O F   B E A M S   A N D      P L A N E   F R A M E S
N O D A L   I N F O R M A T I O N
```

Node	Coordinates		*Nodal			Nodal Loads			In-plane	Type
of support			Restraints						moment	
	X	Y	x	y theta(z)		Fx	Fy		Mz	
1	0.0	0.0	0	0	0	0.0	0.0	0.0	Fixed support	
2	0.0	4.0	1	1	1	0.0	0.0	0.0		
3	3.0	4.0	1	1	1	0.0	0.0	0.0		
4	3.0	0.0	0	0	0	0.0	0.0	0.0	Fixed support	

```
* NOTE: 1 indicates an unrestrained degree of freedom
        0 indicates a  restrained   degree of freedom
The vector of restrained degrees of freedom for the structure is
1 2 3 10 11 12
```

```
E L E M E N T    I N F O R M A T I O N
```

Element	first node	second node	Area (A)	Moment of Inertia(Iz)	Young's modulus(E)	Element load*	w1	x1	w2	x2
1	1	2	0.060	4.500e-04	2.00e+08	PL	-100.0	2.00	-100.0	2.00
2	2	3	0.060	4.500e-04	2.00e+08	DL	-20.0	0.00	-20.0	3.00
3	3	4	0.060	4.500e-04	2.00e+08		0.0	0.00	0.00	0.00

```
*NOTE: DL = Distributed load w on the element
            from a distance of d1 to a distance of d2 from left end
```

```
DETAILED COMPUTATIONS FOR EACH ELEMENT ARE PRESENTED BELOW
 the vector of equivalent nodal forces for each element along local axes
```

Element	Vector of Element Forces					
1	0.000	-50.000	-50.000	0.000	-50.000	50.000
2	0.000	-30.000	-15.000	0.000	-30.000	15.000
3	0.000	0.000	0.000	0.000	0.000	0.000

```
N O D A L   D I S P L A C E M E N T S
```

Node	X- displacement	Y- displacement	Z-rotation
1	0.00000000000000e+00	0.00000000000000e+00	0.00000000000000e+00
2	1.73707099240440e-03	-5.07631708517972e-06	-6.03469637718693e-05
3	1.73139008314678e-03	-1.49236829148203e-05	-1.92402092522238e-04
4	0.00000000000000e+00	0.00000000000000e+00	0.00000000000000e+00

```
F O R C E S    A N D    M O M E N T S    I N    E L E M E N T S
-------------------------------------------------------------------------
ELEMENT  Axial        Shear      Bending        Axial       Shear      Bending
         Force        Force      Moment         Force       Force      Moment
         -----------------------------          ---------------------------
              at FIRST NODE                          at SECOND NODE
-------------------------------------------------------------------------
   1     15.229       77.276    -105.911       -15.229     -22.724       3.195
   2     22.724       15.229       3.195       -22.724     -44.771     -41.118
   3     44.771       22.724     -41.118       -44.771      22.724      49.776

S U P P O R T    R E A C T I O N S
Support node      Rx            Ry           Mz
-------------------------------------------------------------
     1          -77.276       15.229      105.911
     4          -22.724       44.771       49.776
```

A.4. ANALYSIS OF SPACE FRAMES

```matlab
% Finite Element Analysis of Space Frame
% A program by Dr.M.V.Rama Rao
clc
clear
%-------------------------------------------------------
%First, the input and output files are defined
filename = 'SpaceFrame';
%Here we provide the name of the input/output files to be used
infile = strcat(filename,'.inp');
%Input file name  specified by concatenating '.inp' to filename
inp= fopen(infile,'r');
% Input file is opened and file identifier 'inp' is associated with it
outfile =strcat(filename,'.out');
%Output file name specified by concatenating '.out' to filename
out = fopen(outfile,'w');
% Output file is opened and file identifier 'out' is associated with it
%________________________________________________________
%P R E P R O C E S S O R
%________________________________________________________
%R E A D I N G   N O D A L   I N F O R M A T I O N
[row] = fscanf(inp,'%d',2);
% this line reads number of nodes and number of elements in the truss
nNode = row(1);   %number of nodes
nElem = row(2);   %number of elements
ndof=6*nNode;
% 'ndof' is the total number of degrees of freedom for the Space Truss
%Read nodal information from the input file node by node
[mat] = fscanf(inp,'%d %f %f %f  %d %d %d %d %d %d %e %e %e %e %e %e  ',
[16,nNode]);
mat = mat'; %Transposed for future use
Xval = mat(:,2);
%reading x coordinates of nodes(column 2 of input as above)
Yval = mat(:,3);
%reading y coordinates of nodes(column 3 of input as above)
Zval = mat(:,4);%reading y coordinates of nodes(column 4 of input as
above)
% restraints at a node
resx  = mat(:,5);   %reading restraint along x-direction from column 5
resy  = mat(:,6);   %reading restraint along y-direction from column 6
resz  = mat(:,7);   %reading restraint along z-direction from column 7
resmx = mat(:,8);   %reading restraint about x-direction from column 8
resmy = mat(:,9);   %reading restraint about y-direction from column 9
resmz = mat(:,10); %reading restraint about z-direction from column 10
%-------------------------------------------------------------------
%Generation of  nodal force vector (F) (of size ndof) along global axes
%-------------------------------------------------------------------
fx = mat(:,11); %reading fx from column 8
fy = mat(:,12); %reading fy from column 9
```

```
fz = mat(:,13); %reading fy from column 10
Mx = mat(:,14); %reading Mx from column 8
My = mat(:,15); %reading My from column 9
Mz = mat(:,16); %reading My from column 10
F= zeros(ndof,1);
%Creating global nodal force vector and initializing all %elements to
%zero
for i=1:nNode
  F(6*i-5)    = fx(i);
  F(6*i-4)    = fy(i);
  F(6*i-3)    = fz(i);
  F(6*i-2)    = Mx(i);
  F(6*i-1)    = My(i);
  F(6*i)      = Mz(i);
end
%bn is a counter for the total number of restraints in the Space Truss
bn = 0; %initialized to zero
%------------------------------------------------------------------------
%Notation used to indicate restraint associated with a given degree of
%freedom
    % 0 indicates RESTRAINED degree of freedom
    % 1 indicates UNRESTRAINED degree of freedom
%------------------------------------------------------------------------
for i=1:nNode
    if(resx(i)==0)
        %if this degree of freedom along x-direction is restrained
        bn = bn+1;
        ifix(bn) = 6*i-5;
        %add to the list of restrained degrees of freedom
    end
    if(resy(i)==0)
        %if this degree of freedom along y-direction is restrained
        bn = bn+1;
        ifix(bn) = 6*i-4;
        %add to the list of restrained degrees of freedom
    end
    if(resz(i)==0)
        %if this degree of freedom along z-direction is restrained
        bn = bn+1;
        ifix(bn) = 6*i-3;
        %add to the list of restrained degrees of freedom
    end
    if(resx(i)==0)
      %if this degree of freedom about theta_x-direction is restrained
        bn = bn+1;
        ifix(bn) = 6*i-2;
        %add to the list of restrained degrees of freedom
    end
    if(resy(i)==0)
        %if this degree of freedom about theta_y-direction is restrained
```

```matlab
        bn = bn+1;
        ifix(bn) = 6*i-1;
        %add to the list of restrained degrees of freedom
      end
      if(resz(i)==0)
        %if this degree of freedom about theta_z-direction is restrained
        bn = bn+1;
        ifix(bn) = 6*i;
        %add to the list of restrained degrees of freedom
      end
end
fprintf(1,'A N A L Y S I S    O F    A    S P A C E    F R A M   E \n\n');
fprintf(1,'N O D A L    I N F O R M A T I O N \n');
fprintf(1,'__________________________________________________\n');
fprintf(1,'Node        Coordinates    Restraints*    Forces      Moments \n');
fprintf(1,'                                              Theta \n');
fprintf(1,'                                           ---------\n');
fprintf(1,'  X Y Z  X  Y  Z  X  Y  Z    Fx    Fy    Fz Mx  My    Mz\n');
fprintf(1,'-----------------------------------------------------------\n');
for i=1:nNode
     fprintf(1,'%2d %4.1f    %4.1f    %4.1f    %d %d %d %d %d %d
               %5.1f %5.1f   %5.1f %5.1f %5.1f %5.1f\n',...
i,Xval(i),Yval(i),Zval(i),resx(i),resy(i),resz(i),resmx(i),resmy(i),
resmz(i),F(6*i-5),F(6*i-4),F(6*i-3),...
          F(6*i-2),F(6*i-1),F(6*i));
end
fprintf(1,'* NOTE: 1 indicates an unrestrained degree of freedom\n');
fprintf(1,'        0 indicates a  restrained   degree of freedom\n');
%-------------------------------------------------------------------
%READING ELEMENT INFORMATION
[mat] = fscanf(inp,'%d %d %d %f %e %e %e %e %e %d %d %f %f %f
                   %c',[15,nElem]);
mat = mat'; %transposed for future sure
%reading first and second nodes of each element from columns 2 and 3
FirstNode  = mat(:,2);
SecondNode = mat(:,3);
A   = mat(:,4);%reading cross sectional areas of elements from column 4
E   = mat(:,5);%reading Modulus of Elasticity of elements from column 5
G    = mat(:,6);%reading Modulus of Rigidity of elements from column 6
Ix   = mat(:,7);%reading Ix of elements from column 7
Iy   = mat(:,8); %reading Iy of elements from column 8
Iz   = mat(:,9); %reading Iz of elements from column 9
kNode= mat(:,10);
Ltype= mat(:,11);
w1 = mat(:,12);
d1 = mat(:,13);
d2 = mat(:,14);
Lflag = mat(:,15);
eforce = zeros(12,nElem);
fprintf(1,'*NOTE: DL = Distributed load w on the element\n');
```

```matlab
fprintf(1,'          from a distance of d1 to a distance of d2 from left
end\n');
fprintf(1,'________________________________________________\n');
fprintf(1,'DETAILED  COMPUTATIONS  FOR  EACH  ELEMENT  ARE  PRESENTED
BELOW\n');
fprintf(1,'{eforce} the vector of equivalent nodal forces for each
element along local axes\n');
fprintf(1,'Element      Vector of Element Forces  {eforce}\n');
%Computation of equivalent nodal loads owing to uniformly distributed
%load over some portion of the span
fprintf(1,'          Fx      Fy      Fz      Mx      My      Mz
Fx       Fy      Fz      Mx      My       Mz\n');
for e=1:nElem
 x1=Xval(FirstNode(e));y1=Yval(FirstNode(e)); z1=Zval(FirstNode(e));
 %x,y and z coordinates of first node
 x2=Xval(SecondNode(e));y2=Yval(SecondNode(e));z2=Zval(SecondNode(e));
 %x,y and z coordinates of second node
 le =sqrt((x2-x1)^2+(y2-y1)^2+(z2-z1)^2); %length of element e
 if(Ltype(e)==1) %for distributed loading
     %computation of equivalent nodal loads owing to uniformly
     %distributed load over some portion of the span
     c1 = d2(e)-d1(e);
     a = d1(e)+0.5*c1;
     b = 0.5*c1+le-d2(e);
     p1 = (-w1(e)*c1/(12*le^2))*(12*a*b^2+c1^2*(le-3*b));
     p2 = (w1(e)*c1/(12*le^2))* (12*a^2*b+c1^2*(le-3*a));
     p3 = (w1(e)*c1*b/le)-(p1+p2)/le;
     p4 = (w1(e)*c1*a/le)+(p1+p2)/le;
     if(Lflag(e)=='Y')
         eforce(:,e) = eforce(:,e)+ [0 p3 0 0 0 -p1 0 p4 0 0 0 -p2]';
     elseif(Lflag(e)=='Z')
             eforce(:,e)=eforce(:,e)+ [ 0 0 p3 0 p1 0 0 0 p4 0 p2 0]';
     end
     %Negative of vector of fixed end forces in added to the element
     %force vector
 elseif(Ltype(e)==2) %for point loads
     a = d1(e); b = d2(e);
     if((a+b)~=le)
         fprintf(1,'Wrong input for point load\n');
         exit();
     end
     p2 = w1(e)*b/le - w1(e)*a*b*(a-b)/le^3 ;
     p3 = w1(e)*a*b^2/le^2;
     p5 = w1(e)*a/le+w1(e)*a*b*(a-b)/le^3;
     p6 = -w1(e)*a^2*b/le^2;
     if(Lflag(e)=='Y')
         eforce(:,e) = eforce(:,e)+ [0 p2 0 0 0 p3 0 p5 0 0 0 p6]';
     elseif(Lflag(e)=='Z')
         eforce(:,e) = eforce(:,e)+ [0 0 p2 0 -p3 0 0 0 p5 0 -p6 0]';
     end
```

```matlab
  end
  fprintf(1,'  %2d  ',e);
  for row=1:12
      fprintf(1,'%9.3f ',eforce(row,e));
  end
  fprintf(1,'\n');
end
fprintf(1,'_________________________________________________\n');

fprintf(1,'E L E M E N T   I N F O R M A T I O N\n');
fprintf(1,'Element   first   second    Area of c/s Young''s    Rigidity
Ix        Iy        Iz \n');
fprintf(1,'                node       node              (A)        modulus(E)
modulus(G)\n');
fprintf(1,'---------------------------------------------------------\n');
for e=1:nElem
    fprintf(1,'%2d   %2d   %2d   %f   %.3e    %.3e %.3e %.3e %.3e\n',...
        e,FirstNode(e),SecondNode(e),A(e),E(e),G(e)Ix(e),Iy(e),Iz(e));
end
K=zeros(ndof,ndof); %initialize the global stiffness matrix of size ndof
T=zeros(6,6);
Feq = zeros(ndof,1);
for e=1:nElem
%     fprintf(1,'Element = %d\n',e);
    %x and y coordinates of first node
    x1=Xval(FirstNode(e));
    y1=Yval(FirstNode(e));
    z1=Zval(FirstNode(e));
    %x and y coordinates of second node
    x2=Xval(SecondNode(e));
    y2=Yval(SecondNode(e));
    z2=Zval(SecondNode(e));
    % (x1,y1) &  (x2,y2) are coordinates of first and second nodes
    %of element e respectively
    L(e)=sqrt((x2-x1)^2+(y2-y1)^2+(z2-z1)^2); %length of element e
    %direction cosines of element e
    cx =(x2-x1)/L(e);
    cy =(y2-y1)/L(e);
    cz =(z2-z1)/L(e);
    cxz = sqrt(cx^2+cz^2);
    %Storing values of direction cosines c and s in vectors
    % {Cx} and {Cy} for subsequent use
    Cx(e)=cx;   Cy(e)=cy;       Cz(e)=cz;
    %Degrees of freedom for element e with nodes m and n are
    % 2m-1, 2m , 2n-1 and 2n
    %These degrees of freedom are stored in the vector {dof} as shown
    %below
    dof1=[6*FirstNode(e)-5 6*FirstNode(e)-4 6*FirstNode(e)-3
        6*FirstNode(e)-2 6*FirstNode(e)-1 6*FirstNode(e)];
```

```matlab
dof2=[6*SecondNode(e)-5 6*SecondNode(e)-4 6*SecondNode(e)-3
      6*SecondNode(e)-2 6*SecondNode(e)-1 6*SecondNode(e) ];
dof = [dof1 dof2];
xk = Xval(kNode(e));  yk = Yval(kNode(e));  zk = Zval(kNode(e));
xki = xk-x1; yki=yk-y1; zki =zk-z1;
%Rotation transformation matrix for element e
% rotation transformation matrix
if(cxz~=0) %when member is NOT vertical
 xkg = cx*xki + cy*yki + cz*zki;
 ykg = (-cx*cy/cxz)*xki+cxz*yki -(cy*cz/cxz)*zki;
 zkg = (-cz/cxz)*xki+(cx/cxz)*zki;
 salpha = zkg/sqrt(ykg^2+zkg^2);
 calpha = ykg/sqrt(ykg^2+zkg^2);
 T1 =[             cx                  cy                  cz
   (-cx*cy*calpha-cz*salpha)/cxz cxz*calpha (-cy*cz*calpha+cx*salpha)/cxz
   (cx*cy*salpha-cz*calpha)/cxz -cxz*salpha (cy*cz*salpha+cx*calpha)/cxz];
else
     %when member is vertical
     salpha =  zki/sqrt(xki^2+zki^2);
     calpha = (-xki/sqrt(xki^2+zki^2))*cy;
     T1 = [ 0            cy  0
            -cy*calpha    0  salpha
            cy*salpha     0  calpha];
end
O3 = zeros(3);
T = [ T1  O3  O3  O3
      O3  T1  O3  O3
      O3  O3  T1  O3
      O3  O3  O3  T1];
%Element stiffness matrix with reference to local axes
ae = A(e)*E(e);
le=L(e);
eiy = E(e)*Iy(e);
eiz = E(e)*Iz(e);
gix = G(e)*Ix(e);
v1 = ae/le; v2z =12*eiz/le^3;v3z=6*eiz/le^2;
v4z=4*eiz/le;v5z = 2*eiz/le; v6x = gix/le;
v2y =12*eiy/le^3 ; v3y = 6*eiy/le^2; v4y = 4*eiy/le; v5y = 2*eiy/le;

estif = [ v1   0    0    0    0    0   -v1   0    0    0    0    0
          0   v2z   0    0    0   v3z   0  -v2z   0    0    0   v3z
          0    0   v2y   0  -v3y   0    0    0  -v2y   0  -v3y  0
          0    0    0   v6x   0    0    0    0    0  -v6x  0    0
          0    0  -v3y   0   v4y   0    0    0   v3y   0   v5y  0
          0   v3z   0    0    0   v4z   0  -v3z   0    0    0   v5z
         -v1   0    0    0    0    0   v1    0    0    0    0    0
          0  -v2z   0    0    0  -v3z   0   v2z   0    0    0  -v3z
          0    0  -v2y   0   v3y   0    0    0   v2y   0   v3y  0
          0    0    0  -v6x   0    0    0    0    0   v6x  0    0
          0    0  -v3y   0   v5y   0    0    0   v3y   0   v4y  0
          0   v3z   0    0    0   v5z   0  -v3z   0    0    0  v4z];
```

```matlab
    ke = T'*estif*T;
    %element stiffness matrix with reference global axes
    K(dof,dof) = K(dof,dof)+ke; %assembly
    Feq(dof) = Feq(dof) + T'*eforce(:,e);
end
%Assembly of global stiffness matirx
%store the global stiffness matrix for future use
%to compute support reactions
K1 = K;
Fc = F+Feq;
%combined force vector {Fc} is obtained by adding vector
%of nodal loads {F} to {Feq}
%applying boundary conditions
%(specified by restraints along X and Y axes)
%on global stiffness matrix and global force vector
K(ifix,:)=0    ;
%Elements in rows with d.o.f corresponding to ifix are made 0
K(:,ifix)=0    ;
%Elements in columns with d.o.f corresponding to ifix are made 0
K(ifix,ifix)=eye(length(ifix));
%Diagonal elements with d.o.f corresponding to ifix are made 1
Fc(ifix) = 0;
%corresponding columns of Global force vector are also made zero
%obtaining vector of nodal displacements along global axes
%by solving [K]{u} = {Fc}
u=K\Fc;
fprintf(1,'_____________________________________________________\n');
fprintf(1,'N O D A L    D I S P L A C E M E N T S\n');
fprintf(1,'Node        u                    v                   w            thetaX
thetay          thetaz\n');
fprintf(1,'-----------------------------------------------------------------\n');
for i=1:nNode
    fprintf(1,'%2d %14.6e %14.6e %14.6e %14.6e %14.6e %14.6e\n',...
            i,u(6*i-5),u(6*i-4),u(6*i-3),u(6*i-2),u(6*i-1),u(6*i));
end
%-----------------------------------------------------------------
%COMPUTATION OF INTERNAL FORCES IN SPACE FRAME ELEMENTS

for e=1:nElem
%      fprintf(1,'Element = %d\n',e);
    %degrees of freedom corresponding to element e
    dof1=[6*FirstNode(e)-5  6*FirstNode(e)-4 6*FirstNode(e)-3
            6*FirstNode(e)-2 6*FirstNode(e)-1 6*FirstNode(e)];
    dof2=[6*SecondNode(e)-5 6*SecondNode(e)-4 6*SecondNode(e)-3
            6*SecondNode(e)-2 6*SecondNode(e)-1 6*SecondNode(e) ];
    dof = [dof1 dof2];
    %rotation transformation matrix
    cx = Cx(e);
    cy = Cy(e);
```

```matlab
cz = Cz(e);
cxz = sqrt(cx^2+cz^2);
x1 = Xval(FirstNode(e));y1=Yval(FirstNode(e));
z1= Zval(FirstNode(e));
x2 = Xval(SecondNode(e));y2=Yval(SecondNode(e));
z2= Zval(SecondNode(e));
xk = Xval(kNode(e)); yk = Yval(kNode(e));zk = Zval(kNode(e));
xki= xk-x1;      yki= yk-y1;       zki =zk-z1;
%direction cosines
if(cxz~=0) %when member is NOT vertical
    xkg = cx*xki + cy*yki + cz*zki;
    ykg = (-cx*cy/cxz)*xki+cxz*yki -(cy*cz/cxz)*zki;
    zkg = (-cz/cxz)*xki+(cx/cxz)*zki;
    salpha =zkg/sqrt(ykg^2+zkg^2);
    calpha = ykg/sqrt(ykg^2+zkg^2);
        T1 = [cx             cy        cz
(-cx*cy*calpha-cz*salpha)/cxz   cxz*calpha   (-cy*cz*calpha+cx*salpha)/cxz
(cx*cy*salpha-cz*calpha)/cxz  -cxz*salpha    (cy*cz*salpha+cx*calpha)/cxz];
    else
        %when member is vertical
        salpha =  zki/sqrt(xki^2+zki^2);
        calpha =  -xki/sqrt(xki^2+zki^2);
        T1 = [ 0              cy  0
                -cy*calpha     0  salpha
                 cy*salpha     0  calpha];
end
O3 = zeros(3);
T = [ T1   O3   O3   O3
        O3   T1   O3   O3
        O3   O3   T1   O3
        O3   O3   O3   T1];
ae = A(e)*E(e);
le=L(e);
eiy = E(e)*Iy(e);
eiz = E(e)*Iz(e);
gix = G(e)*Ix(e);
v1 = ae/le; v2z =12*eiz/le^3 ; v3z = 6*eiz/le^2; v4z = 4*eiz/le;
v5z = 2*eiz/le; v6x = gix/le;
v2y =12*eiy/le^3 ; v3y = 6*eiy/le^2; v4y = 4*eiy/le; v5y = 2*eiy/le;
estif = [ v1   0     0     0     0     0    -v1   0     0     0    0     0
           0   v2z   0     0     0    v3z    0  -v2z   0     0    0    v3z
           0    0    v2y   0   -v3y    0     0    0   -v2y    0  -v3y   0
           0    0     0   v6x    0     0     0    0     0   -v6x   0     0
           0    0   -v3y    0    v4y   0     0    0    v3y    0   v5y   0
           0   v3z    0     0     0    v4z    0  -v3z   0     0    0    v5z
          -v1   0     0     0     0     0    v1   0     0     0    0     0
           0  -v2z    0     0     0   -v3z    0   v2z   0     0    0   -v3z
           0    0   -v2y    0    v3y   0     0    0    v2y    0   v3y   0
           0    0     0   -v6x    0     0     0    0     0    v6x   0     0
           0    0   -v3y    0    v5y   0     0    0    v3y    0   v4y   0
           0   v3z    0     0     0    v5z    0  -v3z   0     0    0    v4z];
```

```matlab
    lforce(:,e) = estif*T*u(dof)-eforce(:,e);
    %Vector of internal forces of element w.r.t. local axes
end
fprintf(1,'_______________________________________________\n');
fprintf(1,'F O R C E S   AND   M O M E N T S   IN   E L E M E N T S \n');
fprintf(1,'-----------------------------------------------------------\n');
%Axial force   : Tension positive
%Shear force   : Clockwise shear positive
%Bending moment: Sagging moment positive
fprintf(1,'ELEMENT   Node          Fx         Fy          Fz         Mx
My        Mz     \n');
for e=1:nElem
    n1 = FirstNode(e);
    f1= lforce(1,e);f2= lforce(2,e);f3= lforce(3,e);f4= lforce(4,e);
    f5= lforce(5,e);f6= lforce(6,e);
    fprintf(1,'%2d       %2d %10.5f %10.5f %10.5f %10.5f %10.5f
             %10.5f\n',e,n1,f1,f2,f3,f4,f5,f6);
    n2 = SecondNode(e);
    f7= lforce(7,e);f8= lforce(8,e);f9= lforce(9,e);f10= lforce(10,e);
    f11= lforce(11,e);f12= lforce(12,e);
    fprintf(1,'          %2d  %10.5f %10.5f %10.5f %10.5f %10.5f
                  %10.5f\n',n2,f7,f8,f9,f10,f11,f12);
end
K = K1;
ifix=ifix';
codevec = ones(ndof,1); %code vector to indicate restrained dof
for row=1:size(ifix,1)
   codevec(ifix(row)) = 0;
end
fdof = [1:ndof]'; %list of all degrees of freedom
pdof = ifix(:,1); %list of restrained degrees of freedom
fdof(pdof)=[]; %list of unrestrained degrees of freedom
KRF = K(pdof,fdof);
%reduced stiffness matrix with rows corresponding to restrained dof
%and columns corresponding to unrestrained dof
[nrow ncol] = size(KRF);
uF = u(fdof);
%extracting {uF}, the vector of displacements at free dof
supp = KRF*uF;%vector of support reactions
fprintf(1,'_______________________________________________\n');
fprintf(1,'S U P P O R T   R E A C T I O N S\n');
fprintf(1,'Support node      Rx             Ry           Rz         Mx
My        Mz\n');
fprintf(1,'-----------------------------------------------------------\n');
count= 0;
for j=1:nNode
    j1 = 6*j-5;
    j2 = 6*j-4;
    j3 = 6*j-3;
    j4 = 6*j-2;
```

```matlab
    j5 = 6*j-1;
    j6 = 6*j;
    if(   (codevec(j1)==0) || (codevec(j2)==0) || (codevec(j3)==0)
       || (codevec(j4)==0) || (codevec(j4)==0) || (codevec(j6)==0))
        if(codevec(j1)==0)
           count = count+1;
           Rx = supp(count);
        else
           Rx = 0;
        end
        if(codevec(j2)==0)
           count = count+1;
           Ry = supp(count);
        else
           Ry = 0;
        end
        if(codevec(j3)==0)
           count = count+1;
           Rz = supp(count);
        else
           Rz = 0;
        end
        if(codevec(j4)==0)
           count = count+1;
           Mx = supp(count);
        else
           Mx = 0;
        end
        if(codevec(j5)==0)
           count = count+1;
           My = supp(count);
        else
           My = 0;
        end
        if(codevec(j6)==0)
           count = count+1;
           Mz = supp(count);
        else
           Mz = 0;
        end
        fprintf(1,'   %2d    %10.3f %10.3f   %10.3f    %10.3f %10.3f
                %10.3f  \n',j,Rx,Ry,Rz,Mx,My,Mz);
    end
end
fprintf(1,'_____________________________________________________\n');
%close the file identifier for output file
```

A.4.1. Input and output for numerical examples

Numerical Example 12
Input Data:
```
5   3
1   0.0 3.0 0.0    1    1    1    1    1    1 120.0    0.0 0.0 0.0 0.0     0.0
2   6.0 3.0 0.0    1    1    1    1    1    1    0.0 -60.0 0.0 0.0 0.0 -180.0
3   0.0 0.0 0.0    0    0    0    0    0    0    0.0    0.0 0.0 0.0 0.0     0.0
4   9.0 0.0 3.0    0    0    0    0    0    0    0.0    0.0 0.0 0.0 0.0     0.0
5   5.0 5.0 0.0    0    0    0    0    0    0    0.0    0.0 0.0 0.0 0.0     0.0
1   1    2   0.01 2.0e8 0.8e8   0.002   0.001 0.001   5 2 240.0 3.0   3.0   Z
2   3    1   0.01 2.0e8 0.8e8   0.002   0.001 0.001   5 3   0.0 0.0   0.0   N
3   2    4   0.01 2.0e8 0.8e8   0.002   0.001 0.001   5 3   0.0 0.0   0.0   N
%Numerical example 6.12
```
Output:
```
A N A L Y S I S    O F    A    S P A C E    F R A M    E
N O D A L    I N F O R M A T I O N
```

Node	Coordinates			Restraints*						Forces			Moments		
	X	Y	Z	X	Y	Z	X	Y	Z	Fx	Fy	Fz	Mx	My	Mz
1	0.0	3.0	0.0	1	1	1	1	1	1	120.0	0.0	0.0	0.0	0.0	0.0
2	6.0	3.0	0.0	1	1	1	1	1	1	0.0	-60.0	0.0	0.0	0.0	-180.0
3	0.0	0.0	0.0	0	0	0	0	0	0	0.0	0.0	0.0	0.0	0.0	0.0
4	9.0	0.0	3.0	0	0	0	0	0	0	0.0	0.0	0.0	0.0	0.0	0.0
5	5.0	5.0	0.0	0	0	0	0	0	0	0.0	0.0	0.0	0.0	0.0	0.0

```
* NOTE: 1 indicates an unrestrained degree of freedom
        0 indicates a  restrained   degree of freedom
*NOTE: DL = Distributed load w on the element
       PL = Distributed load w on the element
            from a distance of d1 to a distance of d2 from left end
```

```
DETAILED COMPUTATIONS FOR EACH ELEMENT ARE PRESENTED BELOW
the vector of equivalent nodal forces for each element along local axes
Element                   Vector of Element Forces
```

Element	Fx	Fy	Fz	Mx	My	Mz	Fx	Fy	Fz	Mx	My	Mz
1	0.00	0.00	120.00	0.00	-180.00	0.00	0.00	0.00	120.00	0.00	80.00	0.00
2	0.00	0.00	0.00	0.00	0.000	0.00	0.00	0.00	0.00	0.00	0.00	0.00
3	0.00	0.00	0.00	0.00	0.000	0.00	0.00	0.00	0.00	0.00	0.00	0.00

```
E L E M E N T    I N F O R M A T I O N
```

Element	first node	second node	Area of c/s (A)	Young's modulus (E)	Rigidity modulus (G)	Ix	Iy	Iz
1	1	2	0.010	2.0e+08	8.0e+07	2.0e-03	1.0e-03	1.0e-03
2	3	1	0.010	2.0e+08	8.0e+07	2.0e-03	1.0e-03	1.0e-03

```
 3  2      4      0.010     2.0e+08    8.0e+07    2.0e-03   1.0e-03   1.0e-03
```

N O D A L   D I S P L A C E M E N T S

Node	u	v	w	thetaX	thetay	thetaz
1	-8.594097e-04	5.776353e-05	5.007645e-03	2.393332e-03	-1.623169e-03	6.813313e-04
2	-1.176053e-03	3.253162e-03	5.255517e-03	1.288428e-03	1.720936e-03	-7.714689e-04
3	0.000000e+00	0.000000e+00	0.000000e+00	0.000000e+00	0.000000e+00	0.000000e+00
4	0.000000e+00	0.000000e+00	0.000000e+00	0.000000e+00	0.000000e+00	0.000000e+00
5	0.000000e+00	0.000000e+00	0.000000e+00	0.000000e+00	0.000000e+00	0.000000e+00

F O R C E S   A N D   M O M E N T S   I N   E L E M E N T S

ELEMENT	Node	Fx	Fy	Fz	Mx	My	Mz
1	1	105.548	-38.509	-126.013	29.464	86.568	-67.100
	2	-105.548	38.509	-113.986	-29.464	-50.490	-163.953
2	3	-38.509	14.452	-126.013	86.568	348.575	-23.743
	1	38.509	-14.452	126.013	-86.568	29.464	67.100
3	2	183.6227	9.192	5.967	-21.404	46.703	-32.180
	4	-183.6227	-9.192	-5.967	21.404	-77.710	79.945

S U P P O R T   R E A C T I O N S

Support node	Rx	Ry	Rz	Mx	My	Mz
3	-14.452	-38.509	-126.013	-348.575	86.569	-23.744
4	-105.548	98.509	-113.987	-75.898	-75.808	37.163

Numerical Example 13
Input Data:

```
12 16
1    3.0    0.0    0.0    0 0 0 0 0 0    0.0    0.0    0.0 0.0 0.0    0.0
2    3.0    3.0    0.0    0 0 0 0 0 0    0.0    0.0    0.0 0.0 0.0    0.0
3    0.0    3.0    0.0    0 0 0 0 0 0    0.0    0.0    0.0 0.0 0.0    0.0
4    0.0    0.0    0.0    0 0 0 0 0 0    0.0    0.0    0.0 0.0 0.0    0.0
5    3.0    0.0    3.0    1 1 1 1 1 1    0.0   12.0    0.0 0.0 0.0    0.0
6    3.0    3.0    3.0    1 1 1 1 1 1    0.0    0.0    0.0 0.0 0.0    0.0
7    0.0    3.0    3.0    1 1 1 1 1 1    0.0    0.0    0.0 0.0 0.0    0.0
8    0.0    0.0    3.0    1 1 1 1 1 1    0.0    0.0    0.0 0.0 0.0    0.0
9    3.0    0.0    6.0    1 1 1 1 1 1    0.0    0.0    0.0 0.0 0.0    0.0
10   3.0    3.0    6.0    1 1 1 1 1 1    0.0    0.0    0.0 0.0 0.0    0.0
11   0.0    3.0    6.0    1 1 1 1 1 1    0.0    0.0    0.0 0.0 0.0    0.0
12   0.0    0.0    6.0    1 1 1 1 1 1    0.0    7.5    0.0 0.0 0.0    0.0
1  1 5  0.003233 2.1e8 0.80769e8 10.0e-8 150.0e-8 2235.0e-8 2  3 0.0 0.0   0.0 N
2  5 9  0.003233 2.1e8 0.80769e8 10.0e-8 150.0e-8 2235.0e-8 2  3 0.0 0.0   0.0 N
3  2 6  0.003233 2.1e8 0.80769e8 10.0e-8 150.0e-8 2235.0e-8 1  3 0.0 0.0   0.0 N
4  6 10 0.003233 2.1e8 0.80769e8 10.0e-8 150.0e-8 2235.0e-8 1  3 0.0 0.0   0.0 N
5  3 7  0.003233 2.1e8 0.80769e8 10.0e-8 150.0e-8 2235.0e-8 4  3 0.0 0.0   0.0 N
6  7 11 0.003233 2.1e8 0.80769e8 10.0e-8 150.0e-8 2235.0e-8 4  3 0.0 0.0   0.0 N
7  4 8  0.003233 2.1e8 0.80769e8 10.0e-8 150.0e-8 2235.0e-8 3  3 0.0 0.0   0.0 N
8  8 12 0.003233 2.1e8 0.80769e8 10.0e-8 150.0e-8 2235.0e-8 3  3 0.0 0.0   0.0 N
9  5 6  0.0019   2.1e8 0.80769e8 3.24e-8  52.6e-8 726.6e-8 10  3 0.0 0.0   0.0 N
10 6 7  0.0019   2.1e8 0.80769e8 3.24e-8  52.6e-8 726.6e-8 10  3 0.0 0.0   0.0 N
11 7 8  0.0019   2.1e8 0.80769e8 3.24e-8  52.6e-8 726.6e-8 12  3 0.0 0.0   0.0 N
```

```
12 5 8  0.0019 2.1e8 0.80769e8 3.24e-8   52.6e-8 726.6e-8  12 3 0.0 0.0   0.0 N
13 9 10  0.0019 2.1e8 0.80769e8 3.24e-8  52.6e-8 726.6e-8   6 3 0.0 0.0   0.0 N
14 10 11 0.0019 2.1e8 0.80769e8 3.24e-8  52.6e-8 726.6e-8   6 3 0.0 0.0   0.0 N
15 11 12 0.0019 2.1e8 0.80769e8 3.24e-8  52.6e-8 726.6e-8   8 3 0.0 0.0   0.0 N
16 9  12 0.0019 2.1e8 0.80769e8 3.24e-8  52.6e-8 726.6e-8   8 3 0.0 0.0   0.0 N
%Numerical Example 6.13
```

Output:

```
A N A L Y S I S   O F   A   S P A C E   F R A M E
N O D A L   I N F O R M A T I O N
```

Node	Coordinates			Restraints*						Forces			Moments		
	X	Y	Z	X	Y	Z	X	Y	Z	Fx	Fy	Fz	Mx	My	Mz
1	3.0	0.0	0.0	0	0	0	0	0	0	0.0	0.0	0.0	0.0	0.0	0.0
2	3.0	3.0	0.0	0	0	0	0	0	0	0.0	0.0	0.0	0.0	0.0	0.0
3	0.0	3.0	0.0	0	0	0	0	0	0	0.0	0.0	0.0	0.0	0.0	0.0
4	0.0	0.0	0.0	0	0	0	0	0	0	0.0	0.0	0.0	0.0	0.0	0.0
5	3.0	0.0	3.0	1	1	1	1	1	1	0.0	12.0	0.0	0.0	0.0	0.0
6	3.0	3.0	3.0	1	1	1	1	1	1	0.0	0.0	0.0	0.0	0.0	0.0
7	0.0	3.0	3.0	1	1	1	1	1	1	0.0	0.0	0.0	0.0	0.0	0.0
8	0.0	0.0	3.0	1	1	1	1	1	1	0.0	0.0	0.0	0.0	0.0	0.0
9	3.0	0.0	6.0	1	1	1	1	1	1	0.0	0.0	0.0	0.0	0.0	0.0
10	3.0	3.0	6.0	1	1	1	1	1	1	0.0	0.0	0.0	0.0	0.0	0.0
11	0.0	3.0	6.0	1	1	1	1	1	1	0.0	0.0	0.0	0.0	0.0	0.0
12	0.0	0.0	6.0	1	1	1	1	1	1	0.0	7.5	0.0	0.0	0.0	0.0

```
* NOTE: 1 indicates an unrestrained degree of freedom
        0 indicates a  restrained   degree of freedom
*NOTE: DL = Distributed load w on the element
            from a distance of d1 to a distance of d2 from left end
```

```
DETAILED COMPUTATIONS FOR EACH ELEMENT ARE PRESENTED BELOW
the vector of equivalent nodal forces for each element along local axes
Element                      Vector of Element Forces
```

Element	Fx	Fy	Fz	Mx	My	Mz	Fx	Fy	Fz	Mx	My	Mz
1	0.000	0.000	0.000	0.000	0.000	0.000	0.000	0.000	0.000	0.000	0.000	0.000
2	0.000	0.000	0.000	0.000	0.000	0.000	0.000	0.000	0.000	0.000	0.000	0.000
3	0.000	0.000	0.000	0.000	0.000	0.000	0.000	0.000	0.000	0.000	0.000	0.000
4	0.000	0.000	0.000	0.000	0.000	0.000	0.000	0.000	0.000	0.000	0.000	0.000
5	0.000	0.000	0.000	0.000	0.000	0.000	0.000	0.000	0.000	0.000	0.000	0.000
6	0.000	0.000	0.000	0.000	0.000	0.000	0.000	0.000	0.000	0.000	0.000	0.000
7	0.000	0.000	0.000	0.000	0.000	0.000	0.000	0.000	0.000	0.000	0.000	0.000
8	0.000	0.000	0.000	0.000	0.000	0.000	0.000	0.000	0.000	0.000	0.000	0.000
9	0.000	0.000	0.000	0.000	0.000	0.000	0.000	0.000	0.000	0.000	0.000	0.000
10	0.000	0.000	0.000	0.000	0.000	0.000	0.000	0.000	0.000	0.000	0.000	0.000
11	0.000	0.000	0.000	0.000	0.000	0.000	0.000	0.000	0.000	0.000	0.000	0.000
12	0.000	0.000	0.000	0.000	0.000	0.000	0.000	0.000	0.000	0.000	0.000	0.000
13	0.000	0.000	0.000	0.000	0.000	0.000	0.000	0.000	0.000	0.000	0.000	0.000
14	0.000	0.000	0.000	0.000	0.000	0.000	0.000	0.000	0.000	0.000	0.000	0.000
15	0.000	0.000	0.000	0.000	0.000	0.000	0.000	0.000	0.000	0.000	0.000	0.000
16	0.000	0.000	0.000	0.000	0.000	0.000	0.000	0.000	0.000	0.000	0.000	0.000

```
E L E M E N T     I N F O R M A T I O N
Element first  second Area       Young's     Rigidity   Ix    Iy    Iz
        node   node   of c/s(A)  modulus(E)  modulus(G)
------------------------------------------------------------------------------
   1        1      5   0.003233  2.100e+08   8.077e+07  1.000e-07 1.500e-06 2.235e-05
   2        5      9   0.003233  2.100e+08   8.077e+07  1.000e-07 1.500e-06 2.235e-05
   3        2      6   0.003233  2.100e+08   8.077e+07  1.000e-07 1.500e-06 2.235e-05
   4        6     10   0.003233  2.100e+08   8.077e+07  1.000e-07 1.500e-06 2.235e-05
   5        3      7   0.003233  2.100e+08   8.077e+07  1.000e-07 1.500e-06 2.235e-05
   6        7     11   0.003233  2.100e+08   8.077e+07  1.000e-07 1.500e-06 2.235e-05
   7        4      8   0.003233  2.100e+08   8.077e+07  1.000e-07 1.500e-06 2.235e-05
   8        8     12   0.003233  2.100e+08   8.077e+07  1.000e-07 1.500e-06 2.235e-05
   9        5      6   0.001900  2.100e+08   8.077e+07  3.240e-08 5.260e-07 7.266e-06
  10        6      7   0.001900  2.100e+08   8.077e+07  3.240e-08 5.260e-07 7.266e-06
  11        7      8   0.001900  2.100e+08   8.077e+07  3.240e-08 5.260e-07 7.266e-06
  12        5      8   0.001900  2.100e+08   8.077e+07  3.240e-08 5.260e-07 7.266e-06
  13        9     10   0.001900  2.100e+08   8.077e+07  3.240e-08 5.260e-07 7.266e-06
  14       10     11   0.001900  2.100e+08   8.077e+07  3.240e-08 5.260e-07 7.266e-06
  15       11     12   0.001900  2.100e+08   8.077e+07  3.240e-08 5.260e-07 7.266e-06
  16        9     12   0.001900  2.100e+08   8.077e+07  3.240e-08 5.260e-07 7.266e-06


N O D A L    D I S P L A C E M E N T S
Node       u            v            w           thetaX          thetay         thetaz
------------------------------------------------------------------------------
 1  0.000000e+00 0.000000e+00 0.000000e+00  0.000000e+00    0.000000e+00    0.00000e+00
 2  0.000000e+00 0.000000e+00 0.000000e+00  0.000000e+00    0.000000e+00    0.00000e+00
 3  0.000000e+00 0.000000e+00 0.000000e+00  0.000000e+00    0.000000e+00    0.00000e+00
 4  0.000000e+00 0.000000e+00 0.000000e+00  0.000000e+00    0.000000e+00    0.00000e+00
 5 -3.971968e-04 5.380984e-03 2.055130e-05 -1.642707e-03   -4.738377e-05   -1.13054e-04
 6  3.971968e-04 5.336322e-03 -2.055130e-05 -1.637754e-03   4.738377e-05   -1.23933e-04
 7  3.971968e-04 5.243158e-03 -3.745838e-05 -2.309601e-03   4.738377e-05   -1.23933e-04
 8 -3.971968e-04 5.243259e-03  3.745838e-05 -2.313731e-03  -4.738377e-05   -1.13054e-04
 9 -9.846131e-04 8.737213e-03  2.600381e-05 -5.794079e-04   -2.420554e-05   -1.11613e-03
10  9.846131e-04 8.737054e-03 -2.600381e-05 -6.006750e-04    2.420554e-05   -1.10929e-03
11  9.846131e-04 1.346646e-02 -5.483649e-05 -1.998327e-03    2.420554e-05   -1.10929e-03
12 -9.84613e-04 1.349461e-02  5.483649e-05 -2.008608e-03   -2.420554e-05   -1.11613e-03


F O R C E S    A N D    M O M E N T S    I N    E L E M E N T S
------------------------------------------------------------------------------
ELEMENT   Node     Fx          Fy          Fz          Mx          My          Mz
   1        1    -4.65097    -6.08470    -0.04566     0.00030     0.07346    -11.69707
            5     4.65097     6.08470     0.04566    -0.00030     0.06351     -6.55704
   2        5    -1.23396    -0.04809    -0.06720     0.00270     0.09837      1.59139
            9     1.23396     0.04809     0.06720    -0.00270     0.10324     -1.73567
   3        2     4.65097     6.00704    -0.04566     0.00033     0.07346     11.57282
            6    -4.65097    -6.00704     0.04566    -0.00033     0.06351      6.44829
   4        6     1.23396     0.08988    -0.06720     0.00265     0.09837     -1.48768
           10    -1.23396    -0.08988     0.06720    -0.00265     0.10324      1.75733
   5        3     8.47721     3.71049    -0.04566     0.00033     0.07346      9.17910
            7    -8.47721    -3.71049     0.04566    -0.00033     0.06351      1.95236
   6        7     3.93284     3.67431    -0.06720     0.00265     0.09837      5.02447
           11    -3.93284    -3.67431     0.06720    -0.00265     0.10324      5.99845
   7        4    -8.47721    -3.69777    -0.04566     0.00030     0.07346     -9.16649
```

	8	8.47721	3.69777	0.04566	-0.00030	0.06351	-1.92683
8	8	-3.93284	-3.68771	-0.06720	0.00270	0.09837	-5.05421
	12	3.93284	3.68771	0.06720	-0.00270	0.10324	-6.00894
9	5	5.93998	-3.30914	-0.02155	-0.00008	0.03272	-4.96623
	6	-5.93998	3.30914	0.02155	0.00008	0.03192	-4.96119
10	6	0.00000	0.10787	0.02283	-0.00059	-0.03424	0.16180
	7	-0.00000	-0.10787	-0.02283	0.00059	-0.03424	0.16180
11	7	0.01335	4.65223	-0.02155	-0.00008	0.03192	6.97625
	8	-0.01335	-4.65223	0.02155	0.00008	0.03272	6.98045
12	5	0.00000	-0.10787	0.02341	-0.00059	-0.03512	-0.16180
	8	-0.00000	0.10787	-0.02341	0.00059	-0.03512	-0.16180
13	9	0.02108	1.16516	-0.06720	-0.00004	0.10106	1.73692
	10	-0.02108	-1.16516	0.06720	0.00004	0.10056	1.75855
14	10	-0.00000	-0.06880	0.06881	-0.00122	-0.10321	-0.10320
	11	0.00000	0.06880	-0.06881	0.00122	-0.10321	-0.10320
15	11	3.74311	-4.00164	-0.06720	-0.00004	0.10056	-5.99723
	12	-3.74311	4.00164	0.06720	0.00004	0.10106	-6.00769
16	9	0.00000	0.06880	0.06917	-0.00125	-0.10376	0.10320
	12	-0.00000	-0.06880	-0.06917	0.00125	-0.10376	0.10320

S U P P O R T   R E A C T I O N S

Support node	Rx	Ry	Rz
1	0.046	-6.085	-4.651
2	11.697	0.073	0.000
3	-0.046	-6.007	4.651
4	11.573	-0.073	0.000

A.5. SOLVING PLANE STRESS AND PLANE STRAIN PROBLEMS OF PLATES USING CST ELEMENTS

```matlab
%FINITE ELEMENT ANALYSIS OF A TRIANGULAR PLATE
clear
clc
format long
inp= fopen('CSTplate.inp','r');
%reading nodal information
[mat] = fscanf(inp,'%f %f  %f %d %d %f %f %f %f %d %d',11);
Lx = mat(1); %Length of plate
Ly = mat(2); %width of plate
t  = mat(3); %thickness of plate
nx = mat(4); %number of divisions along x-direction
ny = mat(5); %number of divisions along y-direction
E  = mat(6);%Modulus of elasticity in kPa
rho= mat(7);
nue= mat(8); %Poisson's ratio
q  = mat(9);   %uniform pressure kN
nLC = mat(10); %number of loaded corners
nLE = mat(11); %number of loaded edges
[x, y ,nodes, nNode, nElem] = Tmesh(Lx,Ly,nx,ny,1);
dx = Lx/nx;
dy = Ly/ny;
figure(1)
triplot(nodes,x,y,'k')
node1 = nodes(:,1);
node2 = nodes(:,2);
node3 = nodes(:,3);
%Show Node Labels
for i = 1:nNode
    h=text(x(i),y(i),int2str(i),'fontsize',8,'fontweight','bold','Color','k');
    h.HorizontalAlignment='right';
    h.VerticalAlignment='bottom';
end
%Show Element Labels
for e = 1:nElem
    xloc = (x(nodes(e,1))+x(nodes(e,2))+x(nodes(e,3)))/3;
    yloc = (y(nodes(e,1))+y(nodes(e,2))+y(nodes(e,3)))/3;
    text(xloc,yloc,int2str(e),'fontsize',8,'fontweight','bold','Color','k');
end
nn = (nx+1)*(ny+1);
%determining edges of the plate
bottom = zeros(nx+1,1);
top    = zeros(nx+1,1);
left   = zeros(ny+1,1);
right  = zeros(ny+1,1);
iB = 0; iT =0; iL = 0;iR = 0;
inner = 0;
```

```matlab
for i=1:nn
    % extracting nodes corresponding to edges
    if(y(i)==0)
        iB = iB+1;
        bottom(iB) = i;
    end
    if(y(i)==Ly)
        iT = iT+1;
        top(iT) = i;
    end
    if(x(i)==0)
        iL = iL+1;
        left(iL) = i;
    end
    if(x(i)==Lx)
        iR = iR+1;
        right(iR) = i;
    end
end
%-------------------------------------------------------------------
%determining the elements with one edge coinciding with the top edge
ntop = size(top,1)-1; % number of elements on top edge
Tedge = zeros(ntop,2);
for i=1:ntop
    Tedge(i,:)=[top(i) top(i+1)];
end
k=0;
for i=1:size(Tedge,1)
    for e=1:nElem
        if(find(nodes(e,:)==Tedge(i,1))~=0 & find(nodes(e,:)==Tedge(i,2))~=0)
            k = k+1;
            Telem(k)=e;
            break
        end
    end
end
Telem = Telem';
%-------------------------------------------------------------------
%determining the elements with one edge coinciding with the left edge
nleft = size(left,1)-1;
Ledge = zeros(nleft,2);
for i=1:nleft
    Ledge(i,:)=[left(i) left(i+1)];
end
k=0;
for i=1:size(Ledge,1)
    for e=1:nElem
        if(find(nodes(e,:)==Ledge(i,1))~=0 & find(nodes(e,:)==Ledge(i,2))~=0)
            k = k+1;
            Lelem(k)=e;
```

```matlab
                    break
            end
        end
end
Lelem = Lelem';
%------------------------------------------------------------------
%determining the elements with one edge coinciding with the bottom edge
nbottom = size(bottom,1)-1;
Bedge = zeros(nbottom,2);
for i=1:nbottom
    Bedge(i,:)=[bottom(i) bottom(i+1)];
end
k=0;
for i=1:size(Bedge,1)
    for e=1:nElem
        if(find(nodes(e,:)==Bedge(i,1))~=0 & find(nodes(e,:)==Bedge(i,2))~=0)
            k = k+1;
            Belem(k)=e;
            break
        end
    end
end
Belem=Belem';
%------------------------------------------------------------------
%determining the elements with one edge coinciding with the left edge
nright = size(right,1)-1;
Redge = zeros(nright,2);
for i=1:nright
    Redge(i,:)=[right(i) right(i+1)];
end
k=0;
for i=1:size(Redge,1)
    for e=1:nElem
        if(find(nodes(e,:)==Redge(i,1))~=0                          &
find(nodes(e,:)==Redge(i,2))~=0)
            k = k+1;
            Relem(k)=e;
            break
        end
    end
end
Relem = Relem';
%------------------------------------------------------------------
ne = 1:nElem;
optionS = input('Enter 1 for Plane Stress or  2 for Plane Strain:');
if(optionS==1)
    D=(E/(1-nue^2))*[1    nue    0;
                   nue   1     0;
                    0    0   (1-nue)/2];
else
```

```matlab
    D = (E/((1+nue)*(1-2*nue)))*[(1-nue)     nue          0
                                   nue      (1-nue)       0
                                    0          0      (1-2*nue)/2];
end
nodes = [node1 node2 node3];

ndof = 2*nNode;
K = zeros(ndof,ndof); %initialize global stiffness matrix
for i=1:nElem
    x1 = x(node1(i)); y1 = y(node1(i));
    x2 = x(node2(i)); y2 = y(node2(i));
    x3 = x(node3(i)); y3 = y(node3(i));
    mat = [ 1 x1 y1
            1 x2 y2
            1 x3 y3];
    area = 0.5*det(mat);
    i1 =2*node1(i)-1; j1 = 2*node1(i);
    i2 =2*node2(i)-1; j2 = 2*node2(i);
    i3 =2*node3(i)-1; j3 = 2*node3(i);
    dof = [i1 j1 i2 j2 i3 j3];
    b1 = y2-y3; b2 = y3-y1; b3 = y1-y2;
    a1 = x3-x2; a2 = x1-x3; a3 = x2-x1;
    B = (1/(2*area))*[b1 0     b2    0 b3    0
                       0  a1   0    a2 0    a3
                      a1 b1   a2    b2 a3   b3];
    Ke = t*area*B'*D*B;
    K(dof,dof) = K(dof,dof)+Ke;
end
K;
F = zeros(2*nn,1);
%applying load at top right corner
tleft = intersect(top,left);    %TL
tright = intersect(top,right); %TR
bleft = intersect(bottom,left); %BL
bright = intersect(bottom,right); %BR
cleft = 0.5*(tleft+bleft); %ML
cright = 0.5*(tright+bright); %MR
if(nLE~=0)% if distributed loading is present on edges
    F = zeros(ndof,1); %initialize the force vector
    fprintf(1,'Distributed load is present on the edges\n');
    [mat1] = fscanf(inp,'%d %f %f',3);
     Ecode = mat1(1); % code for the edge
     fx = mat1(2);
     fy = mat1(3);
     switch Ecode
         case 1 % loading on top edge
             fprintf(1,'Top edge is loaded\n');
             for i=1:ntop
                 n1 = Tedge(i,1);
                 n2 = Tedge(i,2);
```

```matlab
            F(2*n1-1)=F(2*n1-1)+0.5*fx*dy*t;
            F(2*n1)  =F(2*n1)  +0.5*fy*dx*t;
            F(2*n2-1)=F(2*n2-1)+0.5*fx*dy*t;
            F(2*n2)  =F(2*n2)  +0.5*fy*dx*t;
        end
    case 2 % loading on left edge
        fprintf(1,'Left edge is loaded\n');
        for i=1:nleft
            n1 = Ledge(i,1);
            n2 = Ledge(i,2);
            F(2*n1-1)=F(2*n1-1)+0.5*fx*dx;
            F(2*n1)=F(2*n1)+0.5*fy*dy;
            F(2*n2-1)=F(2*n2-1)+0.5*fx*dx;
            F(2*n2)=F(2*n2)++0.5*fy*dy;
        end
    case 3 % loading on bottom edge
        fprintf(1,'Bottom edge is loaded\n');
        for i=1:nbottom
            n1 = Bedge(i,1);
            n2 = Bedge(i,2);
            F(2*n1-1)=F(2*n1-1)+0.5*fx*dx;
            F(2*n1)=F(2*n1)+0.5*fy*dy;
            F(2*n2-1)=F(2*n2-1)+0.5*fx*dx;
            F(2*n2)=F(2*n2)++0.5*fy*dy;
        end
    case 4 % loading on right edge
        fprintf(1,'Right edge is loaded\n');
        for i=1:nright
            n1 = Redge(i,1);
            n2 = Redge(i,2);
            F(2*n1-1)=F(2*n1-1)+0.5*fx*dx;
            F(2*n1)=F(2*n1)+0.5*fy*dy;
            F(2*n2-1)=F(2*n2-1)+0.5*fx*dx;
            F(2*n2)=F(2*n2)++0.5*fy*dy;
        end
    end
end

if(nLC~=0) % if corners are loaded by point loads
    F = zeros(ndof,1); %initialize the force vector
    for i=1:nLC
        fprintf(1,'Point loads are present at specified nodes\n');
        [mat1] = fscanf(inp,'%d %f %f',3);
        Ncode = mat1(1); %code for the node
        Fx = mat1(2);
        Fy = mat1(3);
        switch Ncode
            case 1  %Top left corner
                F(2*tleft-1)=F(2*tleft-1)+Fx;
                F(2*tleft)=F(2*tleft)+Fy;
```

```matlab
        case 2  % Centre left corner
            F(2*cleft-1) = F(2*cleft-1)+Fx;
            F(2*cleft) = F(2*cleft)+Fy;
        case 3 %Bottom left corner
            F(2*bleft-1)=F(2*bleft-1)+Fx;
            F(2*bleft) = F(2*bleft)+Fy;
        case 4  %Bottom right corner
            F(2*tright-1)=F(2*tright-1)+Fx;
            F(2*tright) = F(2*tright)+Fy;
        case 5  %Centre right corner
            F(2*cright-1)=F(2*cright-1)+Fx
            F(2*cright) = F(2*cright)+Fy;
        case 6  %Bottom right corner
            F(2*bright-1)=F(2*bright-1)+Fx;
            F(2*bright) = F(2*bright)+Fy;
        end
    end
end
choice = input('Enter 1 for Edge restraint OR 2 for Corner restraint:');
if(choice==1) %edges are restrained
    fprintf(1,'enter  edge  restraints  in  the  order  [Bottom  Top  Left
Right]\n');
    fprintf(1,'enter  0  for  restrained  edge  and  1  for  unrestrained
edge\n');
    resB=input('Bottom Edge(1 or 0): ');
    resT=input('Top Edge(1 or 0): ');
    resL=input('Left Edge(1 or 0): ');
    resR=input('Right Edge(1 or 0): ');
    edge =[resB resT resL resR];
    %applying boundary conditions
    resx = ones(nn,1);
    resy = ones(nn,1);
    for i=1:nx+1
        if(edge(1)==0)%bottom edge clamped
            j = bottom(i);
            resx(j)=0;
            resy(j)=0; %all dof on bottom edge are restrained
        end
        if(edge(2)==0)%top edge clamped
            j = top(i);
            resx(j)=0;
            resy(j)=0;%all dof on bottom edge are restrained
        end
    end
    for i=1:ny+1
        if(edge(3)==0)%left edge clamped
            j = left(i);
            resx(j)=0;
            resy(j)=0;
        end
```

```matlab
        if(edge(4)==0)%right edge clamped
            j = right(i);
            resx(j)=0;
            resy(j)=0;
        end
    end
else % corners are restrained
    %all nodes are initialized to free dof
    fprintf(1,'enter corner restraints\n');
    fprintf(1,'enter 1 for Free  2 for hinge  3 for roller\n');
    resTL=input('Top-Left Corner(1/2/3)    : ');
    resBL=input('Bottom-Left Corner(1/2/3): ');
    resTR=input('Top-Right Corner(1/2/3)   : ');
    resBR=input('Bottom-Right Corner(1/2/3): ');

    %all dof iniitialzed to free by default
    resx = ones(nn,1);
    resy = ones(nn,1);
    switch resTL  %TOP-LEFT CORNER
        case 2  %hinge
            resx(tleft)=0;
            resy(tleft)=0;
        case 3  %roller
            resy(tleft)=0;
    end
    switch resBL  %BOTTOM-LEFT CORNER
        case 2  %hinge
            resx(bleft)=0;
            resy(bleft)=0;
        case 3  %roller
            resy(bleft)=0;
    end
    switch resTR  %TOP-RIGHT CORNER
        case 2  %hinge
            resx(tright)=0;
            resy(tright)=0;
        case 3  %roller
            resy(tright)=0;
    end
    switch resBR  %BOTTOM-RIGHT CORNER
        case 2  %hinge
            resx(bright)=0;
            resy(bright)=0;
        case 3  %roller
            resy(bright)=0;
    end
end
bn = 0;
```

```matlab
for i=1:nn
    if(resx(i)==0)
        bn = bn+1;
        ifix(bn) = 2*i-1;
    end
    if(resy(i)==0)
        bn = bn+1;
        ifix(bn) = 2*i;
    end
end
ifix = ifix';
K(ifix,:)=zeros(bn,ndof);
K(:,ifix)=zeros(ndof,bn);
K(ifix,ifix)=eye(length(ifix));
F(ifix)=zeros(bn,1);
u =K\F;
sigma = zeros(3,nElem);
for i=1:nElem
    x1 = x(node1(i)); y1 = y(node1(i));
    x2 = x(node2(i)); y2 = y(node2(i));
    x3 = x(node3(i)); y3 = y(node3(i));
    mat = [ 1 x1 y1
            1 x2 y2
            1 x3 y3];
    area = 0.5*det(mat);
    i1 =2*node1(i)-1; j1 = 2*node1(i);
    i2 =2*node2(i)-1; j2 = 2*node2(i);
    i3 =2*node3(i)-1; j3 = 2*node3(i);
    dof = [i1 j1 i2 j2 i3 j3];
    b1 = y2-y3; b2 = y3-y1; b3 = y1-y2;
    a1 = x3-x2; a2 = x1-x3; a3 = x2-x1;
    B = (1/(2*area))*[b1 0    b2    0 b3    0
                       0  a1  0    a2 0    a3
                       a1 b1 a2    b2 a3   b3];
    sigma(:,i) = D*B*u(dof)  ;
end
ux = u(1:2:2*nn-1);
uy = u(2:2:2*nn);
ux1 = reshape(ux,ny+1,nx+1);
uy1 = reshape(uy,ny+1,nx+1);
dx = Lx/nx;
dy = Ly/ny;
x1 = 0:dx:Lx;
y1 = 0:dy:Ly;
optionD = input('Do you want plots on deformed configuration (1 for Yes
0 for No):');
if(optionD==1)
    x2 = x+ux;
    y2 = y+uy;
    xd = reshape(x2,ny+1,nx+1);
```

```matlab
    yd = reshape(y2,ny+1,nx+1);
end
%-----------------------------------
%horizontal displacement
figure(2)
if(optionD==1)
    surf(xd,yd,ux1)
else
    surf(x1,y1,ux1);
end
colormap('jet')
shading interp
cb = colorbar;
val = cb.Limits;
t = linspace(val(1),val(2),8);
cstr_t = cell(length(t),1);
for ii=1:length(t)
    cstr_t{ii}=num2str(t(ii));
end
tickLabelsToUse = cb.TickLabels;
tickLabelsMod = cellfun(@(c) sprintf('%.3E',str2double(c)) ,cstr_t,
'uni', 0);
cb.TickLabels = [];
cb.TickLabels = tickLabelsMod;
set(cb,'YTick',t);
%-----------------------------------
%vertical displacement
figure(3)
if(optionD==1)
    surf(xd,yd,uy1)
else
    surf(x1,y1,uy1);
end
colormap('jet')
shading interp
cb = colorbar;
val = cb.Limits;
t = linspace(val(1),val(2),8);
cstr_t = cell(length(t),1);
for ii=1:length(t)
    cstr_t{ii}=num2str(t(ii));
end
tickLabelsToUse = cb.TickLabels;
tickLabelsMod = cellfun(@(c) sprintf('%.3E',str2double(c)) ,cstr_t,
'uni', 0);
cb.TickLabels = [];
cb.TickLabels = tickLabelsMod;
set(cb,'YTick',t);
%-----------------------------------
%preparing contour plot for stresses
```

```matlab
%generate location matrix
mat = zeros(ny+1,nx+1);
mat(1,:)               = bottom;
for i=2:ny+1
  mat(i,1:nx+1) = bottom(1:nx+1)+i-1;
end
sigmaX = zeros(ny+1,nx+1);
sigmaY = zeros(ny+1,nx+1);
tauXY  = zeros(ny+1,nx+1);
count = zeros(ny+1,nx+1);
for e=1:nElem
    n1 = node1(e); n2 = node2(e); n3 = node3(e)  ;
    [row1,  col1] = find(mat == n1);
    [row2,  col2] = find(mat == n2);
    [row3,  col3] = find(mat == n3);
    sigmaX(row1,col1) = sigmaX(row1,col1)+sigma(1,e);
    sigmaX(row2,col2) = sigmaX(row2,col2)+sigma(1,e);
    sigmaX(row3,col3) = sigmaX(row3,col3)+sigma(1,e);

    sigmaY(row1,col1) = sigmaY(row1,col1)+sigma(2,e);
    sigmaY(row2,col2) = sigmaY(row2,col2)+sigma(2,e);
    sigmaY(row3,col3) = sigmaY(row3,col3)+sigma(2,e);

    tauXY(row1,col1)  = tauXY(row1,col1)  +sigma(3,e);
    tauXY(row2,col2)  = tauXY(row2,col2)  +sigma(3,e);
    tauXY(row3,col3)  = tauXY(row3,col3)  +sigma(3,e);
    count(row1,col1)  = count(row1,col1)+1;
    count(row2,col2)  = count(row2,col2)+1;
    count(row3,col3)  = count(row3,col3)+1;
end
sigmaX = sigmaX./count;
sigmaY = sigmaY./count;
tauXY = tauXY./count;
%sigma-x
figure(4)
if(optionD==1)
    surf(xd,yd,sigmaX)
else
    surf(x1,y1,sigmaX);
end
colormap('jet')
shading interp
cb = colorbar;
val = cb.Limits;
t = linspace(val(1),val(2),8);
cstr_t = cell(length(t),1);
for ii=1:length(t)
    cstr_t{ii}=num2str(t(ii));
end
```

```matlab
tickLabelsToUse = cb.TickLabels;
tickLabelsMod = cellfun(@(c) sprintf('%.3E',str2double(c)) ,cstr_t,
'uni', 0);
cb.TickLabels = [];
cb.TickLabels = tickLabelsMod;
set(cb,'YTick',t);
%-------------------------------
%sigma-y
figure(5)
if(optionD==1)
    surf(xd,yd,sigmaY)
else
    surf(x1,y1,sigmaY);
end
colormap('jet')
shading interp
cb = colorbar;
val = cb.Limits;
t = linspace(val(1),val(2),8);
cstr_t = cell(length(t),1);
for ii=1:length(t)
    cstr_t{ii}=num2str(t(ii));
end
tickLabelsToUse = cb.TickLabels;
tickLabelsMod = cellfun(@(c) sprintf('%.3E',str2double(c)) ,cstr_t,
'uni', 0);
cb.TickLabels = [];
cb.TickLabels = tickLabelsMod;
set(cb,'YTick',t);
%-------------------------------
%tau-xy
figure(6)
if(optionD==1)
    surf(xd,yd,tauXY)
else
    surf(x1,y1,tauXY);
end
colormap('jet')
shading interp
cb = colorbar;
val = cb.Limits;
t = linspace(val(1),val(2),8);
cstr_t = cell(length(t),1);
for ii=1:length(t)
    cstr_t{ii}=num2str(t(ii));
end
tickLabelsToUse = cb.TickLabels;
tickLabelsMod = cellfun(@(c) sprintf('%.3E',str2double(c)) ,cstr_t,
'uni', 0);
cb.TickLabels = [];
```

```matlab
cb.TickLabels = tickLabelsMod;
set(cb,'YTick',t);
%------------------------------------------------------------------------
% This function generates a mesh of three noded triangular elements
function [x, y, nodes, nNode, nElem] =  Tmesh(Lx,Ly,nx,ny,pattern)
k = 0;
dx = Lx/nx;
dy = Ly/ny;
xo=0;
yo=0;
for i = 1:nx
    for j=1:ny
        k = k + 1;
        n1 = j + (i-1)*(ny + 1);
        geom(n1,:) = [(i-1)*dx - xo (j-1)*dy - yo ];
        n2 = j + i*(ny+1);
        geom(n2,:) = [i*dx - xo (j-1)*dy - yo ];
        n3 = n1 + 1;
        geom(n3,:) = [(i-1)*dx - xo j*dy - yo ];
        n4 = n2 + 1;
        geom(n4,:) = [i*dx- xo j*dy - yo ];
        nel = 2*k;
        m = nel -1;
        if(pattern==1)
            nodes(m,:) = [n3 n1 n4];
            nodes(nel,:) = [n2 n4 n1];
        else
            nodes(m,:) = [n1 n2 n3];
            nodes(nel,:) = [n2 n4 n3];
        end
    end
end
x = geom(:,1);
y = geom(:,2);
nNode = size(x,1);
nElem = size(nodes,1);
%------------------------------------------------------------------------
```

A.6. SOLVING PLANE STRESS AND PLANE STRAIN PROBLEMS OF PLATES USING LST ELEMENTS

```
clear
clc
format long
inp= fopen('LSTplate.inp','r');
%reading nodal information
[mat] = fscanf(inp,'%f %f  %f %d %d %f %f %f %f %d %d',11);
Lx = mat(1); %Length of plate
Ly = mat(2); %width of plate
thick  = mat(3); %thickness of plate
nx = mat(4); %number of divisions along x-direction
ny = mat(5); %number of divisions along y-direction
E  = mat(6);%Modulus of elasticity in kPa
rho= mat(7);
nue= mat(8); %Poisson's ratio
q  = mat(9);  %uniform pressure kN
nLC = mat(10); %number of loaded corners
nLE = mat(11); %number of loaded edges
[xgrid,ygrid,x, y ,nodes, nNode, nElem] = T6mesh(Lx,Ly,nx,ny,2);
dx = Lx/nx;
dy = Ly/ny;
node1 = nodes(:,1); node2 = nodes(:,2); node3 = nodes(:,3);
node4 = nodes(:,4); node5 = nodes(:,5); node6 = nodes(:,6);
%Show Node Labels
for i = 1:nNode
    h=text(x(i),y(i),int2str(i),'fontsize',7,'fontweight','bold','Color','k');
    h.HorizontalAlignment='right';
    h.VerticalAlignment='bottom';
end
%Show Element Labels
for e = 1:nElem
    xloc = (x(nodes(e,1))+x(nodes(e,3))+x(nodes(e,5)))/3;
    yloc = (y(nodes(e,1))+y(nodes(e,3))+y(nodes(e,5)))/3;
    text(xloc,yloc,int2str(e),'fontsize',8,'fontweight','bold','Color','k');
end
nn = (2*nx+1)*(2*ny+1);
%determining edges of the plate
bottom = zeros(2*nx+1,1);
top    = zeros(2*nx+1,1);
left   = zeros(2*ny+1,1);
right  = zeros(2*ny+1,1);
iB = 0; iT =0; iL = 0;iR = 0;
inner = 0;
for i=1:nn
    % extracting nodes corresponding to edges
    if(y(i)==0)
        iB = iB+1;
        bottom(iB) = i;
```

```matlab
    end
    if(y(i)==Ly)
        iT = iT+1;
        top(iT) = i;
    end
    if(x(i)==0)
        iL = iL+1;
        left(iL) = i;
    end
    if(x(i)==Lx)
        iR = iR+1;
        right(iR) = i;
    end
end
%-------------------------------------------------------------------
%determining the elements with one edge coinciding with the top edge
ntop = 0.5*(size(top,1)-1); % number of elements on top edge
Tedge = zeros(ntop,3);
for i=1:ntop
    Tedge(i,:)=[top(2*i-1) top(2*i) top(2*i+1)];
end
%-------------------------------------------------------------------
%determining the elements with one edge coinciding with the left edge
nleft = 0.5*(size(left,1)-1);
Ledge = zeros(nleft,3);
for i=1:nleft
    Ledge(i,:)=[left(2*i-1) left(2*i) left(2*i+1)];
end
%-------------------------------------------------------------------
%determining the elements with one edge coinciding with the bottom edge
nbottom = 0.5*(size(bottom,1)-1);
Bedge = zeros(nbottom,3);
for i=1:nbottom
    Bedge(i,:)=[bottom(2*i-1) bottom(2*i) bottom(2*i+1)];
end
%-------------------------------------------------------------------
%determining the elements with one edge coinciding with the left edge
nright = 0.5*(size(right,1)-1);
Redge = zeros(nright,3);
for i=1:nright
    Redge(i,:)=[right(2*i-1) right(2*i) right(2*i+1)];
end
%-------------------------------------------------------------------

ne = 1:nElem;
optionS = input('Enter 1 for Plane Stress or  2 for Plane Strain:');
if(optionS==1)
    D=(E/(1-nue^2))*[1    nue     0;
                   nue    1      0;
                    0     0    (1-nue)/2];
```

```matlab
else
    D = (E/((1+nue)*(1-2*nue)))*[(1-nue)    nue          0
                                  nue       (1-nue)      0
                                  0          0          (1-2*nue)/2];
end
ndof = 2*nNode;
K = zeros(ndof,ndof); %initialize global stiffness matrix

nhp = 3 ; %Number of Hammer points
samp = hammer1(nhp);
for i=1:nElem
    Ke = zeros(12,12);
    B  = zeros(3,12);
    x1 = x(node1(i)); y1 = y(node1(i));
    x2 = x(node2(i)); y2 = y(node2(i));
    x3 = x(node3(i)); y3 = y(node3(i));
    x4 = x(node4(i)); y4 = y(node4(i));
    x5 = x(node5(i)); y5 = y(node5(i));
    x6 = x(node6(i)); y6 = y(node6(i));
    coord = [x1 x2 x3 x4 x5 x6
             y1 y2 y3 y4 y5 y6]';
    i1 =2*node1(i)-1; j1 = 2*node1(i);
    i2 =2*node2(i)-1; j2 = 2*node2(i);
    i3 =2*node3(i)-1; j3 = 2*node3(i);
    i4 =2*node4(i)-1; j4 = 2*node4(i);
    i5 =2*node5(i)-1; j5 = 2*node5(i);
    i6 =2*node6(i)-1; j6 = 2*node6(i);
    dof = [i1 j1 i2 j2 i3 j3 i4 j4 i5 j5 i6 j6];
    for j=1:nhp  % loop over Hammer points
        wi = samp(j,3);
        [der,fun]=fmT6_quad(samp,j);
        J =der*coord;
        detJ = det(J);
        deriv = J\der ; %inv(J)*der;
        B(1,1:2:12) = deriv(1,:);
        B(2,2:2:12) = deriv(2,:);
        B(3,1:2:12) = deriv(2,:);B(3,2:2:12) = deriv(1,:);
        Ke = Ke+detJ*thick*wi*B'*D*B;
        %element stiffness matrix by numerical intergration
    end
    K(dof,dof) = K(dof,dof)+Ke;
end
K;
F = zeros(2*nn,1);
%applying load at top right corner
tleft = intersect(top,left);   %TL
tright = intersect(top,right); %TR
bleft = intersect(bottom,left); %BL
bright = intersect(bottom,right); %BR
cleft = 0.5*(tleft+bleft); %ML
```

```matlab
cright = 0.5*(tright+bright); %MR
if(nLE~=0)% if distributed loading is present on edges
    F = zeros(ndof,1); %initialize the force vector
    fprintf(1,'Distributed load is present on the edges\n');
    [mat1] = fscanf(inp,'%d %f %f',3);
     Ecode = mat1(1); % code for the edge
     fx = mat1(2);
     fy = mat1(3);
     switch Ecode
        case 1 % loading on top edge
            fprintf(1,'Top edge is loaded\n');
            for i=1:ntop
                n1 = Tedge(i,1);
                n2 = Tedge(i,2);
                n3 = Tedge(i,3);
                F(2*n1-1)=F(2*n1-1)+fx*dy*thick/6;
                F(2*n1)  =F(2*n1)  +fy*dx*thick/6;
                F(2*n2-1)=F(2*n2-1)+2*fx*dy*thick/3;
                F(2*n2)  =F(2*n2)  +2*fy*dx*thick/3;
                F(2*n3-1)=F(2*n3-1)+fx*dy*thick/6;
                F(2*n3)  =F(2*n3)  +fy*dx*thick/6;
            end
        case 2 % loading on left edge
            fprintf(1,'Left edge is loaded\n');
            for i=1:nleft
                n1 = Ledge(i,1);
                n2 = Ledge(i,2);
                F(2*n1-1)=F(2*n1-1)+0.5*fx*dx*thick;
                F(2*n1)=F(2*n1)+0.5*fy*dy*thick;
                F(2*n2-1)=F(2*n2-1)+0.5*fx*dx*thick;
                F(2*n2)=F(2*n2)++0.5*fy*dy*thick;
            end
        case 3 % loading on bottom edge
            fprintf(1,'Bottom edge is loaded\n');
            for i=1:nbottom
                n1 = Bedge(i,1);
                n2 = Bedge(i,2);
                F(2*n1-1)=F(2*n1-1)+0.5*fx*dx*thick;
                F(2*n1)=F(2*n1)+0.5*fy*dy*thick;
                F(2*n2-1)=F(2*n2-1)+0.5*fx*dx*thick;
                F(2*n2)=F(2*n2)++0.5*fy*dy*thick;
            end
        case 4 % loading on right edge
            fprintf(1,'Right edge is loaded\n');
            for i=1:nright
                n1 = Redge(i,1);
                n2 = Redge(i,2);
                F(2*n1-1)=F(2*n1-1)+0.5*fx*dx*thick;
                F(2*n1)=F(2*n1)+0.5*fy*dy*thick;
                F(2*n2-1)=F(2*n2-1)+0.5*fx*dx*thick;
```

```matlab
                        F(2*n2)=F(2*n2)++0.5*fy*dy*thick;
                end
        end
end

if(nLC~=0) % if corners are loaded by point loads
    F = zeros(ndof,1); %initialize the force vector
    for i=1:nLC
        fprintf(1,'Point loads are present at specified nodes\n');
        [mat1] = fscanf(inp,'%d %f %f',3);
        Ncode = mat1(1); %code for the node
        Fx = mat1(2);
        Fy = mat1(3);
        switch Ncode
            case 1  %Top left corner
                F(2*tleft-1)=F(2*tleft-1)+Fx;
                F(2*tleft)   =F(2*tleft)+Fy;
            case 2  % Centre left corner
                F(2*cleft-1) = F(2*cleft-1)+Fx;
                F(2*cleft) = F(2*cleft)+Fy;
            case 3 %Bottom left corner
                F(2*bleft-1)=F(2*bleft-1)+Fx;
                F(2*bleft) = F(2*bleft)+Fy;
            case 4  %Bottom right corner
                F(2*tright-1)=F(2*tright-1)+Fx;
                F(2*tright) = F(2*tright)+Fy;
            case 5  %Centre right corner
                F(2*cright-1)=F(2*cright-1)+Fx;
                F(2*cright) = F(2*cright)+Fy;
            case 6  %Bottom right corner
                F(2*bright-1)=F(2*bright-1)+Fx;
                F(2*bright) = F(2*bright)+Fy;
        end
    end
end
% F
choice = input('Enter 1 for Edge restraint OR 2 for Corner restraint:');
if(choice==1) %edges are restrained
    fprintf(1,'enter edge restraints in the order [Bottom Top Left Right]\n');
    fprintf(1,'enter 0 for restrained edge and 1 for unrestrained edge\n');
    resB=input('Bottom Edge(1 or 0): ');
    resT=input('Top Edge(1 or 0): ');
    resL=input('Left Edge(1 or 0): ');
    resR=input('Right Edge(1 or 0): ');
    edge =[resB resT resL resR];
    %applying boundary conditions
    resx = ones(nn,1);
    resy = ones(nn,1);
    for i=1:2*nx+1
        if(edge(1)==0)%bottom edge clamped
```

```matlab
                j = bottom(i);
                resx(j)=0;
                resy(j)=0; %all dof on bottom edge are restrained
            end
            if(edge(2)==0)%top edge clamped
                j = top(i);
                resx(j)=0;
                resy(j)=0;%all dof on bottom edge are restrained
            end
        end
        for i=1:2*ny+1
            if(edge(3)==0)%left edge clamped
                j = left(i);
                resx(j)=0;
                resy(j)=0;
            end
            if(edge(4)==0)%right edge clamped
                j = right(i);
                resx(j)=0;
                resy(j)=0;
            end
        end
    end
else % corners are restrained
    %all nodes are initialized to free dof
    fprintf(1,'enter corner restraints\n');
    fprintf(1,'enter 1 for Free  2 for hinge   3 for roller\n');
    resTL=input('Top-Left Corner(1/2/3)    : ');
    resBL=input('Bottom-Left Corner(1/2/3): ');
    resTR=input('Top-Right Corner(1/2/3)   : ');
    resBR=input('Bottom-Right Corner(1/2/3): ');

    %all dof iniitialzed to free by default
    resx = ones(nn,1);
    resy = ones(nn,1);
    switch resTL   %TOP-LEFT CORNER
        case 2   %hinge
            resx(tleft)=0;
            resy(tleft)=0;
        case 3   %roller
            resy(tleft)=0;
    end
    switch resBL   %BOTTOM-LEFT CORNER
        case 2   %hinge
            resx(bleft)=0;
            resy(bleft)=0;
        case 3   %roller
            resy(bleft)=0;
    end
```

```matlab
        switch resTR  %TOP-RIGHT CORNER
            case 2  %hinge
                resx(tright)=0;
                resy(tright)=0;
            case 3  %roller
                resy(tright)=0;
        end
        switch resBR  %BOTTOM-RIGHT CORNER
            case 2  %hinge
                resx(bright)=0;
                resy(bright)=0;
            case 3  %roller
                resy(bright)=0;
        end
end
bn = 0;
for i=1:nn
    if(resx(i)==0)
        bn = bn+1;
        ifix(bn)  = 2*i-1;
    end
    if(resy(i)==0)
        bn = bn+1;
        ifix(bn)  = 2*i;
    end
end
ifix = ifix';
K(ifix,:)=zeros(bn,ndof);
K(:,ifix)=zeros(ndof,bn);
K(ifix,ifix)=eye(length(ifix));
F(ifix)=zeros(bn,1);
u =K\F;
nhp=3; % for computing stresses at centroid of each element
samp = hammer1(nhp);
sigma = zeros(3,nElem,nhp);
for i=1:nElem
    x1 = x(node1(i)); y1 = y(node1(i));
    x2 = x(node2(i)); y2 = y(node2(i));
    x3 = x(node3(i)); y3 = y(node3(i));
    x4 = x(node4(i)); y4 = y(node4(i));
    x5 = x(node5(i)); y5 = y(node5(i));
    x6 = x(node6(i)); y6 = y(node6(i));
    coord = [x1 x2 x3 x4 x5 x6
             y1 y2 y3 y4 y5 y6]';
    i1 =2*node1(i)-1; j1 = 2*node1(i);
    i2 =2*node2(i)-1; j2 = 2*node2(i);
    i3 =2*node3(i)-1; j3 = 2*node3(i);
    i4 =2*node4(i)-1; j4 = 2*node4(i);
    i5 =2*node5(i)-1; j5 = 2*node5(i);
    i6 =2*node6(i)-1; j6 = 2*node6(i);
```

```matlab
        dof = [i1 j1 i2 j2 i3 j3 i4 j4 i5 j5 i6 j6];
        for j=1:nhp  % loop over Hammer points
            [der,fun]=fmT6_quad(samp,j);
            J =der*coord;
            detJ = det(J);
            deriv = J\der ;
            B(1,1:2:12) = deriv(1,:);
            B(2,2:2:12) = deriv(2,:);
            B(3,1:2:12) = deriv(2,:);B(3,2:2:12) = deriv(1,:);
            sigma(:,i,j) =D*B*u(dof);
        end
end
%generate location matrix
mat = zeros(2*ny+1,2*nx+1);
mat(1,:)            = bottom;
for i=2:2*ny+1
  mat(i,1:2:2*nx+1) = bottom(1:2:2*nx+1)+i-1;
end
k = 0;
for i=3:2:2*ny+1
      k = k+2;
      mat(i,2:2:2*nx+1) = bottom(2:2:2*nx+1)+k;
end
ux = u(1:2:2*nn-1);
uy = u(2:2:2*nn);
ux1 = reshape(ux,2*ny+1,2*nx+1);
uy1 = reshape(uy,2*ny+1,2*nx+1);
dx = Lx/nx;
dy = Ly/ny;
x1 = 0:0.5*dx:Lx;
y1 = 0:0.5*dy:Ly;
optionD = input('Do you want plots on deformed configuration (1 for Yes
0 for No):');
if(optionD==1)
    x2 = x+ux;
    y2 = y+uy;
    xd = reshape(x2,2*ny+1,2*nx+1);
    yd = reshape(y2,2*ny+1,2*nx+1);
end
%--------------------------------
%horizontal displacement
figure(2)
if(optionD==1)
    surf(xd,yd,ux1)
else
    surf(x1,y1,ux1);
end
colormap('jet')
shading interp
cb = colorbar;
```

```matlab
val = cb.Limits;
t = linspace(val(1),val(2),8);
cstr_t = cell(length(t),1);
for ii=1:length(t)
    cstr_t{ii}=num2str(t(ii));
end
tickLabelsToUse = cb.TickLabels;
tickLabelsMod  = cellfun(@(c)  sprintf('%.3E',str2double(c))  ,cstr_t,
'uni', 0);
cb.TickLabels = [];
cb.TickLabels = tickLabelsMod;
set(cb,'YTick',t);
%------------------------------------
%vertical displacement
figure(3)
if(optionD==1)
    surf(xd,yd,uy1)
else
    surf(x1,y1,uy1);
end
colormap('jet')
shading interp
cb = colorbar;
val = cb.Limits;
t = linspace(val(1),val(2),8);
cstr_t = cell(length(t),1);
for ii=1:length(t)
    cstr_t{ii}=num2str(t(ii));
end
tickLabelsToUse = cb.TickLabels;
tickLabelsMod  = cellfun(@(c)  sprintf('%.3E',str2double(c))  ,cstr_t,
'uni', 0);
cb.TickLabels = [];
cb.TickLabels = tickLabelsMod;
set(cb,'YTick',t);
%------------------------------------
%preparing contour plot for stresses
Tmat = genT6mat;
sigmaX = zeros(2*ny+1,2*nx+1);
sigmaY = zeros(2*ny+1,2*nx+1);
tauXY  = zeros(2*ny+1,2*nx+1);
count  = zeros(2*ny+1,2*nx+1);
for e=1:nElem
    n1 = node1(e); n2 = node2(e); n3 = node3(e) ;
    n4 = node4(e); n5 = node5(e); n6 = node6(e);
    sigx =  Tmat*[sigma(1,e,1) sigma(1,e,2) sigma(1,e,3) ]';
    sigy =  Tmat*[sigma(2,e,1) sigma(2,e,2) sigma(2,e,3) ]';
    tau  =  Tmat*[sigma(3,e,1) sigma(3,e,2) sigma(3,e,3) ]';
    [row1, col1] = find(mat == n1);   [row2, col2] = find(mat == n2);
    [row3, col3] = find(mat == n3);   [row4, col4] = find(mat == n4);
```

```matlab
    [row5, col5] = find(mat == n5);     [row6, col6] = find(mat == n6);
    sigmaX(row1,col1) = sigmaX(row1,col1)+sigx(1);
    sigmaX(row2,col2) = sigmaX(row2,col2)+sigx(2);
    sigmaX(row3,col3) = sigmaX(row3,col3)+sigx(3);
    sigmaX(row4,col4) = sigmaX(row4,col4)+sigx(4);
    sigmaX(row5,col5) = sigmaX(row5,col5)+sigx(5);
    sigmaX(row6,col6) = sigmaX(row6,col6)+sigx(6);

    sigmaY(row1,col1) = sigmaY(row1,col1)+sigy(1);
    sigmaY(row2,col2) = sigmaY(row2,col2)+sigy(2);
    sigmaY(row3,col3) = sigmaY(row3,col3)+sigy(3);
    sigmaY(row4,col4) = sigmaY(row4,col4)+sigy(4);
    sigmaY(row5,col5) = sigmaY(row5,col5)+sigy(5);
    sigmaY(row6,col6) = sigmaY(row6,col6)+sigy(6);

    tauXY(row1,col1)  = tauXY(row1,col1)  +tau(1);
    tauXY(row2,col2)  = tauXY(row2,col2)  +tau(2);
    tauXY(row3,col3)  = tauXY(row3,col3)  +tau(3);
    tauXY(row4,col4)  = tauXY(row4,col4)  +tau(4);
    tauXY(row5,col5)  = tauXY(row5,col5)  +tau(5);
    tauXY(row6,col6)  = tauXY(row6,col6)  +tau(6);
    count(row1,col1)  = count(row1,col1)+1;
    count(row2,col2)  = count(row2,col2)+1;
    count(row3,col3)  = count(row3,col3)+1;
    count(row4,col4)  = count(row4,col4)+1;
    count(row5,col5)  = count(row5,col5)+1;
    count(row6,col6)  = count(row6,col6)+1;
end
sigmaX = sigmaX./count;
sigmaY = sigmaY./count;
tauXY = tauXY./count;

[m1, n1] = find(isnan(sigmaX));

for e=1:0.5*nElem
    i=m1(e);
    j=n1(e);
    sigmaX(i,j) = (sigmaX(i-1,j)+sigmaX(i+1,j)+sigmaX(i,j-1)+sigmaX(i,j+1))/4;
    sigmaY(i,j) = (sigmaY(i-1,j)+sigmaY(i+1,j)+sigmaY(i,j-1)+sigmaY(i,j+1))/4;
    tauXY(i,j) = (tauXY(i-1,j)+tauXY(i+1,j)+tauXY(i,j-1)+tauXY(i,j+1))/4;
end
figure(4)
surf(xgrid,ygrid,sigmaX);
colormap('jet')
shading interp
cb = colorbar;
val = cb.Limits;
t = linspace(val(1),val(2),8);
set(cb,'YTick',t);
```

```
figure(5)
surf(xgrid,ygrid,sigmaY);
colormap('jet')
shading interp
cb = colorbar;
val = cb.Limits;
t = linspace(val(1),val(2),8);
set(cb,'YTick',t);

figure(6)
surf(xgrid,ygrid,tauXY);
colormap('jet')
shading interp
cb = colorbar;
val = cb.Limits;
t = linspace(val(1),val(2),8);
set(cb,'YTick',t);

%---------------------------------------------------------------------
function       [xgrid,ygrid,x,    y,      nodes,     nNode,     nElem]     =
T6mesh(Lx,Ly,nx,ny,pattern)
% This function generates a mesh of the linear strain triangular element
%
% global nnd nel geom nodes x y
% global Length Width nx ny xo yo dhx dhy
%
%
nNode = 0;
k = 0;
dhx = Lx/nx;
dhy = Ly/ny;
xo = 0.0;
yo = 0.0;
figure(1)
for i = 1:nx
    for j=1:ny
        k = k + 1;
        n1 = (2*j-1) + (2*i-2)*(2*ny+1)  ;
        n2 = (2*j-1) + (2*i-1)*(2*ny+1);
        n3 = (2*j-1) + (2*i)*(2*ny+1);
        n4 = n1 + 1;
        n5 = n2 + 1;
        n6 = n3 + 1 ;
        n7 = n1 + 2;
        n8 = n2 + 2;
        n9 = n3 + 2;
        %
        geom(n1,:) = [(i-1)*dhx - xo (j-1)*dhy - yo];
        geom(n2,:) = [((2*i-1)/2)*dhx - xo (j-1)*dhy - yo ];
        geom(n3,:) = [i*dhx - xo (j-1)*dhy - yo ];
```

```matlab
geom(n4,:) = [(i-1)*dhx - xo ((2*j-1)/2)*dhy - yo ];
geom(n5,:) = [((2*i-1)/2)*dhx - xo ((2*j-1)/2)*dhy - yo ];
geom(n6,:) = [i*dhx - xo ((2*j-1)/2)*dhy - yo ];
geom(n7,:) = [(i-1)*dhx - xo j*dhy - yo];
geom(n8,:) = [((2*i-1)/2)*dhx - xo j*dhy - yo];
geom(n9,:) = [i*dhx - xo j*dhy - yo];
%
if(pattern==1) %diagonals right-down
     nodes(2*k-1,:) = [n1 n2 n3 n5 n7 n4];
     nodes(2*k,:)   = [n3 n6 n9 n8 n7 n5];
elseif(pattern==2)%diagonals left-down
      nodes(2*k-1,:) = [n1 n2 n3 n6 n9 n5];
      nodes(2*k,:)   = [n9 n8 n7 n4 n1 n5];
end
max_n = max([n1 n2 n3 n4 n5 n6 n7 n8 n9]);
if(nNode <= max_n)
     nNode = max_n;
end;
xgrid(2*i-1) = geom(n1,1);
xgrid(2*i) = geom(n2,1);
xgrid(2*i+1) = geom(n3,1);
ygrid(2*j-1) = geom(n1,2);
ygrid(2*j) = geom(n4,2);
ygrid(2*j+1) = geom(n7,2);
if(pattern==1)
 h=line([geom(n1,1) geom(n2,1)],[geom(n1,2) geom(n2,2)]);
 h.Color = 'k';h.Marker='.';h.MarkerFaceColor='k'; h.MarkerSize=10;
h=line([geom(n2,1) geom(n3,1)],[geom(n2,2) geom(n3,2)]);
h.Color = 'k';h.Marker='.';h.MarkerFaceColor='k';h.MarkerSize=10;
h=line([geom(n3,1) geom(n5,1)],[geom(n3,2) geom(n5,2)]);
h.Color = 'k';h.Marker='.';h.MarkerFaceColor='k';h.MarkerSize=10;
h=line([geom(n5,1) geom(n7,1)],[geom(n5,2) geom(n7,2)]);
h.Color = 'k';h.Marker='.';h.MarkerFaceColor='k';h.MarkerSize=10;
h=line([geom(n7,1) geom(n4,1)],[geom(n7,2) geom(n4,2)]);
h.Color = 'k';h.Marker='.';h.MarkerFaceColor='k';h.MarkerSize=10;
h=line([geom(n4,1) geom(n1,1)],[geom(n4,2) geom(n1,2)]);
h.Color = 'k';h.Marker='.';h.MarkerFaceColor='k';h.MarkerSize=10;
h=line([geom(n3,1) geom(n6,1)],[geom(n3,2) geom(n6,2)]);
h.Color = 'k';h.Marker='.';h.MarkerFaceColor='k';h.MarkerSize=10;
h=line([geom(n6,1) geom(n9,1)],[geom(n6,2) geom(n9,2)]);
h.Color = 'k';h.Marker='.';h.MarkerFaceColor='k';h.MarkerSize=10;
h=line([geom(n9,1) geom(n8,1)],[geom(n9,2) geom(n8,2)]);
h.Color = 'k';h.Marker='.';h.MarkerFaceColor='k';h.MarkerSize=10;
h=line([geom(n8,1) geom(n7,1)],[geom(n8,2) geom(n7,2)]);
h.Color = 'k';h.Marker='.';h.MarkerFaceColor='k';h.MarkerSize=10;
else
 h=line([geom(n1,1) geom(n2,1)],[geom(n1,2) geom(n2,2)]);
 h.Color = 'k';h.Marker='.';h.MarkerFaceColor='k'; h.MarkerSize=10;
 h=line([geom(n2,1) geom(n3,1)],[geom(n2,2) geom(n3,2)]);
 h.Color = 'k';h.Marker='.';h.MarkerFaceColor='k';h.MarkerSize=10;
 h=line([geom(n3,1) geom(n6,1)],[geom(n3,2) geom(n6,2)]);
h.Color = 'k';h.Marker='.';h.MarkerFaceColor='k';h.MarkerSize=10;
```

```matlab
            h=line([geom(n6,1) geom(n9,1)],[geom(n6,2) geom(n9,2)]);
            h.Color = 'k';h.Marker='.';h.MarkerFaceColor='k';h.MarkerSize=10;
            h=line([geom(n9,1) geom(n5,1)],[geom(n9,2) geom(n5,2)]);
            h.Color = 'k';h.Marker='.';h.MarkerFaceColor='k';h.MarkerSize=10;
            h=line([geom(n5,1) geom(n1,1)],[geom(n5,2) geom(n1,2)]);
            h.Color = 'k';h.Marker='.';h.MarkerFaceColor='k';h.MarkerSize=10;
            h=line([geom(n9,1) geom(n8,1)],[geom(n9,2) geom(n8,2)]);
            h.Color = 'k';h.Marker='.';h.MarkerFaceColor='k';h.MarkerSize=10;
            h=line([geom(n8,1) geom(n7,1)],[geom(n8,2) geom(n7,2)]);
            h.Color = 'k';h.Marker='.';h.MarkerFaceColor='k';h.MarkerSize=10;
            h=line([geom(n7,1) geom(n4,1)],[geom(n7,2) geom(n4,2)]);
            h.Color = 'k';h.Marker='.';h.MarkerFaceColor='k';h.MarkerSize=10;
            h=line([geom(n4,1) geom(n1,1)],[geom(n4,2) geom(n1,2)]);
            h.Color = 'k';h.Marker='.';h.MarkerFaceColor='k';h.MarkerSize=10;
        end
    end
end
x = geom(:,1);
y = geom(:,2);
% nNode = size(x,1);
nElem = size(nodes,1);
%-------------------------------------------------------------------------

function[samp]=hammer(npt)
%
% This function returns the abscissae and weights of the
% integration points for npt equal up to 7

samp=zeros(npt,3);
%
switch npt
    case 1
        samp=[1/3 1/3 1/2];
    case {2 , 3}
        %npt =3;
        samp=[
            1/2  1/2  1/6
            0    1/2  1/6
            1/2  0    1/6
            ];
    case {4 , 5}
        %npt=4;
        samp= [1/3 1/3 -27/96
               1/5 1/5  25/96
               3/5 1/5  25/96
               1/5 3/5  25/96];
    case 6
        a = 0.445948490915965; b = 0.091576213509771;
        samp= [ a        a       0.111690794839005
               1-2*a     a       0.111690794839005
                a      1-2*a     0.111690794839005
```

```matlab
                b        b      0.054975871827661
              1-2*b      b      0.054975871827661
                b      1-2*b    0.054975871827661];
    case 7
        a = (6+sqrt(15))/21 ; b = 4/7 -a;
        A = (155+sqrt(15))/2400; B = (31/240 -A);
    samp= [ 1/3        1/3        9/80
              a          a          A
            1-2*a        a          A
              a        1-2*a        A
              b          b          B
            1-2*b        b          B
              b        1-2*b        B];
end
%-------------------------------------------------------------------

function[der,fun] = fmT6_quad(samp, ig)
%
% This function returns the vector of the shape function and their
% derivatives with respect to xi and eta at the Hammer points for
% an 6-noded triangular element
%
% xi=samp(ig,1);
% eta=samp(ig,2);
%NOTEL the three columns of samp matrix correspond to xi, eta and Weight
%ig is the Hammer point
% samp(ig,1)
% samp(ig,2)
% pause
syms xi eta lamda;
lamda = 1-xi-eta;
N1 = lamda*(2*lamda-1);
N2 = 4*xi*lamda;
N3 = xi*(2*xi-1);
N4 = 4*xi*eta;
N5 = eta*(2*eta-1);
N6 = 4*eta*lamda;
f = [N1 N2 N3 N4 N5 N6];
fun = double(subs(f,{xi eta},{samp(ig,1) samp(ig,2)}));

dermat = [diff(N1,xi)  diff(N2,xi)  diff(N3,xi) diff(N4,xi) diff(N5,xi)
diff(N6,xi)
      diff(N1,eta)        diff(N2,eta)        diff(N3,eta)    diff(N4,eta)
diff(N5,eta) diff(N6,eta)];
der = double(subs(dermat,{xi eta},{samp(ig,1) samp(ig,2)}));

%
% end function fmT6_quad
%-------------------------------------------------------------------
```

```
function [Tmat] = genT6mat
syms xi eta lamda;
N2 = xi;
N3 = eta;
N1 = 1-xi-eta;

Tmat(1,:)= [
double(subs(N1,{xi,eta},[1,1]))
double(subs(N2,{xi,eta},[1,1]))
double(subs(N3,{xi,eta},[1,1]))
];

Tmat(2,:)= [
double(subs(N1,{xi,eta},[0,1]))
double(subs(N2,{xi,eta},[0,1]))
double(subs(N3,{xi,eta},[0,1]))
];

Tmat(3,:)= [
double(subs(N1,{xi,eta},[-1,1]))
double(subs(N2,{xi,eta},[-1,1]))
double(subs(N3,{xi,eta},[-1,1]))
];

Tmat(4,:)= [
double(subs(N1,{xi,eta},[0,0]))
double(subs(N2,{xi,eta},[0,0]))
double(subs(N3,{xi,eta},[0,0]))
];

Tmat(5,:)= [
double(subs(N1,{xi,eta},[1,-1]))
double(subs(N2,{xi,eta},[1,-1]))
double(subs(N3,{xi,eta},[1,-1]))
];
Tmat(6,:)= [
double(subs(N1,{xi,eta},[1,0]))
double(subs(N2,{xi,eta},[1,0]))
double(subs(N3,{xi,eta},[1,0]))
];
```

A.7. SOLVING PLANE STRESS AND PLANE STRAIN PROBLEMS OF PLATES USING 4-NODE QUADRILATERAL ELEMENTS

```matlab
%FINITE ELEMENT ANALYSIS OF A PLATE USING 4-NODE QUADRILATERAL ELEMENTS
clear
clc
format long
inp= fopen('Q4plate.inp','r');
%reading nodal information
[mat] = fscanf(inp,'%f %f  %f %d %d %f %f %f %f %d %d',11);
Lx = mat(1); %Length of plate
Ly = mat(2); %width of plate
thick  = mat(3); %thickness of plate
nx = mat(4); %number of divisions along x-direction
ny = mat(5); %number of divisions along y-direction
E  = mat(6);%Modulus of elasticity in kPa
rho= mat(7);
nue= mat(8); %Poisson's ratio
q  = mat(9);   %uniform pressure
nLC = mat(10); %number of loaded corners
nLE = mat(11); %number of loaded edges
alpha = 5.0;
[xgrid,ygrid,x, y ,nodes, nNode, nElem] = Q4skewmesh(Lx,Ly,nx,ny,alpha);
node1 = nodes(:,1);  node2 = nodes(:,2);  node3 = nodes(:,3);node4 =
nodes(:,4);
%Show Node Labels
for i = 1:nNode
    h=text(x(i),y(i),int2str(i),'fontsize',7,'fontweight','bold','Color','k');
    h.HorizontalAlignment='right';
    h.VerticalAlignment='bottom';
end
%Show Element Labels
for e = 1:nElem
    xloc = (x(nodes(e,1))+x(nodes(e,3)))/2;
    yloc = (y(nodes(e,1))+y(nodes(e,3)))/2;
    text(xloc,yloc,int2str(e),'fontsize',8,'fontweight','bold','Color','k');
end
%determining edges of the plate
bottom = zeros(nx+1,1);
top    = zeros(nx+1,1);
left   = zeros(ny+1,1);
right  = zeros(ny+1,1);
iB = 0; iT =0; iL = 0;iR = 0;
inner = 0;
H = Lx*tand(alpha);
dky = H/nx;
for i=1:nNode
    % extracting nodes corresponding to edges
    if (y(i)==0||(rem(y(i),dky)==0))
        iB = iB+1;
```

```matlab
            bottom(iB) = i;
        end
        if(x(i)==0)
            iL = iL+1;
            left(iL) = i;
        end
        if(x(i)==Lx)
            iR = iR+1;
            right(iR) = i;
        end
end
for i=1:size(bottom,1)
    for j=1:nNode
      if(y(j) == (Ly+y(bottom(i))))
            top(i) = j;
      end
    end
end
%-------------------------------------------------------------------
%determining the elements with one edge coinciding with the top edge
ntop = size(top,1)-1; % number of elements on top edge
Tedge = zeros(ntop,2);
for i=1:ntop
    Tedge(i,:)=[top(i) top(i+1)];
end
k=0;
for i=1:size(Tedge,1)
    for e=1:nElem
        if(find(nodes(e,:)==Tedge(i,1))~=0 &
            find(nodes(e,:)==Tedge(i,2))~=0)
            k = k+1;
            Telem(k)=e;
            break
        end
    end
end
Telem = Telem';
%-------------------------------------------------------------------
%determining the elements with one edge coinciding with the left edge
nleft = (size(left,1)-1);
Ledge = zeros(nleft,2);
for i=1:nleft
    Ledge(i,:)=[left(i) left(i+1)];
end
k=0;
for i=1:size(Ledge,1)
    for e=1:nElem
        if(find(nodes(e,:)==Ledge(i,1))~=0 &
            find(nodes(e,:)==Ledge(i,2))~=0)
            k = k+1;
```

```matlab
                Lelem(k)=e;
                break
            end
        end
end
Lelem = Lelem';
%----------------------------------------------------------------------
%determining the elements with one edge coinciding with the bottom edge
nbottom = (size(bottom,1)-1);
Bedge = zeros(nbottom,2);
for i=1:nbottom
    Bedge(i,:)=[bottom(i) bottom(i+1)];
end
k=0;
for i=1:size(Bedge,1)
    for e=1:nElem
        if(find(nodes(e,:)==Bedge(i,1))~=0                              &
find(nodes(e,:)==Bedge(i,2))~=0)
            k = k+1;
            Belem(k)=e;
            break
        end
    end
end
Belem=Belem';
%----------------------------------------------------------------------
%determining the elements with one edge coinciding with the left edge
nright = (size(right,1)-1);
Redge = zeros(nright,2);
for i=1:nright
    Redge(i,:)=[right(i) right(i+1)];
end
k=0;
for i=1:size(Redge,1)
    for e=1:nElem
        if(find(nodes(e,:)==Redge(i,1))~=0 & find(nodes(e,:)==Redge(i,2))~=0)
            k = k+1;
            Relem(k)=e;
            break
        end
    end
end
Relem = Relem';
ne = 1:nElem;
optionS = input('Enter 1 for Plane Stress or  2 for Plane Strain:');

if(optionS==1)
    D=(E/(1-nue^2))*[1    nue     0;
                 nue   1      0;
                 0     0    (1-nue)/2];
```

```matlab
else
    D = (E/((1+nue)*(1-2*nue)))*[(1-nue)    nue         0
                                  nue       (1-nue)     0
                                  0          0         (1-2*nue)/2];
end
ndof = 2*nNode;
K = zeros(ndof,ndof); %initialize global stiffness matrix

ngp = 2 ; %Number of Gauss points
samp = gauss(ngp);
for i=1:nElem
    Ke = zeros(8,8);
    B  = zeros(3,8);
    x1 = x(node1(i)); y1 = y(node1(i));
    x2 = x(node2(i)); y2 = y(node2(i));
    x3 = x(node3(i)); y3 = y(node3(i));
    x4 = x(node4(i)); y4 = y(node4(i));
    coord = [x1 x2 x3 x4
             y1 y2 y3 y4 ]';
    i1 =2*node1(i)-1; j1 = 2*node1(i);
    i2 =2*node2(i)-1; j2 = 2*node2(i);
    i3 =2*node3(i)-1; j3 = 2*node3(i);
    i4 =2*node4(i)-1; j4 = 2*node4(i);
    dof = [i1 j1 i2 j2 i3 j3 i4 j4 ];
    for ig=1:ngp  % loop over Gauss points
        wi = samp(ig,2);
        for jg=1:ngp
            wj = samp(jg,2);
            [der,fun]=fmlin(samp,ig,jg);
            J =der*coord;
            detJ = det(J);
            deriv = J\der ; %inv(J)*der;
            B(1,1:2:8) = deriv(1,:);
            B(2,2:2:8) = deriv(2,:);
            B(3,1:2:8) = deriv(2,:);B(3,2:2:8) = deriv(1,:);
            Ke = Ke+detJ*thick*wi*wj*B'*D*B;
            %element stiffness matrix by numerical intergration
        end
    end
    K(dof,dof) = K(dof,dof)+Ke;
end
K;
F = zeros(2*nNode,1);
%applying load at top right corner
tleft = intersect(top,left);   %TL
tright = intersect(top,right); %TR
bleft = intersect(bottom,left); %BL
bright = intersect(bottom,right); %BR
cleft = 0.5*(tleft+bleft); %ML
cright = 0.5*(tright+bright); %MR
```

```matlab
if(nLE~=0)% if distributed loading is present on edges
    F = zeros(ndof,1); %initialize the force vector
    fprintf(1,'Distributed load is present on the edges\n');
    [mat1] = fscanf(inp,'%d %f %f',3);
    Ecode = mat1(1); % code for the edge
    qx = mat1(2);
    qy = mat1(3);
    dx = Lx/nx;
    dy = Ly/ny;
    ds = dx*secd(alpha);
    switch Ecode
        case 1 % loading on top edge
            fprintf(1,'Top edge is loaded\n');
            for i=1:ntop
                n1 = Tedge(i,1);
                n2 = Tedge(i,2);
                F(2*n1-1)=F(2*n1-1)-0.5*qy*ds*thick*sind(alpha);
                F(2*n1)  =F(2*n1)  +0.5*qy*ds*thick*cosd(alpha);
                F(2*n2-1)=F(2*n2-1)-0.5*qy*ds*thick*sind(alpha);
                F(2*n2)  =F(2*n2)  +0.5*qy*ds*thick*cosd(alpha);
            end
        case 2 % loading on left edge
            fprintf(1,'Left edge is loaded\n');
            for i=1:nleft
                n1 = Ledge(i,1);
                n2 = Ledge(i,2);
                F(2*n1-1)=F(2*n1-1)+0.5*fx*dhy*thick;
                F(2*n1)=F(2*n1)+0.5*fy*dhx*thick;
                F(2*n2-1)=F(2*n2-1)+0.5*fx*dhy*thick;
                F(2*n2)=F(2*n2)++0.5*fy*dhx*thick;
            end
        case 3 % loading on bottom edge
            fprintf(1,'Bottom edge is loaded\n');
            for i=1:nbottom
                n1 = Bedge(i,1);
                n2 = Bedge(i,2);
                F(2*n1-1)=F(2*n1-1)-0.5*qy*ds*thick*sind(alpha);
                F(2*n1)  =F(2*n1)  +0.5*qy*ds*thick*cosd(alpha);
                F(2*n2-1)=F(2*n2-1)-0.5*qy*ds*thick*sind(alpha);
                F(2*n2)  =F(2*n2)  +0.5*qy*ds*thick*cosd(alpha);
            end
        case 4 % loading on right edge
            fprintf(1,'Right edge is loaded\n');
            for i=1:nright
                n1 = Redge(i,1);
                n2 = Redge(i,2);
                F(2*n1-1)=F(2*n1-1)+0.5*fx*dx*thick;
                F(2*n1)=F(2*n1)+0.5*fy*dy*thick;
                F(2*n2-1)=F(2*n2-1)+0.5*fx*dx*thick;
                F(2*n2)=F(2*n2)++0.5*fy*dy*thick;
```

```matlab
            end
        end
end

if(nLC~=0) % if corners are loaded by point loads
    F = zeros(ndof,1); %initialize the force vector
    for i=1:nLC
        fprintf(1,'Point loads are present at specified nodes\n');
        [mat1] = fscanf(inp,'%d %f %f',3);
        Ncode = mat1(1); %code for the node
        Fx = mat1(2);
        Fy = mat1(3);
        switch Ncode
            case 1  %Top left corner
                F(2*tleft-1)=F(2*tleft-1)+Fx;
                F(2*tleft)=F(2*tleft)+Fy;
            case 2  % Centre left corner
                F(2*cleft-1) = F(2*cleft-1)+Fx;
                F(2*cleft) = F(2*cleft)+Fy;
            case 3 %Bottom left corner
                F(2*bleft-1)=F(2*bleft-1)+Fx;
                F(2*bleft) = F(2*bleft)+Fy;
            case 4  %Bottom right corner
                F(2*tright-1)=F(2*tright-1)+Fx;
                F(2*tright) = F(2*tright)+Fy;
            case 5  %Centre right corner
                F(2*cright-1)=F(2*cright-1)+Fx;
                F(2*cright) = F(2*cright)+Fy;
            case 6  %Bottom right corner
                F(2*bright-1)=F(2*bright-1)+Fx;
                F(2*bright) = F(2*bright)+Fy;
        end
    end
end
choice = input('Enter 1 for Edge restraint OR 2 for Corner restraint:');
if(choice==1) %edges are restrained
    fprintf(1,'enter edge restraints in the order [Bottom Top Left Right]\n');
    fprintf(1,'enter 0 for restrained edge and 1 for unrestrained edge\n');
    resB=input('Bottom Edge(1 or 0): ');
    resT=input('Top Edge(1 or 0): ');
    resL=input('Left Edge(1 or 0): ');
    resR=input('Right Edge(1 or 0): ');
    edge =[resB resT resL resR];
    %applying boundary conditions
    resx = ones(nNode,1);
    resy = ones(nNode,1);
```

```matlab
for i=1:nx+1
    if(edge(1)==0)%bottom edge clamped
        j = bottom(i);
        resx(j)=0;
        resy(j)=0; %all dof on bottom edge are restrained
    end
    if(edge(2)==0)%top edge clamped
        j = top(i);
        resx(j)=0;
        resy(j)=0;%all dof on bottom edge are restrained
    end
end
for i=1:ny+1
    if(edge(3)==0)%left edge clamped
        j = left(i);
        resx(j)=0;
        resy(j)=0;
    end
    if(edge(4)==0)%right edge clamped
        j = right(i);
        resx(j)=0;
        resy(j)=0;
    end
end
else % corners are restrained
    %all nodes are initialized to free dof
    fprintf(1,'enter corner restraints\n');
    fprintf(1,'enter 1 for Free  2 for hinge   3 for roller\n');
    resTL=input('Top-Left Corner(1/2/3)    : ');
    resBL=input('Bottom-Left Corner(1/2/3): ');
    resTR=input('Top-Right Corner(1/2/3)   : ');
    resBR=input('Bottom-Right Corner(1/2/3): ');

    %all dof iniitialzed to free by default
    resx = ones(nNode,1);
    resy = ones(nNode,1);
    switch resTL  %TOP-LEFT CORNER
        case 2  %hinge
            resx(tleft)=0;
            resy(tleft)=0;
        case 3  %roller
            resy(tleft)=0;
    end
    switch resBL   %BOTTOM-LEFT CORNER
        case 2  %hinge
            resx(bleft)=0;
            resy(bleft)=0;
        case 3  %roller
            resy(bleft)=0;
    end
```

```matlab
        switch resTR  %TOP-RIGHT CORNER
            case 2   %hinge
                resx(tright)=0;
                resy(tright)=0;
            case 3   %roller
                resy(tright)=0;
        end
        switch resBR   %BOTTOM-RIGHT CORNER
            case 2   %hinge
                resx(bright)=0;
                resy(bright)=0;
            case 3   %roller
                resy(bright)=0;
        end
end
bn = 0;
for i=1:nNode
    if(resx(i)==0)
        bn = bn+1;
        ifix(bn)  = 2*i-1;
    end
    if(resy(i)==0)
        bn = bn+1;
        ifix(bn)  = 2*i;
    end
end
ifix = ifix';
K(ifix,:)=zeros(bn,ndof);
K(:,ifix)=zeros(ndof,bn);
K(ifix,ifix)=eye(length(ifix));
F(ifix)=zeros(bn,1);
u =K\F;
ngp=2; % for computing stresses at Gauss points
sigma = zeros(3,nElem,ngp^2);
samp = gauss(ngp);
for i=1:nElem
    x1 = x(node1(i)); y1 = y(node1(i));
    x2 = x(node2(i)); y2 = y(node2(i));
    x3 = x(node3(i)); y3 = y(node3(i));
    x4 = x(node4(i)); y4 = y(node4(i));
    coord = [x1 x2 x3 x4
             y1 y2 y3 y4]';
    i1 =2*node1(i)-1; j1 = 2*node1(i);
    i2 =2*node2(i)-1; j2 = 2*node2(i);
    i3 =2*node3(i)-1; j3 = 2*node3(i);
    i4 =2*node4(i)-1; j4 = 2*node4(i);
    dof = [i1 j1 i2 j2 i3 j3 i4 j4 ];
    k= 0 ;
    for ig=1:ngp  % loop over Gauss points
        wi = samp(ig,2);
```

```matlab
        for jg=1:ngp
            wj = samp(jg,2);
            if(samp(ig,1)<0 & samp(jg,1)<0)
                k=1;
            elseif(samp(ig,1)>0 & samp(jg,1)<0)
                k=2;
            elseif(samp(ig,1)>0 & samp(jg,1)>0)
                k=3;
            elseif(samp(ig,1)<0 & samp(jg,1)>0)
                k=4;
            end
            [der,fun]=fmlin(samp,ig,jg);
            J =der*coord;
            detJ = det(J);
            deriv = J\der ; %inv(J)*der;
            B(1,1:2:8) = deriv(1,:);
            B(2,2:2:8) = deriv(2,:);
            B(3,1:2:8) = deriv(2,:);B(3,2:2:8) = deriv(1,:);
            sigma(:,i,k) =D*B*u(dof);
        end
    end
end
ux = u(1:2:2*nNode-1);
uy = u(2:2:2*nNode);
ux1 = reshape(ux,ny+1,nx+1);
uy1 = reshape(uy,ny+1,nx+1);
dx = Lx/nx;
dy = Ly/ny;
x1 =reshape(x,ny+1,nx+1);
y1 =reshape(y,ny+1,nx+1);
optionD = input('Do you want plots on deformed configuration (1 for Yes
0 for No):');
if(optionD==1)
    x2 = x1+ux1;
    y2 = y1+uy1;
    xd = reshape(x2,ny+1,nx+1);
    yd = reshape(y2,ny+1,nx+1);
end
%---------------------------------
%horizontal displacement
figure(2)
if(optionD==1)
    surf(xd,yd,ux1);
else
    surf(x1,y1,ux1);
end
title('Variation                        of                        horizontal
displacement','fontweight','bold','fontsize',16);
xlabel('Length of plate (mm)','fontweight','bold','fontsize',10);
ylabel('Depth of plate (mm)','fontweight','bold','fontsize',10);
```

```matlab
ax = gca;
ax.FontSize = 11;
ax.FontWeight = 'bold';
colormap('jet')
shading interp
cb = colorbar;
val = cb.Limits;
t = linspace(val(1),val(2),8);
cstr_t = cell(length(t),1);
for ii=1:length(t)
    cstr_t{ii}=num2str(t(ii));
end
tickLabelsToUse = cb.TickLabels;
tickLabelsMod = cellfun(@(c)  sprintf('%.3E',str2double(c))  ,cstr_t,
'uni', 0);
cb.TickLabels = [];
cb.TickLabels = tickLabelsMod;
set(cb,'YTick',t);
%vertical displacement
figure(3)
if(optionD==1)
    surf(xd,yd,uy1)
else
    surf(x1,y1,uy1);
end
title('Variation                    of                    vertical
displacement','fontweight','bold','fontsize',16);
xlabel('Length of plate (mm)','fontweight','bold','fontsize',10);
ylabel('Depth of plate (mm)','fontweight','bold','fontsize',10);
ax = gca;
ax.FontSize = 11;
ax.FontWeight = 'bold';
colormap('jet')
shading interp
cb = colorbar;
val = cb.Limits;
t = linspace(val(1),val(2),8);
cstr_t = cell(length(t),1);
for ii=1:length(t)
    cstr_t{ii}=num2str(t(ii));
end
tickLabelsToUse = cb.TickLabels;
tickLabelsMod = cellfun(@(c)  sprintf('%.3E',str2double(c))  ,cstr_t,
'uni', 0);
cb.TickLabels = [];
cb.TickLabels = tickLabelsMod;
set(cb,'YTick',t);
%-----------------------------------
%preparing contour plot for stresses
```

```matlab
for i=1:ny+1
     nodemat(i,:) = left(i):ny+1:right(i);
end
Tmat = [1+sqrt(3)/2      -1/2        1-sqrt(3)/2          -1/2
          -1/2        1+sqrt(3)/2     -1/2          1-sqrt(3)/2
        1-sqrt(3)/2      -1/2       1+sqrt(3)/2         -1/2
          -1/2        1-sqrt(3)/2    -1/2          1+sqrt(3)/2];
sigmaX = zeros(ny+1,nx+1);
sigmaY = zeros(ny+1,nx+1);
tauXY  = zeros(ny+1,nx+1);
for e=1:nElem
   n1 = node1(e); n2 = node2(e); n3 = node3(e) ; n4 = node4(e);
   sigx =  Tmat*[sigma(1,e,1) sigma(1,e,2) sigma(1,e,3) sigma(1,e,4)]';
   sigy =  Tmat*[sigma(2,e,1) sigma(2,e,2) sigma(2,e,3) sigma(2,e,4)]';
   tau  =  Tmat*[sigma(3,e,1) sigma(3,e,2) sigma(3,e,3) sigma(3,e,4)]';
   [row1, col1] = find(nodemat == n1);
   [row2, col2] = find(nodemat == n2);
   [row3, col3] = find(nodemat == n3);
   [row4, col4] = find(nodemat == n4);
   sigmaX(row1,col1) = sigmaX(row1,col1)+sigx(1);
   sigmaX(row2,col2) = sigmaX(row2,col2)+sigx(2);
   sigmaX(row3,col3) = sigmaX(row3,col3)+sigx(3);
   sigmaX(row4,col4) = sigmaX(row4,col4)+sigx(4);
   sigmaY(row1,col1) = sigmaY(row1,col1)+sigy(1);
   sigmaY(row2,col2) = sigmaY(row2,col2)+sigy(2);
   sigmaY(row3,col3) = sigmaY(row3,col3)+sigy(3);
   sigmaY(row4,col4) = sigmaY(row4,col4)+sigy(4);
   tauXY(row1,col1)  = tauXY(row1,col1) +tau(1);
   tauXY(row2,col2)  = tauXY(row2,col2) +tau(2);
   tauXY(row3,col3)  = tauXY(row3,col3) +tau(3);
   tauXY(row4,col4)  = tauXY(row4,col4) +tau(4);
end
%to count number of elements contributing to a node
for i=1:ny+1
    for j=1:nx+1
        gridnode = nodemat(i,j);
        m = 0;
        for i1=1:nElem
            for j1=1:4
                if(nodes(i1,j1)==gridnode)
                    m = m+1;
                end
            end
        end
        sigmaX(i,j) = sigmaX(i,j)/m;
        sigmaY(i,j) = sigmaY(i,j)/m;
        tauXY(i,j)  = tauXY(i,j)/m;
    end
end
```

```matlab
figure(4)
if(optionD==1)
    surf(xd,yd,sigmaX)
else
    surf(x1,y1,sigmaX);
end
title('Variation of \sigma_{X}','fontweight','bold','fontsize',16);
xlabel('Length of plate (mm)','fontweight','bold','fontsize',10);
ylabel('Depth of plate (mm)','fontweight','bold','fontsize',10);
ax = gca;
ax.FontSize = 11;
ax.FontWeight = 'bold';
colormap('jet')
shading interp
cb = colorbar;
val = cb.Limits;
t = linspace(val(1),val(2),8);
set(cb,'YTick',t);
cb.FontWeight = 'bold';
for ii=1:length(t)
    cstr_t{ii}=num2str(t(ii));
end
tickLabelsToUse = cb.TickLabels;
tickLabelsMod = cellfun(@(c) sprintf('%.3e',str2double(c))  ,cstr_t,
'uni', 0);
cb.TickLabels = [];
cb.TickLabels = tickLabelsMod;
set(cb,'YTick',t);
figure(5)
if(optionD==1)
    surf(xd,yd,sigmaY)
else
    surf(x1,y1,sigmaY);
end
title('Variation of \sigma_{Y}','fontweight','bold','fontsize',16);
xlabel('Length of plate (mm)','fontweight','bold','fontsize',10);
ylabel('Depth of plate (mm)','fontweight','bold','fontsize',10);
ax = gca;
ax.FontSize = 11;
ax.FontWeight = 'bold';
colormap('jet')
shading interp
cb = colorbar;
val = cb.Limits;
t = linspace(val(1),val(2),8);
set(cb,'YTick',t);
cb.FontWeight = 'bold';
for ii=1:length(t)
    cstr_t{ii}=num2str(t(ii));
end
```

```matlab
tickLabelsToUse = cb.TickLabels;
tickLabelsMod = cellfun(@(c) sprintf('%.3e',str2double(c)) ,cstr_t, 'uni', 0);
cb.TickLabels = [];
cb.TickLabels = tickLabelsMod;
set(cb,'YTick',t);
figure(6)
if(optionD==1)
    surf(xd,yd,tauXY)
else
    surf(x1,y1,tauXY);
end
title('Variation of \tau_{XY}','fontweight','bold','fontsize',16);
xlabel('Length of plate (mm)','fontweight','bold','fontsize',10);
ylabel('Depth of plate (mm)','fontweight','bold','fontsize',10);
ax = gca;
ax.FontSize = 11;
ax.FontWeight = 'bold';
colormap('jet')
shading interp
cb = colorbar;
val = cb.Limits;
t = linspace(val(1),val(2),8);
cb.FontWeight = 'bold';
for ii=1:length(t)
    cstr_t{ii}=num2str(t(ii));
end
tickLabelsToUse = cb.TickLabels;
tickLabelsMod = cellfun(@(c) sprintf('%.3e',str2double(c)) ,cstr_t, 'uni', 0);
cb.TickLabels = [];
cb.TickLabels = tickLabelsMod;
set(cb,'YTick',t);


%---------------------------------------------------------------------------
% This module generates a mesh of linear quadrilateral elements for a
%skew plate
function [xgrid,ygrid,x, y,nodes,nNode,nElem] =  Q4skewmesh(Lx,Ly,nx,ny,alpha)
nNode = 0;
k = 0;
dhx = Lx/nx;
dhy = Ly/ny;
h = Lx*tand(alpha);
dky = h/nx;
for i = 1:nx
    for j=1:ny
        k = k + 1;
        n1 = j + (i-1)*(ny + 1);
        geom(n1,:) = [(i-1)*dhx    (i-1)*dky+(j-1)*dhy ];
        n2 = j + i*(ny+1);
        geom(n2,:) = [i*dhx         i*dky+(j-1)*dhy ];
        n3 = n1 + 1;
```

```matlab
        geom(n3,:) = [(i-1)*dhx    (i-1)*dky+j*dhy ];
        n4 = n2 + 1;
        geom(n4,:) = [i*dhx         i*dky+j*dhy ];
        nel = k;
        nodes(nel,:) = [n1 n2 n4 n3];
        nNode = n4;
        xgrid(2*i-1) = geom(n1,1); xgrid(2*i) = geom(n2,1);
        ygrid(2*j-1) = geom(n1,2); ygrid(2*j) = geom(n3,2);
        figure (1)
        h=line([geom(n1,1) geom(n2,1)],[geom(n1,2) geom(n2,2)]);
        h.Color = 'k';h.Marker='.';h.MarkerFaceColor='k'; h.MarkerSize=10;
        h=line([geom(n2,1) geom(n4,1)],[geom(n2,2) geom(n4,2)]);
        h.Color = 'k';h.Marker='.';h.MarkerFaceColor='k'; h.MarkerSize=10;
        h=line([geom(n4,1) geom(n3,1)],[geom(n4,2) geom(n3,2)]);
        h.Color = 'k';h.Marker='.';h.MarkerFaceColor='k'; h.MarkerSize=10;
        h=line([geom(n3,1) geom(n1,1)],[geom(n3,2) geom(n1,2)]);
        h.Color = 'k';h.Marker='.';h.MarkerFaceColor='k'; h.MarkerSize=10;
    end
end
x = geom(:,1);
y = geom(:,2);
nElem = size(nodes,1);
%-----------------------------------------------------------------------
function[samp]=gauss(ngp)
% This function returns the abscissas and weights of the Gauss
% points for ngp equal up to 4
samp=zeros(ngp,2);
if ngp==1
    samp=[0 2];
elseif ngp==2
    samp=[-1/sqrt(3) 1.0
          1/sqrt(3) 1.0];
elseif ngp==3
    samp= [-0.2*sqrt(15.)   5/9
             0.0            8/9
           0.2*sqrt(15.)    5/9];
elseif ngp==4
    samp= [-0.861136311594053    0.347854845137454
           -0.339981043584856    0.652145154862546
            0.339981043584856    0.652145154862546
            0.861136311594053    0.347854845137454];
end
```

```
function[der,fun] = fmlin(samp, ig,jg)
% This function returns the vector of the shape function and their
% derivatives with respect to xi and eta
syms xi eta ;
N1 = 0.25*(1-xi)*(1-eta);
N2 = 0.25*(1+xi)*(1-eta);
N3 = 0.25*(1+xi)*(1+eta);
N4 = 0.25*(1-xi)*(1+eta);
f = [N1 N2 N3 N4];
fun = double(subs(f,{xi eta},{samp(ig,1) samp(ig,2)}));
dermat = [diff(N1,xi)  diff(N2,xi)  diff(N3,xi) diff(N4,xi)
        diff(N1,eta)  diff(N2,eta)  diff(N3,eta) diff(N4,eta)];
der = double(subs(dermat,{xi eta},{samp(ig,1) samp(jg,1)}));
```

A.8. SOLVING PLANE STRESS AND PLANE STRAIN PROBLEMS OF PLATES USING 8-NODE QUADRILATERAL ELEMENTS

```matlab
%FINITE ELEMENT ANALYSIS OF A SKEW PLATE USING QUADRADIC QUADRILATERAL
ELEMENTS
clear
clc
format long
inp= fopen('Q8plate.inp','r');
%reading nodal information
[mat] = fscanf(inp,'%f %f  %f %d %d %f %f %f %f %d %d',11);
Lx = mat(1); %Length of plate
Ly = mat(2); %width of plate
thick  = mat(3); %thickness of plate
nx = mat(4); %number of divisions along x-direction
ny = mat(5); %number of divisions along y-direction
E  = mat(6);%Modulus of elasticity in kPa
rho= mat(7);
nue= mat(8); %Poisson's ratio
q  = mat(9);   %uniform pressure
nLC = mat(10); %number of loaded corners
nLE = mat(11); %number of loaded edges
alpha = 5.0;
[xgrid,ygrid,x, y ,nodes, nNode, nElem] = Q8skewmesh(Lx,Ly,nx,ny,alpha);
node1 = nodes(:,1); node2 = nodes(:,2); node3 = nodes(:,3);node4 =
nodes(:,4);
node5 = nodes(:,5); node6 = nodes(:,6); node7 = nodes(:,7);node8 =
nodes(:,8);
%Show Node Labels
for i = 1:nNode
    h=text(x(i),y(i),int2str(i),'fontsize',7,'fontweight','bold','Color','k');
    h.HorizontalAlignment='right';
    h.VerticalAlignment='bottom';
end
%Show Element Labels
for e = 1:nElem
    xloc = (x(nodes(e,1))+x(nodes(e,5)))/2;
    yloc = (y(nodes(e,1))+y(nodes(e,5)))/2;
    text(xloc,yloc,int2str(e),'fontsize',8,'fontweight','bold','Color','k');
end
%determining edges of the plate
bottom = zeros(2*nx+1,1);
top     = zeros(2*nx+1,1);
left    = zeros(2*ny+1,1);
right   = zeros(2*ny+1,1);
iB = 0; iT =0; iL = 0;iR = 0;
H = Lx*tand(alpha);
dky = H/nx;
```

```matlab
for i=1:nNode
    % extracting nodes corresponding to edges
    if (y(i)==0||(rem(y(i),0.5*dky)==0))
        iB = iB+1;
        bottom(iB) = i;
    end
    if(x(i)==0)
        iL = iL+1;
        left(iL) = i;
    end
    if(x(i)==Lx)
        iR = iR+1;
        right(iR) = i;
    end
end
for i=1:size(bottom,1)
    for j=1:nNode
      if(y(j) == (Ly+y(bottom(i))))
          top(i) = j;
      end
    end
end
%-----------------------------------------------------------------
%determining the elements with one edge coinciding with the top edge
ntop = size(top,1)-1; % number of elements on top edge
Tedge = zeros(ntop/2,3);
for i=1:ntop/2
    Tedge(i,:)=[top(2*i-1) top(2*i)  top(2*i+1)];
end
%-----------------------------------------------------------------
%determining the elements with one edge coinciding with the left edge
nleft = (size(left,1)-1);
Ledge = zeros(nleft/2,3);
for i=1:nleft/2
    Ledge(i,:)=[left(2*i-1) left(2*i)  left(2*i+1)];
end
%-----------------------------------------------------------------
%determining the elements with one edge coinciding with the bottom edge
nbottom = (size(bottom,1)-1);
Bedge = zeros(nbottom/2,3);
for i=1:nbottom/2
    Bedge(i,:)=[bottom(2*i-1)  bottom(2*i)  bottom(2*i+1)];
end
%-----------------------------------------------------------------
%determining the elements with one edge coinciding with the left edge
nright = (size(right,1)-1);
Redge = zeros(nright/2,3);
for i=1:nright/2
    Redge(i,:)=[right(2*i-1) right(2*i) right(2*i+1)];
end
```

```matlab
%-------------------------------------------------------------------------
ne = 1:nElem;
optionS = input('Enter 1 for Plane Stress or  2 for Plane Strain:');
if(optionS==1)
    D=(E/(1-nue^2))*[1    nue     0;
                nue   1      0;
                0     0     (1-nue)/2];
else
    D = (E/((1+nue)*(1-2*nue)))*[(1-nue)   nue        0
                                nue     (1-nue)     0
                                 0         0      (1-2*nue)/2];
end
ndof = 2*nNode;
K = zeros(ndof,ndof); %initialize global stiffness matrix

ngp = 3 ; %Number of Gauss points
samp = gauss(ngp);
for i=1:nElem
    Ke = zeros(16,16);
    B  = zeros(3,16);
    x1 = x(node1(i)); y1 = y(node1(i));
    x2 = x(node2(i)); y2 = y(node2(i));
    x3 = x(node3(i)); y3 = y(node3(i));
    x4 = x(node4(i)); y4 = y(node4(i));
    x5 = x(node5(i)); y5 = y(node5(i));
    x6 = x(node6(i)); y6 = y(node6(i));
    x7 = x(node7(i)); y7 = y(node7(i));
    x8 = x(node8(i)); y8 = y(node8(i));
    coord = [x1 x2 x3 x4 x5 x6 x7 x8
             y1 y2 y3 y4 y5 y6 y7 y8 ]';
    i1 =2*node1(i)-1; j1 = 2*node1(i);
    i2 =2*node2(i)-1; j2 = 2*node2(i);
    i3 =2*node3(i)-1; j3 = 2*node3(i);
    i4 =2*node4(i)-1; j4 = 2*node4(i);
    i5 =2*node5(i)-1; j5 = 2*node5(i);
    i6 =2*node6(i)-1; j6 = 2*node6(i);
    i7 =2*node7(i)-1; j7 = 2*node7(i);
    i8 =2*node8(i)-1; j8 = 2*node8(i);
    dof = [i1 j1 i2 j2 i3 j3 i4 j4 i5 j5 i6 j6 i7 j7 i8 j8];
    for ig=1:ngp  % loop over Gauss points
        wi = samp(ig,2);
        for jg=1:ngp
            wj = samp(jg,2);
            [der,fun]=fmquad(samp,ig,jg);
            J =der*coord;
            detJ = det(J);
            deriv = J\der ; %inv(J)*der;
            B(1,1:2:16) = deriv(1,:);
            B(2,2:2:16) = deriv(2,:);
            B(3,1:2:16) = deriv(2,:);B(3,2:2:16) = deriv(1,:);
```

```matlab
            Ke = Ke+detJ*thick*wi*wj*B'*D*B;
            %element stiffness matrix by numerical intergration
        end
    end
    K(dof,dof) = K(dof,dof)+Ke;
end
K;
F = zeros(2*nNode,1);
%applying load at top right corner
tleft = intersect(top,left);    %TL
tright = intersect(top,right);  %TR
bleft = intersect(bottom,left); %BL
bright = intersect(bottom,right); %BR
cleft = 0.5*(tleft+bleft); %ML
cright = 0.5*(tright+bright); %MR
if(nLE~=0)% if distributed loading is present on edges
    fprintf(1,'Distributed load is present on the edges\n');
    [mat1] = fscanf(inp,'%d %f %f',3);
     Ecode = mat1(1); % code for the edge
     qx = mat1(2);
     qy = mat1(3);
     dx = Lx/nx;
     ds = dx*secd(alpha);
     switch Ecode
        case 1 % loading on top edge
            fprintf(1,'Top edge is loaded\n');
            for i=1:ntop/2
                n1 = Tedge(i,1);
                n2 = Tedge(i,2);
                n3 = Tedge(i,3);
                F(2*n1-1)=F(2*n1-1)-qy*ds*thick*sind(alpha)/6;
                F(2*n1)  =F(2*n1)  +qy*ds*thick*cosd(alpha)/6;
                F(2*n2-1)=F(2*n2-1)-qy*ds*thick*sind(alpha)*(2/3);
                F(2*n2)  =F(2*n2)  +qy*ds*thick*cosd(alpha)*(2/3);
                F(2*n3-1)=F(2*n3-1)-qy*ds*thick*sind(alpha)/6;
                F(2*n3)  =F(2*n3)  +qy*ds*thick*cosd(alpha)/6;
            end
        case 2 % loading on left edge
            fprintf(1,'Left edge is loaded\n');
            for i=1:nleft
                n1 = Ledge(i,1);
                n2 = Ledge(i,2);
                F(2*n1-1)=F(2*n1-1)+0.5*fx*dx*thick;
                F(2*n1)=F(2*n1)+0.5*fy*dy*thick;
                F(2*n2-1)=F(2*n2-1)+0.5*fx*dx*thick;
                F(2*n2)=F(2*n2)++0.5*fy*dy*thick;
            end
        case 3 % loading on bottom edge
            fprintf(1,'Bottom edge is loaded\n');
            for i=1:nbottom
```

```matlab
                    n1 = Bedge(i,1);
                    n2 = Bedge(i,2);
                    F(2*n1-1)=F(2*n1-1)-0.5*qy*ds*thick*sind(alpha);
                    F(2*n1)  =F(2*n1)  +0.5*qy*ds*thick*cosd(alpha);
                    F(2*n2-1)=F(2*n2-1)-0.5*qy*ds*thick*sind(alpha);
                    F(2*n2)  =F(2*n2)  +0.5*qy*ds*thick*cosd(alpha);
                end
            case 4 % loading on right edge
                fprintf(1,'Right edge is loaded\n');
                for i=1:nright
                    n1 = Redge(i,1);
                    n2 = Redge(i,2);
                    F(2*n1-1)=F(2*n1-1)+0.5*fx*dx*thick;
                    F(2*n1)=F(2*n1)+0.5*fy*dy*thick;
                    F(2*n2-1)=F(2*n2-1)+0.5*fx*dx*thick;
                    F(2*n2)=F(2*n2)++0.5*fy*dy*thick;
                end
        end
end

if(nLC~=0) % if corners are loaded by point loads
    for i=1:nLC
        fprintf(1,'Point loads are present at specified nodes\n');
        [mat1] = fscanf(inp,'%d %f %f',3);
        Ncode = mat1(1); %code for the node
        Fx = mat1(2);
        Fy = mat1(3);
        switch Ncode
            case 1  %Top left corner
                F(2*tleft-1)=F(2*tleft-1)+Fx;
                F(2*tleft)=F(2*tleft)+Fy;
            case 2  % Centre left corner
                F(2*cleft-1) = F(2*cleft-1)+Fx;
                F(2*cleft) = F(2*cleft)+Fy;
            case 3 %Bottom left corner
                F(2*bleft-1)=F(2*bleft-1)+Fx;
                F(2*bleft) = F(2*bleft)+Fy;
            case 4  %Bottom right corner
                F(2*tright-1)=F(2*tright-1)+Fx;
                F(2*tright) = F(2*tright)+Fy;
            case 5  %Centre right corner
                F(2*cright-1)=F(2*cright-1)+Fx;
                F(2*cright) = F(2*cright)+Fy;
            case 6  %Bottom right corner
                F(2*bright-1)=F(2*bright-1)+Fx;
                F(2*bright) = F(2*bright)+Fy;
        end
    end
end
```

```matlab
choice = input('Enter 1 for Edge restraint OR 2 for Corner restraint:');
if(choice==1) %edges are restrained
    fprintf(1,'enter edge restraints in the order [Bottom Top Left Right]\n');
    fprintf(1,'enter 0 for restrained edge and 1 for unrestrained edge\n');
    resB=input('Bottom Edge(1 or 0): ');
    resT=input('Top Edge(1 or 0): ');
    resL=input('Left Edge(1 or 0): ');
    resR=input('Right Edge(1 or 0): ');
    edge =[resB resT resL resR];
    %applying boundary conditions
    resx = ones(nNode,1);
    resy = ones(nNode,1);
    for i=1:2*nx+1
        if(edge(1)==0)%bottom edge clamped
            j = bottom(i);
            resx(j)=0;
            resy(j)=0; %all dof on bottom edge are restrained
        end
        if(edge(2)==0)%top edge clamped
            j = top(i);
            resx(j)=0;
            resy(j)=0;%all dof on bottom edge are restrained
        end
    end
    for i=1:2*ny+1
        if(edge(3)==0)%left edge clamped
            j = left(i);
            resx(j)=0;
            resy(j)=0;
        end
        if(edge(4)==0)%right edge clamped
            j = right(i);
            resx(j)=0;
            resy(j)=0;
        end
    end
else % corners are restrained
    %all nodes are initialized to free dof
    fprintf(1,'enter corner restraints\n');
    fprintf(1,'enter 1 for Free  2 for hinge   3 for roller\n');
    resTL=input('Top-Left Corner(1/2/3)    : ');
    resBL=input('Bottom-Left Corner(1/2/3): ');
    resTR=input('Top-Right Corner(1/2/3)   : ');
    resBR=input('Bottom-Right Corner(1/2/3): ');

    %all dof iniitialzed to free by default
    resx = ones(nNode,1);
    resy = ones(nNode,1);
    switch resTL  %TOP-LEFT CORNER
        case 2  %hinge
```

```matlab
                resx(tleft)=0;
                resy(tleft)=0;
          case 3  %roller
                resy(tleft)=0;
    end
    switch resBL  %BOTTOM-LEFT CORNER
        case 2  %hinge
              resx(bleft)=0;
              resy(bleft)=0;
        case 3  %roller
              resy(bleft)=0;
    end
    switch resTR   %TOP-RIGHT CORNER
        case 2  %hinge
              resx(tright)=0;
              resy(tright)=0;
        case 3  %roller
              resy(tright)=0;
    end
    switch resBR   %BOTTOM-RIGHT CORNER
        case 2  %hinge
              resx(bright)=0;
              resy(bright)=0;
        case 3  %roller
              resy(bright)=0;
    end
end
bn = 0;
for i=1:nNode
    if(resx(i)==0)
        bn = bn+1;
        ifix(bn) = 2*i-1;
    end
    if(resy(i)==0)
        bn = bn+1;
        ifix(bn) = 2*i;
    end
end
ifix = ifix';
K(ifix,:)=zeros(bn,ndof);
K(:,ifix)=zeros(ndof,bn);
K(ifix,ifix)=eye(length(ifix));
F(ifix)=zeros(bn,1);
u =K\F;
sigma = zeros(3,nElem,ngp^2);
sigmac = zeros(3,nElem);
ngp=3; % for computing stresses at Gauss points
samp = gauss(ngp);
ngp1 = 1; %for computing stresses at centroid of each element
sampc = gauss(ngp1);
```

```matlab
for i=1:nElem
    x1 = x(node1(i)); y1 = y(node1(i));
    x2 = x(node2(i)); y2 = y(node2(i));
    x3 = x(node3(i)); y3 = y(node3(i));
    x4 = x(node4(i)); y4 = y(node4(i));
    x5 = x(node5(i)); y5 = y(node5(i));
    x6 = x(node6(i)); y6 = y(node6(i));
    x7 = x(node7(i)); y7 = y(node7(i));
    x8 = x(node8(i)); y8 = y(node8(i));
    coord = [x1 x2 x3 x4 x5 x6 x7 x8
             y1 y2 y3 y4 y5 y6 y7 y8 ]';
    i1 =2*node1(i)-1; j1 = 2*node1(i);
    i2 =2*node2(i)-1; j2 = 2*node2(i);
    i3 =2*node3(i)-1; j3 = 2*node3(i);
    i4 =2*node4(i)-1; j4 = 2*node4(i);
    i5 =2*node5(i)-1; j5 = 2*node5(i);
    i6 =2*node6(i)-1; j6 = 2*node6(i);
    i7 =2*node7(i)-1; j7 = 2*node7(i);
    i8 =2*node8(i)-1; j8 = 2*node8(i);
    dof = [i1 j1 i2 j2 i3 j3 i4 j4 i5 j5 i6 j6 i7 j7 i8 j8];
    k=0;
    for ig=1:ngp  % loop over Gauss points
        wi = samp(ig,2);
        for jg=1:ngp
            wj = samp(jg,2);
            k=k+1;
            [der,fun]=fmquad(samp,ig,jg);
            J =der*coord;
            detJ = det(J);
            deriv = J\der ;
            B(1,1:2:16) = deriv(1,:);
            B(2,2:2:16) = deriv(2,:);
            B(3,1:2:16) = deriv(2,:);B(3,2:2:16) = deriv(1,:);
            sigma(:,i,k)  = D*B*u(dof);
        end
    end
end
ux = u(1:2:2*nNode-1);
uy = u(2:2:2*nNode);
%generate location matrix
mat = zeros(2*ny+1,2*nx+1);
mat(1,:)            = bottom;
for i=2:2*ny+1
  mat(i,1:2:2*nx+1) = bottom(1:2:2*nx+1)+i-1;
end
k = 0;
for i=3:2:2*ny+1
      k = k+1;
    mat(i,2:2:2*nx+1) = bottom(2:2:2*nx+1)+k;
end
```

```matlab
ux1 = zeros(2*ny+1,2*nx+1);
uy1 = zeros(2*ny+1,2*nx+1);
for i=1:2*ny+1
    for j=1:2*nx+1
        nodval = mat(i,j);
        if(nodval==0)
            x1(i,j) = 0.5*(x(mat(i,j-1))+x(mat(i,j+1)));
            y1(i,j) = 0.5*(y(mat(i,j-1))+y(mat(i,j+1)));
            ux1(i,j)=0.25*(ux(mat(i,j-1))+ux(mat(i,j+1))+...
                        ux(mat(i-1,j))+ux(mat(i+1,j)));
            uy1(i,j)=0.25*(uy(mat(i,j-1))+uy(mat(i,j+1))+...
                        uy(mat(i-1,j))+uy(mat(i+1,j)));
        else
            x1(i,j) = x(nodval);
            y1(i,j) = y(nodval);
            ux1(i,j)=ux(nodval);
            uy1(i,j)=uy(nodval);
        end
    end
end
optionD=input('Do you want plots on deformed configuration(1 for Yes 0 for
No):');
if(optionD==1)
    x2 = x1+ux1;
    y2 = y1+uy1;
end
%-----------------------------------
%horizontal displacement
figure(2)
if(optionD==1)
    surf(x2,y2,ux1)
else
    surf(x1,y1,ux1);
end
colormap('jet')
shading interp
cb = colorbar;
val = cb.Limits;
t = linspace(val(1),val(2),8);
cstr_t = cell(length(t),1);
for ii=1:length(t)
    cstr_t{ii}=num2str(t(ii));
end
tickLabelsToUse = cb.TickLabels;
tickLabelsMod = cellfun(@(c) sprintf('%.3E',str2double(c)) ,cstr_t,
'uni', 0);
cb.TickLabels = [];
cb.TickLabels = tickLabelsMod;
set(cb,'YTick',t);
%-----------------------------------
```

```matlab
%vertical displacement
figure(3)
if(optionD==1)
    surf(x2,y2,uy1)
else
    surf(x1,y1,uy1);
end
colormap('jet')
shading interp
cb = colorbar;
val = cb.Limits;
t = linspace(val(1),val(2),8);
cstr_t = cell(length(t),1);
for ii=1:length(t)
    cstr_t{ii}=num2str(t(ii));
end
tickLabelsToUse = cb.TickLabels;
tickLabelsMod  = cellfun(@(c)  sprintf('%.3E',str2double(c))   ,cstr_t,
'uni', 0);
cb.TickLabels = [];
cb.TickLabels = tickLabelsMod;
set(cb,'YTick',t);
%-----------------------------------
%preparing contour plot for stresses
Tmat = genTmatA;
sigmaX = zeros(2*ny+1,2*nx+1);
sigmaY = zeros(2*ny+1,2*nx+1);
tauXY  = zeros(2*ny+1,2*nx+1);
for e=1:nElem
%     fprintf(1,'Element = %d\n',e);
   n1 = node1(e); n2 = node2(e); n3 = node3(e) ; n4 = node4(e);
   n5 = node5(e); n6 = node6(e); n7 = node7(e) ; n8 = node8(e);
   sigx =  Tmat*[sigma(1,e,1) sigma(1,e,2) sigma(1,e,3) sigma(1,e,4)...
                 sigma(1,e,5) sigma(1,e,6) sigma(1,e,7) sigma(1,e,8)
                 sigma(1,e,9)]';
   sigy =  Tmat*[sigma(2,e,1) sigma(2,e,2) sigma(2,e,3) sigma(2,e,4)...
                 sigma(2,e,5) sigma(2,e,6) sigma(2,e,7) sigma(2,e,8)
                 sigma(2,e,9)]';
   tau  =  Tmat*[sigma(3,e,1) sigma(3,e,2) sigma(3,e,3) sigma(3,e,4)...
                 sigma(3,e,5) sigma(3,e,6) sigma(3,e,7) sigma(3,e,8)
                 sigma(3,e,9)]';
   [row1, col1] = find(mat == n1);    [row2, col2] = find(mat == n2);
   [row3, col3] = find(mat == n3);    [row4, col4] = find(mat == n4);
   [row5, col5] = find(mat == n5);    [row6, col6] = find(mat == n6);
   [row7, col7] = find(mat == n7);    [row8, col8] = find(mat == n8);
   sigmaX(row1,col1) = sigmaX(row1,col1)+sigx(1);
   sigmaX(row2,col2) = sigmaX(row2,col2)+sigx(2);
   sigmaX(row3,col3) = sigmaX(row3,col3)+sigx(3);
   sigmaX(row4,col4) = sigmaX(row4,col4)+sigx(4);
   sigmaX(row5,col5) = sigmaX(row5,col5)+sigx(5);
```

```matlab
    sigmaX(row6,col6)  = sigmaX(row6,col6)+sigx(6);
    sigmaX(row7,col7)  = sigmaX(row7,col7)+sigx(7);
    sigmaX(row8,col8)  = sigmaX(row8,col8)+sigx(8);

    sigmaY(row1,col1)  = sigmaY(row1,col1)+sigy(1);
    sigmaY(row2,col2)  = sigmaY(row2,col2)+sigy(2);
    sigmaY(row3,col3)  = sigmaY(row3,col3)+sigy(3);
    sigmaY(row4,col4)  = sigmaY(row4,col4)+sigy(4);
    sigmaY(row5,col5)  = sigmaY(row5,col5)+sigy(5);
    sigmaY(row6,col6)  = sigmaY(row6,col6)+sigy(6);
    sigmaY(row7,col7)  = sigmaY(row7,col7)+sigy(7);
    sigmaY(row8,col8)  = sigmaY(row8,col8)+sigy(8);

    tauXY(row1,col1)   = tauXY(row1,col1)  +tau(1);
    tauXY(row2,col2)   = tauXY(row2,col2)  +tau(2);
    tauXY(row3,col3)   = tauXY(row3,col3)  +tau(3);
    tauXY(row4,col4)   = tauXY(row4,col4)  +tau(4);
    tauXY(row5,col5)   = tauXY(row5,col5)  +tau(5);
    tauXY(row6,col6)   = tauXY(row6,col6)  +tau(6);
    tauXY(row7,col7)   = tauXY(row7,col7)  +tau(7);
    tauXY(row8,col8)   = tauXY(row8,col8)  +tau(8);
end
%to count number of elements contributing to a node
for i=1:2*ny+1
    for j=1:2*nx+1
        gridnode = mat(i,j);
        m = 0;
        for i1=1:nElem
            for j1=1:8
                if(nodes(i1,j1)==gridnode)
                    m = m+1;
                end
            end
        end
        sigmaX(i,j) = sigmaX(i,j)/m;
        sigmaY(i,j) = sigmaY(i,j)/m;
        tauXY(i,j)  = tauXY(i,j)/m;
    end
end
[m1, n1] = find(isnan(sigmaX));
for e=1:nElem
    i=m1(e);
    j=n1(e);
sigmaX(i,j) = (sigmaX(i-1,j)+sigmaX(i+1,j)+sigmaX(i,j-1)+sigmaX(i,j+1)...
        +sigmaX(i-1,j-1)+sigmaX(i+1,j+1)+sigmaX(i-1,j+1)+sigmaX(i+1,j-1))/8;
    sigmaY(i,j) = (sigmaY(i-1,j)+sigmaY(i+1,j)+sigmaY(i,j-1)+sigmaY(i,j+1)...
        +sigmaY(i-1,j-1)+sigmaY(i+1,j+1)+sigmaY(i-1,j+1)+sigmaY(i+1,j-1))/8;
    tauXY(i,j) = (tauXY(i-1,j)+tauXY(i+1,j)+tauXY(i,j-1)+tauXY(i,j+1)...
        +tauXY(i-1,j-1)+tauXY(i+1,j+1)+tauXY(i-1,j+1)+tauXY(i+1,j-1))/8;
end
```

```matlab
figure(4)
if(optionD==1)
    surf(x2,y2,sigmaX)
else
    surf(x1,y1,sigmaX);
end
colormap('jet')
shading interp
cb = colorbar;
val = cb.Limits;
t = linspace(val(1),val(2),8);
t = linspace(val(1),val(2),8);
cb.FontWeight = 'bold';
for ii=1:length(t)
    cstr_t{ii}=num2str(t(ii));
end
tickLabelsToUse = cb.TickLabels;
tickLabelsMod = cellfun(@(c) sprintf('%.1f',str2double(c)) ,cstr_t, 'uni', 0);
cb.TickLabels = [];
cb.TickLabels = tickLabelsMod;
set(cb,'YTick',t);

figure(5)
if(optionD==1)
    surf(x2,y2,sigmaY)
else
    surf(x1,y1,sigmaY);
end
colormap('jet')
shading interp
cb = colorbar;
val = cb.Limits;
t = linspace(val(1),val(2),8);
cb.FontWeight = 'bold';
for ii=1:length(t)
    cstr_t{ii}=num2str(t(ii));
end
tickLabelsToUse = cb.TickLabels;
tickLabelsMod = cellfun(@(c) sprintf('%.1f',str2double(c)) ,cstr_t, 'uni', 0);
cb.TickLabels = [];
cb.TickLabels = tickLabelsMod;
set(cb,'YTick',t);

figure(6)
if(optionD==1)
    surf(x2,y2,tauXY)
else
    surf(x1,y1,tauXY);
end
colormap('jet')
```

```matlab
shading interp
cb = colorbar;
val = cb.Limits;
t = linspace(val(1),val(2),8);
t = linspace(val(1),val(2),8);
cb.FontWeight = 'bold';
for ii=1:length(t)
    cstr_t{ii}=num2str(t(ii));
end
tickLabelsToUse = cb.TickLabels;
tickLabelsMod = cellfun(@(c) sprintf('%.1f',str2double(c)) ,cstr_t, 'uni', 0);
cb.TickLabels = [];
cb.TickLabels = tickLabelsMod;
set(cb,'YTick',t);
%----------------------------------------------------------------
% This module generates a mesh of linear quadrilateral elements for a
skew plate
function     [xgrid,ygrid,x,     y,      nodes,     nNode,     nElem]     =
Q8skewmesh(Lx,Ly,nx,ny,alpha)
nNode = 0;
k = 0;
dhx = Lx/nx;
dhy = Ly/ny;
h = Lx*tand(alpha);
dky = h/nx;
for i = 1:nx
    for j=1:ny
      k = k + 1;
      n1 = (i-1)*(3*ny+2)+2*j - 1;
      n2 = i*(3*ny+2)+j - ny - 1;
      n3 = i*(3*ny+2)+2*j-1;
      n4 = n3 + 1;
      n5 = n3 + 2;
      n6 = n2 + 1;
      n7 = n1 + 2;
      n8 = n1 + 1;
      geom(n1,:) = [(i-1)*dhx   (i-1)*dky+(j-1)*dhy   ];
      geom(n3,:) = [ i*dhx        i*dky+(j-1)*dhy   ];
      geom(n2,:) = [(geom(n1,1)+geom(n3,1))/2 (geom(n1,2)+geom(n3,2))/2];
      geom(n5,:) = [i*dhx    i*dky+j*dhy   ];
      geom(n4,:) = [(geom(n3,1)+ geom(n5,1))/2 (geom(n3,2)+ geom(n5,2))/2];
      geom(n7,:) = [(i-1)*dhx   (i-1)*dky+j*dhy ];
      geom(n6,:) = [(geom(n5,1)+ geom(n7,1))/2 (geom(n5,2)+ geom(n7,2))/2];
      geom(n8,:) = [(geom(n1,1)+ geom(n7,1))/2 (geom(n1,2)+ geom(n7,2))/2];
        nel = k;
        nNode = n5;
        nodes(k,:) = [n1 n2 n3 n4 n5 n6 n7 n8];
        xgrid(2*i-1) = geom(n1,1); xgrid(2*i) = geom(n3,1);
        ygrid(2*j-1) = geom(n1,2); ygrid(2*j) = geom(n3,2);
        figure (1)
```

```matlab
        h=line([geom(n1,1) geom(n2,1)],[geom(n1,2) geom(n2,2)]);
        h.Color = 'k';h.Marker='.';h.MarkerFaceColor='k'; h.MarkerSize=10;
        h=line([geom(n2,1) geom(n3,1)],[geom(n2,2) geom(n3,2)]);
        h.Color = 'k';h.Marker='.';h.MarkerFaceColor='k'; h.MarkerSize=10;
        h=line([geom(n3,1) geom(n4,1)],[geom(n3,2) geom(n4,2)]);
        h.Color = 'k';h.Marker='.';h.MarkerFaceColor='k'; h.MarkerSize=10;
        h=line([geom(n4,1) geom(n4,1)],[geom(n4,2) geom(n5,2)]);
        h.Color = 'k';h.Marker='.';h.MarkerFaceColor='k'; h.MarkerSize=10;
        h=line([geom(n5,1) geom(n6,1)],[geom(n5,2) geom(n6,2)]);
        h.Color = 'k';h.Marker='.';h.MarkerFaceColor='k'; h.MarkerSize=10;
        h=line([geom(n6,1) geom(n7,1)],[geom(n6,2) geom(n7,2)]);
        h.Color = 'k';h.Marker='.';h.MarkerFaceColor='k'; h.MarkerSize=10;
        h=line([geom(n7,1) geom(n8,1)],[geom(n7,2) geom(n8,2)]);
        h.Color = 'k';h.Marker='.';h.MarkerFaceColor='k'; h.MarkerSize=10;
        h=line([geom(n8,1) geom(n1,1)],[geom(n8,2) geom(n1,2)]);
        h.Color = 'k';h.Marker='.';h.MarkerFaceColor='k'; h.MarkerSize=10;
    end
end
x = geom(:,1);
y = geom(:,2);
nElem = size(nodes,1);
%-------------------------------------------------------------------
function[samp]=gauss(ngp)
% This function returns the abscissas and weights of the Gauss
% points for ngp equal up to 4
samp=zeros(ngp,2);
if ngp==1
    samp=[0 2];
elseif ngp==2
    samp=[-1/sqrt(3) 1.0
           1/sqrt(3) 1.0];
elseif ngp==3
    samp= [-0.2*sqrt(15.)   5/9
             0.0            8/9
           0.2*sqrt(15.)   5/9];
elseif ngp==4
    samp= [-0.861136311594053      0.347854845137454
           -0.339981043584856      0.652145154862546
            0.339981043584856      0.652145154862546
            0.861136311594053      0.347854845137454];
end
%-------------------------------------------------------------------
function[der,fun] = fmquad(samp, ig,jg)
%
% This function returns the vector of the shape function and their
% derivatives with respect to xi and eta
%
% xi=samp(ig,1);
% eta=samp(jg,1);
%
```

```
syms xi eta ;
N1 = -0.25*(1-xi)*(1-eta)*(1+xi+eta) ;
N2 =  0.5*(1-xi^2)*(1-eta) ;
N3 = -0.25*(1+xi)*(1-eta)*(1-xi+eta) ;
N4 =  0.5*(1+xi)*(1-eta^2) ;
N5 = -0.25*(1+xi)*(1+eta)*(1-xi-eta) ;
N6 =  0.5*(1-xi^2)*(1+eta) ;
N7 = -0.25*(1-xi)*(1+eta)*(1+xi-eta) ;
N8 =  0.5*(1-xi)*(1-eta^2) ;

f = [N1 N2 N3 N4 N5 N6 N7 N8];
fun = double(subs(f,{xi eta},{samp(ig,1) samp(ig,2)}));
dermat = [diff(N1,xi)   diff(N2,xi)   diff(N3,xi) diff(N4,xi)  ...
          diff(N5,xi)   diff(N6,xi)   diff(N7,xi) diff(N8,xi)
          diff(N1,eta)  diff(N2,eta)   diff(N3,eta) diff(N4,eta)...
          diff(N5,eta)  diff(N6,eta)   diff(N7,eta) diff(N8,eta) ];
der = double(subs(dermat,{xi eta},{samp(ig,1) samp(jg,1)}));
%------------------------------------------------------------------------
function [Tmat] =  genTmatA
syms xi eta ;
f1 = 0.5*xi*(xi-1); f2 = 1-xi^2 ; f3 = 0.5*xi*(xi+1);
g1 = 0.5*eta*(eta-1); g2 = 1-eta^2 ; g3 = 0.5*eta*(eta+1);

N1 = f1*g1; N2 = f1*g2 ; N3 = f1*g3;
N4 = f2*g1; N5 = f2*g2 ; N6 = f2*g3;
N7 = f3*g1; N8 = f3*g2 ; N9 = f3*g3;
Tmat = zeros(8,9) ;
v = sqrt(5/3) ;
Tmat(1,:)= [
double(subs(N1,{xi,eta},[-v,-v]))
double(subs(N2,{xi,eta},[-v, -v]))
double(subs(N3,{xi,eta},[-v, -v]))
double(subs(N4,{xi,eta},[-v,-v]))
double(subs(N5,{xi,eta},[-v,-v]))
double(subs(N6,{xi,eta},[-v,-v]))
double(subs(N7,{xi,eta},[-v,-v]))
double(subs(N8,{xi,eta},[-v,-v]))
double(subs(N9,{xi,eta},[-v,-v]))];
Tmat(2,:)= [
double(subs(N1,{xi,eta},[-v,0]))
double(subs(N2,{xi,eta},[-v,0]))
double(subs(N3,{xi,eta},[-v,0]))
double(subs(N4,{xi,eta},[-v,0]))
double(subs(N5,{xi,eta},[-v,0]))
double(subs(N6,{xi,eta},[-v,0]))
double(subs(N7,{xi,eta},[-v,0]))
double(subs(N8,{xi,eta},[-v,0]))
double(subs(N9,{xi,eta},[-v,0]))];
Tmat(3,:)= [
```

```
double(subs(N1,{xi,eta},[v,-v]))
double(subs(N2,{xi,eta},[v,-v]))
double(subs(N3,{xi,eta},[v,-v]))
double(subs(N4,{xi,eta},[v,-v]))
double(subs(N5,{xi,eta},[v,-v]))
double(subs(N6,{xi,eta},[v,-v]))
double(subs(N7,{xi,eta},[v,-v]))
double(subs(N8,{xi,eta},[v,-v]))
double(subs(N9,{xi,eta},[v,-v]))];
Tmat(4,:)= [
double(subs(N1,{xi,eta},[ v,  0]))
double(subs(N2,{xi,eta},[ v,  0]))
double(subs(N3,{xi,eta},[ v,  0]))
double(subs(N4,{xi,eta},[ v,  0]))
double(subs(N5,{xi,eta},[ v,  0]))
double(subs(N6,{xi,eta},[ v,  0]))
double(subs(N7,{xi,eta},[ v,  0]))
double(subs(N8,{xi,eta},[ v,  0]))
double(subs(N9,{xi,eta},[ v,  0]))];
Tmat(5,:)= [
double(subs(N1,{xi,eta},[v,v]))
double(subs(N2,{xi,eta},[v,v]))
double(subs(N3,{xi,eta},[v,v]))
double(subs(N4,{xi,eta},[v,v]))
double(subs(N5,{xi,eta},[v,v]))
double(subs(N6,{xi,eta},[v,v]))
double(subs(N7,{xi,eta},[v,v]))
double(subs(N8,{xi,eta},[v,v]))
double(subs(N9,{xi,eta},[v,v]))];


Tmat(6,:)= [
double(subs(N1,{xi,eta},[0,v]))
double(subs(N2,{xi,eta},[0,v]))
double(subs(N3,{xi,eta},[0,v]))
double(subs(N4,{xi,eta},[0,v]))
double(subs(N5,{xi,eta},[0,v]))
double(subs(N6,{xi,eta},[0,v]))
double(subs(N7,{xi,eta},[0,v]))
double(subs(N8,{xi,eta},[0,v]))
double(subs(N9,{xi,eta},[0,v]))];
Tmat(7,:)= [
double(subs(N1,{xi,eta},[-v,v]))
double(subs(N2,{xi,eta},[-v,v]))
double(subs(N3,{xi,eta},[-v,v]))
double(subs(N4,{xi,eta},[-v,v]))
double(subs(N5,{xi,eta},[-v,v]))
double(subs(N6,{xi,eta},[-v,v]))
double(subs(N7,{xi,eta},[-v,v]))
double(subs(N8,{xi,eta},[-v,v]))
```

```
double(subs(N9,{xi,eta},[-v,v]))];
Tmat(8,:)= [
double(subs(N1,{xi,eta},[-v, 0]))
double(subs(N2,{xi,eta},[-v, 0]))
double(subs(N3,{xi,eta},[-v, 0]))
double(subs(N4,{xi,eta},[-v, 0]))
double(subs(N5,{xi,eta},[-v, 0]))
double(subs(N6,{xi,eta},[-v, 0]))
double(subs(N7,{xi,eta},[-v, 0]))
double(subs(N8,{xi,eta},[-v, 0]))
double(subs(N9,{xi,eta},[-v, 0]))];
%----------------------------------------------------------------------
```

A.9. ANALYSIS OF AXISYMMETRIC DOMAIN USING 3-NODE TRIANGULAR ELEMENTS

```
%FINITE ELEMENT ANALYSIS OF AXI-SYMMETRIC DOMAIN USING 3 noded TRIANGULAR
ELEMENTS
clear
clc
format long
inp= fopen('axiCST.inp','r');
%reading nodal information
[mat] = fscanf(inp,'%f %f  %d %d %f %f %f %d ',8);
Lx = mat(1); %Length of plate
Ly = mat(2); %width of plate
nx = mat(3); %number of divisions along x-direction
ny = mat(4); %number of divisions along y-direction
E  = mat(5);%Modulus of elasticity in kPa
nue= mat(6); %Poisson's ratio
q  = mat(7);  %uniform pressure kN
nLE= mat(8); %number of loaded edges
[x, y ,nodes, nNode, nElem] = Tmesh(Lx,Ly,nx,ny,1);
dx = Lx/nx;
dy = Ly/ny;
figure(1)
triplot(nodes,x,y,'k');
node1 = nodes(:,1); node2 = nodes(:,2); node3 = nodes(:,3);
%Show Node Labels
for i = 1:nNode

h=text(x(i),y(i),int2str(i),'fontsize',7,'fontweight','bold','Color','k');
    h.HorizontalAlignment='right';
    h.VerticalAlignment='bottom';
end
%Show Element Labels
for e = 1:nElem
  xloc = (x(nodes(e,1))+x(nodes(e,2))+x(nodes(e,3)))/3;
  yloc = (y(nodes(e,1))+y(nodes(e,2))+y(nodes(e,3)))/3;
  text(xloc,yloc,int2str(e),'fontsize',8,'fontweight','bold','Color','k');
end
pbaspect([Lx/Ly 1 1]);
nn = (nx+1)*(ny+1);
%determining edges of the plate
bottom = zeros(nx+1,1);
top    = zeros(nx+1,1);
left   = zeros(ny+1,1);
right  = zeros(ny+1,1);
iB = 0; iT =0; iL = 0;iR = 0;
inner = 0;
for i=1:nn
    % extracting nodes corresponding to edges
    if(y(i)==0)
        iB = iB+1;
        bottom(iB) = i;
    end
    if(y(i)==Ly)
        iT = iT+1;
        top(iT) = i;
    end
    if(x(i)==0)
        iL = iL+1;
        left(iL) = i;
    end
```

```matlab
    if(x(i)==Lx)
        iR = iR+1;
        right(iR) = i;
    end
end
%-------------------------------------------------------------------
%determining the elements with one edge coinciding with the top edge
ntop = (size(top,1)-1); % number of elements on top edge
Tedge = zeros(ntop,2);
for i=1:ntop
    Tedge(i,:)=[top(i) top(i+1)];
end
%-------------------------------------------------------------------
%determining the elements with one edge coinciding with the left edge
nleft = size(left,1)-1;
Ledge = zeros(nleft,2);
for i=1:nleft
    Ledge(i,:)=[left(i)  left(i+1)];
end
%-------------------------------------------------------------------
%determining the elements with one edge coinciding with the bottom edge
nbottom = (size(bottom,1)-1);
Bedge = zeros(nbottom,2);
for i=1:nbottom
    Bedge(i,:)=[bottom(i) bottom(i+1)];
end
%-------------------------------------------------------------------
%determining the elements with one edge coinciding with the left edge
nright = (size(right,1)-1);
Redge = zeros(nright,2);
for i=1:nright
    Redge(i,:)=[right(i)  right(i+1)];
end
%-------------------------------------------------------------------
D =(E/((1+nue)*(1-2*nue)))* [ 1-nue    nue      nue         0
                              nue      1-nue    nue      0
                              nue      nue      1-nue    0
                              0        0        0        (1-2*nue)/2];
ndof = 2*nNode;
K = zeros(ndof,ndof); %initialize global stiffness matrix
%Computing stiffness matrix
nhp = 1 ; %Hammer point is located at the centroid of the
samp = hammer(nhp);
for i=1:nElem
    Ke = zeros(6,6);
    B  = zeros(4,6);
    x1 = x(node1(i)); y1 = y(node1(i));
    x2 = x(node2(i)); y2 = y(node2(i));
    x3 = x(node3(i)); y3 = y(node3(i));
    coord = [x1 x2 x3
             y1 y2 y3 ]';
    i1 =2*node1(i)-1; j1 = 2*node1(i);
    i2 =2*node2(i)-1; j2 = 2*node2(i);
    i3 =2*node3(i)-1; j3 = 2*node3(i);
    dof = [i1 j1 i2 j2 i3 j3];
    for j=1:nhp  % loop over Hammer points
        wi = samp(j,3);
        [der,fun]=fmT3lin(samp,j);
        radius = dot(fun,coord(:,1));
        J =der*coord;
        detJ = det(J);
        deriv = J\der ;
```

```matlab
            B(1,1:2:6) = deriv(1,:);
            B(4,2:2:6) = deriv(1,:);
            B(2,2:2:6) = deriv(2,:);
            B(4,1:2:6) = deriv(2,:);
            B(3,1:2:6) = fun/radius;
            Ke = Ke+detJ*wi*B'*D*B*radius;
            %element stiffness matrix by numerical intergration
        end
        K(dof,dof) = K(dof,dof)+Ke;
end
F = zeros(2*nn,1);
if(nLE~=0)% if distributed loading is present on edges
    F = zeros(ndof,1); %initialize the force vector
    fprintf(1,'Distributed load is present on the edges\n');
    [mat1] = fscanf(inp,'%d %f %f %f',4);
     Ecode = mat1(1); % code for the edge
     fx = mat1(2);
     fy = mat1(3);
     dist=mat1(4);
     switch Ecode
        case 1 % loading on top edge
            fprintf(1,'Top edge is loaded\n');
            for i=1:ntop
                n1 = Tedge(i,1);
                n2 = Tedge(i,2);
                if(x(n2)<=dist)
                    r0 =x(n1); r1=x(n2);
                    F1 = fy*(r1^2+r0*r1-2*r0^2)/6;
                    F2 = fy*(2*r1^2-r0*r1-r0^2)/6;
                    F(2*n1)   =F(2*n1)+F1;
                    F(2*n2)   =F(2*n2)+F2;
                end
            end
        case 2 % loading on left edge
            fprintf(1,'Left edge is loaded\n');
            for i=1:nleft
                n1 = Ledge(i,1);
                n2 = Ledge(i,2);
                r0 =y(n1); r1=y(n2);
                F1 = fx*(r1^2+r0*r1-2*r0^2)/6;
                F2 = fx*(2*r1^2-r0*r1-r0^2)/6;
                F(2*n1-1)   =F(2*n1-1)+F1;
                F(2*n2-1)   =F(2*n2-1)+F2;
            end
        case 3 % loading on bottom edge
            fprintf(1,'Bottom edge is loaded\n');
            for i=1:nbottom
                n1 = Bedge(i,1);
                n2 = Bedge(i,2);
                r0 =x(n1); r1=x(n2);
                F1 = fy*(r1^2+r0*r1-2*r0^2)/6;
                F2 = fy*(2*r1^2-r0*r1-r0^2)/6;
                F(2*n1)   =F(2*n1)+F1;
                F(2*n2)   =F(2*n2)+F2;
            end
        case 4 % loading on right edge
            fprintf(1,'Right edge is loaded\n');
            for i=1:nright
                n1 = Redge(i,1);
                n2 = Redge(i,2);
                r0 =y(n1); r1=y(n2);
                F1 = fx*(r1^2+r0*r1-2*r0^2)/6;
```

```matlab
                        F2 = fx*(2*r1^2-r0*r1-r0^2)/6;
                        F(2*n1-1)   =F(2*n1-1)+F1;
                        F(2*n2-1)   =F(2*n2-1)+F2;
                end
        end
end
fprintf(1,'enter edge restraints in the order [Bottom Top Left Right]\n');
fprintf(1,'enter 2 for roller edge/1 for unrestrained edge/0 for hinged edge\n');
resB=input('Bottom Edge(2 or 1 or 0): ');
resT=input('Top Edge(2 or 1 or 0): ');
resL=input('Left Edge(2 or 1 or 0): ');
resR=input('Right Edge(2 or 1 or 0): ');
edge =[resB resT resL resR];
%applying boundary conditions
resx = ones(nn,1);
resy = ones(nn,1);
for i=1:ny+1
    if(edge(3)==0)%left edge hinged
        j = left(i);
        resx(j)=0;
        resy(j)=0;
    elseif (edge(3)==2)%left edge roller (on vertical plane)
        j = left(i);
        resx(j)=0;
        resy(j)=1;
    end
    if(edge(4)==0)%right edge hinged
        j = right(i);
        resx(j)=0;
        resy(j)=0;
    elseif (edge(4)==2)%right edge roller (on vertical plane)
        j = right(i);
        resx(j)=0;
        resy(j)=1;
    end
end
for i=1:nx+1
    if(edge(1)==0)%bottom edge hinged
        j = bottom(i);
        resx(j)=0;
        resy(j)=0; %all dof on bottom edge are restrained
    elseif (edge(1)==2)%bottom edge roller (on horizontal plane)
        j = bottom(i);
        resx(j)=1;
        resy(j)=0;
    end
    if(edge(2)==0)%top edge hinged
        j = top(i);
        resx(j)=0;
        resy(j)=0;%all dof on bottom edge are restrained
    elseif (edge(2)==2)%top edge roller (on horizontal plane)
        j = top(i);
        resx(j)=1;
        resy(j)=0;
    end
end

bn = 0;
for i=1:nn
    if(resx(i)==0)
```

```matlab
            bn = bn+1;
            ifix(bn) = 2*i-1;
        end
        if(resy(i)==0)
            bn = bn+1;
            ifix(bn) = 2*i;
        end
    end
end
ifix = ifix';
K(ifix,:)=zeros(bn,ndof);
K(:,ifix)=zeros(ndof,bn);
K(ifix,ifix)=eye(length(ifix));
F(ifix)=zeros(bn,1);
u =K\F;
%Computation of stresses at centroid of element
nhp=1;
samp = hammer(nhp);
sigma = zeros(4,nElem);
for i=1:nElem
    B   = zeros(4,6);
    x1 = x(node1(i)); y1 = y(node1(i));
    x2 = x(node2(i)); y2 = y(node2(i));
    x3 = x(node3(i)); y3 = y(node3(i));
    coord = [x1 x2 x3
             y1 y2 y3 ]';
    i1 =2*node1(i)-1; j1 = 2*node1(i);
    i2 =2*node2(i)-1; j2 = 2*node2(i);
    i3 =2*node3(i)-1; j3 = 2*node3(i);
    dof = [i1 j1 i2 j2 i3 j3];
    for j=1:nhp  % loop over Hammer points
        [der,fun]=fmT3lin(samp,j);
        radius = dot(fun,coord(:,1));
        J =der*coord;
        detJ = det(J);
        deriv = J\der ;
        B(1,1:2:6) = deriv(1,:);
        B(4,2:2:6) = deriv(1,:);
        B(2,2:2:6) = deriv(2,:);
        B(4,1:2:6) = deriv(2,:);
        B(3,1:2:6) = fun/radius;
        sigma(:,i,j) =D*B*u(dof);
    end
end
%generate location matrix
mat = zeros(ny+1,nx+1);
mat(1,:)              = bottom;
for i=2:ny+1
  mat(i,1:nx+1) = bottom(1:nx+1)+i-1;
end
k = 0;
for i=3:2:ny+1
      k = k+2;
      mat(i,2:2:nx+1) = bottom(2:2:nx+1)+k;
end
ux = u(1:2:2*nn-1);
uy = u(2:2:2*nn);
ux1 = reshape(ux,ny+1,nx+1);
uy1 = reshape(uy,ny+1,nx+1);
dx = Lx/nx;
dy = Ly/ny;
x1 = 0:dx:Lx;
y1 = 0:dy:Ly;
```

```matlab
optionD = input('Do you want plots on deformed configuration (1 for Yes
0 for No):');
if(optionD==1)
    x2 = x+ux;
    y2 = y+uy;
    xd = reshape(x2,ny+1,nx+1);
    yd = reshape(y2,ny+1,nx+1);
end
%-----------------------------------
%horizontal displacement
figure(2)
if(optionD==1)
    surf(xd,yd,ux1)
else
    surf(x1,y1,ux1);
end
colormap('jet')
shading interp
cb = colorbar;
val = cb.Limits;
t = linspace(val(1),val(2),8);
cstr_t = cell(length(t),1);
for ii=1:length(t)
    cstr_t{ii}=num2str(t(ii));
end
tickLabelsToUse = cb.TickLabels;
tickLabelsMod = cellfun(@(c) sprintf('%.3E',str2double(c)) ,cstr_t,
'uni', 0);
cb.TickLabels = [];
cb.TickLabels = tickLabelsMod;
set(cb,'YTick',t);
pbaspect([Lx/Ly 1 1]);
%-----------------------------------
%vertical displacement
figure(3)
if(optionD==1)
    surf(xd,yd,uy1)
else
    surf(x1,y1,uy1);
end
colormap('jet')
shading interp
cb = colorbar;
val = cb.Limits;
t = linspace(val(1),val(2),8);
cstr_t = cell(length(t),1);
for ii=1:length(t)
    cstr_t{ii}=num2str(t(ii));
end
tickLabelsToUse = cb.TickLabels;
tickLabelsMod = cellfun(@(c) sprintf('%.3E',str2double(c)) ,cstr_t,
'uni', 0);
cb.TickLabels = [];
cb.TickLabels = tickLabelsMod;
set(cb,'YTick',t);
pbaspect([Lx/Ly 1 1]);
%preparing contour plot for stresses
sigmaR = zeros(ny+1,nx+1);
sigmaZ = zeros(ny+1,nx+1);
sigmaT = zeros(ny+1,nx+1);
tauRZ  = zeros(ny+1,nx+1);
count  = zeros(ny+1,nx+1);
```

```matlab
for e=1:nElem
    n1 = node1(e) ; n2 = node2(e) ; n3 = node3(e)  ;
    sigr = sigma(1,e) ;sigz = sigma(2,e) ;sigt = sigma(3,e) ;tau=sigma(4,e) ;
    [row1, col1] = find(mat == n1) ;    [row2, col2] = find(mat == n2) ;
    [row3, col3] = find(mat == n3) ;
    sigmaR(row1,col1) =sigmaR(row1,col1)+sigr;
    sigmaR(row2,col2) =sigmaR(row2,col2)+sigr;
    sigmaR(row3,col3) =sigmaR(row3,col3)+sigr;

    sigmaZ(row1,col1) =sigmaZ(row1,col1)+sigz;
    sigmaZ(row2,col2) =sigmaZ(row2,col2)+sigz;
    sigmaZ(row3,col3) =sigmaZ(row3,col3)+sigz;

    sigmaT(row1,col1) =sigmaT(row1,col1)+sigt;
    sigmaT(row2,col2) =sigmaT(row2,col2)+sigt;
    sigmaT(row3,col3) =sigmaT(row3,col3)+sigt;

    tauRZ(row1,col1)   = tauRZ(row1,col1)  +tau;
    tauRZ(row2,col2)   = tauRZ(row2,col2)  +tau;
    tauRZ(row3,col3)   = tauRZ(row3,col3)  +tau;

    count(row1,col1)   = count(row1,col1)+1;
    count(row2,col2)   = count(row2,col2)+1;
    count(row3,col3)   = count(row3,col3)+1;
end
sigmaR =sigmaR./count;
sigmaZ =sigmaZ./count;
sigmaT =sigmaT./count;
tauRZ = tauRZ./count;

xgrid = 0:dx:Lx;
ygrid = 0:dy:Ly;

figure(4)
surf(xgrid,ygrid,sigmaR) ;
colormap('jet')
shading interp
cb = colorbar;
val = cb.Limits;
t = linspace(val(1),val(2),8) ;
set(cb,'YTick',t) ;
pbaspect([Lx/Ly 1 1]) ;

figure(5)
surf(xgrid,ygrid,sigmaZ) ;
colormap('jet')
shading interp
cb = colorbar;
val = cb.Limits;
t = linspace(val(1),val(2),8) ;
set(cb,'YTick',t) ;
pbaspect([Lx/Ly 1 1]) ;

figure(6)
surf(xgrid,ygrid,sigmaT) ;
colormap('jet')
shading interp
cb = colorbar;
val = cb.Limits;
t = linspace(val(1),val(2),8) ;
set(cb,'YTick',t) ;
pbaspect([Lx/Ly 1 1]) ;
```

```matlab
figure(7)
surf(xgrid,ygrid,tauRZ);
colormap('jet')
shading interp
cb = colorbar;
val = cb.Limits;
t = linspace(val(1),val(2),8);
set(cb,'YTick',t);
pbaspect([Lx/Ly 1 1]);

%-----------------------------------------------------------------------

% This function generates a mesh of three noded triangular elements
function [x, y, nodes, nNode, nElem] =  Tmesh(Lx,Ly,nx,ny,pattern)
k = 0;
dx = Lx/nx;
dy = Ly/ny;
xo=0;
yo=0;
for i = 1:nx
    for j=1:ny
        k = k + 1;
        n1 = j + (i-1)*(ny + 1);
        geom(n1,:) = [(i-1)*dx - xo (j-1)*dy - yo ];
        n2 = j + i*(ny+1);
        geom(n2,:) = [i*dx - xo (j-1)*dy - yo ];
        n3 = n1 + 1;
        geom(n3,:) = [(i-1)*dx - xo j*dy - yo ];
        n4 = n2 + 1;
        geom(n4,:) = [i*dx- xo j*dy - yo ];
        nel = 2*k;
        m = nel -1;
        if(pattern==1)
            nodes(m,:) = [n3 n1 n4];
            nodes(nel,:) = [n2 n4 n1];
        else
            nodes(m,:) = [n1 n2 n3];
            nodes(nel,:) = [n2 n4 n3];
        end
    end
end
x = geom(:,1);
y = geom(:,2);
nNode = size(x,1);
nElem = size(nodes,1);

%-----------------------------------------------------------------------
function[der,fun] = fmT3lin(samp, ig)
%
% This function returns the vector of the shape function and their
% derivatives with respect to xi and eta at the Hammer points for
% an 3-noded triangular element
```

```matlab
%
% xi=samp(ig,1);
% eta=samp(ig,2);
%NOTE: the three columns of samp matrix correspond to xi, eta and Weight
%ig is the Hammer point

syms xi eta;
N2 = xi;
N3 = eta;
N1 = 1-xi-eta;
f = [N1 N2 N3];
fun = double(subs(f,{xi eta},{samp(ig,1) samp(ig,2)}));
dermat = [diff(N1,xi)   diff(N2,xi)   diff(N3,xi)
        diff(N1,eta)   diff(N2,eta)   diff(N3,eta)];
der = double(subs(dermat,{xi eta},{samp(ig,1) samp(ig,2)}));


%
% end function fmT3_lin
%--------------------------------------------------------------------------
function[samp]=hammer(npt)
%
% This function returns the abscissae and weights of the
% integration points for npt equal up to 7
samp=zeros(npt,3);
switch npt
    case 1
        samp=[1/3 1/3 1/2];
    case {2 , 3}
        %npt =3;
        samp=[1/6   1/6 1/6
              2/3   1/6   1/6
              1/6   2/3 1/6];
    case {4 , 5}
        %npt=4;
        samp= [1/3 1/3 -27/96
               1/5 1/5   25/96
               3/5 1/5   25/96
               1/5 3/5   25/96];
    case 6
        a = 0.445948490915965; b = 0.091576213509771;
        samp= [ a        a        0.111690794839005
               1-2*a     a        0.111690794839005
                a      1-2*a      0.111690794839005
                b        b        0.054975871827661
               1-2*b     b        0.054975871827661
                b      1-2*b      0.054975871827661];
```

```
case 7
      a = (6+sqrt(15))/21 ; b = 4/7 -a;
      A = (155+sqrt(15))/2400; B = (31/240 -A);
samp= [ 1/3        1/3       9/80
          a          a        A
        1-2*a        a        A
          a        1-2*a      A
          b          b        B
        1-2*b        b        B
          b        1-2*b     B];
end
%-----------------------------------------------------------------------------
```

A.10. ANALYSIS OF AXISYMMETRIC DOMAIN USING 6-NODE TRIANGULAR AXISYMMETRIC ELEMENTS

```matlab
%FINITE ELEMENT ANALYSIS OF AXI-SYMMETRIC DOMAIN
clear
clc
format long
inp= fopen('axiLST.inp','r');
%reading nodal information
[mat] = fscanf(inp,'%f %f  %d %d %f %f %f %d ',8);
Lx = mat(1); %Length of plate
Ly = mat(2); %width of plate
nx = mat(3); %number of divisions along x-direction
ny = mat(4); %number of divisions along y-direction
E  = mat(5);%Modulus of elasticity in kPa
nue= mat(6); %Poisson's ratio
q  = mat(7);   %uniform pressure kN
nLE= mat(8); %number of loaded edges
nLC =0;
[xgrid,ygrid,x, y ,nodes, nNode, nElem] = T6mesh(Lx,Ly,nx,ny,2);
dx = Lx/nx;
dy = Ly/ny;
node1 = nodes(:,1); node2 = nodes(:,2); node3 = nodes(:,3);
node4 = nodes(:,4); node5 = nodes(:,5); node6 = nodes(:,6);
%Show Node Labels
for i = 1:nNode
    h=text(x(i),y(i),int2str(i),'fontsize',7,'fontweight','bold','Color','k');
    h.HorizontalAlignment='right';
    h.VerticalAlignment='bottom';
end
%Show Element Labels
for e = 1:nElem
    xloc = (x(nodes(e,1))+x(nodes(e,3))+x(nodes(e,5)))/3;
    yloc = (y(nodes(e,1))+y(nodes(e,3))+y(nodes(e,5)))/3;
    text(xloc,yloc,int2str(e),'fontsize',8,'fontweight','bold','Color','k');
end
pbaspect([Lx/Ly 1 1]);
nn = (2*nx+1)*(2*ny+1);
%determining edges of the plate
bottom = zeros(2*nx+1,1);
top    = zeros(2*nx+1,1);
left   = zeros(2*ny+1,1);
right  = zeros(2*ny+1,1);
iB = 0; iT =0; iL = 0;iR = 0;
inner = 0;
```

```matlab
for i=1:nn
    % extracting nodes corresponding to edges
    if(y(i)==0)
        iB = iB+1;
        bottom(iB) = i;
    end
    if(y(i)==Ly)
        iT = iT+1;
        top(iT) = i;
    end
    if(x(i)==0)
        iL = iL+1;
        left(iL) = i;
    end
    if(x(i)==Lx)
        iR = iR+1;
        right(iR) = i;
    end
end
%-------------------------------------------------------------------
%determining the elements with one edge coinciding with the top edge
ntop = 0.5*(size(top,1)-1); % number of elements on top edge
Tedge = zeros(ntop,3);
for i=1:ntop
    Tedge(i,:)=[top(2*i-1) top(2*i) top(2*i+1)];
end
%-------------------------------------------------------------------
%determining the elements with one edge coinciding with the left edge
nleft = 0.5*(size(left,1)-1);
Ledge = zeros(nleft,3);
for i=1:nleft
    Ledge(i,:)=[left(2*i-1) left(2*i) left(2*i+1)];
end
%-------------------------------------------------------------------
%determining the elements with one edge coinciding with the bottom edge
nbottom = 0.5*(size(bottom,1)-1);
Bedge = zeros(nbottom,3);
for i=1:nbottom
    Bedge(i,:)=[bottom(2*i-1) bottom(2*i) bottom(2*i+1)];
end
%determining the elements with one edge coinciding with the left edge
nright = 0.5*(size(right,1)-1);
Redge = zeros(nright,3);
for i=1:nright
    Redge(i,:)=[right(2*i-1) right(2*i) right(2*i+1)];
end
```

```matlab
D =(E/((1+nue)*(1-2*nue)))* [ 1-nue    nue      nue        0
                              nue      1-nue    nue        0
                              nue      nue      1-nue      0
                              0        0        0        (1-2*nue)/2];
ndof = 2*nNode;
K = zeros(ndof,ndof); %initialize global stiffness matrix

nhp = 3 ; %Number of Hammer points
samp = hammer(nhp);
for i=1:nElem
    Ke = zeros(12,12);
    B  = zeros(4,12);
    x1 = x(node1(i)); y1 = y(node1(i));
    x2 = x(node2(i)); y2 = y(node2(i));
    x3 = x(node3(i)); y3 = y(node3(i));
    x4 = x(node4(i)); y4 = y(node4(i));
    x5 = x(node5(i)); y5 = y(node5(i));
    x6 = x(node6(i)); y6 = y(node6(i));
    coord = [x1 x2 x3 x4 x5 x6
             y1 y2 y3 y4 y5 y6]';
    i1 =2*node1(i)-1; j1 = 2*node1(i);
    i2 =2*node2(i)-1; j2 = 2*node2(i);
    i3 =2*node3(i)-1; j3 = 2*node3(i);
    i4 =2*node4(i)-1; j4 = 2*node4(i);
    i5 =2*node5(i)-1; j5 = 2*node5(i);
    i6 =2*node6(i)-1; j6 = 2*node6(i);
    dof = [i1 j1 i2 j2 i3 j3 i4 j4 i5 j5 i6 j6];
    for j=1:nhp  % loop over Hammer points
        wi = samp(j,3);
        [der,fun]=fmT6_quad(samp,j);
        radius = dot(fun,coord(:,1));
        J =der*coord;
        detJ = det(J);
        deriv = J\der ;
        B(1,1:2:12) = deriv(1,:);
        B(4,2:2:12) = deriv(1,:);
        B(2,2:2:12) = deriv(2,:);
        B(4,1:2:12) = deriv(2,:);
        B(3,1:2:12) = fun/radius;
        Ke = Ke+detJ*wi*B'*D*B*radius;
        %element stiffness matrix by numerical intergration
    end
    K(dof,dof) = K(dof,dof)+Ke;
end
K;
F = zeros(2*nn,1);
```

```matlab
%applying load at top right corner
tleft = intersect(top,left);  %TL
tright = intersect(top,right); %TR
bleft = intersect(bottom,left); %BL
bright = intersect(bottom,right); %BR
cleft = 0.5*(tleft+bleft); %ML
cright = 0.5*(tright+bright); %MR
if(nLE~=0)% if distributed loading is present on edges
    F = zeros(ndof,1); %initialize the force vector
    fprintf(1,'Distributed load is present on the edges\n');
    [mat1] = fscanf(inp,'%d %f %f %f',4);
    Ecode = mat1(1); % code for the edge
    fx = mat1(2);
    fy = mat1(3);
    dist=mat1(4);
    switch Ecode
        case 1 % loading on top edge
            fprintf(1,'Top edge is loaded\n');
            for i=1:ntop
                n1 = Tedge(i,1);
                n2 = Tedge(i,2);
                n3 = Tedge(i,3);
                if(x(n3)<=dist)
                    r0 =x(n1); r1=x(n3);
                    F1 = fy*(r1-r0)*r0/6;
                    F2 = fy*(r1^2-r0^2)/3;
                    F3 = fy*(r1-r0)*r1/6;
                    F(2*n1)   =F(2*n1)+F1;
                    F(2*n2)   =F(2*n2)   +F2;
                    F(2*n3)   =F(2*n3)   +F3;
                end
            end
        case 2 % loading on left edge
            fprintf(1,'Left edge is loaded\n');
            for i=1:nleft
                n1 = Ledge(i,1);
                n2 = Ledge(i,2);
                r0 =x(n1); r1=x(n3);
                F1 = fx*(r1-r0)*r0/6;
                F2 = fx*(r1^2-r0^2)/3;
                F3 = fx*(r1-r0)*r1/6;
                F(2*n1-1)   =F(2*n1-1)+F1;
                F(2*n2-1)   =F(2*n2-1)+F2;
                F(2*n3-1)   =F(2*n3-1)+F3;
            end
        case 3 % loading on bottom edge
```

```matlab
            fprintf(1,'Bottom edge is loaded\n');
            for i=1:nbottom
                n1 = Bedge(i,1);
                n2 = Bedge(i,2);
                r0 =x(n1); r1=x(n3);
                F1 = fy*(r1-r0)*r0/6;
                F2 = fy*(r1^2-r0^2)/3;
                F3 = fy*(r1-r0)*r1/6;
                F(2*n1)   =F(2*n1)+F1;
                F(2*n2)   =F(2*n2)+F2;
                F(2*n3)   =F(2*n3)+F3;
            end
        case 4 % loading on right edge
            fprintf(1,'Right edge is loaded\n');
            for i=1:nright
                n1 = Redge(i,1);
                n2 = Redge(i,2);
                r0 =x(n1); r1=x(n3);
                F1 = fx*(r1-r0)*r0/6;
                F2 = fx*(r1^2-r0^2)/3;
                F3 = fx*(r1-r0)*r1/6;
                F(2*n1-1)   =F(2*n1-1)+F1;
                F(2*n2-1)   =F(2*n2-1)+F2;
                F(2*n3-1)   =F(2*n3-1)+F3;
            end
    end
end
fprintf(1,'enter edge restraints in the order [Bottom Top Left Right]\n');
fprintf(1,'enter 2 for roller edge/1 for unrestrained edge/0 for hinged edge\n');
resB=input('Bottom Edge(2 or 1 or 0): ');
resT=input('Top Edge(2 or 1 or 0): ');
resL=input('Left Edge(2 or 1 or 0): ');
resR=input('Right Edge(2 or 1 or 0): ');
edge =[resB resT resL resR];
%applying boundary conditions
resx = ones(nn,1);
resy = ones(nn,1);
for i=1:2*ny+1
    if(edge(3)==0)%left edge hinged
        j = left(i);
        resx(j)=0;
        resy(j)=0;
    elseif (edge(3)==2)%left edge roller (on vertical plane)
        j = left(i);
        resx(j)=0;
        resy(j)=1;
```

```matlab
        end
    if(edge(4)==0)%right edge hinged
        j = right(i);
        resx(j)=0;
        resy(j)=0;
    elseif (edge(4)==2)%right edge roller (on vertical plane)
        j = right(i);
        resx(j)=0;
        resy(j)=1;
    end
end
for i=1:2*nx+1
    if(edge(1)==0)%bottom edge hinged
        j = bottom(i);
        resx(j)=0;
        resy(j)=0; %all dof on bottom edge are restrained
    elseif (edge(1)==2)%bottom edge roller (on horizontal plane)
        j = bottom(i);
        resx(j)=1;
        resy(j)=0;
    end
    if(edge(2)==0)%top edge hinged
        j = top(i);
        resx(j)=0;
        resy(j)=0;%all dof on bottom edge are restrained
    elseif (edge(2)==2)%top edge roller (on horizontal plane)
        j = top(i);
        resx(j)=1;
        resy(j)=0;
    end
end
bn = 0;
for i=1:nn
    if(resx(i)==0)
        bn = bn+1;
        ifix(bn) = 2*i-1;
    end
    if(resy(i)==0)
        bn = bn+1;
        ifix(bn) = 2*i;
    end
end
ifix = ifix';
K(ifix,:)=zeros(bn,ndof);
K(:,ifix)=zeros(ndof,bn);
K(ifix,ifix)=eye(length(ifix));
```

```matlab
F(ifix)=zeros(bn,1);
u =K\F;
nhp=3;
samp = hammer(nhp);
sigma = zeros(4,nElem,nhp);
for i=1:nElem
    x1 = x(node1(i)); y1 = y(node1(i));
    x2 = x(node2(i)); y2 = y(node2(i));
    x3 = x(node3(i)); y3 = y(node3(i));
    x4 = x(node4(i)); y4 = y(node4(i));
    x5 = x(node5(i)); y5 = y(node5(i));
    x6 = x(node6(i)); y6 = y(node6(i));
    coord = [x1 x2 x3 x4 x5 x6
             y1 y2 y3 y4 y5 y6]';
    i1 =2*node1(i)-1; j1 = 2*node1(i);
    i2 =2*node2(i)-1; j2 = 2*node2(i);
    i3 =2*node3(i)-1; j3 = 2*node3(i);
    i4 =2*node4(i)-1; j4 = 2*node4(i);
    i5 =2*node5(i)-1; j5 = 2*node5(i);
    i6 =2*node6(i)-1; j6 = 2*node6(i);
    dof = [i1 j1 i2 j2 i3 j3 i4 j4 i5 j5 i6 j6];
    for j=1:nhp  % loop over Hammer points
        [der,fun]=fmT6_quad(samp,j);
        radius = dot(fun,coord(:,1));
        J =der*coord;
        detJ = det(J);
        deriv = J\der ;
        B(1,1:2:12) = deriv(1,:);
        B(4,2:2:12) = deriv(1,:);
        B(2,2:2:12) = deriv(2,:);
        B(4,1:2:12) = deriv(2,:);
        B(3,1:2:12) = fun/radius;
        sigma(:,i,j) =D*B*u(dof);
    end
end
%generate location matrix
mat = zeros(2*ny+1,2*nx+1);
mat(1,:)          = bottom;
for i=2:2*ny+1
  mat(i,1:2:2*nx+1) = bottom(1:2:2*nx+1)+i-1;
end
k = 0;
for i=3:2:2*ny+1
    k = k+2;
    mat(i,2:2:2*nx+1) = bottom(2:2:2*nx+1)+k;
end
```

```matlab
ux = u(1:2:2*nn-1);
uy = u(2:2:2*nn);
ux1 = reshape(ux,2*ny+1,2*nx+1);
uy1 = reshape(uy,2*ny+1,2*nx+1);
dx = Lx/nx;
dy = Ly/ny;
x1 = 0:0.5*dx:Lx;
y1 = 0:0.5*dy:Ly;
optionD = input('Do you want plots on deformed configuration (1 for Yes
0 for No):');
if(optionD==1)
    x2 = x+ux;
    y2 = y+uy;
    xd = reshape(x2,2*ny+1,2*nx+1);
    yd = reshape(y2,2*ny+1,2*nx+1);
end
%horizontal displacement
figure(2)
if(optionD==1)
    surf(xd,yd,ux1)
else
    surf(x1,y1,ux1);
end
colormap('jet')
shading interp
cb = colorbar;
val = cb.Limits;
t = linspace(val(1),val(2),8);
cstr_t = cell(length(t),1);
for ii=1:length(t)
    cstr_t{ii}=num2str(t(ii));
end
tickLabelsToUse = cb.TickLabels;
tickLabelsMod = cellfun(@(c)  sprintf('%.3E',str2double(c))  ,cstr_t,
'uni', 0);
cb.TickLabels = [];
cb.TickLabels = tickLabelsMod;
set(cb,'YTick',t);
pbaspect([Lx/Ly 1 1]);
%vertical displacement
figure(3)
if(optionD==1)
    surf(xd,yd,uy1)
else
    surf(x1,y1,uy1);
end
```

```matlab
colormap('jet')
shading interp
cb = colorbar;
val = cb.Limits;
t = linspace(val(1),val(2),8);
cstr_t = cell(length(t),1);
for ii=1:length(t)
    cstr_t{ii}=num2str(t(ii));
end
tickLabelsToUse = cb.TickLabels;
tickLabelsMod  = cellfun(@(c) sprintf('%.3E',str2double(c))  ,cstr_t,
'uni', 0);
cb.TickLabels = [];
cb.TickLabels = tickLabelsMod;
set(cb,'YTick',t);
pbaspect([Lx/Ly 1 1]);
%-------------------------------------
%preparing contour plot for stresses
Tmat = genT6AXI;
sigmaR = zeros(2*ny+1,2*nx+1);
sigmaZ = zeros(2*ny+1,2*nx+1);
sigmaT = zeros(2*ny+1,2*nx+1);
tauRZ  = zeros(2*ny+1,2*nx+1);
count  = zeros(2*ny+1,2*nx+1);
for e=1:nElem
   n1 = node1(e); n2 = node2(e); n3 = node3(e)  ;
   n4 = node4(e); n5 = node5(e); n6 = node6(e);
   sigr  =  Tmat*[sigma(1,e,1) sigma(1,e,2) sigma(1,e,3) ]';
   sigz  =  Tmat*[sigma(2,e,1) sigma(2,e,2) sigma(2,e,3) ]';
   sigt  =  Tmat*[sigma(3,e,1) sigma(3,e,2) sigma(3,e,3) ]';
   tau   =  Tmat*[sigma(4,e,1) sigma(4,e,2) sigma(4,e,3) ]';
   [row1, col1] = find(mat == n1);    [row2, col2] = find(mat == n2);
   [row3, col3] = find(mat == n3);    [row4, col4] = find(mat == n4);
   [row5, col5] = find(mat == n5);    [row6, col6] = find(mat == n6);

   sigmaR(row1,col1) =sigmaR(row1,col1)+sigr(1);
   sigmaR(row2,col2) =sigmaR(row2,col2)+sigr(2);
   sigmaR(row3,col3) =sigmaR(row3,col3)+sigr(3);
   sigmaR(row4,col4) =sigmaR(row4,col4)+sigr(4);
   sigmaR(row5,col5) =sigmaR(row5,col5)+sigr(5);
   sigmaR(row6,col6) =sigmaR(row6,col6)+sigr(6);

   sigmaZ(row1,col1) =sigmaZ(row1,col1)+sigz(1);
   sigmaZ(row2,col2) =sigmaZ(row2,col2)+sigz(2);
   sigmaZ(row3,col3) =sigmaZ(row3,col3)+sigz(3);
   sigmaZ(row4,col4) =sigmaZ(row4,col4)+sigz(4);
```

```matlab
    sigmaZ(row5,col5) =sigmaZ(row5,col5)+sigz(5);
    sigmaZ(row6,col6) =sigmaZ(row6,col6)+sigz(6);

    sigmaT(row1,col1) =sigmaT(row1,col1)+sigt(1);
    sigmaT(row2,col2) =sigmaT(row2,col2)+sigt(2);
    sigmaT(row3,col3) =sigmaT(row3,col3)+sigt(3);
    sigmaT(row4,col4) =sigmaT(row4,col4)+sigt(4);
    sigmaT(row5,col5) =sigmaT(row5,col5)+sigt(5);
    sigmaT(row6,col6) =sigmaT(row6,col6)+sigt(6);

    tauRZ(row1,col1)  = tauRZ(row1,col1) +tau(1);
    tauRZ(row2,col2)  = tauRZ(row2,col2) +tau(2);
    tauRZ(row3,col3)  = tauRZ(row3,col3) +tau(3);
    tauRZ(row4,col4)  = tauRZ(row4,col4) +tau(4);
    tauRZ(row5,col5)  = tauRZ(row5,col5) +tau(5);
    tauRZ(row6,col6)  = tauRZ(row6,col6) +tau(6);
    count(row1,col1)  = count(row1,col1)+1;
    count(row2,col2)  = count(row2,col2)+1;
    count(row3,col3)  = count(row3,col3)+1;
    count(row4,col4)  = count(row4,col4)+1;
    count(row5,col5)  = count(row5,col5)+1;
    count(row6,col6)  = count(row6,col6)+1;
end
sigmaR =sigmaR./count;
sigmaZ =sigmaZ./count;
sigmaT =sigmaT./count;
tauRZ = tauRZ./count;

[m1, n1] = find(isnan(sigmaR));

for e=1:0.5*nElem
    i=m1(e);
    j=n1(e);
    sigmaR(i,j) = (sigmaR(i-1,j)+sigmaR(i+1,j)+sigmaR(i,j-1)+sigmaR(i,j+1))/4;
    sigmaZ(i,j) = (sigmaZ(i-1,j)+sigmaZ(i+1,j)+sigmaZ(i,j-1)+sigmaZ(i,j+1))/4;
    sigmaT(i,j) = (sigmaT(i-1,j)+sigmaT(i+1,j)+sigmaT(i,j-1)+sigmaT(i,j+1))/4;
    tauRZ(i,j) = (tauRZ(i-1,j)+tauRZ(i+1,j)+tauRZ(i,j-1)+tauRZ(i,j+1))/4;
end
figure(4)
surf(xgrid,ygrid,sigmaR);
colormap('jet')
shading interp
cb = colorbar;
val = cb.Limits;
t = linspace(val(1),val(2),8);
set(cb,'YTick',t);
```

```matlab
pbaspect([Lx/Ly 1 1]);

figure(5)
surf(xgrid,ygrid,sigmaZ);
colormap('jet')
shading interp
cb = colorbar;
val = cb.Limits;
t = linspace(val(1),val(2),8);
set(cb,'YTick',t);
pbaspect([Lx/Ly 1 1]);

figure(6)
surf(xgrid,ygrid,sigmaT);
colormap('jet')
shading interp
cb = colorbar;
val = cb.Limits;
t = linspace(val(1),val(2),8);
set(cb,'YTick',t);
pbaspect([Lx/Ly 1 1]);

figure(7)
surf(xgrid,ygrid,tauRZ);
colormap('jet')
shading interp
cb = colorbar;
val = cb.Limits;
t = linspace(val(1),val(2),8);
set(cb,'YTick',t);
pbaspect([Lx/Ly 1 1]);

%-------------------------------------------------------------------
function [xgrid,ygrid,x, y, nodes, nNode, nElem] =  T6mesh(Lx,Ly,nx,ny,pattern)
% This function generates a mesh of the linear strain triangular element
% global nnd nel geom nodes x y
% global Length Width nx ny xo yo dhx dhy
nNode = 0;
k = 0;
dhx = Lx/nx;
dhy = Ly/ny;
xo = 0.0;
yo = 0.0;
figure(1)
for i = 1:nx
    for j=1:ny
```

```matlab
k = k + 1;
n1 = (2*j-1) + (2*i-2)*(2*ny+1)  ;
n2 = (2*j-1) + (2*i-1)*(2*ny+1);
n3 = (2*j-1) + (2*i)*(2*ny+1);
n4 = n1 + 1;
n5 = n2 + 1;
n6 = n3 + 1 ;
n7 = n1 + 2;
n8 = n2 + 2;
n9 = n3 + 2;
%
geom(n1,:) = [(i-1)*dhx - xo (j-1)*dhy - yo];
geom(n2,:) = [((2*i-1)/2)*dhx - xo (j-1)*dhy - yo ];
geom(n3,:) = [i*dhx - xo (j-1)*dhy - yo ];
geom(n4,:) = [(i-1)*dhx - xo ((2*j-1)/2)*dhy - yo ];
geom(n5,:) = [((2*i-1)/2)*dhx - xo ((2*j-1)/2)*dhy - yo ];
geom(n6,:) = [i*dhx - xo ((2*j-1)/2)*dhy - yo ];
geom(n7,:) = [(i-1)*dhx - xo j*dhy - yo];
geom(n8,:) = [((2*i-1)/2)*dhx - xo j*dhy - yo];
geom(n9,:) = [i*dhx - xo j*dhy - yo];
%

if(pattern==1) %diagonals right-down
    nodes(2*k-1,:) = [n1 n2 n3 n5 n7 n4];
    nodes(2*k,:)   = [n3 n6 n9 n8 n7 n5];
elseif(pattern==2) %diagonals left-down
    nodes(2*k-1,:) = [n1 n2 n3 n6 n9 n5];
    nodes(2*k,:)   = [n9 n8 n7 n4 n1 n5];
end
max_n = max([n1 n2 n3 n4 n5 n6 n7 n8 n9]);
if(nNode <= max_n)
    nNode = max_n;
end;
xgrid(2*i-1) = geom(n1,1); xgrid(2*i) = geom(n2,1);
xgrid(2*i+1) = geom(n3,1);
ygrid(2*j-1) = geom(n1,2); ygrid(2*j) = geom(n4,2);
ygrid(2*j+1) = geom(n7,2);
if(pattern==1)
 h=line([geom(n1,1) geom(n2,1)],[geom(n1,2) geom(n2,2)]);
 h.Color = 'k';h.Marker='.';h.MarkerFaceColor='k'; h.MarkerSize=10;
 h=line([geom(n2,1) geom(n3,1)],[geom(n2,2) geom(n3,2)]);
 h.Color = 'k';h.Marker='.';h.MarkerFaceColor='k';h.MarkerSize=10;
 h=line([geom(n3,1) geom(n5,1)],[geom(n3,2) geom(n5,2)]);
 h.Color = 'k';h.Marker='.';h.MarkerFaceColor='k';h.MarkerSize=10;
 h=line([geom(n5,1) geom(n7,1)],[geom(n5,2) geom(n7,2)]);
 h.Color = 'k';h.Marker='.';h.MarkerFaceColor='k';h.MarkerSize=10;
```

```matlab
            h=line([geom(n7,1) geom(n4,1)],[geom(n7,2) geom(n4,2)]);
            h.Color = 'k';h.Marker='.';h.MarkerFaceColor='k';h.MarkerSize=10;
            h=line([geom(n4,1) geom(n1,1)],[geom(n4,2) geom(n1,2)]);
            h.Color = 'k';h.Marker='.';h.MarkerFaceColor='k';h.MarkerSize=10;
            h=line([geom(n3,1) geom(n6,1)],[geom(n3,2) geom(n6,2)]);
            h.Color = 'k';h.Marker='.';h.MarkerFaceColor='k';h.MarkerSize=10;
            h=line([geom(n6,1) geom(n9,1)],[geom(n6,2) geom(n9,2)]);
            h.Color = 'k';h.Marker='.';h.MarkerFaceColor='k';h.MarkerSize=10;
            h=line([geom(n9,1) geom(n8,1)],[geom(n9,2) geom(n8,2)]);
            h.Color = 'k';h.Marker='.';h.MarkerFaceColor='k';h.MarkerSize=10;
            h=line([geom(n8,1) geom(n7,1)],[geom(n8,2) geom(n7,2)]);
            h.Color = 'k';h.Marker='.';h.MarkerFaceColor='k';h.MarkerSize=10;
        else
            h=line([geom(n1,1) geom(n2,1)],[geom(n1,2) geom(n2,2)]);
            h.Color = 'k';h.Marker='.';h.MarkerFaceColor='k'; h.MarkerSize=10;
            h=line([geom(n2,1) geom(n3,1)],[geom(n2,2) geom(n3,2)]);
            h.Color = 'k';h.Marker='.';h.MarkerFaceColor='k';h.MarkerSize=10;
            h=line([geom(n3,1) geom(n6,1)],[geom(n3,2) geom(n6,2)]);
            h.Color = 'k';h.Marker='.';h.MarkerFaceColor='k';h.MarkerSize=10;
            h=line([geom(n6,1) geom(n9,1)],[geom(n6,2) geom(n9,2)]);
            h.Color = 'k';h.Marker='.';h.MarkerFaceColor='k';h.MarkerSize=10;
            h=line([geom(n9,1) geom(n5,1)],[geom(n9,2) geom(n5,2)]);
            h.Color = k';h.Marker='.';h.MarkerFaceColor='k';h.MarkerSize=10;
            h=line([geom(n5,1) geom(n1,1)],[geom(n5,2) geom(n1,2)]);
            h.Color = 'k';h.Marker='.';h.MarkerFaceColor='k';h.MarkerSize=10;
            h=line([geom(n9,1) geom(n8,1)],[geom(n9,2) geom(n8,2)]);
            h.Color = 'k';h.Marker='.';h.MarkerFaceColor='k';h.MarkerSize=10;
            h=line([geom(n8,1) geom(n7,1)],[geom(n8,2) geom(n7,2)]);
            h.Color = 'k';h.Marker='.';h.MarkerFaceColor='k';h.MarkerSize=10;
            h=line([geom(n7,1) geom(n4,1)],[geom(n7,2) geom(n4,2)]);
            h.Color = 'k';h.Marker='.';h.MarkerFaceColor='k';h.MarkerSize=10;
            h=line([geom(n4,1) geom(n1,1)],[geom(n4,2) geom(n1,2)]);
            h.Color = 'k';h.Marker='.';h.MarkerFaceColor='k';h.MarkerSize=10;
        end
    end
end
x = geom(:,1);
y = geom(:,2);
% nNode = size(x,1);
nElem = size(nodes,1);


%-----------------------------------------------------------------------
function[der,fun] = fmT6_quad(samp, ig)
%
% This function returns the vector of the shape function and their
% derivatives with respect to xi and eta at the Hammer points for
% an 6-noded triangular element
%
% xi=samp(ig,1);
```

```matlab
% eta=samp(ig,2);
%NOTE the three columns of samp matrix correspond to xi, eta and Weight
%ig is the Hammer point
% samp(ig,1)
% samp(ig,2)
syms xi eta lamda;
lamda = 1-xi-eta;
N1 = lamda*(2*lamda-1);
N2 = 4*xi*lamda;
N3 = xi*(2*xi-1);
N4 = 4*xi*eta;
N5 = eta*(2*eta-1);
N6 = 4*eta*lamda;
f = [N1 N2 N3 N4 N5 N6];
fun = double(subs(f,{xi eta},{samp(ig,1) samp(ig,2)}));
dermat =
[diff(N1,xi)  diff(N2,xi)  diff(N3,xi) diff(N4,xi) diff(N5,xi) diff(N6,xi)
 diff(N1,eta) diff(N2,eta) diff(N3,eta) diff(N4,eta) diff(N5,eta)
diff(N6,eta)];
der = double(subs(dermat,{xi eta},{samp(ig,1) samp(ig,2)}));
% end function fmT6_quad
%-------------------------------------------------------------------

function [Tmat] = genT6AXI
syms xi eta ;
N2 = xi;
N3 = eta;
N1 = 1-xi-eta;

Tmat(1,:)= [
double(subs(N1,{xi,eta},[-1/3,-1/3]))
double(subs(N2,{xi,eta},[-1/3,-1/3]))
double(subs(N3,{xi,eta},[-1/3,-1/3]))
];

Tmat(2,:)= [
double(subs(N1,{xi,eta},[2/3,-1/3]))
double(subs(N2,{xi,eta},[2/3,-1/3]))
double(subs(N3,{xi,eta},[2/3,-1/3]))
];

Tmat(3,:)= [
double(subs(N1,{xi,eta},[5/3,-1/3]))
double(subs(N2,{xi,eta},[5/3,-1/3]))
double(subs(N3,{xi,eta},[5/3,-1/3]))
];
```

```
Tmat(4,:)= [
double(subs(N1,{xi,eta},[2/3,2/3]))
double(subs(N2,{xi,eta},[2/3,2/3]))
double(subs(N3,{xi,eta},[2/3,2/3]))
];

Tmat(5,:)= [
double(subs(N1,{xi,eta},[-1/3,5/3]))
double(subs(N2,{xi,eta},[-1/3,5/3]))
double(subs(N3,{xi,eta},[-1/3,5/3]))
];

Tmat(6,:)= [
double(subs(N1,{xi,eta},[-1/3,2/3]))
double(subs(N2,{xi,eta},[-1/3,2/3]))
double(subs(N3,{xi,eta},[-1/3,2/3]))
];
%-----------------------------------------------------------------------
```

A.11. ANALYSIS OF AXISYMMETRIC DOMAIN USING 8-NODE QUADRILATERAL AXISYMMETRIC ELEMENTS

```
clear
clc
format long
inp= fopen('axiQ8.inp','r');
%reading nodal information
[mat] = fscanf(inp,'%f %f  %d %d %f %f %f %d ',8);
Lx = mat(1); %Length of plate
Ly = mat(2); %width of plate
nx = mat(3); %number of divisions along x-direction
ny = mat(4); %number of divisions along y-direction
E  = mat(5);%Modulus of elasticity in kPa
nue= mat(6); %Poisson's ratio
q  = mat(7);  %uniform pressure
nLE = mat(8); %number of loaded edges
[xgrid,ygrid,x, y ,nodes, nNode, nElem] = Q8skewmesh(Lx,Ly,nx,ny,0.0);
node1=nodes(:,1);node2=nodes(:,2);node3= nodes(:,3);node4 = nodes(:,4);
node5=nodes(:,5);node6=nodes(:,6);node7= nodes(:,7);node8 = nodes(:,8);
dx = Lx/nx;
dy = Ly/ny;
%Show Node Labels
for i = 1:nNode
  h=text(x(i),y(i),int2str(i),'fontsize',7,'fontweight','bold','Color','k');
    h.HorizontalAlignment='right';
    h.VerticalAlignment='bottom';
end
%Show Element Labels
for e = 1:nElem
    xloc = (x(nodes(e,1))+x(nodes(e,5)))/2;
    yloc = (y(nodes(e,1))+y(nodes(e,5)))/2;
    text(xloc,yloc,int2str(e),'fontsize',8,'fontweight','bold','Color','k');
end
%determining edges of the plate
bottom = zeros(2*nx+1,1);
top    = zeros(2*nx+1,1);
left   = zeros(2*ny+1,1);
right  = zeros(2*ny+1,1);
iB = 0; iT =0; iL = 0;iR = 0;
for i=1:nNode
    % extracting nodes corresponding to edges
    if(y(i)==0)
        iB = iB+1;
        bottom(iB) = i;
    end
    if(y(i)==Ly)
```

```matlab
        iT = iT+1;
        top(iT) = i;
    end
    if(x(i)==0)
        iL = iL+1;
        left(iL) = i;
    end
    if(x(i)==Lx)
        iR = iR+1;
        right(iR) = i;
    end
end
%----------------------------------------------------------------
%determining the elements with one edge coinciding with the top edge
ntop = size(top,1)-1; % number of elements on top edge
Tedge = zeros(ntop/2,3);
for i=1:ntop/2
    Tedge(i,:)=[top(2*i-1) top(2*i)  top(2*i+1)];
end
%----------------------------------------------------------------
%determining the elements with one edge coinciding with the left edge
nleft = (size(left,1)-1);
Ledge = zeros(nleft/2,3);
for i=1:nleft/2
    Ledge(i,:)=[left(2*i-1) left(2*i)  left(2*i+1)];
end
%----------------------------------------------------------------
%determining the elements with one edge coinciding with the bottom edge
nbottom = (size(bottom,1)-1);
Bedge = zeros(nbottom/2,3);
for i=1:nbottom/2
    Bedge(i,:)=[bottom(2*i-1)  bottom(2*i)  bottom(2*i+1)];
end


%----------------------------------------------------------------
%determining the elements with one edge coinciding with the right edge
nright = (size(right,1)-1);
Redge = zeros(nright/2,3);
for i=1:nright/2
    Redge(i,:)=[right(2*i-1) right(2*i) right(2*i+1)];
end
%----------------------------------------------------------------
D =(E/((1+nue)*(1-2*nue)))* [ 1-nue     nue     nue        0
                                nue     1-nue   nue        0
                                nue      nue    1-nue    0
                                0         0       0      (1-2*nue)/2];
ndof = 2*nNode;
```

```matlab
K = zeros(ndof,ndof); %initialize global stiffness matrix
ngp = 3 ; %Number of Gauss points
samp = gauss(ngp);
for i=1:nElem
    Ke = zeros(16,16);
    B  = zeros(4,16);
    x1 = x(node1(i)); y1 = y(node1(i));
    x2 = x(node2(i)); y2 = y(node2(i));
    x3 = x(node3(i)); y3 = y(node3(i));
    x4 = x(node4(i)); y4 = y(node4(i));
    x5 = x(node5(i)); y5 = y(node5(i));
    x6 = x(node6(i)); y6 = y(node6(i));
    x7 = x(node7(i)); y7 = y(node7(i));
    x8 = x(node8(i)); y8 = y(node8(i));
    coord = [x1 x2 x3 x4 x5 x6 x7 x8
             y1 y2 y3 y4 y5 y6 y7 y8 ]';
    i1 =2*node1(i)-1; j1 = 2*node1(i);
    i2 =2*node2(i)-1; j2 = 2*node2(i);
    i3 =2*node3(i)-1; j3 = 2*node3(i);
    i4 =2*node4(i)-1; j4 = 2*node4(i);
    i5 =2*node5(i)-1; j5 = 2*node5(i);
    i6 =2*node6(i)-1; j6 = 2*node6(i);
    i7 =2*node7(i)-1; j7 = 2*node7(i);
    i8 =2*node8(i)-1; j8 = 2*node8(i);
    dof = [i1 j1 i2 j2 i3 j3 i4 j4 i5 j5 i6 j6 i7 j7 i8 j8];
    for ig=1:ngp
        wi = samp(ig,2);
        for jg=1:ngp
            wj = samp(jg,2);
            [der,fun]=fmquad(samp,ig,jg);
            radius = dot(fun,coord(:,1));
            J =der*coord;
            detJ = det(J);
            deriv = J\der ;
            B(1,1:2:16) = deriv(1,:);
            B(4,2:2:16) = deriv(1,:);
            B(2,2:2:16) = deriv(2,:);
            B(4,1:2:16) = deriv(2,:);
            B(3,1:2:16) = fun/radius;
            Ke = Ke+detJ*wi*wj*B'*D*B*radius;
            %element stiffness matrix by numerical intergration
        end
    end
    K(dof,dof) = K(dof,dof)+Ke;
end
K;
F = zeros(2*nNode,1);
```

```matlab
%applying load at top right corner
tleft = intersect(top,left);   %TL
tright = intersect(top,right); %TR
bleft = intersect(bottom,left); %BL
bright = intersect(bottom,right); %BR
cleft = 0.5*(tleft+bleft); %ML
cright = 0.5*(tright+bright); %MR
if(nLE~=0)
    fprintf(1,'Distributed load is present on the edges\n');
    [mat1] = fscanf(inp,'%d %f %f %f',4);
     Ecode = mat1(1); % code for the edge
     fx = mat1(2);
     fy = mat1(3);
     dist = mat1(4);
     switch Ecode
        case 1 % loading on top edge
            fprintf(1,'Top edge is loaded\n');
            for i=1:ntop/2
                n1 = Tedge(i,1);
                n2 = Tedge(i,2);
                n3 = Tedge(i,3);
                if(x(n3)<=dist)
                    r0 =x(n1);  r1=x(n3);
                    F1 = fy*(r1-r0)*r0/6;
                    F2 = fy*(r1^2-r0^2)/3;
                    F3 = fy*(r1-r0)*r1/6;
                    F(2*n1)   =F(2*n1)+F1;
                    F(2*n2)   =F(2*n2)   +F2;
                    F(2*n3)   =F(2*n3)   +F3;
                end
            end
        case 2 % loading on left edge
            fprintf(1,'Left edge is loaded\n');
            for i=1:nleft
                n1 = Ledge(i,1);
                n2 = Ledge(i,2);
                r0 =x(n1);  r1=x(n3);
                F1 = fx*(r1-r0)*r0/6;
                F2 = fx*(r1^2-r0^2)/3;
                F3 = fx*(r1-r0)*r1/6;
                F(2*n1-1)   =F(2*n1-1)+F1;
                F(2*n2-1)   =F(2*n2-1)+F2;
                F(2*n3-1)   =F(2*n3-1)+F3;
            end
        case 3 % loading on bottom edge
            fprintf(1,'Bottom edge is loaded\n');
            for i=1:nbottom
```

```matlab
            n1 = Bedge(i,1);
            n2 = Bedge(i,2);
            r0 =x(n1); r1=x(n3);
            F1 = fy*(r1-r0)*r0/6;
            F2 = fy*(r1^2-r0^2)/3;
            F3 = fy*(r1-r0)*r1/6;
            F(2*n1)   =F(2*n1)+F1;
            F(2*n2)   =F(2*n2)   +F2;
            F(2*n3)   =F(2*n3)   +F3;
        end
    case 4 % loading on right edge
        fprintf(1,'Right edge is loaded\n');
        for i=1:nright
            n1 = Redge(i,1);
            n2 = Redge(i,2);
            r0 =x(n1); r1=x(n3);
            F1 = fx*(r1-r0)*r0/6;
            F2 = fx*(r1^2-r0^2)/3;
            F3 = fx*(r1-r0)*r1/6;
            F(2*n1-1)   =F(2*n1-1)+F1;
            F(2*n2-1)   =F(2*n2-1)+F2;
            F(2*n3-1)   =F(2*n3-1)+F3;
        end
    end
end

fprintf(1,'enter edge restraints in the order [Bottom Top Left Right]\n');
fprintf(1,'enter 2 for roller edge/1 for unrestrained edge/0 for hinged edge\n');
resB=input('Bottom Edge(2 or 1 or 0): ');
resT=input('Top Edge(2 or 1 or 0): ');
resL=input('Left Edge(2 or 1 or 0): ');
resR=input('Right Edge(2 or 1 or 0): ');
edge =[resB resT resL resR];
%applying boundary conditions
resx = ones(nNode,1);
resy = ones(nNode,1);
for i=1:2*ny+1
    if(edge(3)==0)%left edge hinged
        j = left(i);
        resx(j)=0;
        resy(j)=0;
    elseif (edge(3)==2)%left edge roller (on vertical plane)
        j = left(i);
        resx(j)=0;
        resy(j)=1;
    end
    if(edge(4)==0)%right edge hinged
```

```matlab
        j = right(i);
        resx(j)=0;
        resy(j)=0;
    elseif (edge(4)==2)%right edge roller (on vertical plane)
        j = right(i);
        resx(j)=0;
        resy(j)=1;
    end
end
for i=1:2*nx+1
    if(edge(1)==0)%bottom edge hinged
        j = bottom(i);
        resx(j)=0;
        resy(j)=0; %all dof on bottom edge are restrained
    elseif (edge(1)==2)%bottom edge roller (on horizontal plane)
        j = bottom(i);
        resx(j)=1;
        resy(j)=0;
    end
    if(edge(2)==0)%top edge hinged
        j = top(i);
        resx(j)=0;
        resy(j)=0;%all dof on bottom edge are restrained
    elseif (edge(2)==2)%top edge roller (on horizontal plane)
        j = top(i);
        resx(j)=1;
        resy(j)=0;
    end
end
bn = 0;
for i=1:nNode
    if(resx(i)==0)
        bn = bn+1;
        ifix(bn) = 2*i-1;
    end
    if(resy(i)==0)
        bn = bn+1;
        ifix(bn) = 2*i;
    end
end
ifix = ifix';
K(ifix,:)=zeros(bn,ndof);
K(:,ifix)=zeros(ndof,bn);
K(ifix,ifix)=eye(length(ifix));
F(ifix)=zeros(bn,1);
u =K\F;
sigma = zeros(4,nElem,ngp^2);
```

```matlab
ngp=3; % for computing stresses at Gauss points
samp = gauss(ngp);
for i=1:nElem
    x1 = x(node1(i)); y1 = y(node1(i));
    x2 = x(node2(i)); y2 = y(node2(i));
    x3 = x(node3(i)); y3 = y(node3(i));
    x4 = x(node4(i)); y4 = y(node4(i));
    x5 = x(node5(i)); y5 = y(node5(i));
    x6 = x(node6(i)); y6 = y(node6(i));
    x7 = x(node7(i)); y7 = y(node7(i));
    x8 = x(node8(i)); y8 = y(node8(i));
    coord = [x1 x2 x3 x4 x5 x6 x7 x8
             y1 y2 y3 y4 y5 y6 y7 y8 ]';
    i1 =2*node1(i)-1; j1 = 2*node1(i);
    i2 =2*node2(i)-1; j2 = 2*node2(i);
    i3 =2*node3(i)-1; j3 = 2*node3(i);
    i4 =2*node4(i)-1; j4 = 2*node4(i);
    i5 =2*node5(i)-1; j5 = 2*node5(i);
    i6 =2*node6(i)-1; j6 = 2*node6(i);
    i7 =2*node7(i)-1; j7 = 2*node7(i);
    i8 =2*node8(i)-1; j8 = 2*node8(i);
    dof = [i1 j1 i2 j2 i3 j3 i4 j4 i5 j5 i6 j6 i7 j7 i8 j8];
    k=0;
    for ig=1:ngp
        for jg=1:ngp  % loop over Gauss points
            k=k+1;
            [der,fun]=fmquad(samp,ig,jg);
            radius = dot(fun,coord(:,1));
            J =der*coord;
            detJ = det(J);
            deriv = J\der ;
            B(1,1:2:16)  = deriv(1,:);
            B(4,2:2:16)  = deriv(1,:);
            B(2,2:2:16)  = deriv(2,:);
            B(4,1:2:16)  = deriv(2,:);
            B(3,1:2:16)  = fun/radius;
            sigma(:,i,k)   = D*B*u(dof);
        end
    end
end
ux = u(1:2:2*nNode-1);
uy = u(2:2:2*nNode);
%generate location matrix
mat = zeros(2*ny+1,2*nx+1);
mat(1,:)            = bottom;
```

```matlab
for i=2:2*ny+1
   mat(i,1:2:2*nx+1) = bottom(1:2:2*nx+1)+i-1;
end
k = 0;
for i=3:2:2*ny+1
      k = k+1;
      mat(i,2:2:2*nx+1) = bottom(2:2:2*nx+1)+k;
end
ux1 = zeros(2*ny+1,2*nx+1);
uy1 = zeros(2*ny+1,2*nx+1);
for i=1:2*ny+1
    for j=1:2*nx+1
        nodval = mat(i,j);
        if(nodval==0)
            x1(i,j) = 0.5*(x(mat(i,j-1))+x(mat(i,j+1)));
            y1(i,j) = 0.5*(y(mat(i,j-1))+y(mat(i,j+1)));
            ux1(i,j)= 0.25*(ux(mat(i,j-1))+ux(mat(i,j+1))
                            +ux(mat(i-1,j))+ux(mat(i+1,j)));
            uy1(i,j)=0.25*(uy(mat(i,j-1))+uy(mat(i,j+1))
                            +uy(mat(i-1,j))+uy(mat(i+1,j)));
        else
            x1(i,j) = x(nodval);
            y1(i,j) = y(nodval);
            ux1(i,j)=ux(nodval);
            uy1(i,j)=uy(nodval);
        end
    end
end
optionD = input('Do you want plots on deformed configuration (1 for Yes
0 for No):');
if(optionD==1)
    x2 = x1+ux1;
    y2 = y1+uy1;
end
%---------------------------------
%horizontal displacement
figure(2)
if(optionD==1)
    surf(x2,y2,ux1)
else
    surf(x1,y1,ux1);
end
colormap('jet')
shading interp
cb = colorbar;
val = cb.Limits;
t = linspace(val(1),val(2),8);
```

```matlab
cstr_t = cell(length(t),1);
for ii=1:length(t)
    cstr_t{ii}=num2str(t(ii));
end
tickLabelsToUse = cb.TickLabels;
tickLabelsMod = cellfun(@(c)
sprintf('%.3E',str2double(c)),cstr_t,'uni', 0);
cb.TickLabels = [];
cb.TickLabels = tickLabelsMod;
set(cb,'YTick',t);
pbaspect([Lx/Ly 1 1]);
%------------------------------------
%vertical displacement
figure(3)
if(optionD==1)
    surf(x2,y2,uy1)
else
    surf(x1,y1,uy1);
end
colormap('jet')
shading interp
cb = colorbar;
val = cb.Limits;
t = linspace(val(1),val(2),8);
cstr_t = cell(length(t),1);
for ii=1:length(t)
    cstr_t{ii}=num2str(t(ii));
end
tickLabelsToUse = cb.TickLabels;
tickLabelsMod = cellfun(@(c)
sprintf('%.3E',str2double(c)),cstr_t,'uni', 0);
cb.TickLabels = [];
cb.TickLabels = tickLabelsMod;
set(cb,'YTick',t);
pbaspect([Lx/Ly 1 1]);

%------------------------------------
%preparing contour plot for stresses
Tmat = genTmatA;
sigmaR = zeros(2*ny+1,2*nx+1);
sigmaZ = zeros(2*ny+1,2*nx+1);
sigmaT = zeros(2*ny+1,2*nx+1);
tauRZ  = zeros(2*ny+1,2*nx+1);
for e=1:nElem
%     fprintf(1,'Element = %d\n',e);
 n1 = node1(e); n2 = node2(e); n3 = node3(e) ; n4 = node4(e);
 n5 = node5(e); n6 = node6(e); n7 = node7(e) ; n8 = node8(e);
```

```matlab
sigx =  Tmat*[sigma(1,e,1) sigma(1,e,2) sigma(1,e,3) sigma(1,e,4)...
              sigma(1,e,5) sigma(1,e,6) sigma(1,e,7) sigma(1,e,8)
              sigma(1,e,9)]';
sigy=Tmat*[sigma(2,e,1) sigma(2,e,2) sigma(2,e,3) sigma(2,e,4)...
            sigma(2,e,5) sigma(2,e,6) sigma(2,e,7) sigma(2,e,8)
            sigma(2,e,9)]';
sigt=Tmat*[sigma(3,e,1) sigma(3,e,2) sigma(3,e,3) sigma(3,e,4)...
            sigma(3,e,5) sigma(3,e,6) sigma(3,e,7) sigma(3,e,8)
            sigma(3,e,9)]';
tau=Tmat*[sigma(4,e,1) sigma(4,e,2) sigma(4,e,3) sigma(4,e,4)...
           sigma(4,e,5) sigma(4,e,6) sigma(4,e,7) sigma(4,e,8)
           sigma(4,e,9)]';

[row1, col1] = find(mat == n1);    [row2, col2] = find(mat == n2);
[row3, col3] = find(mat == n3);    [row4, col4] = find(mat == n4);
[row5, col5] = find(mat == n5);    [row6, col6] = find(mat == n6);
[row7, col7] = find(mat == n7);    [row8, col8] = find(mat == n8);
sigmaR(row1,col1) = sigmaR(row1,col1)+sigx(1);
sigmaR(row2,col2) = sigmaR(row2,col2)+sigx(2);
sigmaR(row3,col3) = sigmaR(row3,col3)+sigx(3);
sigmaR(row4,col4) = sigmaR(row4,col4)+sigx(4);
sigmaR(row5,col5) = sigmaR(row5,col5)+sigx(5);
sigmaR(row6,col6) = sigmaR(row6,col6)+sigx(6);
sigmaR(row7,col7) = sigmaR(row7,col7)+sigx(7);
sigmaR(row8,col8) = sigmaR(row8,col8)+sigx(8);

sigmaZ(row1,col1) = sigmaZ(row1,col1)+sigy(1);
sigmaZ(row2,col2) = sigmaZ(row2,col2)+sigy(2);
sigmaZ(row3,col3) = sigmaZ(row3,col3)+sigy(3);
sigmaZ(row4,col4) = sigmaZ(row4,col4)+sigy(4);
sigmaZ(row5,col5) = sigmaZ(row5,col5)+sigy(5);
sigmaZ(row6,col6) = sigmaZ(row6,col6)+sigy(6);
sigmaZ(row7,col7) = sigmaZ(row7,col7)+sigy(7);
sigmaZ(row8,col8) = sigmaZ(row8,col8)+sigy(8);

sigmaT(row1,col1) = sigmaT(row1,col1)+sigt(1);
sigmaT(row2,col2) = sigmaT(row2,col2)+sigt(2);
sigmaT(row3,col3) = sigmaT(row3,col3)+sigt(3);
sigmaT(row4,col4) = sigmaT(row4,col4)+sigt(4);
sigmaT(row5,col5) = sigmaT(row5,col5)+sigt(5);
sigmaT(row6,col6) = sigmaT(row6,col6)+sigt(6);
sigmaT(row7,col7) = sigmaT(row7,col7)+sigt(7);
sigmaT(row8,col8) = sigmaT(row8,col8)+sigt(8);

tauRZ(row1,col1)  = tauRZ(row1,col1) +tau(1);
tauRZ(row2,col2)  = tauRZ(row2,col2) +tau(2);
tauRZ(row3,col3)  = tauRZ(row3,col3) +tau(3);
```

```
 tauRZ(row4,col4)    = tauRZ(row4,col4) +tau(4);
 tauRZ(row5,col5)    = tauRZ(row5,col5) +tau(5);
 tauRZ(row6,col6)    = tauRZ(row6,col6) +tau(6);
 tauRZ(row7,col7)    = tauRZ(row7,col7) +tau(7);
 tauRZ(row8,col8)    = tauRZ(row8,col8) +tau(8);
end

%to count number of elements contributing to a node
for i=1:2*ny+1
    for j=1:2*nx+1
        gridnode = mat(i,j);
        m = 0;
        for i1=1:nElem
            for j1=1:8
                if(nodes(i1,j1)==gridnode)
                    m = m+1;
                end
            end
        end
        sigmaR(i,j)  = sigmaR(i,j)/m;
        sigmaZ(i,j)  = sigmaZ(i,j)/m;
        sigmaT(i,j)  = sigmaT(i,j)/m;
        tauRZ(i,j)   = tauRZ(i,j)/m;
    end
end
[m1, n1] = find(isnan(sigmaR));
for e=1:nElem
    i=m1(e);
    j=n1(e);
    sigmaR(i,j)=(sigmaR(i-1,j)+sigmaR(i+1,j)+sigmaR(i,j-1)+sigmaR(i,j+1)...
        +sigmaR(i-1,j-1)+sigmaR(i+1,j+1)+sigmaR(i-1,j+1)+sigmaR(i+1,j-1))/8;
    sigmaZ(i,j)=(sigmaZ(i-1,j)+sigmaZ(i+1,j)+sigmaZ(i,j-1)+sigmaZ(i,j+1)...
        +sigmaZ(i-1,j-1)+sigmaZ(i+1,j+1)+sigmaZ(i-1,j+1)+sigmaZ(i+1,j-1))/8;
    sigmaT(i,j)=(sigmaT(i-1,j)+sigmaT(i+1,j)+sigmaT(i,j-1)+sigmaT(i,j+1)...
        +sigmaT(i-1,j-1)+sigmaT(i+1,j+1)+sigmaT(i-1,j+1)+sigmaT(i+1,j-1))/8;
    tauRZ(i,j)=(tauRZ(i-1,j)+tauRZ(i+1,j)+tauRZ(i,j-1)+tauRZ(i,j+1)...
        +tauRZ(i-1,j-1)+tauRZ(i+1,j+1)+tauRZ(i-1,j+1)+tauRZ(i+1,j-1))/8;
end

figure(4)
if(optionD==1)
    surf(x2,y2,sigmaR)
else
    surf(x1,y1,sigmaR);
end
colormap('jet')
shading interp
```

```matlab
cb = colorbar;
val = cb.Limits;
t = linspace(val(1),val(2),8);
t = linspace(val(1),val(2),8);
cb.FontWeight = 'bold';
for ii=1:length(t)
    cstr_t{ii}=num2str(t(ii));
end
tickLabelsToUse = cb.TickLabels;
tickLabelsMod = cellfun(@(c)sprintf('%.1f',str2double(c)),cstr_t, 'uni', 0);
cb.TickLabels = [];
cb.TickLabels = tickLabelsMod;
set(cb,'YTick',t);
pbaspect([Lx/Ly 1 1]);

figure(5)
if(optionD==1)
    surf(x2,y2,sigmaZ)
else
    surf(x1,y1,sigmaZ);
end
colormap('jet')
shading interp
cb = colorbar;
val = cb.Limits;
t = linspace(val(1),val(2),8);
cb.FontWeight = 'bold';
for ii=1:length(t)
    cstr_t{ii}=num2str(t(ii));
end
tickLabelsToUse = cb.TickLabels;
tickLabelsMod=cellfun(@(c)sprintf('%.1f',str2double(c)),cstr_t,'uni', 0);
cb.TickLabels = [];
cb.TickLabels = tickLabelsMod;
set(cb,'YTick',t);
pbaspect([Lx/Ly 1 1]);

figure(6)
if(optionD==1)
    surf(x2,y2,sigmaT)
else
    surf(x1,y1,sigmaT);
end
colormap('jet')
shading interp
cb = colorbar;
val = cb.Limits;
```

```matlab
t = linspace(val(1),val(2),8);
t = linspace(val(1),val(2),8);
cb.FontWeight = 'bold';
for ii=1:length(t)
    cstr_t{ii}=num2str(t(ii));
end
tickLabelsToUse = cb.TickLabels;
tickLabelsMod=cellfun(@(c)sprintf('%.1f',str2double(c)),cstr_t,'uni', 0);
cb.TickLabels = [];
cb.TickLabels = tickLabelsMod;
set(cb,'YTick',t);
pbaspect([Lx/Ly 1 1]);

figure(7)
if(optionD==1)
    surf(x2,y2,tauRZ)
else
    surf(x1,y1,tauRZ);
end
colormap('jet')
shading interp
cb = colorbar;
val = cb.Limits;
t = linspace(val(1),val(2),8);
t = linspace(val(1),val(2),8);
cb.FontWeight = 'bold';
for ii=1:length(t)
    cstr_t{ii}=num2str(t(ii));
end
tickLabelsToUse = cb.TickLabels;
tickLabelsMod = cellfun(@(c)
sprintf('%.1f',str2double(c)),cstr_t,'uni', 0);
cb.TickLabels = [];
cb.TickLabels = tickLabelsMod;
set(cb,'YTick',t);
pbaspect([Lx/Ly 1 1]);

%-----------------------------------------------------------------------
function [Tmat] = genTmatA
syms xi eta ;

f1 = 0.5*xi*(xi-1); f2 = 1-xi^2 ; f3 = 0.5*xi*(xi+1);
g1 = 0.5*eta*(eta-1); g2 = 1-eta^2 ; g3 = 0.5*eta*(eta+1);

N1 = f1*g1; N2 = f1*g2 ; N3 = f1*g3;
N4 = f2*g1; N5 = f2*g2 ; N6 = f2*g3;
N7 = f3*g1; N8 = f3*g2 ; N9 = f3*g3;
```

```
Tmat = zeros(8,9);
v = sqrt(5/3);

Tmat(1,:)= [
double(subs(N1,{xi,eta},[-v,-v]))
double(subs(N2,{xi,eta},[-v, -v]))
double(subs(N3,{xi,eta},[-v, -v]))
double(subs(N4,{xi,eta},[-v,-v]))
double(subs(N5,{xi,eta},[-v,-v]))
double(subs(N6,{xi,eta},[-v,-v]))
double(subs(N7,{xi,eta},[-v,-v]))
double(subs(N8,{xi,eta},[-v,-v]))
double(subs(N9,{xi,eta},[-v,-v]))];

Tmat(2,:)= [
double(subs(N1,{xi,eta},[-v,0]))
double(subs(N2,{xi,eta},[-v,0]))
double(subs(N3,{xi,eta},[-v,0]))
double(subs(N4,{xi,eta},[-v,0]))
double(subs(N5,{xi,eta},[-v,0]))
double(subs(N6,{xi,eta},[-v,0]))
double(subs(N7,{xi,eta},[-v,0]))
double(subs(N8,{xi,eta},[-v,0]))
double(subs(N9,{xi,eta},[-v,0]))];


Tmat(3,:)= [
double(subs(N1,{xi,eta},[v,-v]))
double(subs(N2,{xi,eta},[v,-v]))
double(subs(N3,{xi,eta},[v,-v]))
double(subs(N4,{xi,eta},[v,-v]))
double(subs(N5,{xi,eta},[v,-v]))
double(subs(N6,{xi,eta},[v,-v]))
double(subs(N7,{xi,eta},[v,-v]))
double(subs(N8,{xi,eta},[v,-v]))
double(subs(N9,{xi,eta},[v,-v]))];


Tmat(4,:)= [
double(subs(N1,{xi,eta},[ v, 0]))
double(subs(N2,{xi,eta},[ v, 0]))
double(subs(N3,{xi,eta},[ v, 0]))
double(subs(N4,{xi,eta},[ v, 0]))
double(subs(N5,{xi,eta},[ v, 0]))
double(subs(N6,{xi,eta},[ v, 0]))
double(subs(N7,{xi,eta},[ v, 0]))
double(subs(N8,{xi,eta},[ v, 0]))
```

```
double(subs(N9,{xi,eta},[ v, 0])))];
Tmat(5,:)= [
double(subs(N1,{xi,eta},[v,v]))
double(subs(N2,{xi,eta},[v,v]))
double(subs(N3,{xi,eta},[v,v]))
double(subs(N4,{xi,eta},[v,v]))
double(subs(N5,{xi,eta},[v,v]))
double(subs(N6,{xi,eta},[v,v]))
double(subs(N7,{xi,eta},[v,v]))
double(subs(N8,{xi,eta},[v,v]))
double(subs(N9,{xi,eta},[v,v])))];

Tmat(6,:)= [
double(subs(N1,{xi,eta},[0,v]))
double(subs(N2,{xi,eta},[0,v]))
double(subs(N3,{xi,eta},[0,v]))
double(subs(N4,{xi,eta},[0,v]))
double(subs(N5,{xi,eta},[0,v]))
double(subs(N6,{xi,eta},[0,v]))
double(subs(N7,{xi,eta},[0,v]))
double(subs(N8,{xi,eta},[0,v]))
double(subs(N9,{xi,eta},[0,v])))];


Tmat(7,:)= [
double(subs(N1,{xi,eta},[-v,v]))
double(subs(N2,{xi,eta},[-v,v]))
double(subs(N3,{xi,eta},[-v,v]))
double(subs(N4,{xi,eta},[-v,v]))
double(subs(N5,{xi,eta},[-v,v]))
double(subs(N6,{xi,eta},[-v,v]))
double(subs(N7,{xi,eta},[-v,v]))
double(subs(N8,{xi,eta},[-v,v]))
double(subs(N9,{xi,eta},[-v,v])))];

Tmat(8,:)= [
double(subs(N1,{xi,eta},[-v, 0]))
double(subs(N2,{xi,eta},[-v, 0]))
double(subs(N3,{xi,eta},[-v, 0]))
double(subs(N4,{xi,eta},[-v, 0]))
double(subs(N5,{xi,eta},[-v, 0]))
double(subs(N6,{xi,eta},[-v, 0]))
double(subs(N7,{xi,eta},[-v, 0]))
double(subs(N8,{xi,eta},[-v, 0]))
double(subs(N9,{xi,eta},[-v, 0])))];
%--------------------------------------------------------------------------
```

A.12. ANALYSIS OF PLATES USING ACM ELEMENTS

```matlab
clear
clc
format long
inp= fopen('ACMplate.inp','r');
%reading nodal information
[mat] = fscanf(inp,'%f %f  %f %d %d %f %f %f %f %d %f',10);
Lx = mat(1); %Length of plate
Ly = mat(2); %width of plate
thick  = mat(3); %thickness of plate
nx = mat(4); %number of divisions along x-direction
ny = mat(5); %number of divisions along y-direction
E  = mat(6);%Modulus of elasticity in kPa
rho= mat(7);
nue= mat(8); %Poisson's ratio
q  = mat(9);  %uniform pressure
nLC = mat(10); %number of loaded nodes
[xgrid,ygrid,x, y ,nodes, nNode, nElem] = Q4mesh(Lx,Ly,nx,ny);
node1=nodes(:,1);node2=nodes(:,2);node3=nodes(:,3);node4 = nodes(:,4);
%Show Node Labels
for i = 1:nNode
    h=text(x(i),y(i),int2str(i),'fontsize',7,'fontweight','bold','Color','k');
    h.HorizontalAlignment='right';
    h.VerticalAlignment='bottom';
end
%Show Element Labels
for e = 1:nElem
    xloc = (x(nodes(e,1))+x(nodes(e,3)))/2;
    yloc = (y(nodes(e,1))+y(nodes(e,3)))/2;
    text(xloc,yloc,int2str(e),'fontsize',8,'fontweight','bold','Color','k');
end
%determining edges of the plate
bottom = zeros(nx+1,1);
top     = zeros(nx+1,1);
left    = zeros(ny+1,1);
right   = zeros(ny+1,1);
iB = 0; iT =0; iL = 0;iR = 0;
inner = 0;
for i=1:nNode
    % extracting nodes corresponding to edges
    if (y(i)==Ly)
        iT = iT+1;
        top(iT) = i;
    end
    if (y(i)==0)
        iB = iB+1;
```

```matlab
            bottom(iB) = i;
        end
        if(x(i)==0)
            iL = iL+1;
            left(iL) = i;
        end
        if(x(i)==Lx)
            iR = iR+1;
            right(iR) = i;
        end
    end
end


%-----------------------------------------------------------------------
ne = 1:nElem;

ndof = 3*nNode; %total dof for the entire structure
K = zeros(ndof,ndof); %initialize global stiffness matrix

ngp = 3 ; %Number of Gauss points
samp = gauss(ngp);
syms xa ya
poly=[1,xa,ya,xa^2,xa*ya,ya^2,xa^3,xa^2*ya,xa*ya^2,ya^3,xa^3*ya,xa*ya^
3];
node = [poly; diff(poly,xa); diff(poly,ya)];
A=ones(12,12);
a = Lx/(2*nx);
b = Ly/(2*ny);
A(1:3,:)   = subs(node,{xa,ya},{-a,-b});
A(4:6,:)   = subs(node,{xa,ya},{ a,-b});
A(7:9,:)   = subs(node,{xa,ya},{ a, b});
A(10:12,:)= subs(node,{xa,ya},{-a, b});

H=sym(ones(3,12));
H(1,:) =    -diff(diff(poly,xa),xa);
H(2,:) =    -diff(diff(poly,ya),ya);
H(3,:) = -2*diff(diff(poly,xa),ya);
D=E*thick^3/(12*(1-nue^2))*[1   nue   0
                            nue 1     0
                            0   0   (1-nue)/2];
bmat = H*inv(A);
nmat = poly*inv(A);
kmat = bmat'*D*bmat;
F = zeros(3*nNode,1);
Fe =double(int((int(nmat'*q,xa,-a,a)),ya,-b,b));
```

```matlab
for i=1:nElem
    B  = zeros(3,12);
    x1 = x(node1(i)); y1 = y(node1(i));
    x2 = x(node2(i)); y2 = y(node2(i));
    x3 = x(node3(i)); y3 = y(node3(i));
    x4 = x(node4(i)); y4 = y(node4(i));
    coord = [x1 x2 x3 x4
             y1 y2 y3 y4 ]';
    % dof are in the order w theta-x theta-y
    i1 =3*node1(i)-2; j1 = 3*node1(i)-1; k1 = 3*node1(i);
    i2 =3*node2(i)-2; j2 = 3*node2(i)-1; k2 = 3*node2(i);
    i3 =3*node3(i)-2; j3 = 3*node3(i)-1; k3 = 3*node3(i);
    i4 =3*node4(i)-2; j4 = 3*node4(i)-1; k4 = 3*node4(i);
    dof = [i1 j1 k1 i2 j2 k2 i3 j3 k3 i4 j4 k4];
    if(i==1)
        %Note: Element stiffness matrix is computed only once as all
        %       elements are identical
        Ke = zeros(12,12);
        for ig=1:ngp  % loop over Gauss points
         wi = samp(ig,2);
         for jg=1:ngp
             wj = samp(jg,2);
              Ke= Ke+a*b*wi*wj*
              double(subs(kmat,{xa,ya},{a*samp(ig,1),b*samp(jg,1)}));
            end
        end
    end
    K(dof,dof) = K(dof,dof)+Ke;
    F(dof)= F(dof)+Fe;
end
tleft = intersect(top,left);   %TL
tright = intersect(top,right); %TR
bleft = intersect(bottom,left); %BL
bright = intersect(bottom,right); %BR
cleft = 0.5*(tleft+bleft); %ML
cright = 0.5*(tright+bright); %MR
center = 0.5*(cleft+cright); %CP
if(nLC~=0)
    for i=1:nLC
        fprintf(1,'Point loads are present at specified nodes\n');
        [mat1] = fscanf(inp,'%d %f',2);
        Ncode = mat1(1); %code for the node
        Fn = mat1(2);
        switch Ncode
            case 1  %Top left corner
                F(3*tleft-2)=F(3*tleft-2)+Fn;
```

```matlab
        case 2  % Centre left corner
            F(3*cleft-2) = F(3*cleft-2)+Fn;
        case 3 %Bottom left corner
            F(3*bleft-2)=F(3*bleft-2)+Fn;
        case 4   %Bottom right corner
            F(3*tright-2)=F(3*tright-2)+Fn;
        case 5   %Centre right corner
            F(3*cright-2)=F(3*cright-2)+Fn;
        case 6   %Bottom right corner
            F(3*bright-2)=F(3*bright-2)+Fn;
        case 7   %centre of the plate
            F(3*center-2)=F(3*center-2)+Fn;
        end
    end
end
fprintf(1,'enter edge restraints in the order [Bottom Top Left Right]\n');
fprintf(1,'enter 0 for clamped edge\n');
fprintf(1,'enter 1 for simply-supported edge\n');
fprintf(1,'enter 2 for symmetric edge condition\n');
resB=input('Bottom Edge(0/1/2): ');
resT=input('Top Edge(0/1/2): ');
resL=input('Left Edge(0/1/2): ');
resR=input('Right Edge(0/1/2): ');
edge =[resB resT resL resR];
%applying boundary conditions
resx = ones(nNode,1);
resy = ones(nNode,1);
resz = ones(nNode,1);
for i=1:nx+1
    j=bottom(i);
    if(edge(1)==0)%bottom edge clamped
        resx(j)=0;%all dof on bottom edge are restrained
        resy(j)=0;
        resz(j)=0;
    elseif(edge(1)==1)% bottom edge is simply-supported
        resz(j) = 0;
        resx(j) = 0;
    elseif(edge(1)==2);%bottom edge symmetric
        resy(j) = 0;
    end
end
for i=1:nx+1
    j = top(i);
    if(edge(2)==0)%top edge clamped
        resx(j)=0;%all dof on bottom edge are restrained
        resy(j)=0;
```

```matlab
            resz(j)=0;
        elseif(edge(2)==1)%top edge simply-supported
            resz(j) = 0;
            resx(j) = 0;
        elseif(edge(2)==2)%top edge symmetric
            resy(j) = 0;
        end
    end
    for i=1:ny+1
        j = left(i);
        if(edge(3)==0)%left edge clamped
            resx(j)=0;
            resy(j)=0;
            resz(j)=0;
        elseif(edge(3)==1)%left edge simply-supported
            resz(j) = 0;
            resy(j) = 0;
        elseif(edge(3)==2)%left edge symmetric
            resx(j) = 0;
        end
    end
    for i=1:ny+1
        j = right(i);
        if(edge(4)==0)%right edge clamped
            resx(j)=0;
            resy(j)=0;
            resz(j)=0;
        elseif(edge(4)==1)%right edge simply-supported
            resz(j) = 0;
            resy(j) = 0;
        elseif(edge(4)==2) %right edge symmetric
            resx(j) = 0;
        end
    end
    bn = 0;
    for i=1:nNode
        if(resz(i)==0)
            bn = bn+1;
            ifix(bn) = 3*i-2;
        end
        if(resx(i)==0)
            bn = bn+1;
            ifix(bn) = 3*i-1;
        end
        if(resy(i)==0)
            bn = bn+1;
```

```matlab
        ifix(bn) = 3*i;
    end
end
ifix = ifix';
K(ifix,:)=zeros(bn,ndof);
K(:,ifix)=zeros(ndof,bn);
K(ifix,ifix)=eye(length(ifix));
F(ifix)=zeros(bn,1);
u =K\F;
ngp=1; % for computing stresses at Gauss points
sigma = zeros(3,nElem,ngp^2);
samp = gauss(ngp);
ab= a*b;
BB =
0.25*[ 0    0    -1/a   0      0    1/a   0    0    1/a   0      0    -1/a
        0   -1/b   0     0    -1/b   0    0   1/b   0     0     1/b    0
       4/ab 1/a   1/b  -4/ab  -1/a  1/b  4/ab -1/a  -1/b -4/ab 1/a   -1/b];
for i=1:nElem
    % dof are in the order w theta-x theta-y
    i1 =3*node1(i)-2; j1 = 3*node1(i)-1; k1 = 3*node1(i);
    i2 =3*node2(i)-2; j2 = 3*node2(i)-1; k2 = 3*node2(i);
    i3 =3*node3(i)-2; j3 = 3*node3(i)-1; k3 = 3*node3(i);
    i4 =3*node4(i)-2; j4 = 3*node4(i)-1; k4 = 3*node4(i);
    dof = [i1 j1 k1 i2 j2 k2 i3 j3 k3 i4 j4 k4];
    k=0;
    for ig=1:ngp  % loop over Gauss points
        for jg=1:ngp
            B = double(subs(bmat,{xa,ya},{a*samp(ig,1),b*samp(jg,1)}));
        end
    end
    sigma(:,i) =D*B*u(dof);
end
w = u(1:3:3*nNode-2);
w1 = reshape(w,ny+1,nx+1);
dx = Lx/nx;
dy = Ly/ny;
x1 =reshape(x,ny+1,nx+1);
y1 =reshape(y,ny+1,nx+1);
%------------------------------------
%z-translation
figure(2)
surf(x1,y1,w1);
title('Variation of vertical displacement','fontweight','bold','fontsize',16);
xlabel('Length of plate (m)','fontweight','bold','fontsize',10);
ylabel('Depth of plate (m)','fontweight','bold','fontsize',10);
ax = gca;
```

```matlab
ax.FontSize = 11;
ax.FontWeight = 'bold';
colormap('jet')
shading interp
cb = colorbar;
val = cb.Limits;
t = linspace(val(1),val(2),8);
cstr_t = cell(length(t),1);
for ii=1:length(t)
    cstr_t{ii}=num2str(t(ii));
end
tickLabelsToUse = cb.TickLabels;
tickLabelsMod = cellfun(@(c) sprintf('%.3E',str2double(c)) ,cstr_t, 'uni', 0);
cb.TickLabels = [];
cb.TickLabels = tickLabelsMod;
set(cb,'YTick',t);
val = abs(min(w))/min(Lx,Ly);
if(val<0.01)
    fprintf(1,'maximum deflection is within limits\n');
else
    fprintf(1,'large deflection. please check again\n');
end
%-----------------------------------
%theta-x
thetax = u(2:3:3*nNode-1);
thetax1 = reshape(thetax,ny+1,nx+1);
figure(3)
surf(x1,y1,thetax1)
title('Variation of {\theta_{x}}','fontweight','bold','fontsize',16);
xlabel('Length of plate (m)','fontweight','bold','fontsize',10);
ylabel('Depth of plate (m)','fontweight','bold','fontsize',10);
ax = gca;
ax.FontSize = 11;
ax.FontWeight = 'bold';
colormap('jet')
shading interp
cb = colorbar;
val = cb.Limits;
t = linspace(val(1),val(2),8);
cstr_t = cell(length(t),1);
for ii=1:length(t)
    cstr_t{ii}=num2str(t(ii));
end
tickLabelsToUse = cb.TickLabels;
tickLabelsMod = cellfun(@(c) sprintf('%.3E',str2double(c)) ,cstr_t, 'uni', 0);
cb.TickLabels = [];
```

```matlab
cb.TickLabels = tickLabelsMod;
set(cb,'YTick',t);
%----------------------------------
%theta-y
thetay = u(3:3:3*nNode);
thetay1 = reshape(thetay,ny+1,nx+1);
figure(4)
surf(x1,y1,thetay1)
title('Variation of {\theta_{y}}','fontweight','bold','fontsize',16);
xlabel('Length of plate (m)','fontweight','bold','fontsize',10);
ylabel('Depth of plate (m)','fontweight','bold','fontsize',10);
ax = gca;
ax.FontSize = 11;
ax.FontWeight = 'bold';
colormap('jet')
shading interp
cb = colorbar;
val = cb.Limits;
t = linspace(val(1),val(2),8);
cstr_t = cell(length(t),1);
for ii=1:length(t)
    cstr_t{ii}=num2str(t(ii));
end
tickLabelsToUse = cb.TickLabels;
tickLabelsMod = cellfun(@(c) sprintf('%.3E',str2double(c)) ,cstr_t, 'uni', 0);
cb.TickLabels = [];
cb.TickLabels = tickLabelsMod;
set(cb,'YTick',t);
%preparing contour plot for stresses
for i=1:ny+1
    nodemat(i,:) = left(i):ny+1:right(i);
end
MXX = zeros(ny+1,nx+1);
MYY = zeros(ny+1,nx+1);
MXY  = zeros(ny+1,nx+1);
count = zeros(ny+1,nx+1);
for e=1:nElem
   n1 = node1(e); n2 = node2(e); n3 = node3(e) ; n4 = node4(e);
   [row1, col1] = find(nodemat == n1);
   [row2, col2] = find(nodemat == n2);
   [row3, col3] = find(nodemat == n3);
   [row4, col4] = find(nodemat == n4);
   MXX(row1,col1) = MXX(row1,col1)+sigma(1,e);
   MXX(row2,col2) = MXX(row2,col2)+sigma(1,e);
   MXX(row3,col3) = MXX(row3,col3)+sigma(1,e);
   MXX(row4,col4) = MXX(row4,col4)+sigma(1,e);
```

```matlab
    MYY(row1,col1)  = MYY(row1,col1)+sigma(2,e);
    MYY(row2,col2)  = MYY(row2,col2)+sigma(2,e);
    MYY(row3,col3)  = MYY(row3,col3)+sigma(2,e);
    MYY(row4,col4)  = MYY(row4,col4)+sigma(2,e);

    MXY(row1,col1)   = MXY(row1,col1) +sigma(3,e);
    MXY(row2,col2)   = MXY(row2,col2) +sigma(3,e);
    MXY(row3,col3)   = MXY(row3,col3) +sigma(3,e);
    MXY(row4,col4)   = MXY(row4,col4) +sigma(3,e);

    count(row1,col1)   = count(row1,col1)+1;
    count(row2,col2)   = count(row2,col2)+1;
    count(row3,col3)   = count(row3,col3)+1;
    count(row4,col4)   = count(row4,col4)+1;
end
MXX = MXX./count;
MYY = MYY./count;
MXY = MXY./count;
figure(5)
surf(x1,y1,MXX);
title('Variation of M_{XX}','fontweight','bold','fontsize',16);
xlabel('Length of plate (m)','fontweight','bold','fontsize',10);
ylabel('Depth of plate (m)','fontweight','bold','fontsize',10);
ax = gca;
ax.FontSize = 11;
ax.FontWeight = 'bold';
colormap('jet')
shading interp
cb = colorbar;
val = cb.Limits;
t = linspace(val(1),val(2),8);
set(cb,'YTick',t);
cb.FontWeight = 'bold';
for ii=1:length(t)
    cstr_t{ii}=num2str(t(ii));
end
tickLabelsToUse = cb.TickLabels;
tickLabelsMod  =  cellfun(@(c)  sprintf('%.1f',str2double(c))  ,cstr_t,
'uni', 0);
cb.TickLabels = [];
cb.TickLabels = tickLabelsMod;
set(cb,'YTick',t);

figure(6)
surf(x1,y1,MYY);
```

```matlab
title('Variation of M_{YY}','fontweight','bold','fontsize',16);
xlabel('Length of plate (mm)','fontweight','bold','fontsize',10);
ylabel('Depth of plate (mm)','fontweight','bold','fontsize',10);
ax = gca;
ax.FontSize = 11;
ax.FontWeight = 'bold';
colormap('jet')
shading interp
cb = colorbar;
val = cb.Limits;
t = linspace(val(1),val(2),8);
set(cb,'YTick',t);
cb.FontWeight = 'bold';
for ii=1:length(t)
    cstr_t{ii}=num2str(t(ii));
end
tickLabelsToUse = cb.TickLabels;
tickLabelsMod = cellfun(@(c) sprintf('%.1f',str2double(c)) ,cstr_t,
'uni', 0);
cb.TickLabels = [];
cb.TickLabels = tickLabelsMod;
set(cb,'YTick',t);
figure(7)
surf(x1,y1,MXY);
title('Variation of M_{XY}','fontweight','bold','fontsize',16);
xlabel('Length of plate (mm)','fontweight','bold','fontsize',10);
ylabel('Depth of plate (mm)','fontweight','bold','fontsize',10);
ax = gca;
ax.FontSize = 11;
ax.FontWeight = 'bold';
colormap('jet')
shading interp
cb = colorbar;
val = cb.Limits;
t = linspace(val(1),val(2),8);
cb.FontWeight = 'bold';
for ii=1:length(t)
    cstr_t{ii}=num2str(t(ii));
end
tickLabelsToUse = cb.TickLabels;
tickLabelsMod = cellfun(@(c) sprintf('%.1f',str2double(c)) ,cstr_t,
'uni', 0);
cb.TickLabels = [];
cb.TickLabels = tickLabelsMod;
set(cb,'YTick',t);
```

A.13. ANALYSIS OF PLATES USING 4-NODE QUADRILATERAL ELEMENTS

```matlab
%FINITE ELEMENT ANALYSIS OF A SKEW PLATE SUBJECTED TO NORMAL PRESSURE
%USING S4R5 ELEMENTS
%This version works also for quarter plates under conditions of symmetry
clear
clc
format long
inp= fopen('S4plate.inp','r');
%reading nodal information
[mat] = fscanf(inp,'%f %f  %f %d %d %f %f %f %f %d %f %f',11);
Lx = mat(1); %Length of plate
Ly = mat(2); %width of plate
thick  = mat(3); %thickness of plate
nx = mat(4); %number of divisions along x-direction
ny = mat(5); %number of divisions along y-direction
E  = mat(6);%Modulus of elasticity in kPa
rho= mat(7);
nue= mat(8); %Poisson's ratio
q  = mat(9);   %uniform pressure
nLC = mat(10); %number of loaded nodes
alpha=mat(11); %skew angle
[xgrid,ygrid,x, y ,nodes, nNode, nElem] = Q4skewmesh(Lx,Ly,nx,ny,alpha);
node1 = nodes(:,1); node2 = nodes(:,2); node3 = nodes(:,3);node4 = nodes(:,4);

%Show Node Labels
for i = 1:nNode
    h=text(x(i),y(i),int2str(i),'fontsize',7,'fontweight','bold','Color','k');
     h.HorizontalAlignment='right';
     h.VerticalAlignment='bottom';
end
%Show Element Labels
for e = 1:nElem
    xloc = (x(nodes(e,1))+x(nodes(e,3)))/2;
    yloc = (y(nodes(e,1))+y(nodes(e,3)))/2;
    text(xloc,yloc,int2str(e),'fontsize',8,'fontweight','bold','Color','k');
end
%determining edges of the plate
bottom = zeros(nx+1,1);
top    = zeros(nx+1,1);
left   = zeros(ny+1,1);
right  = zeros(ny+1,1);
iB = 0; iT =0; iL = 0;iR = 0;
H = Lx*tand(alpha);
dky = H/nx;
```

```matlab
for i=1:nNode
    % extracting nodes corresponding to edges
    if (y(i)==0||(rem(y(i),dky)==0))
        iB = iB+1;
        bottom(iB) = i;
    end
    if(x(i)==0)
        iL = iL+1;
        left(iL) = i;
    end
    if(x(i)==Lx)
        iR = iR+1;
        right(iR) = i;
    end
end
i=0;
for j=1:nNode
  if(y(j) == (Ly+y(bottom(i+1))))
        i = i+1;
        top(i) = j;
  end
end
%-----------------------------------------------------------------------
ne = 1:nElem;

ndof = 3*nNode; %total dof for the entire structure
K = zeros(ndof,ndof); %initialize global stiffness matrix

ngpb = 2 ; %Number of Gauss points
ngps = 1 ; %for reduced integration
sampb = gauss(ngpb);
samps = gauss(ngps);
D=(E*thick^3/(12*(1-nue^2)))*[1   nue   0
                              nue  1    0
                              0    0  (1-nue)/2];
G = E/(2*(1+nue));
Ds = G*[thick 0
          0    thick];
a = Lx/(2*nx);
b = Ly/(2*ny);
F = zeros(3*nNode,1);
for i=1:nElem
    Keb = zeros(12,12); %bending
    Kes = zeros(12,12); %shear
    B  = zeros(3,12); Bs = zeros(2,12);
    Fe = zeros(12,1);
```

```matlab
    x1 = x(node1(i)); y1 = y(node1(i));
    x2 = x(node2(i)); y2 = y(node2(i));
    x3 = x(node3(i)); y3 = y(node3(i));
    x4 = x(node4(i)); y4 = y(node4(i));
    coord = [x1 x2 x3 x4
             y1 y2 y3 y4 ]';
    % dof are in the order w theta-x theta-y
    i1 =3*node1(i)-2; j1 = 3*node1(i)-1; k1 = 3*node1(i);
    i2 =3*node2(i)-2; j2 = 3*node2(i)-1; k2 = 3*node2(i);
    i3 =3*node3(i)-2; j3 = 3*node3(i)-1; k3 = 3*node3(i);
    i4 =3*node4(i)-2; j4 = 3*node4(i)-1; k4 = 3*node4(i);
    dof = [i1 j1 k1 i2 j2 k2  i3 j3 k3 i4 j4 k4];
    for ig=1:ngpb  % loop over Gauss points
        wi = sampb(ig,2);
        for jg=1:ngpb
            wj = sampb(jg,2);
            [der,fun]=smlin(sampb,ig,jg);
            J =der*coord;
            detJ = det(J);
            deriv = J\der ;
            B(1,2:3:11) = -deriv(1,:);
            B(2,3:3:12) = -deriv(2,:);
            B(3,2:3:11) = -deriv(2,:);
            B(3,3:3:12) = -deriv(1,:);
            Keb = Keb+detJ*wi*wj*B'*D*B;
            %element stiffness matrix by numerical intergration
        Fe=Fe+detJ*wi*wj*q*[fun(1)0 0 fun(2) 0 0 fun(3) 0 0 fun(4)0 0]';
        end
    end
    for ig=1:ngps  % loop over Gauss points REDUCED INTEGRATION
        wi = samps(ig,2);
        for jg=1:ngps
            wj = samps(jg,2);
            [der,fun]=smlin(samps,ig,jg);
            J =der*coord;
            detJ = det(J);
            deriv = J\der ;
            Bs(1,1:3:10) = -deriv(2,:); Bs(1,3:3:12) = fun;
            Bs(2,1:3:10) = -deriv(1,:); Bs(2,2:3:11) = fun;
            Kes = Kes+(5/6)*detJ*wi*wj*Bs'*Ds*Bs;
            %element stiffness matrix by numerical intergration
        end
    end
    K(dof,dof) = K(dof,dof)+Keb+Kes;
    F(dof)     = F(dof)+Fe;
end
```

```matlab
tleft = intersect(top,left);   %TL
tright = intersect(top,right); %TR
bleft = intersect(bottom,left); %BL
bright = intersect(bottom,right); %BR
cleft = 0.5*(tleft+bleft); %ML
cright = 0.5*(tright+bright); %MR
center = 0.5*(cleft+cright); %CP
if(nLC~=0)
    for i=1:nLC
        fprintf(1,'Point loads are present at specified nodes\n');
        [mat1] = fscanf(inp,'%d %f',2);
        Ncode = mat1(1); %code for the node
        Fn = mat1(2);
        switch Ncode
            case 1  %Top left corner
                F(3*tleft-2)=F(3*tleft-2)+Fn;
            case 2  % Centre left corner
                F(3*cleft-2) = F(3*cleft-2)+Fn;
            case 3 %Bottom left corner
                F(3*bleft-2)=F(3*bleft-2)+Fn;
            case 4  %Bottom right corner
                F(3*tright-2)=F(3*tright-2)+Fn;
            case 5  %Centre right corner
                F(3*cright-2)=F(3*cright-2)+Fn;
            case 6  %Bottom right corner
                F(3*bright-2)=F(3*bright-2)+Fn;
            case 7  %centre of the plate
                F(3*center-2)=F(3*center-2)+Fn;
        end
    end
end
fprintf(1,'enter edge restraints in the order [Bottom Top Left Right]\n');
fprintf(1,'enter 0 for clamped edge\n');
fprintf(1,'enter 1 for simply-supported edge\n');
fprintf(1,'enter 2 for symmetric edge condition\n');
resB=input('Bottom Edge(0/1/2): ');
resT=input('Top Edge(0/1/2): ');
resL=input('Left Edge(0/1/2): ');
resR=input('Right Edge(0/1/2): ');
edge =[resB resT resL resR];
%applying boundary conditions
resx = ones(nNode,1); %rotation about x-axis
resy = ones(nNode,1); %rotation about y-axis
resz = ones(nNode,1); %translation along z-axis
```

```matlab
for i=1:nx+1
    j=bottom(i);
    if(edge(1)==0)%bottom edge clamped
        resx(j)=0;%all dof on bottom edge are restrained
        resy(j)=0;
        resz(j)=0;
    elseif(edge(1)==1)% bottom edge is simply-supported
        resz(j) = 0;
        resx(j) = 0;
    elseif(edge(1)==2);%bottom edge symmetric
        resy(j) = 0;
    end
end
for i=1:nx+1
    j = top(i);
    if(edge(2)==0)%top edge clamped
        resx(j)=0;%all dof on bottom edge are restrained
        resy(j)=0;
        resz(j)=0;
    elseif(edge(2)==1)%top edge simply-supported
        resz(j) = 0;
        resx(j) = 0;
    elseif(edge(2)==2)%top edge symmetric
        resy(j) = 0;
    end
end
for i=1:ny+1
    j = left(i);
    if(edge(3)==0)%left edge clamped
        resx(j)=0;
        resy(j)=0;
        resz(j)=0;
    elseif(edge(3)==1)%left edge simply-supported
        resz(j) = 0;
        resy(j) = 0;
    elseif(edge(3)==2)%left edge symmetric
        resx(j) = 0;
    end
end
for i=1:ny+1
    j = right(i);
    if(edge(4)==0)%right edge clamped
        resx(j)=0;
        resy(j)=0;
        resz(j)=0;
    elseif(edge(4)==1)%right edge simply-supported
```

```
            resz(j) = 0;
            resy(j) = 0;
        elseif(edge(4)==2) %right edge symmetric
            resx(j) = 0;
        end
    end
end
bn = 0;
for i=1:nNode
    if(resz(i)==0)
        bn = bn+1;
        ifix(bn) = 3*i-2;
    end
    if(resx(i)==0)
        bn = bn+1;
        ifix(bn) = 3*i-1;
    end
    if(resy(i)==0)
        bn = bn+1;
        ifix(bn) = 3*i;
    end
end
ifix = ifix';
K(ifix,:)=zeros(bn,ndof);
K(:,ifix)=zeros(ndof,bn);
K(ifix,ifix)=eye(length(ifix));
F(ifix)=zeros(bn,1);
u =K\F;
%NOTE: MXX,MYY and MXY are computed at the centroid of the element
%thus, integration is performed using a single Gauss point
ngpb=1; % for bending
ngps=1; %for shear
sigma = zeros(3,nElem,ngpb^2);
sigmas = zeros(2,nElem);
sampb = gauss(ngpb);
samps = gauss(ngps);
B  = zeros(3,12);
for i=1:nElem
    B  = zeros(3,12); Bs = zeros(2,12);
    x1 = x(node1(i)); y1 = y(node1(i));
    x2 = x(node2(i)); y2 = y(node2(i));
    x3 = x(node3(i)); y3 = y(node3(i));
    x4 = x(node4(i)); y4 = y(node4(i));
    coord = [x1 x2 x3 x4
             y1 y2 y3 y4 ]';
    % dof are in the order w theta-x theta-y
    i1 =3*node1(i)-2; j1 = 3*node1(i)-1; k1 = 3*node1(i);
```

```matlab
    i2 =3*node2(i)-2; j2 = 3*node2(i)-1; k2 = 3*node2(i);
    i3 =3*node3(i)-2; j3 = 3*node3(i)-1; k3 = 3*node3(i);
    i4 =3*node4(i)-2; j4 = 3*node4(i)-1; k4 = 3*node4(i);
    dof = [i1 j1 k1 i2 j2 k2  i3 j3 k3 i4 j4 k4];
    %because all elements are identical, it is enough to compute Kb and
Ks once
    for ig=1:ngpb  % loop over Gauss points
        wi = sampb(ig,2);
        for jg=1:ngpb
            wj = sampb(jg,2);
            [der,fun]=smlin(sampb,ig,jg);
            J =der*coord;
            detJ = det(J);
            deriv = J\der ;
            B(1,2:3:11) = -deriv(1,:); B(2,3:3:12) = -deriv(2,:);
            B(3,2:3:11) = -deriv(2,:); B(3,3:3:12) = -deriv(1,:);
            Bs(1,1:3:10) = -deriv(2,:); Bs(1,3:3:12) = fun;
            Bs(2,1:3:10) = -deriv(1,:); Bs(2,2:3:11) = fun;
            sigma(:,i)  = sigma(:,i)+ D*B*u(dof);
            sigmas(:,i) = sigmas(:,i)+Ds*Bs*u(dof);
        end
    end
end
w = u(1:3:3*nNode-2);
w1 = reshape(w,ny+1,nx+1);
dx = Lx/nx;
dy = Ly/ny;
x1 =reshape(x,ny+1,nx+1);
y1 =reshape(y,ny+1,nx+1);
%--------------------------------
%z-translation
figure(2)
surf(x1,y1,w1);
% axis tight
if(Lx==Ly)
  axis square
end
title('Variation of vertical displacement','fontweight','bold','fontsize',16);
xlabel('Length of plate (m)','fontweight','bold','fontsize',10);
ylabel('Depth of plate (m)','fontweight','bold','fontsize',10);
ax = gca;
ax.FontSize = 11;
ax.FontWeight = 'bold';
colormap('jet')
shading interp
cb = colorbar;
```

```matlab
val = cb.Limits;
t = linspace(val(1),val(2),8);
cstr_t = cell(length(t),1);
for ii=1:length(t)
    cstr_t{ii}=num2str(t(ii));
end
tickLabelsToUse = cb.TickLabels;
tickLabelsMod = cellfun(@(c) sprintf('%.3E',str2double(c))  ,cstr_t, 'uni', 0);
cb.TickLabels = [];
cb.TickLabels = tickLabelsMod;
set(cb,'YTick',t);
val = abs(min(w))/min(Lx,Ly);
if(val<0.01)
    fprintf(1,'maximum deflection is within limits\n');
else
    fprintf(1,'large deflection. please check again\n');
end
%------------------------------------
%preparing contour plot for stresses
for i=1:ny+1
    nodemat(i,:) = left(i):ny+1:right(i);
end
MXX = zeros(ny+1,nx+1);
MYY = zeros(ny+1,nx+1);
MXY = zeros(ny+1,nx+1);
SXX = zeros(ny+1,nx+1);
SYY = zeros(ny+1,nx+1);
count = zeros(ny+1,nx+1);
for e=1:nElem
   n1 = node1(e); n2 = node2(e); n3 = node3(e)  ; n4 = node4(e);
   [row1, col1] = find(nodemat == n1);
   [row2, col2] = find(nodemat == n2);
   [row3, col3] = find(nodemat == n3);
   [row4, col4] = find(nodemat == n4);
   MXX(row1,col1) = MXX(row1,col1)+sigma(1,e);
   MXX(row2,col2) = MXX(row2,col2)+sigma(1,e);
   MXX(row3,col3) = MXX(row3,col3)+sigma(1,e);
   MXX(row4,col4) = MXX(row4,col4)+sigma(1,e);

   MYY(row1,col1) = MYY(row1,col1)+sigma(2,e);
   MYY(row2,col2) = MYY(row2,col2)+sigma(2,e);
   MYY(row3,col3) = MYY(row3,col3)+sigma(2,e);
   MYY(row4,col4) = MYY(row4,col4)+sigma(2,e);

   MXY(row1,col1)   = MXY(row1,col1) +sigma(3,e);
   MXY(row2,col2)   = MXY(row2,col2) +sigma(3,e);
```

```matlab
    MXY(row3,col3)   = MXY(row3,col3) +sigma(3,e);
    MXY(row4,col4)   = MXY(row4,col4) +sigma(3,e);

    SXX(row1,col1) = SXX(row1,col1)+sigmas(1,e);
    SXX(row2,col2) = SXX(row2,col2)+sigmas(1,e);
    SXX(row3,col3) = SXX(row3,col3)+sigmas(1,e);
    SXX(row4,col4) = SXX(row4,col4)+sigmas(1,e);

    SYY(row1,col1) = SYY(row1,col1)+sigmas(2,e);
    SYY(row2,col2) = SYY(row2,col2)+sigmas(2,e);
    SYY(row3,col3) = SYY(row3,col3)+sigmas(2,e);
    SYY(row4,col4) = SYY(row4,col4)+sigmas(2,e);
    count(row1,col1)   = count(row1,col1)+1;
    count(row2,col2)   = count(row2,col2)+1;
    count(row3,col3)   = count(row3,col3)+1;
    count(row4,col4)   = count(row4,col4)+1;
end
MXX = MXX./count;
MYY = MYY./count;
MXY = MXY./count;
SXX = SXX./count;
SYY = SYY./count;
figure(3)
surf(x1,y1,MXX);
axis tight
if(Lx==Ly)
  axis square
end
title('Variation of M_{XX}','fontweight','bold','fontsize',16);
xlabel('Length of plate (m)','fontweight','bold','fontsize',10);
ylabel('Depth of plate (m)','fontweight','bold','fontsize',10);
ax = gca;
ax.FontSize = 11;
ax.FontWeight = 'bold';
colormap('jet')
shading interp
cb = colorbar;
val = cb.Limits;
t = linspace(val(1),val(2),8);
set(cb,'YTick',t);
cb.FontWeight = 'bold';
for ii=1:length(t)
    cstr_t{ii}=num2str(t(ii));
end
tickLabelsToUse = cb.TickLabels;
tickLabelsMod = cellfun(@(c) sprintf('%.1f',str2double(c)) ,cstr_t, 'uni', 0);
```

```matlab
cb.TickLabels = [];
cb.TickLabels = tickLabelsMod;
set(cb,'YTick',t);

figure(4)
surf(x1,y1,MYY);
axis tight
if(Lx==Ly)
  axis square
end
title('Variation of M_{YY}','fontweight','bold','fontsize',16);
xlabel('Length of plate (mm)','fontweight','bold','fontsize',10);
ylabel('Depth of plate (mm)','fontweight','bold','fontsize',10);
ax = gca;
ax.FontSize = 11;
ax.FontWeight = 'bold';
colormap('jet')
shading interp
cb = colorbar;
val = cb.Limits;
t = linspace(val(1),val(2),8);
set(cb,'YTick',t);
cb.FontWeight = 'bold';
for ii=1:length(t)
    cstr_t{ii}=num2str(t(ii));
end
tickLabelsToUse = cb.TickLabels;
tickLabelsMod = cellfun(@(c) sprintf('%.1f',str2double(c)) ,cstr_t, 'uni', 0);
cb.TickLabels = [];
cb.TickLabels = tickLabelsMod;
set(cb,'YTick',t);
figure(5)
surf(x1,y1,MXY);
axis tight
if(Lx==Ly)
  axis square
end
title('Variation of M_{XY}','fontweight','bold','fontsize',16);
xlabel('Length of plate (mm)','fontweight','bold','fontsize',10);
ylabel('Depth of plate (mm)','fontweight','bold','fontsize',10);
ax = gca;
ax.FontSize = 11;
ax.FontWeight = 'bold';
colormap('jet')
shading interp
cb = colorbar;
```

```matlab
val = cb.Limits;
t = linspace(val(1),val(2),8);
cb.FontWeight = 'bold';
for ii=1:length(t)
    cstr_t{ii}=num2str(t(ii));
end
tickLabelsToUse = cb.TickLabels;
tickLabelsMod = cellfun(@(c) sprintf('%.1f',str2double(c)) ,cstr_t, 'uni', 0);
cb.TickLabels = [];
cb.TickLabels = tickLabelsMod;
set(cb,'YTick',t);

figure(8)
surf(x1,y1,SXX);
axis tight
if(Lx==Ly)
  axis square
end
title('Variation of S_{XX}','fontweight','bold','fontsize',16);
xlabel('Length of plate (mm)','fontweight','bold','fontsize',10);
ylabel('Depth of plate (mm)','fontweight','bold','fontsize',10);
ax = gca;
ax.FontSize = 11;
ax.FontWeight = 'bold';
colormap('jet')
shading interp
cb = colorbar;
val = cb.Limits;
t = linspace(val(1),val(2),8);
cb.FontWeight = 'bold';
for ii=1:length(t)
    cstr_t{ii}=num2str(t(ii));
end
tickLabelsToUse = cb.TickLabels;
tickLabelsMod = cellfun(@(c) sprintf('%.1f',str2double(c)) ,cstr_t, 'uni', 0);
cb.TickLabels = [];
cb.TickLabels = tickLabelsMod;
set(cb,'YTick',t);

figure(9)
surf(x1,y1,SYY);
axis tight
if(Lx==Ly)
  axis square
end
title('Variation of S_{YY}','fontweight','bold','fontsize',16);
```

```matlab
xlabel('Length of plate (mm)','fontweight','bold','fontsize',10);
ylabel('Depth of plate (mm)','fontweight','bold','fontsize',10);
ax = gca;
ax.FontSize = 11;
ax.FontWeight = 'bold';
colormap('jet')
shading interp
cb = colorbar;
val = cb.Limits;
t = linspace(val(1),val(2),8);
cb.FontWeight = 'bold';
for ii=1:length(t)
    cstr_t{ii}=num2str(t(ii));
end
tickLabelsToUse = cb.TickLabels;
tickLabelsMod = cellfun(@(c) sprintf('%.1f',str2double(c)) ,cstr_t, 'uni', 0);
cb.TickLabels = [];
cb.TickLabels = tickLabelsMod;
set(cb,'YTick',t);
% This module generates a mesh of linear quadrilateral elements
% for a skew plate
function      [xgrid,ygrid,x,    y,      nodes,      nNode,      nElem]    =
Q4skewmesh(Lx,Ly,nx,ny,alpha)
nNode = 0;
k = 0;
dhx = Lx/nx;
dhy = Ly/ny;
h = Lx*tand(alpha);
dky = h/nx;
for i = 1:nx
    for j=1:ny
        k = k + 1;
        n1 = j + (i-1)*(ny + 1);
        geom(n1,:) = [(i-1)*dhx    (i-1)*dky+(j-1)*dhy ];
        n2 = j + i*(ny+1);
        geom(n2,:) = [i*dhx        i*dky+(j-1)*dhy ];
        n3 = n1 + 1;
        geom(n3,:) = [(i-1)*dhx    (i-1)*dky+j*dhy ];
        n4 = n2 + 1;
        geom(n4,:) = [i*dhx        i*dky+j*dhy ];
        nel = k;
        nodes(nel,:) = [n1 n2 n4 n3];
        nNode = n4;

        xgrid(2*i-1) = geom(n1,1); xgrid(2*i) = geom(n2,1);
        ygrid(2*j-1) = geom(n1,2); ygrid(2*j) = geom(n3,2);
```

```matlab
        figure (1)
        h=line([geom(n1,1) geom(n2,1)],[geom(n1,2) geom(n2,2)]);
        h.Color = 'k';h.Marker='.';h.MarkerFaceColor='k'; h.MarkerSize=10;
        h=line([geom(n2,1) geom(n4,1)],[geom(n2,2) geom(n4,2)]);
        h.Color = 'k';h.Marker='.';h.MarkerFaceColor='k'; h.MarkerSize=10;
        h=line([geom(n4,1) geom(n3,1)],[geom(n4,2) geom(n3,2)]);
        h.Color = 'k';h.Marker='.';h.MarkerFaceColor='k'; h.MarkerSize=10;
        h=line([geom(n3,1) geom(n1,1)],[geom(n3,2) geom(n1,2)]);
        h.Color = 'k';h.Marker='.';h.MarkerFaceColor='k'; h.MarkerSize=10;
    end
end
x = geom(:,1);
y = geom(:,2);
nElem = size(nodes,1);


%------------------------------------------------------------------
function[der,fun] = smlin(samp, ig,jg)
%
% This function returns the vector of the shape function and their
% derivatives with respect to xi and eta
%
% xi=samp(ig,1);
% eta=samp(jg,1);
%
syms xi eta ;
N1 = 0.25*(1-xi)*(1-eta);
N2 = 0.25*(1+xi)*(1-eta);
N3 = 0.25*(1+xi)*(1+eta);
N4 = 0.25*(1-xi)*(1+eta);
f = [N1 N2 N3  N4];
fun = double(subs(f,{xi eta},{samp(ig,1) samp(jg,1)}));
dermat = [diff(N1,xi)  diff(N2,xi)  diff(N3,xi)  diff(N4,xi)
          diff(N1,eta) diff(N2,eta) diff(N3,eta) diff(N4,eta)];
der = double(subs(dermat,{xi eta},{samp(ig,1) samp(jg,1)}));
%------------------------------------------------------------------
function[samp]=gauss(ngp)
%
% This function returns the abscissas and weights of the Gauss
% points for ngp equal up to 4
samp=zeros(ngp,2);
if ngp==1
    samp=[0 2];
elseif ngp==2
    samp=[-1/sqrt(3) 1.0
          1/sqrt(3) 1.0];
```

```
elseif ngp==3
    samp= [-0.2*sqrt(15.)    5/9
              0.0            8/9
           0.2*sqrt(15.)     5/9];
elseif ngp==4
    samp= [-0.861136311594053      0.347854845137454
           -0.339981043584856      0.652145154862546
            0.339981043584856      0.652145154862546
            0.861136311594053      0.347854845137454];
End
```

A.14. ANALYSIS OF PLATES USING 8-NODE QUADRILATERAL ELEMENTS

```matlab
%FINITE ELEMENT ANALYSIS OF A SKEW PLATE SUBJECTED TO NORMAL PRESSURE
%USING S8 ELEMENTS
clear
clc
format long
inp= fopen('S8plate.inp','r');
%reading nodal information
[mat] = fscanf(inp,'%f %f  %f %d %d %f %f %f %f %d %f %f',11);
Lx = mat(1); %Length of plate
Ly = mat(2); %width of plate
thick  = mat(3); %thickness of plate
nx = mat(4); %number of divisions along x-direction
ny = mat(5); %number of divisions along y-direction
E  = mat(6);%Modulus of elasticity in kPa
rho= mat(7);
nue= mat(8); %Poisson's ratio
q  = mat(9);   %uniform pressure
nLC = mat(10); %number of loaded nodes
alpha=mat(11); %skew angle
[xgrid,ygrid,x,        y        ,nodes,       nNode,       nElem]        =
Q8skewmesh(Lx,Ly,nx,ny,alpha);
node1=nodes(:,1);node2=nodes(:,2);node3=nodes(:,3);
node4=nodes(:,4);node5=nodes(:,5);node6=nodes(:,6);
node7 = nodes(:,7);node8 = nodes(:,8);
%Show Node Labels
for i = 1:nNode
h=text(x(i),y(i),int2str(i),'fontsize',7,'fontweight','bold','Color','k');
    h.HorizontalAlignment='right';
    h.VerticalAlignment='bottom';
end
%Show Element Labels


for e = 1:nElem
    xloc = (x(nodes(e,1))+x(nodes(e,5)))/2;
    yloc = (y(nodes(e,1))+y(nodes(e,5)))/2;
text(xloc,yloc,int2str(e),'fontsize',8,'fontweight','bold','Color','k');
end
%determining edges of the plate
bottom = zeros(2*nx+1,1);
top    = zeros(2*nx+1,1);
left   = zeros(2*ny+1,1);
right  = zeros(2*ny+1,1);
iB = 0; iT =0; iL = 0;iR = 0;
H = Lx*tand(alpha);
dky = H/nx;
```

```matlab
for i=1:nNode
    % extracting nodes corresponding to edges
    if (y(i)==0||(rem(y(i),0.5*dky)==0))
        iB = iB+1;
        bottom(iB) = i;
    end
    if(x(i)==0)
        iL = iL+1;
        left(iL) = i;
    end
    if(x(i)==Lx)
        iR = iR+1;
        right(iR) = i;
    end
end
i=0;
for j=1:nNode
  if(y(j) == (Ly+y(bottom(i+1))))
      i = i+1;
      top(i) = j;
  end
end
%----------------------------------------------------------------------
ndof = 3*nNode; %total dof for the entire structure
K = zeros(ndof,ndof); %initialize global stiffness matrix

ngpb = 3 ; %Number of Gauss points
ngps = 2 ; %for reduced integration
sampb = gauss(ngpb);
samps=  gauss(ngps);
D=(E*thick^3/(12*(1-nue^2)))*[1   nue   0
                             nue 1    0
                             0    0 (1-nue)/2];
G = E/(2*(1+nue));
Ds = G*[thick 0
          0    thick];
a = Lx/(2*nx);
b = Ly/(2*ny);
F = zeros(3*nNode,1);
for i=1:nElem
    Keb = zeros(24,24); %bending
    Kes = zeros(24,24); %shear
    Fe= zeros(24,1);
    B   = zeros(3,24); Bs = zeros(2,24);
    x1 = x(node1(i)); y1 = y(node1(i));
    x2 = x(node2(i)); y2 = y(node2(i));
    x3 = x(node3(i)); y3 = y(node3(i));
```

```matlab
x4 = x(node4(i)); y4 = y(node4(i));
x5 = x(node5(i)); y5 = y(node5(i));
x6 = x(node6(i)); y6 = y(node6(i));
x7 = x(node7(i)); y7 = y(node7(i));
x8 = x(node8(i)); y8 = y(node8(i));
coord = [x1 x2 x3 x4 x5 x6 x7 x8
         y1 y2 y3 y4 y5 y6 y7 y8 ]';
% dof are in the order w theta-x theta-y
i1 =3*node1(i)-2; j1 = 3*node1(i)-1; k1 = 3*node1(i);
i2 =3*node2(i)-2; j2 = 3*node2(i)-1; k2 = 3*node2(i);
i3 =3*node3(i)-2; j3 = 3*node3(i)-1; k3 = 3*node3(i);
i4 =3*node4(i)-2; j4 = 3*node4(i)-1; k4 = 3*node4(i);
i5 =3*node5(i)-2; j5 = 3*node5(i)-1; k5 = 3*node5(i);
i6 =3*node6(i)-2; j6 = 3*node6(i)-1; k6 = 3*node6(i);
i7 =3*node7(i)-2; j7 = 3*node7(i)-1; k7 = 3*node7(i);
i8 =3*node8(i)-2; j8 = 3*node8(i)-1; k8 = 3*node8(i);
dof = [i1 j1 k1 i2 j2 k2  i3 j3 k3 i4 j4 k4 ...
       i5 j5 k5 i6 j6 k6  i7 j7 k7 i8 j8 k8];
k=0;
for ig=1:ngpb  % loop over Gauss points
    wi = sampb(ig,2);
    for jg=1:ngpb
        k = k+1;
        wj = sampb(jg,2);
        [der,fun]=fmquad(sampb,ig,jg);
        J =der*coord;
        detJ = det(J);
        deriv = J\der ;
        B(1,2:3:23) = -deriv(1,:);
        B(2,3:3:24) = -deriv(2,:);
        B(3,2:3:23) = -deriv(2,:);
        B(3,3:3:24) = -deriv(1,:);
        Keb = Keb+detJ*wi*wj*B'*D*B;
        %element stiffness matrix by numerical intergration
        Fe  = Fe + detJ*wi*wj*q*
          [fun(1) 0 0 fun(2) 0  0  fun(3)  0  0 fun(4) 0 0 ...
           fun(5) 0 0 fun(6) 0  0  fun(7)  0  0 fun(8) 0 0]';
    end
end
kk=0;
for ig=1:ngps  % loop over Gauss points REDUCED INTEGRATION
    wi = samps(ig,2);
    for jg=1:ngps
        kk = kk+1;
        wj = samps(jg,2);
        [der,fun]=fmquad(samps,ig,jg);
        J =der*coord;
```

```matlab
                    detJ = det(J);
                    deriv = J\der ;
                    Bs(1,1:3:22) = -deriv(2,:); Bs(1,3:3:24) = fun;
                    Bs(2,1:3:22) = -deriv(1,:); Bs(2,2:3:23) = fun;
                    Kes = Kes+(5/6)*detJ*wi*wj*Bs'*Ds*Bs;
                    %element stiffness matrix by numerical intergration
                end
            end
            K(dof,dof) = K(dof,dof)+Keb+Kes;
            F(dof)     = F(dof)+Fe;
    end
    K;
    tleft = intersect(top,left);   %TL
    tright = intersect(top,right); %TR
    bleft = intersect(bottom,left); %BL
    bright = intersect(bottom,right); %BR
    cleft = 0.5*(tleft+bleft); %ML
    cright = 0.5*(tright+bright); %MR
    center = 0.5*(cleft+cright); %CP
    if(nLC~=0)
        for i=1:nLC
            fprintf(1,'Point loads are present at specified nodes\n');
            [mat1] = fscanf(inp,'%d %f',2);
            Ncode = mat1(1); %code for the node
            Fn = mat1(2);
            switch Ncode
                case 1   %Top left corner
                    F(3*tleft-2)=F(3*tleft-2)+Fn;
                case 2   % Centre left corner
                    F(3*cleft-2) = F(3*cleft-2)+Fn;
                case 3 %Bottom left corner
                    F(3*bleft-2)=F(3*bleft-2)+Fn;
                case 4   %Bottom right corner
                    F(3*tright-2)=F(3*tright-2)+Fn;
                case 5   %Centre right corner
                    F(3*cright-2)=F(3*cright-2)+Fn;
                case 6   %Bottom right corner
                    F(3*bright-2)=F(3*bright-2)+Fn;
                case 7   %centre of the plate
                    F(3*center-2)=F(3*center-2)+Fn;
            end
        end
    end
    fprintf(1,'enter edge restraints in the order [Bottom Top Left Right]\n');
    fprintf(1,'enter 0 for restrained edge and 1 for simply-supported edge\n');
    resB=input('Bottom Edge(1 or 0): ');
    resT=input('Top Edge(1 or 0): ');
```

```
resL=input('Left Edge(1 or 0): ');
resR=input('Right Edge(1 or 0): ');
edge =[resB resT resL resR];
%applying boundary conditions
resx = ones(nNode,1);
resy = ones(nNode,1);
resz = ones(nNode,1);
for i=1:2*nx+1
    j=bottom(i);
    if(edge(1)==0)%bottom edge clamped
        resx(j)=0;%all dof on bottom edge are restrained
        resy(j)=0;
        resz(j)=0;
    elseif(edge(1)==1)% bottom edge is simply-supported
        resz(j) = 0;
        resx(j) = 0;
    elseif(edge(1)==2);%bottom edge symmetric
        resy(j) = 0;
    end
end
for i=1:2*nx+1
    j = top(i);
    if(edge(2)==0)%top edge clamped
        resx(j)=0;%all dof on bottom edge are restrained
        resy(j)=0;
        resz(j)=0;
    elseif(edge(2)==1)%top edge simply-supported
        resz(j) = 0;
        resx(j) = 0;
    elseif(edge(2)==2)%top edge symmetric
        resy(j) = 0;
    end
end

for i=1:2*ny+1
    j = left(i);
    if(edge(3)==0)%left edge clamped
        resx(j)=0;
        resy(j)=0;
        resz(j)=0;
    elseif(edge(3)==1)%left edge simply-supported
        resz(j) = 0;
        resy(j) = 0;
    elseif(edge(3)==2)%left edge symmetric
        resx(j) = 0;
    end
end
```

```matlab
for i=1:2*ny+1
    j = right(i);
    if(edge(4)==0)%right edge clamped
        resx(j)=0;
        resy(j)=0;
        resz(j)=0;
    elseif(edge(4)==1)%right edge simply-supported
        resz(j) = 0;
        resy(j) = 0;
    elseif(edge(4)==2) %right edge symmetric
        resx(j) = 0;
    end
end
bn = 0;
for i=1:nNode
    if(resz(i)==0)
        bn = bn+1;
        ifix(bn) = 3*i-2;
    end
    if(resx(i)==0)
        bn = bn+1;
        ifix(bn) = 3*i-1;
    end
    if(resy(i)==0)
        bn = bn+1;
        ifix(bn) = 3*i;
    end
end
ifix = ifix';
K(ifix,:)=zeros(bn,ndof);
K(:,ifix)=zeros(ndof,bn);
K(ifix,ifix)=eye(length(ifix));
F(ifix)=zeros(bn,1);
u =K\F;
%NOTE: MXX,MYY and MXY are computed at the integration points of the
element
ngpb=2; % for bending
sigma = zeros(3,nElem,ngpb^2);
sampb = gauss(ngpb);
B   = zeros(3,24);
for i=1:nElem
    x1 = x(node1(i)); y1 = y(node1(i));
    x2 = x(node2(i)); y2 = y(node2(i));
    x3 = x(node3(i)); y3 = y(node3(i));
    x4 = x(node4(i)); y4 = y(node4(i));
    x5 = x(node5(i)); y5 = y(node5(i));
    x6 = x(node6(i)); y6 = y(node6(i));
```

```matlab
    x7 = x(node7(i)); y7 = y(node7(i));
    x8 = x(node8(i)); y8 = y(node8(i));
    coord = [x1 x2 x3 x4 x5 x6 x7 x8
             y1 y2 y3 y4 y5 y6 y7 y8 ]';
    % dof are in the order w theta-x theta-y
    i1 =3*node1(i)-2; j1 = 3*node1(i)-1; k1 = 3*node1(i);
    i2 =3*node2(i)-2; j2 = 3*node2(i)-1; k2 = 3*node2(i);
    i3 =3*node3(i)-2; j3 = 3*node3(i)-1; k3 = 3*node3(i);
    i4 =3*node4(i)-2; j4 = 3*node4(i)-1; k4 = 3*node4(i);
    i5 =3*node5(i)-2; j5 = 3*node5(i)-1; k5 = 3*node5(i);
    i6 =3*node6(i)-2; j6 = 3*node6(i)-1; k6 = 3*node6(i);
    i7 =3*node7(i)-2; j7 = 3*node7(i)-1; k7 = 3*node7(i);
    i8 =3*node8(i)-2; j8 = 3*node8(i)-1; k8 = 3*node8(i);
    dof = [i1 j1 k1 i2 j2 k2  i3 j3 k3 i4 j4 k4 ....
           i5 j5 k5 i6 j6 k6  i7 j7 k7 i8 j8 k8];
    for ig=1:ngpb  % loop over Gauss points
        for jg=1:ngpb
            [der,fun]=fmquad(sampb,ig,jg);
            J =der*coord;
            detJ = det(J);
            deriv = J\der ;
            B(1,2:3:23) = -deriv(1,:);
            B(2,3:3:24) = -deriv(2,:);
            B(3,2:3:23) = -deriv(2,:);
            B(3,3:3:24) = -deriv(1,:);
            if(ig==1&&jg==1)
                k=1;
            elseif(ig==2&&jg==1)
                k=2;
            elseif(ig==2&&jg==2)
                k=3;
            elseif(ig==1&&jg==2)
                k=4;
            end
            sigma(:,i,k) = D*B*u(dof);
        end
    end
end
w = u(1:3:3*nNode-2);
%generate location matrix
mat = zeros(2*ny+1,2*nx+1);
mat(1,:)           = bottom;
for i=2:2*ny+1
  mat(i,1:2:2*nx+1) = bottom(1:2:2*nx+1)+i-1;
end
k = 0;
```

```matlab
for i=3:2:2*ny+1
      k = k+1;
      mat(i,2:2:2*nx+1) = bottom(2:2:2*nx+1)+k;
end
w1 = zeros(2*ny+1,2*nx+1);
for i=1:2*ny+1
    for j=1:2*nx+1
        nodval = mat(i,j);
        if(nodval==0)
            x1(i,j) = 0.5*(x(mat(i,j-1))+x(mat(i,j+1)));
            y1(i,j) = 0.5*(y(mat(i,j-1))+y(mat(i,j+1)));

            w1(i,j)=0.25*(w(mat(i,j-1))+w(mat(i,j+1))+w(mat(i-1,j))
                          +w(mat(i+1,j)));
        else
            x1(i,j) = x(nodval);
            y1(i,j) = y(nodval);
            w1(i,j) = w(nodval);
        end
    end
end
%z-translation
figure(2)
surf(x1,y1,w1);
if(Lx==Ly)
  axis square
end
title('Variation of vertical
              displacement','fontweight','bold','fontsize',16);
xlabel('Length of plate (m)','fontweight','bold','fontsize',10);
ylabel('Depth of plate (m)','fontweight','bold','fontsize',10);
ax = gca;
ax.FontSize = 11;
ax.FontWeight = 'bold';
colormap('jet')
shading interp
cb = colorbar;
val = cb.Limits;
t = linspace(val(1),val(2),8);
cstr_t = cell(length(t),1);
for ii=1:length(t)
    cstr_t{ii}=num2str(t(ii));
end
tickLabelsToUse = cb.TickLabels;
tickLabelsMod=cellfun(@(c)  sprintf('%.3E',str2double(c))   ,cstr_t,
'uni', 0);
cb.TickLabels = [];
```

```matlab
cb.TickLabels = tickLabelsMod;
set(cb,'YTick',t);
val = abs(min(w))/min(Lx,Ly);
if(val<0.01)
    fprintf(1,'maximum deflection is within limits\n');
else
    fprintf(1,'large deflection. please check again\n');
end
%-----------------------------------
%preparing contour plot for stresses
Tmat = genT4mat;
MXX = zeros(2*ny+1,2*nx+1);
MYY = zeros(2*ny+1,2*nx+1);
MXY = zeros(2*ny+1,2*nx+1);
count = zeros(2*ny+1,2*nx+1);
for e=1:nElem
    n1 = node1(e); n2 = node2(e); n3 = node3(e)  ; n4 = node4(e);
    n5 = node5(e); n6 = node6(e); n7 = node7(e)  ; n8 = node8(e);
    [row1, col1] = find(mat == n1);    [row2, col2] = find(mat == n2);
    [row3, col3] = find(mat == n3);    [row4, col4] = find(mat == n4);
    [row5, col5] = find(mat == n5);    [row6, col6] = find(mat == n6);
    [row7, col7] = find(mat == n7);    [row8, col8] = find(mat == n8);
    mxx =  Tmat*[sigma(1,e,1) sigma(1,e,2) sigma(1,e,3) sigma(1,e,4)]';
    myy =  Tmat*[sigma(2,e,1) sigma(2,e,2) sigma(2,e,3) sigma(2,e,4)]';
    mxy =  Tmat*[sigma(3,e,1) sigma(3,e,2) sigma(3,e,3) sigma(3,e,4)]';
    MXX(row1,col1) = MXX(row1,col1)+mxx(1);
    MXX(row2,col2) = MXX(row2,col2)+mxx(2);
    MXX(row3,col3) = MXX(row3,col3)+mxx(3);
    MXX(row4,col4) = MXX(row4,col4)+mxx(4);
    MXX(row5,col5) = MXX(row5,col5)+mxx(5);
    MXX(row6,col6) = MXX(row6,col6)+mxx(6);
    MXX(row7,col7) = MXX(row7,col7)+mxx(7);
    MXX(row8,col8) = MXX(row8,col8)+mxx(8);


    MYY(row1,col1) = MYY(row1,col1)+myy(1);
    MYY(row2,col2) = MYY(row2,col2)+myy(2);
    MYY(row3,col3) = MYY(row3,col3)+myy(3);
    MYY(row4,col4) = MYY(row4,col4)+myy(4);
    MYY(row5,col5) = MYY(row5,col5)+myy(5);
    MYY(row6,col6) = MYY(row6,col6)+myy(6);
    MYY(row7,col7) = MYY(row7,col7)+myy(7);
    MYY(row8,col8) = MYY(row8,col8)+myy(8);

    MXY(row1,col1) = MXY(row1,col1)+mxy(1);
    MXY(row2,col2) = MXY(row2,col2)+mxy(2);
    MXY(row3,col3) = MXY(row3,col3)+mxy(3);
```

```matlab
    MXY(row4,col4)  = MXY(row4,col4)+mxy(4);
    MXY(row5,col5)  = MXY(row5,col5)+mxy(5);
    MXY(row6,col6)  = MXY(row6,col6)+mxy(6);
    MXY(row7,col7)  = MXY(row7,col7)+mxy(7);
    MXY(row8,col8)  = MXY(row8,col8)+mxy(8);

    count(row1,col1)   = count(row1,col1)+1;
    count(row2,col2)   = count(row2,col2)+1;
    count(row3,col3)   = count(row3,col3)+1;
    count(row4,col4)   = count(row4,col4)+1;
    count(row5,col5)   = count(row5,col5)+1;
    count(row6,col6)   = count(row6,col6)+1;
    count(row7,col7)   = count(row7,col7)+1;
    count(row8,col8)   = count(row8,col8)+1;
end
MXX = MXX./count;
MYY = MYY./count;
MXY = MXY./count;
[m1, n1] = find(isnan(MXX));

for e=1:nElem
    i=m1(e);
    j=n1(e);
    MXX(i,j) = (MXX(i-1,j)+MXX(i+1,j)+MXX(i,j-1)+MXX(i,j+1)...
        +MXX(i-1,j-1)+MXX(i+1,j+1)+MXX(i-1,j+1)+MXX(i+1,j-1))/8;
    MYY(i,j) = (MYY(i-1,j)+MYY(i+1,j)+MYY(i,j-1)+MYY(i,j+1)...
        +MYY(i-1,j-1)+MYY(i+1,j+1)+MYY(i-1,j+1)+MYY(i+1,j-1))/8;
    MXY(i,j) = (MXY(i-1,j)+MXY(i+1,j)+MXY(i,j-1)+MXY(i,j+1)...
        +MXY(i-1,j-1)+MXY(i+1,j+1)+MXY(i-1,j+1)+MXY(i+1,j-1))/8;
end

figure(3)
surf(x1,y1,MXX);
axis tight
if(Lx==Ly)
  axis square
end
title('Variation of M_{XX}','fontweight','bold','fontsize',16);
xlabel('Length of plate (m)','fontweight','bold','fontsize',10);
ylabel('Depth of plate (m)','fontweight','bold','fontsize',10);
ax = gca;
ax.FontSize = 11;
ax.FontWeight = 'bold';
colormap('jet')
shading interp
cb = colorbar;
val = cb.Limits;
```

```matlab
t = linspace(val(1),val(2),8);
set(cb,'YTick',t);
cb.FontWeight = 'bold';
for ii=1:length(t)
    cstr_t{ii}=num2str(t(ii));
end
tickLabelsToUse = cb.TickLabels;
tickLabelsMod = cellfun(@(c) sprintf('%.1f',str2double(c))  ,cstr_t,
'uni', 0);
cb.TickLabels = [];
cb.TickLabels = tickLabelsMod;
set(cb,'YTick',t);
figure(4)
surf(x1,y1,MYY);
axis tight
if(Lx==Ly)
  axis square
end
title('Variation of M_{YY}','fontweight','bold','fontsize',16);
xlabel('Length of plate (mm)','fontweight','bold','fontsize',10);
ylabel('Depth of plate (mm)','fontweight','bold','fontsize',10);
ax = gca;
ax.FontSize = 11;
ax.FontWeight = 'bold';
colormap('jet')
shading interp
cb = colorbar;
val = cb.Limits;
t = linspace(val(1),val(2),8);
set(cb,'YTick',t);
cb.FontWeight = 'bold';
for ii=1:length(t)
    cstr_t{ii}=num2str(t(ii));
end
tickLabelsToUse = cb.TickLabels;
tickLabelsMod=cellfun(@(c)   sprintf('%.1f',str2double(c))   ,cstr_t,
'uni', 0);
cb.TickLabels = [];
cb.TickLabels = tickLabelsMod;
set(cb,'YTick',t);
figure(5)
surf(x1,y1,MXY);
axis tight
if(Lx==Ly)
  axis square
end
title('Variation of M_{XY}','fontweight','bold','fontsize',16);
```

```matlab
xlabel('Length of plate (mm)','fontweight','bold','fontsize',10);
ylabel('Depth of plate (mm)','fontweight','bold','fontsize',10);
ax = gca;
ax.FontSize = 11;
ax.FontWeight = 'bold';
colormap('jet')
shading interp
cb = colorbar;
val = cb.Limits;
t = linspace(val(1),val(2),8);
cb.FontWeight = 'bold';
for ii=1:length(t)
    cstr_t{ii}=num2str(t(ii));
end
tickLabelsToUse = cb.TickLabels;
tickLabelsMod=cellfun(@(c)   sprintf('%.1f',str2double(c))   ,cstr_t,
'uni', 0);
cb.TickLabels = [];
cb.TickLabels = tickLabelsMod;
set(cb,'YTick',t);
%------------------------------------------------------------------
% This module generates a mesh of linear quadrilateral elements for
a skew %plate
Function [xgrid,ygrid,x,y,nodes,nNode,nElem]= Q8skewmesh(Lx,Ly,nx,ny,alpha)
nNode = 0;
k = 0;
dhx = Lx/nx;
dhy = Ly/ny;
h = Lx*tand(alpha);
dky = h/nx;
for i = 1:nx
    for j=1:ny
        k = k + 1;
        n1 = (i-1)*(3*ny+2)+2*j - 1;
        n2 = i*(3*ny+2)+j - ny - 1;
        n3 = i*(3*ny+2)+2*j-1;
        n4 = n3 + 1;
        n5 = n3 + 2;
        n6 = n2 + 1;
        n7 = n1 + 2;
        n8 = n1 + 1;
        geom(n1,:)= [(i-1)*dhx   (i-1)*dky+(j-1)*dhy   ];
        geom(n3,:)= [ i*dhx        i*dky+(j-1)*dhy   ];
        geom(n2,:) = [(geom(n1,1)+geom(n3,1))/2 (geom(n1,2)+geom(n3,2))/2];
        geom(n5,:)= [i*dhx   i*dky+j*dhy   ];
        geom(n4,:)=[(geom(n3,1)+ geom(n5,1))/2 (geom(n3,2)+ geom(n5,2))/2];
        geom(n7,:)=[(i-1)*dhx   (i-1)*dky+j*dhy ];
```

```matlab
        geom(n6,:)=[(geom(n5,1)+geom(n7,1))/2 (geom(n5,2)+ geom(n7,2))/2];
        geom(n8,:)=[(geom(n1,1)+geom(n7,1))/2 (geom(n1,2)+ geom(n7,2))/2];
        nel = k;
        nNode = n5;
        nodes(k,:) = [n1 n2 n3 n4 n5 n6 n7 n8];

        xgrid(2*i-1) = geom(n1,1); xgrid(2*i) = geom(n3,1);
        ygrid(2*j-1) = geom(n1,2); ygrid(2*j) = geom(n3,2);
        figure (1)
        h=line([geom(n1,1) geom(n2,1)],[geom(n1,2) geom(n2,2)]);
        h.Color = 'k';h.Marker='.';h.MarkerFaceColor='k'; h.MarkerSize=10;
        h=line([geom(n2,1) geom(n3,1)],[geom(n2,2) geom(n3,2)]);
        h.Color = 'k';h.Marker='.';h.MarkerFaceColor='k'; h.MarkerSize=10;
        h=line([geom(n3,1) geom(n4,1)],[geom(n3,2) geom(n4,2)]);
        h.Color = 'k';h.Marker='.';h.MarkerFaceColor='k'; h.MarkerSize=10;
        h=line([geom(n4,1) geom(n4,1)],[geom(n4,2) geom(n5,2)]);
        h.Color = 'k';h.Marker='.';h.MarkerFaceColor='k'; h.MarkerSize=10;
        h=line([geom(n5,1) geom(n6,1)],[geom(n5,2) geom(n6,2)]);
        h.Color = 'k';h.Marker='.';h.MarkerFaceColor='k'; h.MarkerSize=10;
        h=line([geom(n6,1) geom(n7,1)],[geom(n6,2) geom(n7,2)]);
        h.Color = 'k';h.Marker='.';h.MarkerFaceColor='k'; h.MarkerSize=10;
        h=line([geom(n7,1) geom(n8,1)],[geom(n7,2) geom(n8,2)]);
        h.Color = 'k';h.Marker='.';h.MarkerFaceColor='k'; h.MarkerSize=10;
        h=line([geom(n8,1) geom(n1,1)],[geom(n8,2) geom(n1,2)]);
        h.Color = 'k';h.Marker='.';h.MarkerFaceColor='k'; h.MarkerSize=10;
    end
end
x = geom(:,1);
y = geom(:,2);
nElem = size(nodes,1);
%-----------------------------------------------------------------------
function[samp]=gauss(ngp)
% This function returns the abscissas and weights of the Gauss
% points for ngp equal up to 4
samp=zeros(ngp,2);
if ngp==1
    samp=[0 2];
elseif ngp==2
    samp=[-1/sqrt(3) 1.0
           1/sqrt(3) 1.0];
elseif ngp==3
    samp= [-0.2*sqrt(15.)   5/9
            0.0             8/9
            0.2*sqrt(15.)   5/9];
```

```matlab
elseif ngp==4
    samp= [-0.861136311594053      0.347854845137454
           -0.339981043584856      0.652145154862546
            0.339981043584856      0.652145154862546
            0.861136311594053      0.347854845137454];
end
%-----------------------------------------------------------------
function[der,fun] = fmquad(samp, ig,jg)
%
% This function returns the vector of the shape function and their
% derivatives with respect to xi and eta
%
% xi=samp(ig,1);
% eta=samp(jg,1);
%
syms xi eta ;
N1 = -0.25*(1-xi)*(1-eta)*(1+xi+eta);
N2 =  0.5*(1-xi^2)*(1-eta);
N3 = -0.25*(1+xi)*(1-eta)*(1-xi+eta);
N4 =  0.5*(1+xi)*(1-eta^2);
N5 = -0.25*(1+xi)*(1+eta)*(1-xi-eta);
N6 =  0.5*(1-xi^2)*(1+eta);
N7 = -0.25*(1-xi)*(1+eta)*(1+xi-eta);
N8 =  0.5*(1-xi)*(1-eta^2);

f = [N1 N2 N3 N4 N5 N6 N7 N8];
fun = double(subs(f,{xi eta},{samp(ig,1) samp(jg,1)}));
dermat = [diff(N1,xi)   diff(N2,xi)   diff(N3,xi) diff(N4,xi)  ...
          diff(N5,xi)   diff(N6,xi)   diff(N7,xi) diff(N8,xi)
          diff(N1,eta)  diff(N2,eta)  diff(N3,eta) diff(N4,eta)...
          diff(N5,eta)  diff(N6,eta)  diff(N7,eta) diff(N8,eta) ];
der = double(subs(dermat,{xi eta},{samp(ig,1) samp(jg,1)}));


%-----------------------------------------------------------------
function [Tmat] =  genT4mat
syms xi eta ;

N1 = 0.25*(1-xi)*(1-eta);
N2 = 0.25*(1+xi)*(1-eta);
N3 = 0.25*(1+xi)*(1+eta);
N4 = 0.25*(1-xi)*(1+eta);
Tmat = zeros(8,4);
v = sqrt(3);
```

```
Tmat(1,:)= [
double(subs(N1,{xi,eta},[-v,-v]))
double(subs(N2,{xi,eta},[-v, -v]))
double(subs(N3,{xi,eta},[-v, -v]))
double(subs(N4,{xi,eta},[-v,-v]))
];
Tmat(2,:)= [
double(subs(N1,{xi,eta},[-v,0]))
double(subs(N2,{xi,eta},[-v,0]))
double(subs(N3,{xi,eta},[-v,0]))
double(subs(N4,{xi,eta},[-v,0]))
];
Tmat(3,:)= [
double(subs(N1,{xi,eta},[v,-v]))
double(subs(N2,{xi,eta},[v,-v]))
double(subs(N3,{xi,eta},[v,-v]))
double(subs(N4,{xi,eta},[v,-v]))
];
Tmat(4,:)= [
double(subs(N1,{xi,eta},[ v,  0]))
double(subs(N2,{xi,eta},[ v,  0]))
double(subs(N3,{xi,eta},[ v,  0]))
double(subs(N4,{xi,eta},[ v,  0]))
];
Tmat(5,:)= [
double(subs(N1,{xi,eta},[v,v]))
double(subs(N2,{xi,eta},[v,v]))
double(subs(N3,{xi,eta},[v,v]))
double(subs(N4,{xi,eta},[v,v]))
];
Tmat(6,:)= [
double(subs(N1,{xi,eta},[0,v]))
double(subs(N2,{xi,eta},[0,v]))
double(subs(N3,{xi,eta},[0,v]))
double(subs(N4,{xi,eta},[0,v]))
];


Tmat(7,:)= [
double(subs(N1,{xi,eta},[-v,v]))
double(subs(N2,{xi,eta},[-v,v]))
double(subs(N3,{xi,eta},[-v,v]))
double(subs(N4,{xi,eta},[-v,v]))];
Tmat(8,:)= [
double(subs(N1,{xi,eta},[-v, 0]))
double(subs(N2,{xi,eta},[-v, 0]))
double(subs(N3,{xi,eta},[-v, 0]))
double(subs(N4,{xi,eta},[-v, 0]))];
```

A.15. ANALYSIS OF A SOLID BEAM USING 8-NODE HEXAHEDRAL (BRICK) ELEMENTS

```matlab
%FINITE ELEMENT ANALYSIS OF A 3D BEAM using 8-node brick elements
clear
clc
format long
inp= fopen('Beam3D.inp','r');
%reading nodal information
[mat] = fscanf(inp,'%f %f  %f %d %d %f %f %f %f %d %d',11);
Lx = mat(1); %dimension b
Ly = mat(2); %dimension d
Lz  = mat(3); %span of the beam
nx = mat(4); %number of divisions along x-direction
ny = mat(5); %number of divisions along y-direction
nz = mat(6); %number of divisions along y-direction
E  = mat(7);%Modulus of elasticity in kPa
nue= mat(8); %Poisson's ratio
q  = mat(9);  %uniform pressure
nLN = mat(10); %number of loaded nodes
nLF = mat(11); %number of loaded faces
dx=Lx/nx; dy=Ly/ny; dz=Lz/nz;
x1 = 0:dx:Lx; %vectors for 3D grid
y1 = 0:dy:Ly;
z1 = 0:dz:Lz;
[Nodes, Rectangles]=Rectangles_Mesh(x1,y1);
[Nodes3D,Brick] = Mesh2D_to_Mesh3D(Nodes,Rectangles,z1);
Plot_Mesh3D(Nodes3D,Brick);
node1 = Brick(:,1); node2 = Brick(:,2); node3 = Brick(:,3);
node4 = Brick(:,4);node5 = Brick(:,5); node6 = Brick(:,6);
node7 = Brick(:,7);node8 = Brick(:,8);
nNode = size(Nodes3D,1);
nElem = size(Brick,1);
x = Nodes3D(:,1);y = Nodes3D(:,2);z = Nodes3D(:,3);
for i = 1:nNode     %Show Node Labels
h=text(Nodes3D(i,1),Nodes3D(i,2),Nodes3D(i,3),int2str(i),'fontsize',8,'font
weight','bold','Color','red');
    h.HorizontalAlignment='right';
    h.VerticalAlignment='bottom';
end
for e = 1:nElem %Show Element Labels
    xloc = (Nodes3D(Brick(e,1),1)+Nodes3D(Brick(e,7),1))/2;
    yloc = (Nodes3D(Brick(e,1),2)+Nodes3D(Brick(e,7),2))/2;
    zloc = (Nodes3D(Brick(e,1),3)+Nodes3D(Brick(e,7),3))/2;
text(xloc,yloc,zloc,int2str(e),'fontsize',10,'fontweight','bold',…
    'Color','blue');
end
```

```matlab
%determining faces of the plate
bottom = zeros(2*nx+1,1);
top    = zeros(2*nx+1,1);
left   = zeros(2*ny+1,1);
right  = zeros(2*ny+1,1);
iB = 0; iT =0; iL = 0;iR = 0;
for i=1:nNode
    % extracting nodes corresponding to faces
    if (Nodes3D(i,2)==0)
        iB = iB+1;
        bottom(iB) = i;
    end
    if (Nodes3D(i,2)==Ly)
        iT = iT+1;
        top(iT) = i;
    end
    if(Nodes3D(i,1)==0)
        iL = iL+1;
        left(iL) = i;
    end
    if(Nodes3D(i,1)==Lx)
        iR = iR+1;
        right(iR) = i;
    end
end
ne = 1:nElem;
D = E/((1+nue)*(1-2*nue))*...
[1-nue      nue      nue        0           0            0
  nue     1-nue      nue        0           0            0
  nue       nue    1-nue        0           0            0
   0         0       0      (1-2*nue)/2      0            0
   0         0       0          0        (1-2*nue)/2      0
   0         0       0          0           0       (1-2*nue)/2];

ndof = 3*nNode;
K = zeros(ndof,ndof); %initialize global stiffness matrix
ngp =2;
samp = gauss(ngp);
for i=1:nElem
    B = zeros(6,24);
    G = zeros(6,9);
    Kuu=zeros(24,24);Kua=zeros(24,9);
    Kau=zeros(9,24);Kaa=zeros(9,9);
    x1 = x(node1(i)); y1 = y(node1(i));z1 = z(node1(i));
    x2 = x(node2(i)); y2 = y(node2(i));z2 = z(node2(i));
    x3 = x(node3(i)); y3 = y(node3(i));z3 = z(node3(i));
    x4 = x(node4(i)); y4 = y(node4(i));z4 = z(node4(i));
```

```matlab
x5 = x(node5(i)); y5 = y(node5(i));z5 = z(node5(i));
x6 = x(node6(i)); y6 = y(node6(i));z6 = z(node6(i));
x7 = x(node7(i)); y7 = y(node7(i));z7 = z(node7(i));
x8 = x(node8(i)); y8 = y(node8(i));z8 = z(node8(i));
coord = [x1 x2 x3 x4 x5 x6 x7 x8
         y1 y2 y3 y4 y5 y6 y7 y8
         z1 z2 z3 z4 z5 z6 z7 z8]';
i1 =3*node1(i)-2; j1 = 3*node1(i)-1; k1 = 3*node1(i);
i2 =3*node2(i)-2; j2 = 3*node2(i)-1; k2 = 3*node2(i);
i3 =3*node3(i)-2; j3 = 3*node3(i)-1; k3 = 3*node3(i);
i4 =3*node4(i)-2; j4 = 3*node4(i)-1; k4 = 3*node4(i);
i5 =3*node5(i)-2; j5 = 3*node5(i)-1; k5 = 3*node5(i);
i6 =3*node6(i)-2; j6 = 3*node6(i)-1; k6 = 3*node6(i);
i7 =3*node7(i)-2; j7 = 3*node7(i)-1; k7 = 3*node7(i);
i8 =3*node8(i)-2; j8 = 3*node8(i)-1; k8 = 3*node8(i);
dof=[i1 j1 k1 i2 j2 k2 i3 j3 k3 i4 j4 k4 i5 j5 k5 i6 j6 k6...
     i7 j7 k7 i8 j8 k8];
for ig=1:ngp
    wi = samp(ig,2);
    for jg=1:ngp
        wj = samp(jg,2);
        for kg=1:ngp
          wk = samp(kg,2);
          [der,derP]=fmbrick(samp,ig,jg,kg);
          J =der*coord;
          detJ = det(J);
          deriv = J\der ;
          derivP = J\derP;
          B(1,1:3:24) = deriv(1,:);
          B(2,2:3:24) = deriv(2,:);
          B(3,3:3:24) = deriv(3,:);
          B(4,2:3:24) = deriv(3,:);B(4,3:3:24) = deriv(2,:);
          B(5,1:3:24) = deriv(3,:);B(5,3:3:24) = deriv(1,:);
          B(6,1:3:24) = deriv(2,:);B(6,2:3:24) = deriv(1,:);

          G(1,1:3:9)=derivP(1,:);G(2,2:3:9)=derivP(2,:);
          G(3,3:3:9) = derivP(3,:);
          G(4,1:3:9) = derivP(2,:); G(4,2:3:9) = derivP(1,:);
          G(5,2:3:9) = derivP(3,:); G(5,3:3:9) = derivP(2,:);
          G(6,1:3:9) = derivP(3,:); G(6,3:3:9) = derivP(1,:);
          Kuu = Kuu+detJ*wi*wj*wk*B'*D*B;
          Kua = Kua+detJ*wi*wj*wk*B'*D*G;
          Kau = Kau+detJ*wi*wj*wk*G'*D*B;
          Kaa = Kaa+detJ*wi*wj*wk*G'*D*G;
        end
    end
end
```

```matlab
    Ke = Kuu - Kua*inv(Kaa)*Kau;
    K(dof,dof) = K(dof,dof)+Ke;
end
K;
F = zeros(3*nNode,1);
%applying load at top right corner
tleft = intersect(top,left);   %TL
tright = intersect(top,right); %TR
bleft = intersect(bottom,left); %BL
bright = intersect(bottom,right); %BR
mleft = 0.5*(tleft+bleft); %ML
mright = 0.5*(tright+bright); %MR
axleft = mleft(ceil(end/2));
axright = mright(ceil(end/2));
if(nLN~=0) % if nodes are loaded by point loads
    for i=1:nLN
        fprintf(1,'Point loads are present at specified nodes\n');
        [mat1] = fscanf(inp,'%d %f %f %f',4);
        Ncode = mat1(1); %code for the node
        Fx = mat1(2);
        Fy = mat1(3);
        Fz = mat1(4);
        switch Ncode
            case 1  %left axis point
                F(3*axleft-2)=F(3*tleft-1)+Fx;
                F(3*axleft-1)=F(3*axleft-1)+Fy;
                F(3*axleft)  =F(3*axleft)+Fz;
            case 2  % right axis point
                F(3*axright-2)=F(3*axright-2)+Fx;
                F(3*axright-1)=F(3*axright-1)+Fy;
                F(3*axright)  =F(3*axright)  +Fz;
        end
    end
end
fprintf(1,'enter  face  restraints  in  the  order  [Bottom  Top  Left
Right]\n');
fprintf(1,'enter  0  for  restrained  face  and  1  for  unrestrained
face\n');
resB=input('Bottom Face(1 or 0): ');
resT=input('Top Face(1 or 0): ');
resL=input('Left Face(1 or 0): ');
resR=input('Right Face(1 or 0): ');
edge =[resB resT resL resR];
%applying boundary conditions
resx = ones(nNode,1);
resy = ones(nNode,1);
resz = ones(nNode,1);
```

```matlab
for i=1:size(bottom)
    if(edge(1)==0)%bottom edge clamped
        j = bottom(i);
        resx(j)=0;
        resy(j)=0;
        resz(j)=0;
    end
end
for i=1:size(top)
    if(edge(2)==0)%top edge clamped
        j = top(i);
        resx(j)=0;
        resy(j)=0;
        resz(j)=0;
    end
end
for i=1:size(left)
    if(edge(3)==0)%left edge clamped
        j = left(i);
        resx(j)=0;
        resy(j)=0;
        resz(j)=0;
    end
end
for i=1:size(right)
    if(edge(4)==0)%right edge clamped
        j = right(i);
        resx(j)=0;
        resy(j)=0;
        resz(j)=0;
    end
end
bn = 0;
for i=1:nNode
    if(resx(i)==0)
        bn = bn+1;
        ifix(bn) = 3*i-2;
    end
    if(resy(i)==0)
        bn = bn+1;
        ifix(bn) = 3*i-1;
    end
    if(resz(i)==0)
        bn = bn+1;
        ifix(bn) = 3*i;
    end
end
```

```matlab
ifix = ifix';
K(ifix,:)=zeros(bn,ndof);
K(:,ifix)=zeros(ndof,bn);
K(ifix,ifix)=eye(length(ifix));
F(ifix)=zeros(bn,1);
u =K\F;
min(u)
sigma = zeros(3,nElem,ngp^2);
sigmac = zeros(3,nElem);
ngp=3; % for computing stresses at Gauss points
samp = gauss(ngp);
ngp1 = 1; %for computing stresses at centroid of each element
sampc = gauss(ngp1);
for i=1:nElem
    x1 = x(node1(i)); y1 = y(node1(i));
    x2 = x(node2(i)); y2 = y(node2(i));
    x3 = x(node3(i)); y3 = y(node3(i));
    x4 = x(node4(i)); y4 = y(node4(i));
    x5 = x(node5(i)); y5 = y(node5(i));
    x6 = x(node6(i)); y6 = y(node6(i));
    x7 = x(node7(i)); y7 = y(node7(i));
    x8 = x(node8(i)); y8 = y(node8(i));
    coord = [x1 x2 x3 x4 x5 x6 x7 x8
             y1 y2 y3 y4 y5 y6 y7 y8 ]';
    i1 =2*node1(i)-1; j1 = 2*node1(i);
    i2 =2*node2(i)-1; j2 = 2*node2(i);
    i3 =2*node3(i)-1; j3 = 2*node3(i);
    i4 =2*node4(i)-1; j4 = 2*node4(i);
    i5 =2*node5(i)-1; j5 = 2*node5(i);
    i6 =2*node6(i)-1; j6 = 2*node6(i);
    i7 =2*node7(i)-1; j7 = 2*node7(i);
    i8 =2*node8(i)-1; j8 = 2*node8(i);
    dof = [i1 j1 i2 j2 i3 j3 i4 j4 i5 j5 i6 j6 i7 j7 i8 j8];
    k=0;
    for ig=1:ngp  % loop over Gauss points
        wi = samp(ig,2);
        for jg=1:ngp
            wj = samp(jg,2);
            k=k+1;
            [der,fun]=fmquad(samp,ig,jg);
            J =der*coord;
            detJ = det(J);
            deriv = J\der ;
            B(1,1:2:16) = deriv(1,:);
            B(2,2:2:16) = deriv(2,:);
            B(3,1:2:16) = deriv(2,:);B(3,2:2:16) = deriv(1,:);
            sigma(:,i,k)  = D*B*u(dof);
```

```matlab
            end
        end
end
ux = u(1:2:2*nNode-1);
uy = u(2:2:2*nNode);
uymin=min(uy)
fprintf(1,'Maximum displacement of beam = %f\n',uymin);
```

A.16. FREE VIBRATION ANALYSIS OF BEAMS AND FRAMES

```
inp= fopen('dynframe.inp','r');
[mat] = fscanf(inp,'%d %d %d',3); % this line reads number of nodes and
number of elements in the truss
nNode = mat(1);   %number of nodes
nElem = mat(2);   %number of elements
imass = mat(3);    %code for mass matrix : 1 for lumped mass 2 for
consistent mass matrix
[mat1] = fscanf(inp,'%d %f %f %d %d %d  %e %e  %e',[9,nNode]);
mat1 = mat1'; %Transposed for future use
[mat2] = fscanf(inp,'%d %d %d %f %e %e %f %d %f %f %f',[11,nElem]);
mat2 = mat2';%transposed for future sure
ndof=3*nNode;
Xval = mat1(:,2);
Yval = mat1(:,3);
resx = mat1(:,4);
resy = mat1(:,5);
resz = mat1(:,6);
fx = mat1(:,7);
fy = mat1(:,8);
Mz = mat1(:,9);
F= zeros(ndof,1);
bn = 0;
for i=1:nNode
   if(resx(i)==0)
        bn = bn+1;
        ifix(bn) = 3*i-2;
   end
   if(resy(i)==0)
        bn = bn+1;
        ifix(bn) = 3*i-1;
   end
   if(resz(i)==0)
        bn = bn+1;
        ifix(bn) = 3*i;
   end
end
end
%-------------------------------------------------------------------------
%READING ELEMENT INFORMATION
A        = zeros(nElem,1)  ;
Iz       = zeros(nElem,1)  ;
E        = zeros(nElem,1);
Cx       = zeros(nElem,1);
Cy       = zeros(nElem,1);
Le       = zeros(nElem,1);
eforce = zeros(6,nElem);
FirstNode  = mat2(:,2);
```

```matlab
SecondNode = mat2(:,3);
A         = mat2(:,4);
Iz        = mat2(:,5);
E         = mat2(:,6);
rho      = mat2(:,7);
Ltype    = mat2(:,8);
w1 = mat2(:,9);
d1 = mat2(:,10);
d2 = mat2(:,11);
loadflag = zeros(nElem);
%flag used to mark loaded elements
%flag =0 indicates element NOT carrying any load
%flag =1 indicates element carrying load
mbar = rho.*A;

for e=1:nElem
    x1=Xval(FirstNode(e));     y1=Yval(FirstNode(e));
    x2=Xval(SecondNode(e));    y2=Yval(SecondNode(e));
    le =sqrt((x2-x1)^2+(y2-y1)^2);
    c=(x2-x1)/le;
    s=(y2-y1)/le;
    Cx(e)=c;
    Cy(e)=s;
    L(e)=le;
    if(Ltype(e)==1) %for distributed loading
        %computation of equivalent nodal loads owing to uniformly
        %distributed load over some portion of the span
        c1 = d2(e)-d1(e);
        a = d1(e)+0.5*c1;
        b = 0.5*c1+le-d2(e);
        p1 = (-w1(e)*c1/(12*le^2))*(12*a*b^2+c1^2*(le-3*b));
        p2 = (w1(e)*c1/(12*le^2))* (12*a^2*b+c1^2*(le-3*a));
        p3 = (w1(e)*c1*b/le)-(p1+p2)/le;
        p4 = (w1(e)*c1*a/le)+(p1+p2)/le;
        eforce(:,e) = eforce(:,e)+ [ 0.0 p3 -p1 0.0 p4 -p2]';
        %Negative of vector of fixed end forces in added to
        % the element force vector
    elseif(Ltype(e)==2) %for point loads
        a = d1(e); b = d2(e);
        if((a+b)~=le)
            fprintf(1,'Wrong input for point load\n');
            exit();
        end
        p2 = w1(e)*b/le - w1(e)*a*b*(a-b)/le^3 ;
        p3 = w1(e)*a*b^2/le^2;
        p5 = w1(e)*a/le+w1(e)*a*b*(a-b)/le^3;
        p6 = -w1(e)*a^2*b/le^2;
```

```matlab
        eforce(:,e) = eforce(:,e)+[0.0 p2 p3 0.0 p5 p6]';
    end
end
K=zeros(ndof,ndof); %initialize the global stiffness matrix of size ndof
M=zeros(ndof,ndof); %initialize the global mass matrix of size ndof
Feq=zeros(ndof,1); %nodal force vector due to element loads
for e=1:nElem
    c =Cx(e);
    s =Cy(e);
    le = L(e);
    ae = A(e)*E(e); %axial rigidity
    ei = E(e)*Iz(e); %flexural rigidity
    dof=[3*FirstNode(e)-2  3*FirstNode(e)-1    3*FirstNode(e)...
        3*SecondNode(e)-2  3*SecondNode(e)-1 3*SecondNode(e)];
    %rotation transformation matrix for element e
    T=[c     s    0    0    0     0;
      -s     c    0    0    0     0;
       0     0    1    0    0     0;
       0     0    0    c    s     0;
       0     0    0   -s    c     0;
       0     0    0    0    0     1;];
    %element stiffness matrix w.r.t local axes
     [estif]=
    [ae/le      0              0         -ae/le      0             0
        0    12*ei/le^3   6*ei/le^2       0      -12*ei/le^3   6*ei/le^2
        0    6*ei/le^2    4*ei/le         0      -6*ei/le^2    2*ei/le
    -ae/le      0              0         ae/le       0             0
        0    -12*ei/le^3  -6*ei/le^2      0      12*ei/le^3   -6*ei/le^2
        0    6*ei/le^2     2*ei/le        0      -6*ei/le^2    4*ei/le];
     if(imass==1)
        emass = (mbar(e)*le/2)*diag([1 1 0  1 1 0]);
     else
        emass = (mbar(e)*le/420)*...
        [140      0        0        70       0        0
          0      156     22*le       0       54      -13*le
          0     22*le    4*le^2       0      13*le    -3*le^2
         70       0        0        140       0        0
          0      54      13*le        0      156      -22*le
          0    -13*le   -3*le^2        0     -22*le    4*le^2];
     end
    %element mass matrix with reference to local axes
    ke=(T'*estif*T);
    %element stiffness matrix with reference to global axes
    me =(T'*emass*T);
    %element mass matrix with reference to global axes
    K(dof,dof)=K(dof,dof)+ke;
    %assembly of global stiffness matrix K
```

```matlab
        M(dof,dof)=M(dof,dof)+me;
        %assembly of global mass matrix M
        %{Feq} is vector of equivalent nodal forces due to element loads
        Feq(dof) = Feq(dof) + T'*eforce(:,e);
end
Fc = F+Feq;
%combined force vector {Fc}is obtained by adding
%vector of nodal loads {F} to {Feq}
%store the global stiffness matrix for future use to compute support
reactions
K1 = K;
M1 = M;
%-----------------------------------------------------------------
K(ifix,:)=0;
%Elements in rows with d.o.f corresponding to ifix are made 0
K(:,ifix)=0;
%Elements in columns with dof corresponding to ifix are made 0
K(ifix,ifix)=eye(length(ifix));
%Diagonal elements with d.o.f corresponding to ifix are made 1
M(ifix,:)=0;
M(:,ifix)=0;
M(ifix,ifix)=eye(length(ifix));
Fc(ifix) = 0; %corresponding columns of Global force vector are also
made zero
%-----------------------------------------------------------------
%obtaining vector of nodal displacements along global axes
u=K\Fc;
K1(ifix,:)=[];
K1(:,ifix)=[];
M1(ifix,:)=[];
M1(:,ifix)=[];
if(imass==1)%for lumped mass matrix, perform static condensation
    n=0;
    m = size(K1,1);
    SlaveDofs=1:m;
    for i=1:m
        if(find(M1(i,:))~=0)
            n = n+1;
            vec(n)=i;
        end
    end
    SlaveDofs(vec)=[];
    [KR,MR,T,Kss]=FastGuyanReduction(K1,M1,SlaveDofs);
    [V,D] = eig(KR,MR);
else
  [V,D] = eig(K1,M1);
end
```

```
[var1,permutation]=sort(diag(D));
D=D(permutation,permutation);
V=V(:,permutation);
w = sqrt(diag(D));

fprintf(1,'The natural frequencies are\n');
for i=1:size(w)
    fprintf(1,'%d %10.3f\n',i,w(i));
end
```

A.17. FREE VIBRATION ANALYSIS OF BARS AND TRUSSES

```matlab
clear
clc
inp= fopen('dyntruss.inp','r');
[mat] = fscanf(inp,'%d %d',2);
% this line reads number of nodes and number of elements in the truss
nNode = mat(1);   %number of nodes
nElem = mat(2);   %number of elements
[mat1] = fscanf(inp,'%d %f %f %d %d %e  %e',[7,nNode]);
mat1 = mat1'; %Transposed for future use
[mat2] = fscanf(inp,'%d %d %d %f %e %e',[6,nElem]);
mat2 = mat2';%transposed for future sure
ndof=2*nNode;
Xval = mat1(:,2);
Yval = mat1(:,3);
resx = mat1(:,4);
resy = mat1(:,5);
fx = mat1(:,6);
fy = mat1(:,7);
F= zeros(ndof,1);
for i=1:nNode
  F(2*i-1) = fx(i);
  F(2*i)   = fy(i);
end
bn = 0;
for i=1:nNode
   if(resx(i)==0)
        bn = bn+1;
        ifix(bn) = 2*i-1;
   end
   if(resy(i)==0)
        bn = bn+1;
        ifix(bn) = 2*i;
   end
 end
%---------------------------------------------------------------------
%READING ELEMENT INFORMATION
E          = zeros(nElem,1);
Cx         = zeros(nElem,1);
Cy         = zeros(nElem,1);
FirstNode  = mat2(:,2);
SecondNode = mat2(:,3);
A          = mat2(:,4);
E          = mat2(:,5);
rho        = mat2(:,6);
```

```matlab
K=zeros(ndof,ndof); %initialize the global stiffness matrix of size ndof
M=zeros(ndof,ndof); %initialize the global mass matrix of size ndof
for e=1:nElem
    x1 =Xval(FirstNode(e));  y1 =Yval(FirstNode(e));
    x2 =Xval(SecondNode(e)); y2 =Yval(SecondNode(e));
    le = sqrt((x2-x1)^2+(y2-y1)^2);
    c = (x2-x1)/le;
    s = (y2-y1)/le;
    mbar = rho(e)*A(e);
    ae = A(e)*E(e); %axial rigidity
    dof=[2*FirstNode(e)-1  2*FirstNode(e)...
         2*SecondNode(e)-1 2*SecondNode(e)];
    %rotation transformation matrix for element e
    T=[c    s    0    0
      -s    c    0    0
       0    0    c    s
       0    0   -s    c];
    %element stiffness matrix w.r.t local axes
     estif = (ae/le)*  [ 1  0   -1   0
                         0  0    0   0
                        -1  0    1   0
                         0  0    0   0];
     emass = (mbar*le/6)*[2  0  1   0
                          0  2  0   1
                          1  0  2   0
                          0  1  0   2];
    %element mass matrix with reference to local axes
    ke=(T'*estif*T);
    %element stiffness matrix with reference to global axes
    me =(T'*emass*T);
    %element mass matrix with reference to global axes
    K(dof,dof)=K(dof,dof)+ke;  %assembly of global stiffness matrix K
    M(dof,dof)=M(dof,dof)+me;  %assembly of global mass matrix M
end
%store the global stiffness matrix for future use to compute support
reactions
K1 = K;
M1 = M;
%-------------------------------------------------------------------
K(ifix,:)=0;
%Elements in rows with d.o.f corresponding to ifix are made 0
K(:,ifix)=0;
%Elements in columns with dof corresponding to ifix are made 0
K(ifix,ifix)=eye(length(ifix));
%Diagonal elements with d.o.f corresponding to ifix are made 1
M(ifix,:)=0;
M(:,ifix)=0;
```

```
M(ifix,ifix)=eye(length(ifix));
Fc(ifix) = 0;
%corresponding columns of Global force vector are also made zero
K1(ifix,:)=[];
K1(:,ifix)=[];
M1(ifix,:)=[];
M1(:,ifix)=[];
[V,D] = eig(K1,M1);
[var1,permutation]=sort(diag(D));
D=D(permutation,permutation);
V=V(:,permutation);
w = sqrt(diag(D));

fprintf(1,'The natural frequencies are\n');
for i=1:size(w)
    fprintf(1,'%d %10.3f\n',i,w(i));
end
```

Step-by-Step Procedure for Solving Problems on ABAQUS

B.1 AXIALLY LOADED BARS

The following is the procedure for the analysis of an axially loaded bar using ABAQUS for solving Numerical Example 6.1 solved earlier in this chapter.

I.   Start **ABAQUS**. In **Start Session** dialog, choose **Create Model Database with Standard/Explicit Model**. On the main menu, click on **File** and set **Set Work Directory** to choose your working directory. Click on **Save As** and name the file **bar-1.cae**.

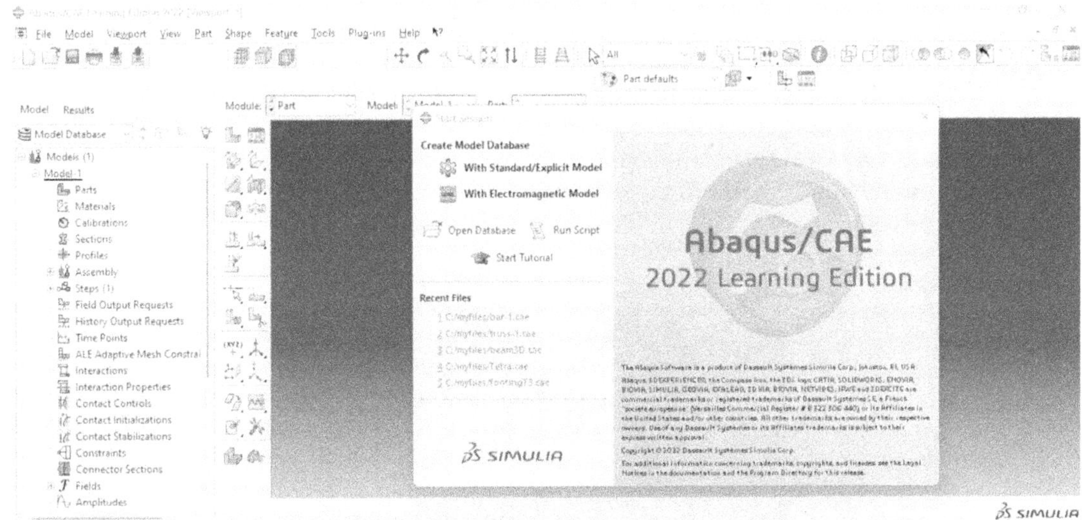

II. Click on *Create Part* icon in the *Module Icon* menu. In the **Create Part** dialog, give name as **bar-A**, choose *Modelling Space* as **2D Planar**, *Type* as **Deformable** and *Base feature* as **Wire** and *Type* as **Planar**. Indicate *Approximate size* as **1** and press **Continue**.

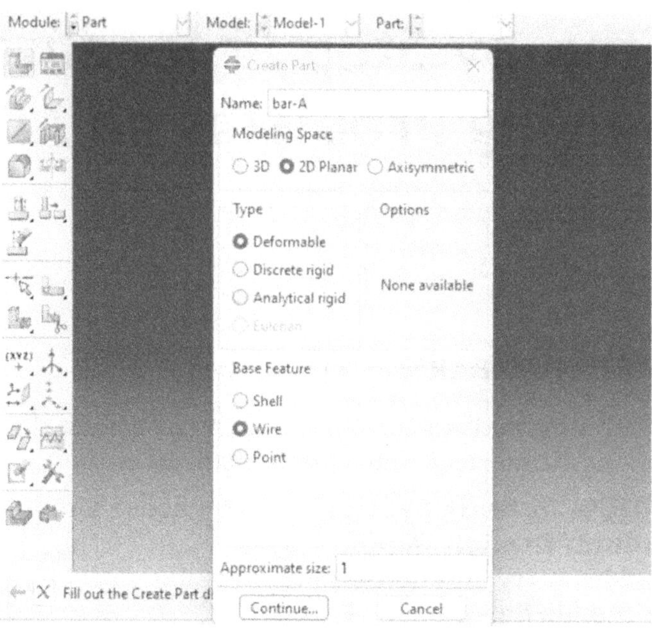

III. In the sketcher menu, choose Create Lines: connected and enter **0.0,0.0** in the window *"Pick a starting point for a line or enter X, Y."* and press **Enter**. When the same dialog appears again, enter **0.5,0.0** and press **Enter**. When the same dialog appears again, enter **0.9,0.0** and press **Enter**. Press × symbol for the next dialog. A prompt *"Sketch the section for the wire"* appears. Press **Done** and complete the line. A grey line appears in the sketcher window.

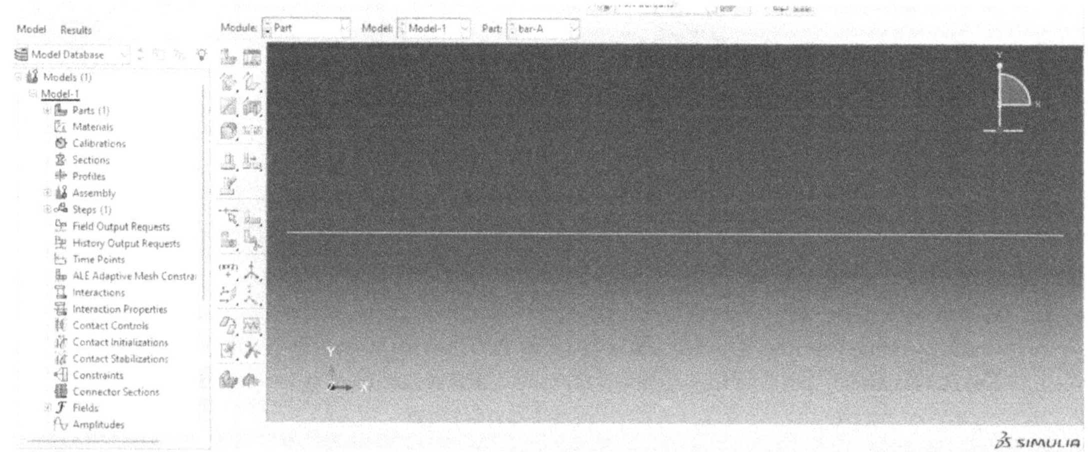

IV. Click the *Material Manager* icon, and click *Create* in the **Material Manager** dialog box that appears. Enter **steel** in place of Material-1. Edit the description box and enter *"**E is in kPa**"*. Enter **Mechanical** menu and

Elastic tab and enter **2.0e8** for Young's modulus and **0.3** for Poisson's ratio. Press ok to confirm the entered values. Press Dismiss on the **Material Manager Dialog** box.

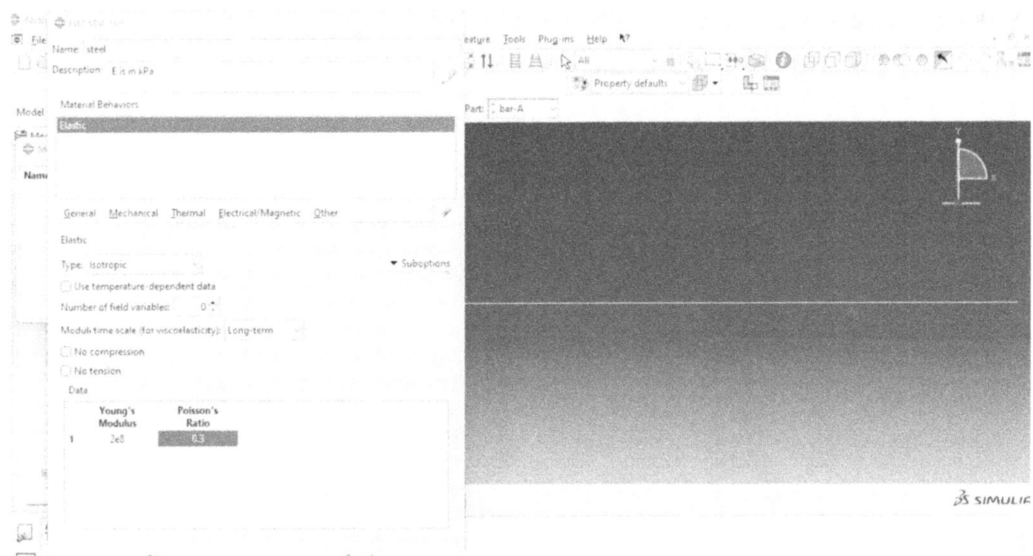

V. Click on the **Create Section** icon. In the dialog box that appears, enter **Section** in place of Section-1. Choose ***Beam*** and ***Truss*** and press *Continue.* In the **Edit Section dialog** box, enter cross sectional area as 0.0026 and click Ok. Repeat the procedure to create Section-2 with cross sectional area 0.0013.

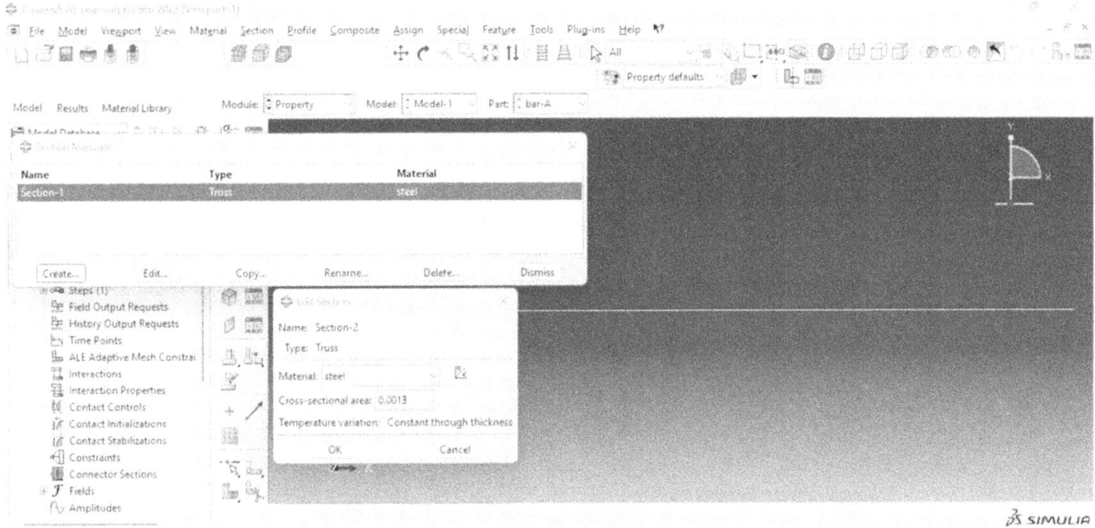

VI. Click on **Assign Section** icon. Use mouse click to select the first part of the line and press *Done.* In the **Edit Section Dialog**, choose Section-1. press Ok. For the prompt: *Select the regions to be assigned a section*, select the line using the mouse click and press *Done.* Similarly, use mouse click to select the second part of line and assign Section-2. Finish by pressing *Done.*

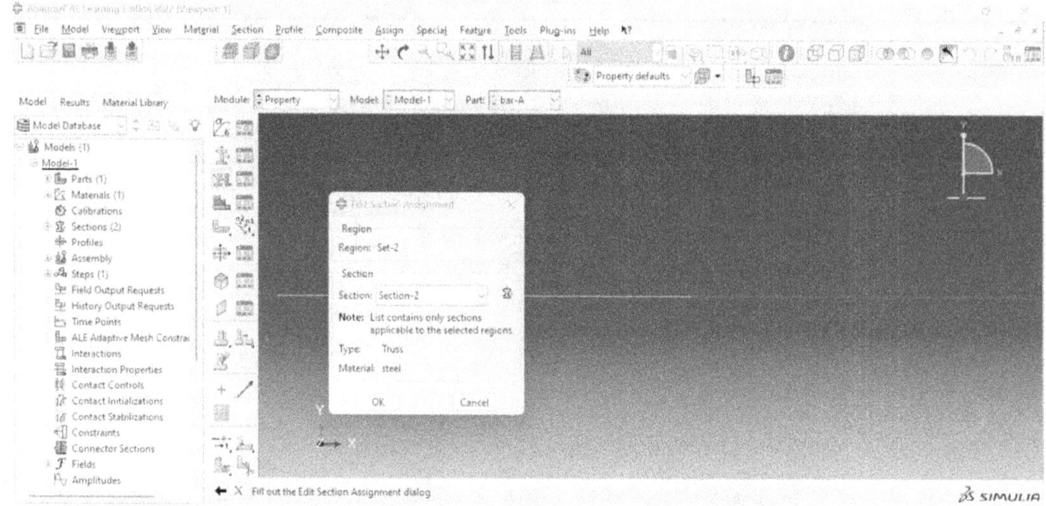

VII. Click assembly in module menu and click on **Create Instance** icon. Choose *Independent (Mesh on instance)* and press Ok. Click *Step* in module menu and click on **Create Step** icon. Rename Step-1 as ApplyLoad, accept *Static, General* in and click *Procedure Type* and click *Continue.* In the description of **Edit step dialog**, type Apply axial load on bar. Click Ok

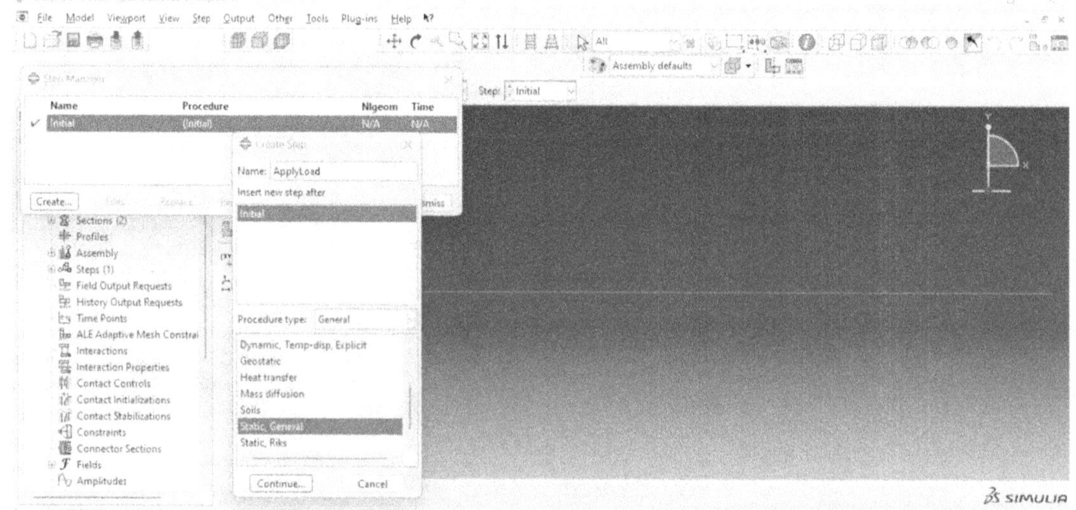

VIII. Click **Load** in module menu and click on **Create Load** icon. In the **Edit Load dialog** box, replace Load-1 by typing P1. Use mouse click to select point for the load and press *Done*. The **Edit Load dialog** appears. Enter CF1 as 250 and click OK. Load P1 is now displayed. Repeat the procedure for load P2 and enter 400 for CF1.

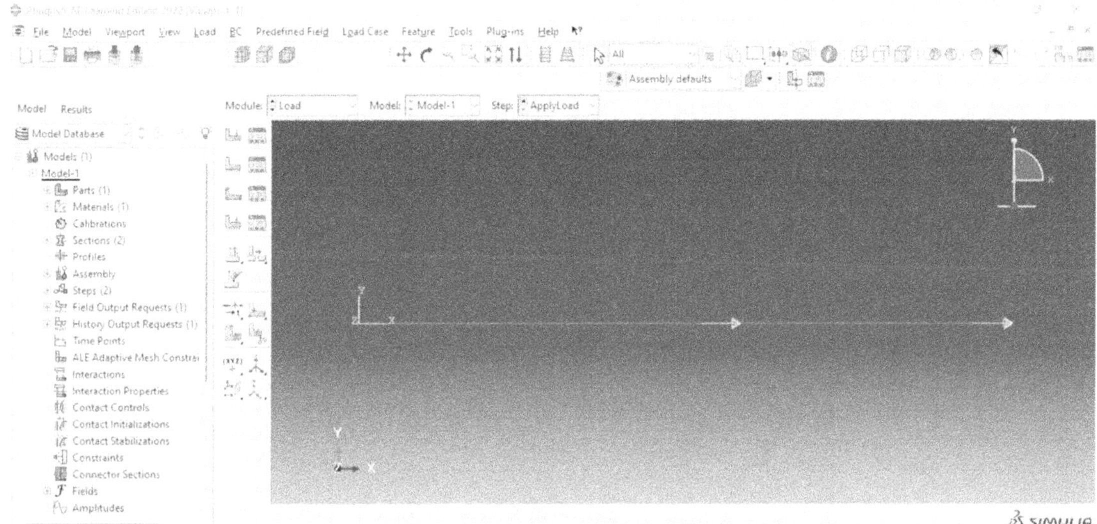

IX. Click **BC** on main menu, and click ***Create***. Type **fixed** for name and choose Category **Mechanical** and Type *Displacement/Rotation* and click *Continue*. With the mouse, choose the leftmost point on bar and click *Done*. In the Edit Boundary condition dialog box, click checkbox for U1,U2,UR3 and click *Ok*.

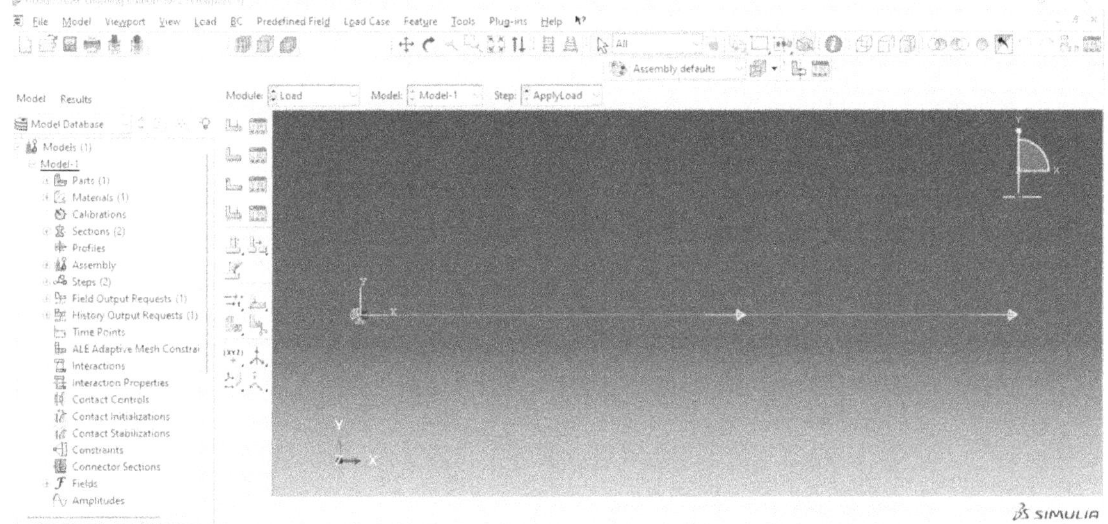

X. Select **Mesh** from Module menu and choose **Mesh** from main menu. Choose **Element Type** and use shift and mouse to click both elements of bar and press *Done*. Select *Standard* for element library, geometric order as *Linear* and choose *Truss. T2D2*: The description : *A 2-node linear 2-D truss* is displayed. Press *Ok* to accept. Press *Done* to end the dialoag. Click **Mesh** and Instance w from main menu.

XI. Go to the **Job** menu under **Module** and create a new job as *ApplyPressure* and click *Continue*. Accept the default options and click Ok. Dismiss the **Job Manager Dialog**. Click *Submit*. After Job is completed, click on *Results* and choose **Visualization** from Module menu. Choose **Plot** from the main menu and choose *Plot on the undeformed shape*. All results can now be seen as per choice.

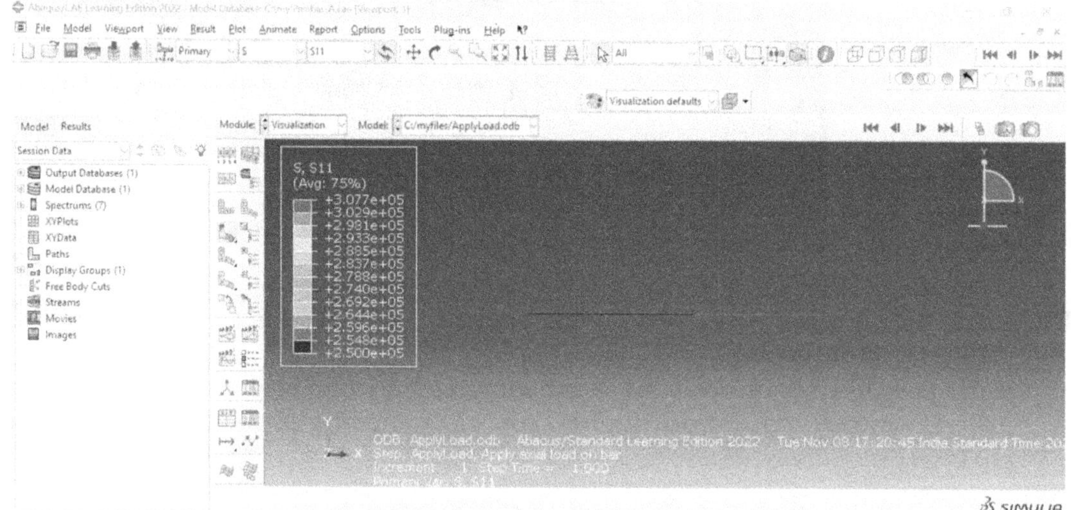

The values of stress in elements 1 and 2 obtained match exactly with the results obtained using MATLAB code.

B.2   PLANE TRUSSES

The following is the procedure for the analysis of a triangular truss using ABAQUS for solving Numerical Example 6.4 solved earlier in this chapter.

A. Start **ABAQUS**. In **Start Session** dialog, choose **Create Model Database with Standard/Explicit Model**. On the main menu, click on **File** and set **Set Work Directory** to choose your working directory. Click on **Save As** and name the file **truss-A.cae**.

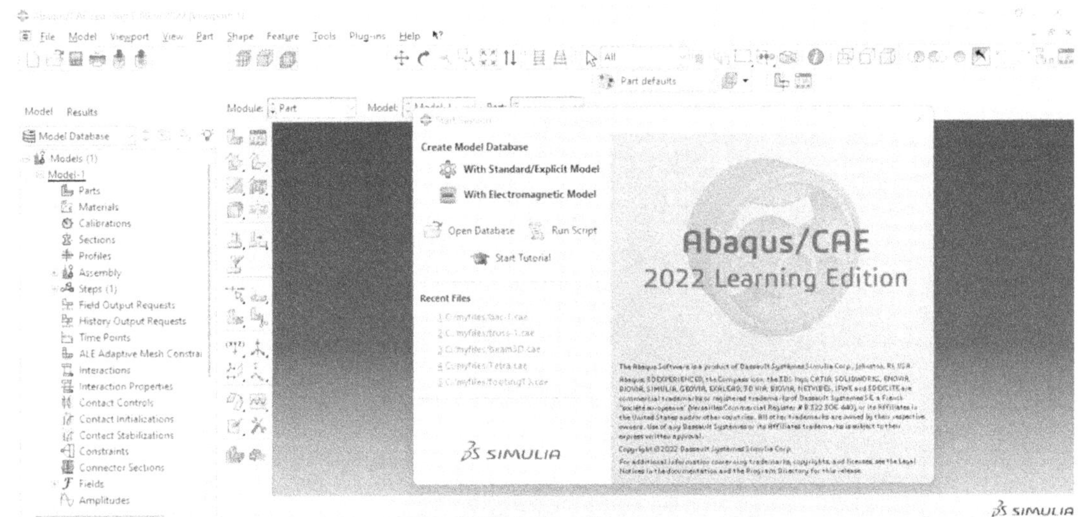

B. Click on ***Create Part*** icon in the *Module Icon* menu. In the **Create Part** dialog, give name as **truss-A**, choose *Modelling Space* as **2D Planar** , *Type* as **Deformable** and *Base feature* as **Wire** and *Type* as **Planar**. Indicate *Approximate size* as **2** and press **Continue.**

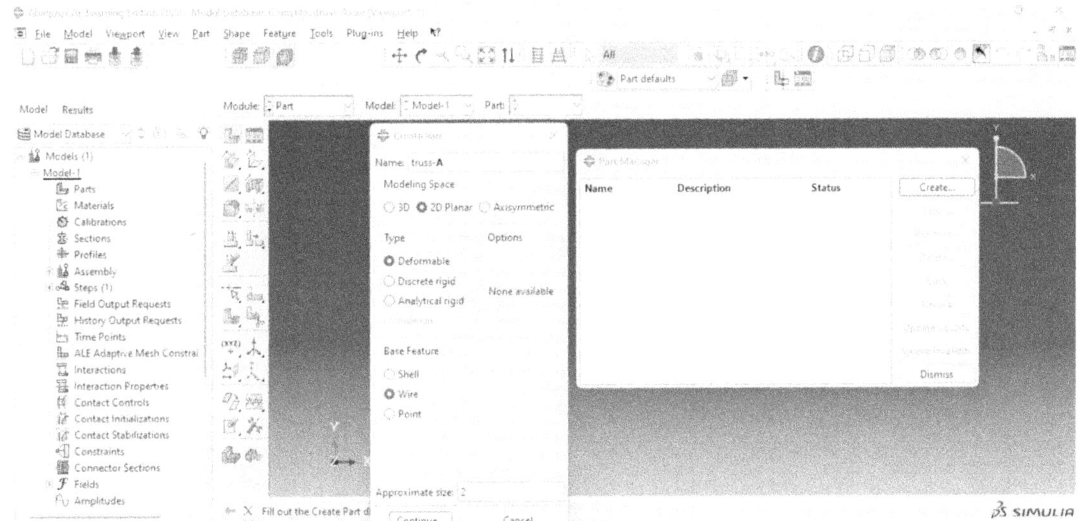

C. In the *sketcher* menu, choose **Create Lines** and enter **0.0,0.0** in the window "*Pick a starting point for a line or enter X, Y.*" and press ***Enter***. When the same dialog appears again, enter **1,1.732** and press ***Enter***. When the same dialog appears again, enter **2,0** and press ***Enter***. When the same dialog appears again, enter **0,0** and press ***Enter***. Press × symbol for the next dialog. A prompt "*Sketch the section for the wire*" appears. Press ***Done*** and complete the line. A grey line appears in the sketcher window.

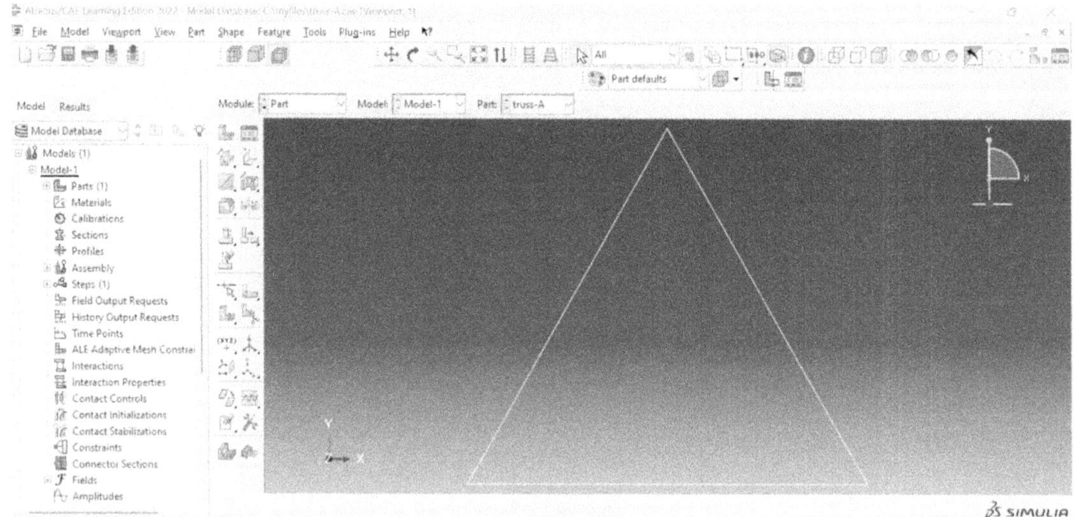

D. Click the *Material Manager* icon, and click *Create* in the Material Manager dialog box that appears. Enter **steel** in place of Material-1. Edit the description box and enter "***E is in kPa***". Enter Mechanical menu and Elastic tab and enter **2.0e8** for Young's modulus and **0.3** for Poisson's ratio. Press ok to confirm the entered values. Press Dismiss on the **Material Manager Dialog** box.

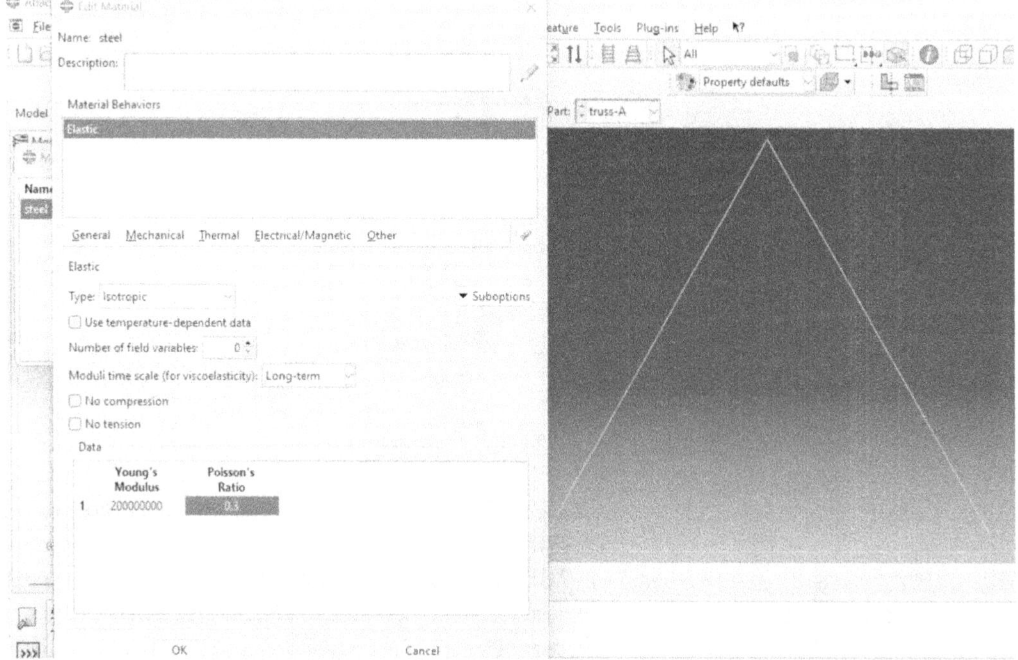

E. Click on the **Create Section** icon. In the dialog box that appears, enter *TrussSection* in place of Section-1. Choose *Beam* and *Truss* and press *Continue.* In the **Edit Section dialog** box, enter cross sectional area as 0.01 and click Ok.

F. Click on **Assign Section** icon. Use mouse click to select the first part of the line and press *Done.* In the **Edit Section Dialog**, choose Section-1. press Ok. For the prompt: *Select the regions to be assigned a section,* select the plate using the mouse click and press *Done.* Similarly, use mouse click to select the second part of line and assign Section-2. Finish by pressing *Done.*

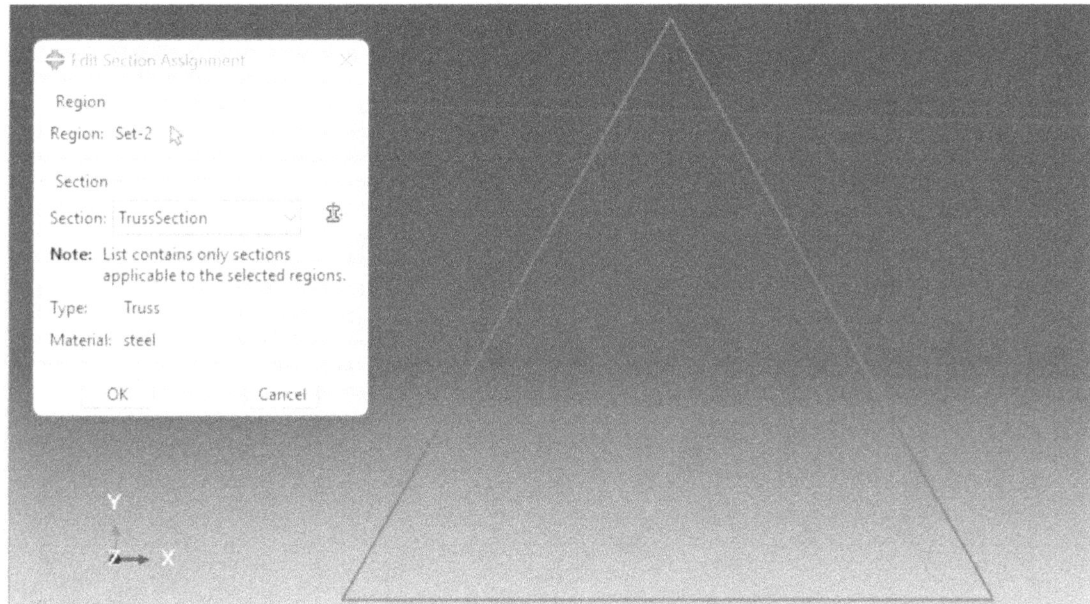

G. Click assembly in module menu and click on **Create Instance** icon. Choose *Independent (Mesh on instance)* and press *Ok.* Click **Step** in **module** menu and click on *Create Step* icon. Rename Step-1 as ApplyLoad, accept *Static, General* in and click *Procedure Type* and click *Continue.* In the description of **Edit step dialog**, type Apply axial load on bar. Click *Ok*

H. Click **Load** in **Module** menu and click on **Create Load** icon. In the **Edit Load dialog** box, replace Load-1 by typing *PointLoad*. Select *Concentrated Load* under choose *Concentrated Force*. Use mouse click to select point for the load and press *Done*. The **Edit Load dialog** appears. Enter CF1 as 100 and CF2 as -150 and click OK. Load is now displayed.

I. Click **BC** on main menu, and click *Create*. Type **Hinge** for name and choose Category Mechanical and Type *Displacement/Rotation* and click *Continue*. With the mouse, choose the node 1 on truss and click *Done*. In the Edit

Boundary condition dialog box, click checkbox for U1,U2 and click *OK*. Type **Roller** for name and choose Category *Mechanical* and Type *Displacement/Rotation* and click *Continue*. With the mouse, choose the node 3 on truss and click *Done*. In the **Edit Boundary condition dialog** box, click checkbox for U2 and click *OK*.

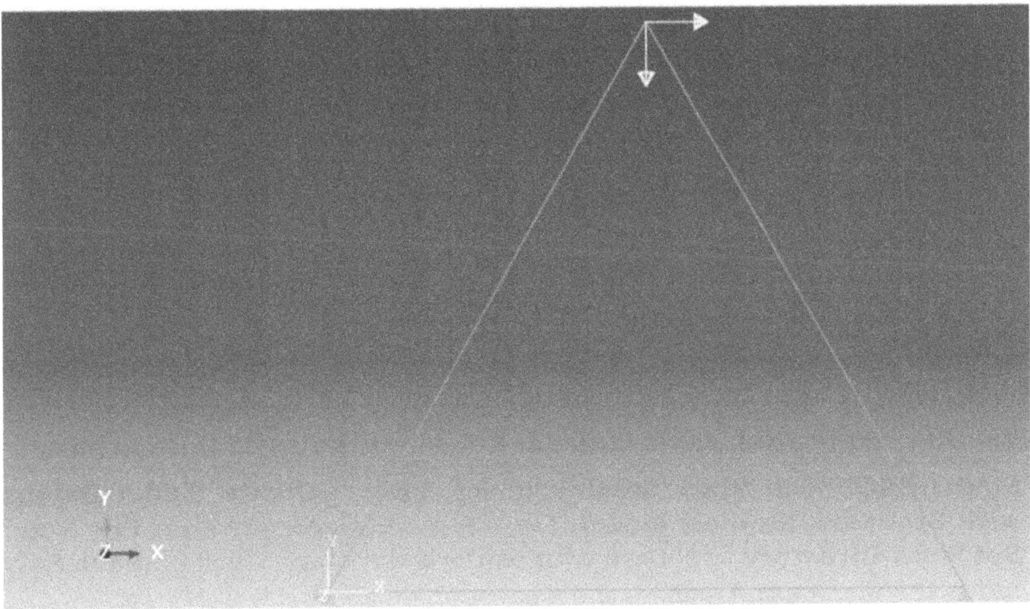

J. Select **Mesh** from **Module** menu and choose **Mesh** from main menu. Choose *Element Type* and use mouse to click all elements of from and press *Done*. Select Standard for element library, geometric order as Linear and choose Truss. T2D2: The description : *A 2-node linear 2-D truss* is displayed. Press Ok to accept. Press *Done* to end the dialoag. Select *Mesh* from the main menu and click *Instance*. After selecting the elements, press *Done*. Select **mesh** and instances from main menu.

K. Go to the **Job** menu under **Module** and create a new job as *ApplyLoad* and click *Continue*. Accept the default options and click OK. Dismiss the **Job Manager Dialog**. Click *Submit*. After Job is completed, click on *Results* and choose **Visualization** from **Module** menu. Choose **Plot** from the main menu and choose Plot on the undeformed shape. All results can now be seen as per choice.

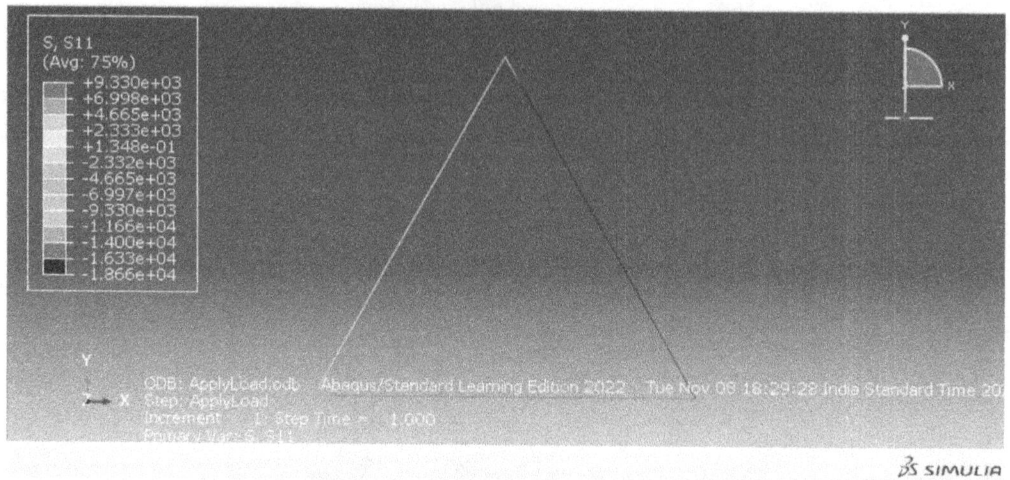

B.3 BEAMS

The following is the procedure for the analysis of continuous beam using ABAQUS for solving Numerical Example 6.9 solved earlier in this chapter.

A. Start **ABAQUS**. In **Start Session** dialog, choose **Create Model Database with Standard/Explicit Model**. On the main menu, click on **File** and set **Set Work Directory** to choose your working directory. Click on **Save As** and name the file **beam-A.cae**.

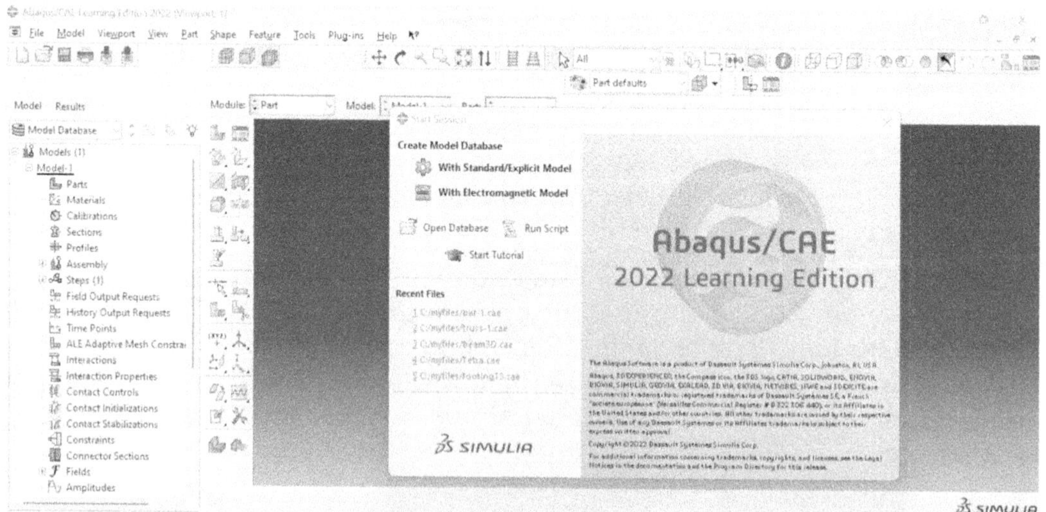

B. Click on ***Create Part*** icon in the *Module Icon* menu. In the **Create Part** dialog, give name as **beam-A**, choose *Modelling Space* as **2D Planar**, *Type* as **Deformable** and *Base feature* as **Wire** and *Type* as **Planar**. Indicate *Approximate size* as **12** and press **Continue**.

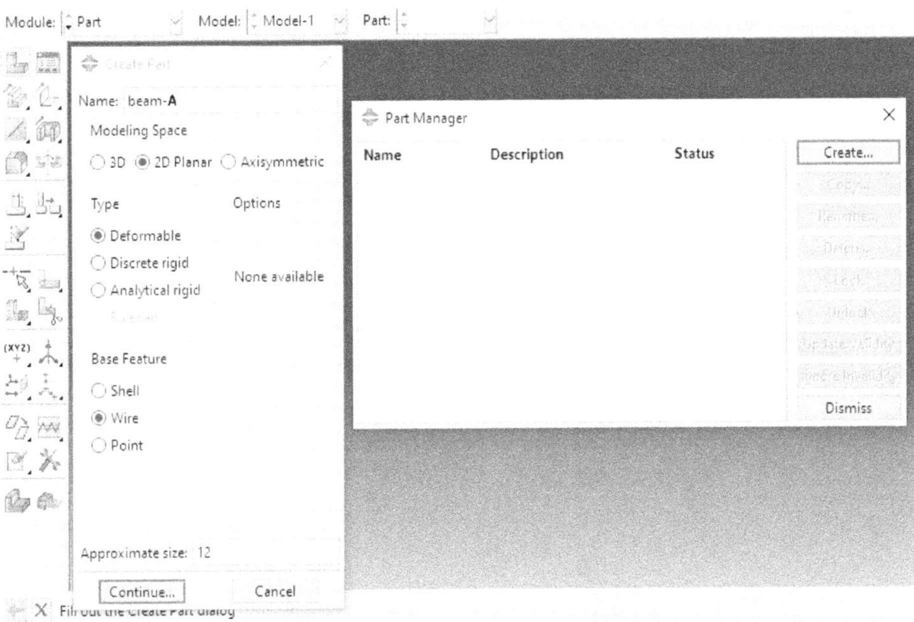

C. In the sketcher menu, choose **Create Lines** and enter **0.0,0.0** in the window *"Pick a starting point for a line or enter X,Y."* and press **Enter**. When the same dialog appears again, enter **5,0** and press **Enter**. When the same dialog appears again, enter **7,0** and press **Enter**. When the same dialog appears again, enter **8,0** and press **Enter**. When the same dialog appears again, enter **12,0** and press **Enter** Press × symbol for the next dialog. A prompt *"Sketch the section for the wire"* appears. Press **Done** and complete the line. A grey line appears in the sketcher window.

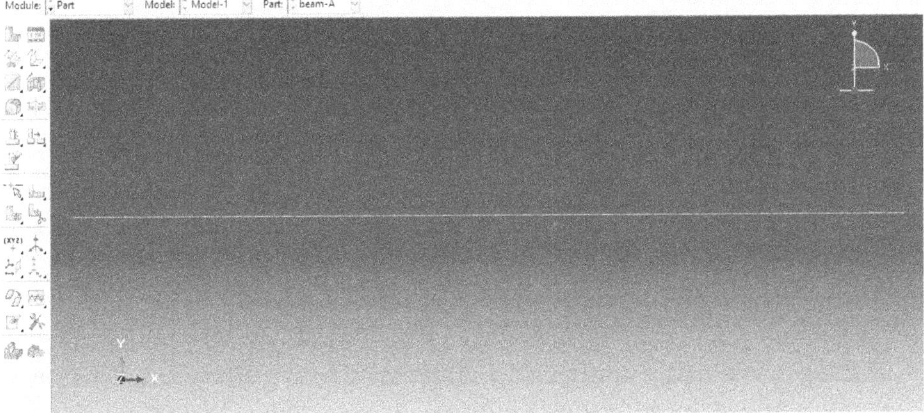

D. Click the **Material Manager** icon, and click **Create** in the Material Manager dialog box that appears. Enter **steel** in place of *Material-1*. Edit the description box and enter "***E is in kPa***". Enter *Mechanical* menu and *Elastic*

tab and enter **2.0e8** for Young's modulus and **0.3** for Poisson's ratio. Press ok to confirm the entered values. Press *Dismiss* on the **Material Manager Dialog** box.

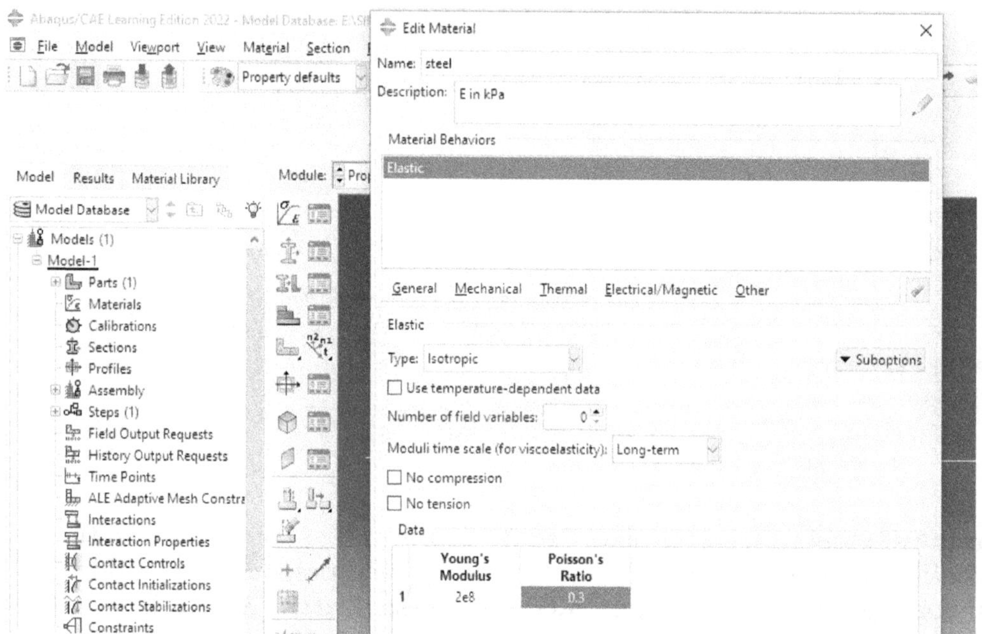

E. Click on the **Create Section** icon. In the dialog box that appears, enter **BeamSection** in place of Section-1. Choose *Beam* and *Beam* and press *Continue.* In the **Edit Section dialog** box, choose rectangular profile and press *Continue.* Enter 0.2 and 0.3 for a and b, and click Ok. Now click Ok to close the **Edit Section dialog box**.

F. Click on **Assign Section** icon. Use mouse click to select the entire line and press *Done.* In the **Edit Section Dialog**, choose *BeamSection.* press Ok. For the prompt: *Select the regions to be assigned a section,* select the line using the mouse click and press *Done.* Finish by pressing *Done.*

G. Click **assembly** in **module** menu and click on **Create Instance** icon. Choose *Independent (Mesh on instance)* and press Ok. Click on **Assign Beam orientation** icon and select the entire beam using Shift+Mouse click. Press then. Accept values of 0.0,0.0,-1.0 in the textbox and click Ok to confirm input.

H. Click **Step** in **module** menu and click on **Create Step** icon. Rename Step-1 as ApplyLoad, accept *Static, General* in and click *Procedure Type* and click *Continue.* In the description of **Edit step dialog**, type *Apply load* on beam. Click Ok.

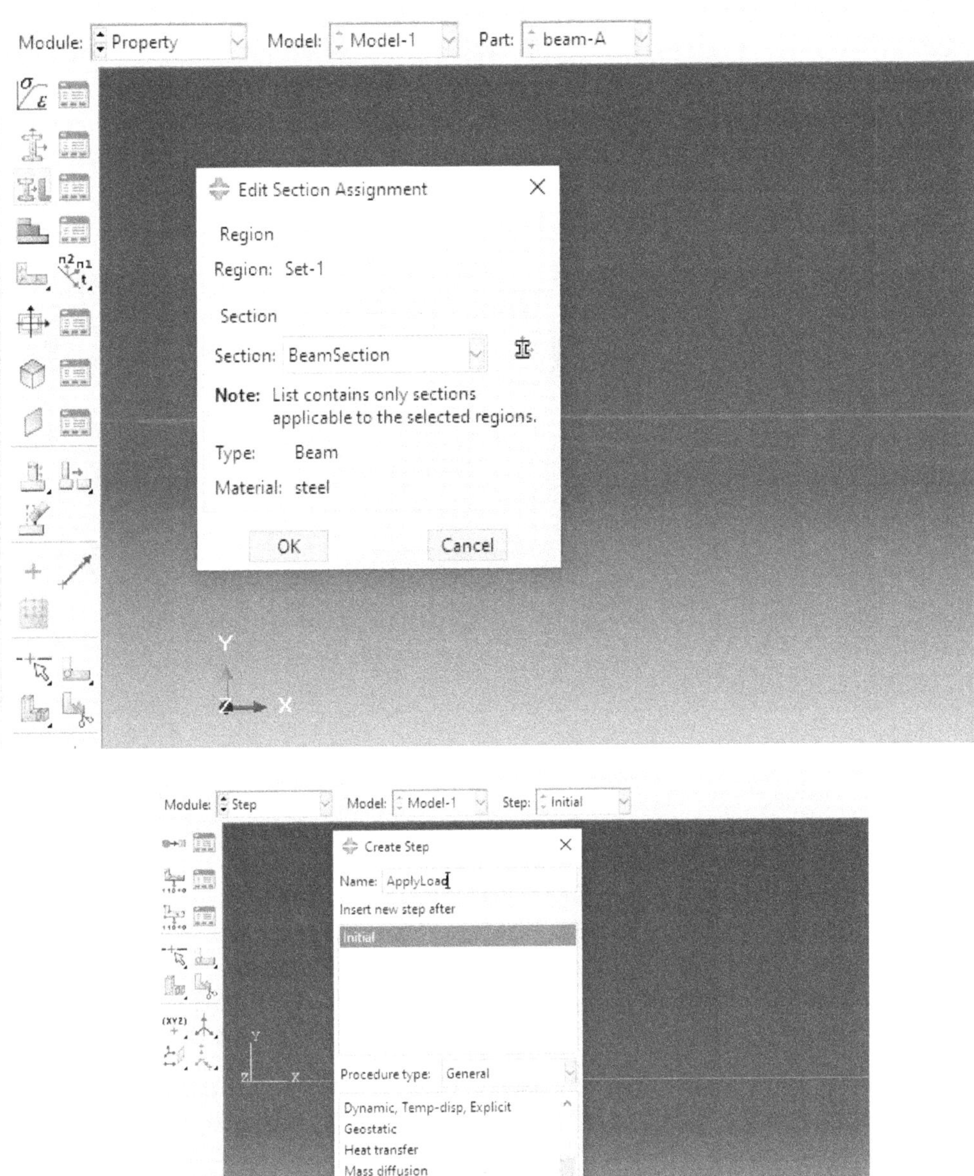

I. Click on **Load** in **Module** menu and press on **CreateLoad** icon. Choose line load under *Types for Selected step* and press *Continue*. Click on the first line segment and click *Done*. Enter **-25.0** under Component-2 and press Ok.

Press on **CreateLoad** icon again. Choose *Concentrated load* under *Types for Selected step* and press *Continue*. With mouse select node at 7m from left end. Enter **-100.0** under CF2 and press Ok. Press on **CreateLoad** icon again. Choose line load under *Types for Selected step* and press *Continue*. Click on the first line segment and click *Done*. Enter **-50.0** under Component-2 and press Ok.

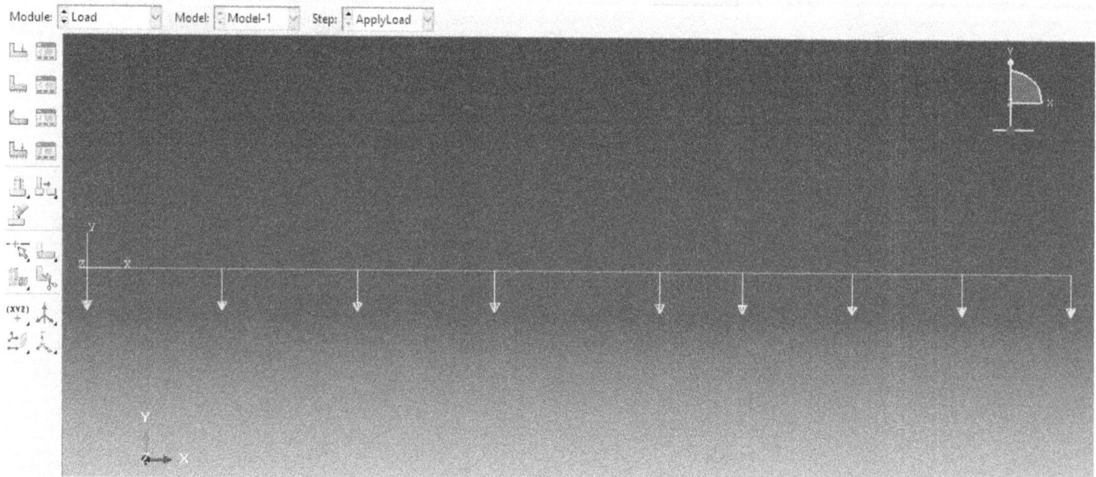

J. Click on **BC** in the main menu and choose **Create**. In the Create Boundary Condition dialog box, type Fixed in place of BC-1, select *Mechanical* under Category and choose *Symmetry/Anti-symmetry/Encastre* under Types for Selected Step. Click *Continue* and choose the left-most node using mouse-click and press *Done*. In the **Edit Boundary Condition dialog** box, choose Encastre and press Ok. Repeat the procedure to create Roller and choose Displacement/Rotation. Click on U2 in the **Edit Boundary Condition dialog** box and press Ok. Repeat the procedure to create Roller-2 by choosing node 4. Create *Hinge* by choosing node 5 and click on U1 and U2.

K. Click **Module** menu and choose **Mesh** from the main menu. Click on *Element type* and using Shift+ Mouse click choose the entire beam and click on *Done*. In the **Element type dialog** box, choose *Beam* and *Linear* to select *B21: A 2-node linear beam in a plane*. Click Ok and click *Done*. Choose **Mesh** from the main menu and click on **Instance**. Click Yes

L. Click **Module** menu and choose Job. Replace Job-1 by beam-A and click *Continue* and press Ok. Click on **Job Manager** icon and click on Submit button and wait till analysis is finished. Click on **Results**

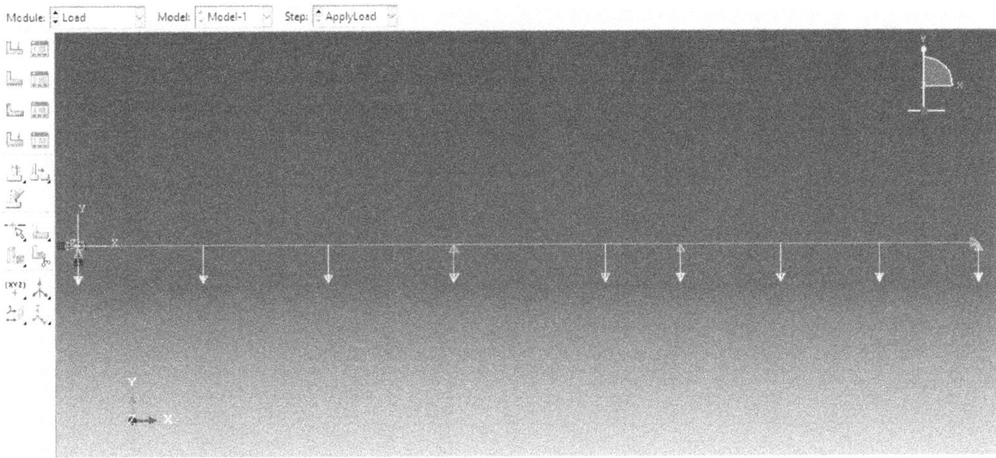

B.4   PLANE STRESS

The following is the common procedure for the analysis of in-plane stress using ABAQUS. It shall be noted that that the selection of element type as given in step M will be different for four node and eight node elements.

A.   Start **ABAQUS**. In **Start Session** dialog, choose **Create Model Database with Standard/Explicit Model**. On the main menu, click on **File** and set **Set Work Directory** to choose your working directory. Click on **Save As** and name the file **Pplate.cae**.

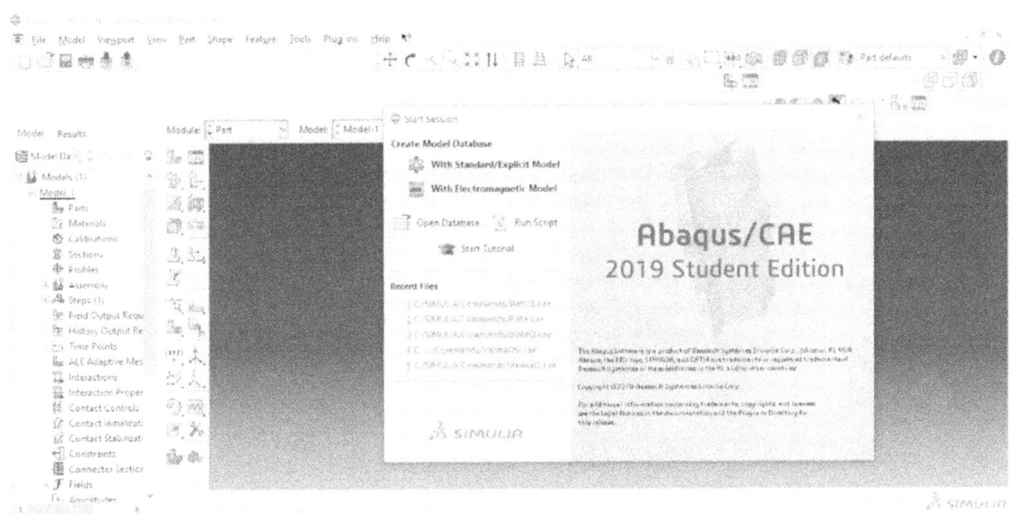

B.   Click on **Create Part** icon in the *Module Icon* menu. In the **Create Part** dialog, give name as **Pplate**, choose *Modelling Space* as **2D Planar** , *Type*

as **Deformable** and *Base feature* as **Shell**. Indicate *Approximate size* as **2** and press **Continue.**

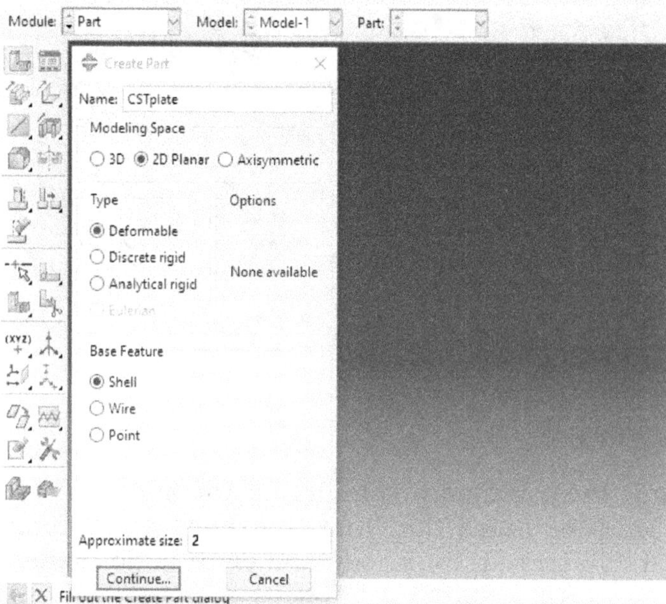

C. In the sketcher menu, choose Create Lines: Rectangle(4 lines) and enter **0,0**in the window *"Pick a starting corner for Rectangle or enter X,Y."* and press ***Enter***. In the window *"Pick the opposite corner for Rectangle or enter X,Y."*, enter **0.5,0.25** and press ***Enter***. Press × symbol for the next dialog. A prompt *"Sketch the section for the planar shell"* appears. Press ***Done*** and complete the rectangle. A grey rectangle appears in the sketcher window.

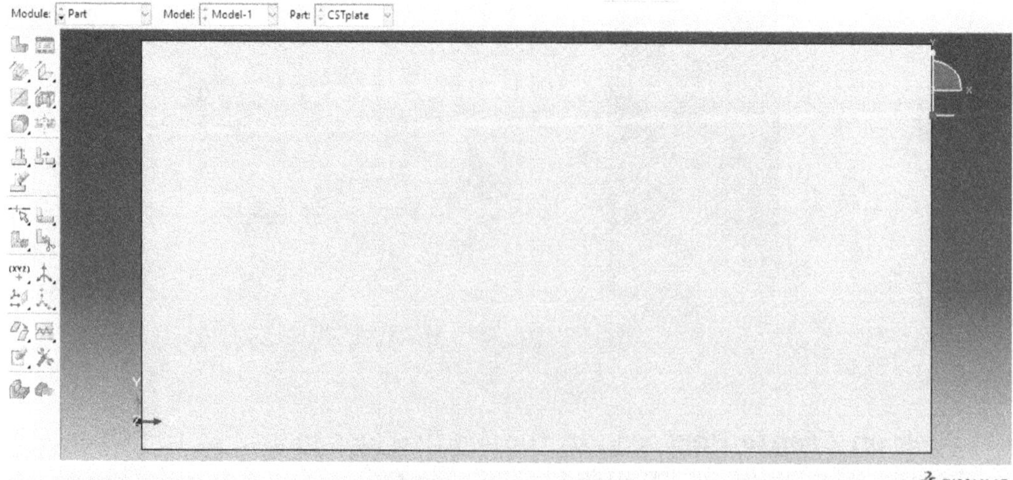

D.   Click the *Material Manager* icon, and click *Create* in the Material Manager dialog box that appears.   Enter **Steel** in place of Material-1. Edit the description box and enter "***E is in kPa***". Enter Mechanical menu and Elastic tab and enter **2.1e8** for Young's modulus and **0.3** for Poisson's ratio. Press ok to confirm the entered values. Press Dismiss on the Material Manager Dialog box.

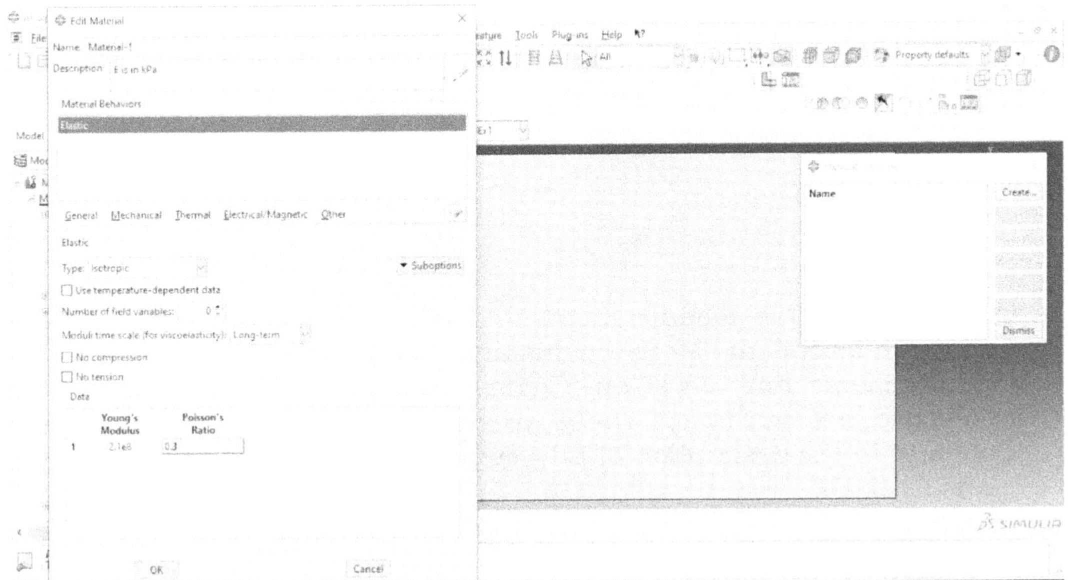

E.   Click on the *Create Section* icon. In the dialog box that appears, enter PlateSection in place of Section-1. Choose *Shell* and *Homogeneous* and press *Continue.* In the Edit Section dialog box, enter thickness of section as 0.025 and click Ok

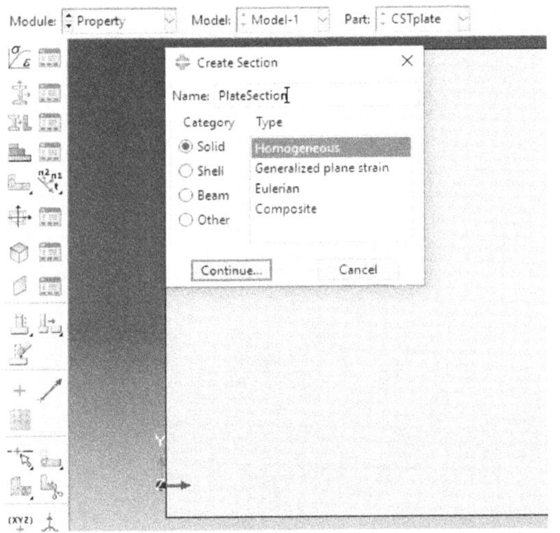
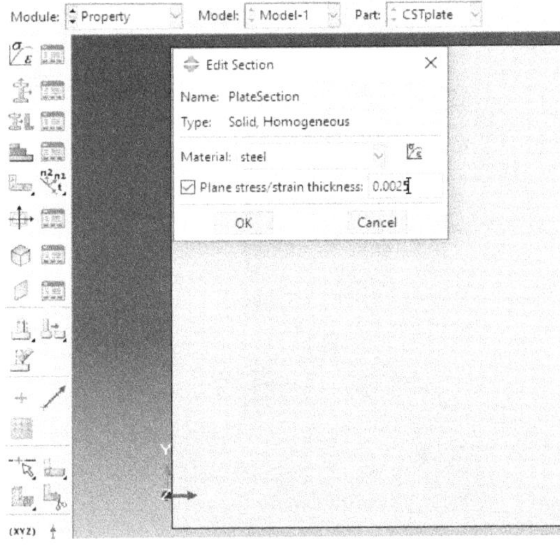

F. Click on *Assign Section* icon. Use mouse click to select the plate and press *Done.* In the Edit Section Dialog, Plate Section appears. Accept and press Ok.

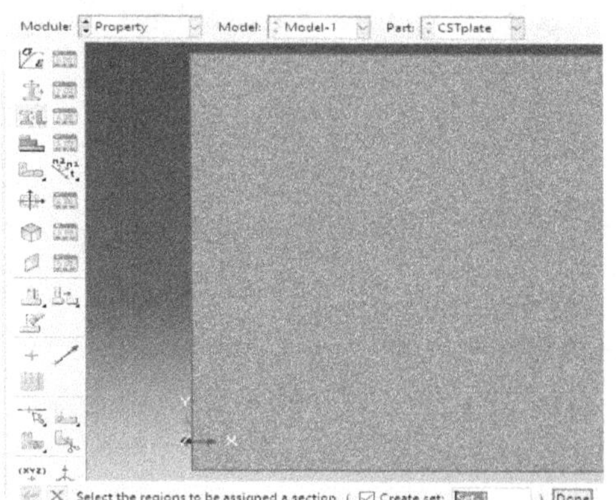 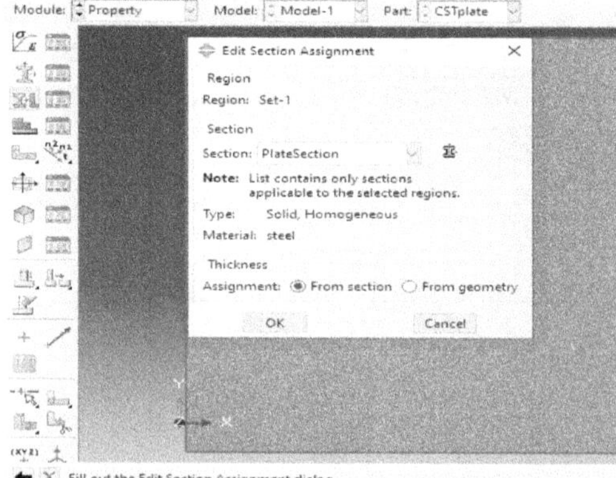

G. Click assembly in module menu and click on *Create Instance* icon. Choose Independent (Mesh on instance) and press Ok. Click *Step* in module menu and click on *Create Step* icon. Rename Step-1 as ApplyLoad, accept *Static, General* in and click *Procedure Type* and click *Continue.* In the description of Edit step dialog, type Apply point load on plate. Click Ok.

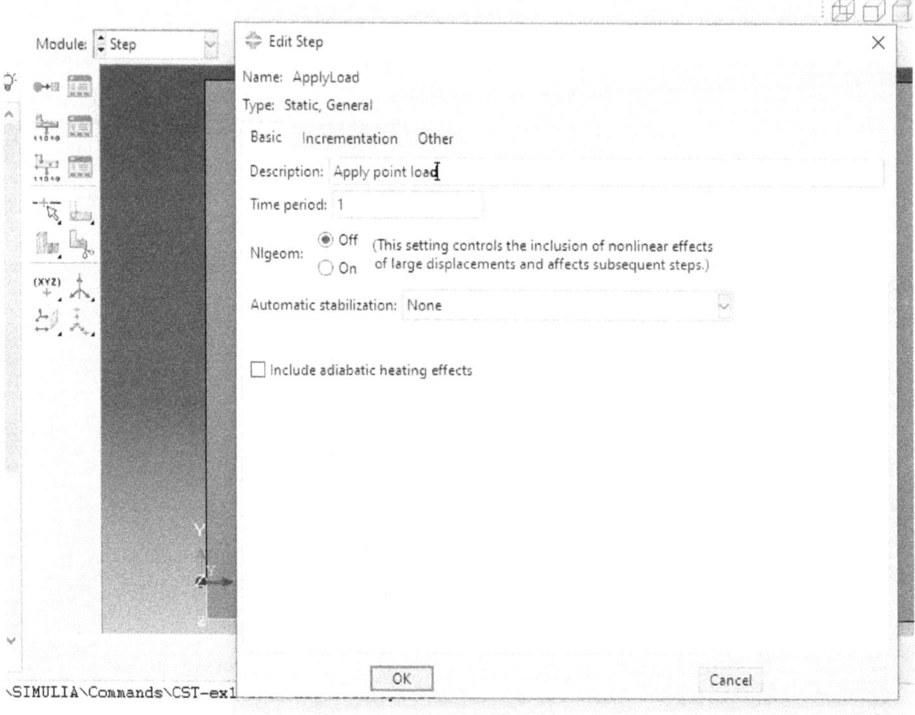

H.   Click *Load* in module menu and click on *Create Load* icon. Replace Load-1 by typing **PointLoad**. Accept Category as *Mechanical* and choose *Concentrated Force* in the *Type of Step* dialog. Press *Continue*. Use mouse click to select top-right corner for the load and click *Done*. The Edit Load dialog appears. Enter -100 for CF2 and click Ok.

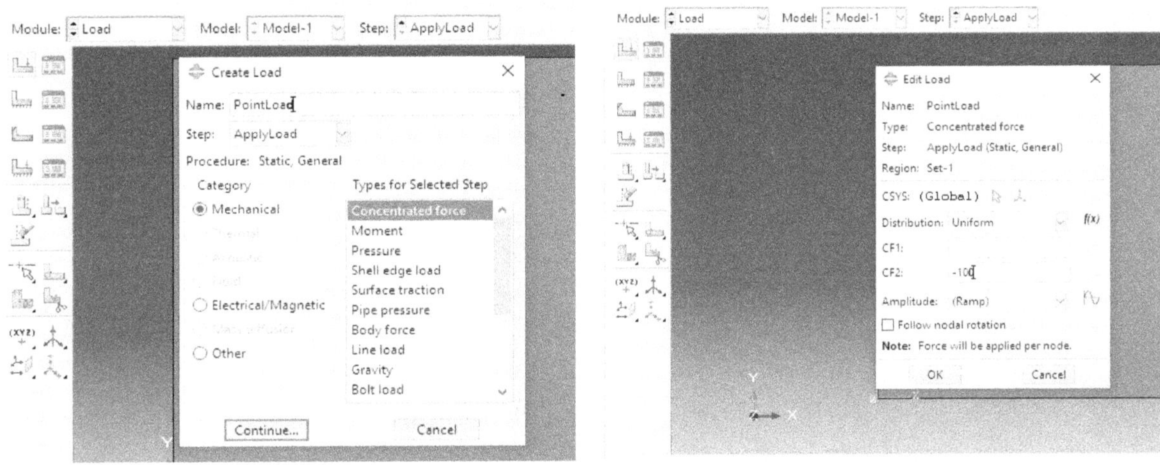

I.   In the *Create Boundary Condition* dialog, type SimpleX and choose *Symmetry/Anti-symmetry/Encastre* from *Types for selected step,*and click *Continue*. Using shift+click, choose left edge and click *Done*. In the Edit Boundary condition dialog, click the checkbox corresponding to Encastre and click Ok.

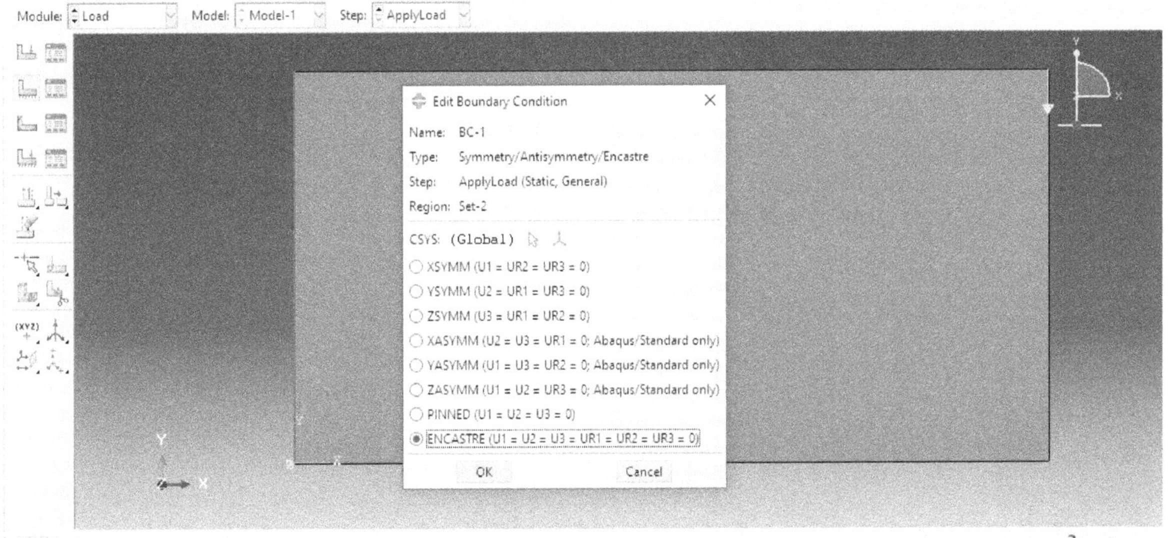

J. Boundary condtions appear as shown below.

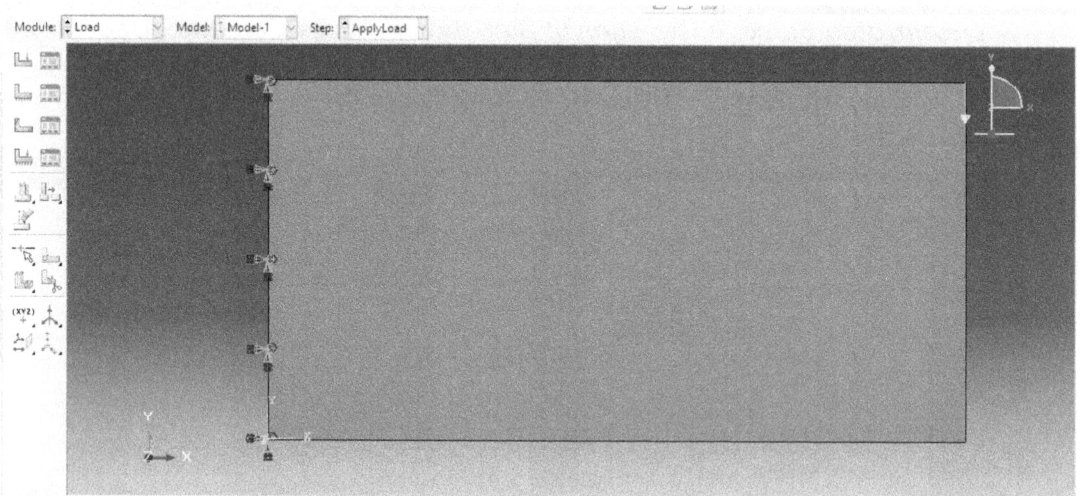

K. Click mesh in the model menu and click on *Controls*.

(a) For CST/LST elements: In the Mesh Controls dialog box, choose *Element type* as **Tri** and *Technique* as **Structured** and click Ok.

(b) For 4-node/8-node quadrilateral elements: In the Mesh Controls dialog box, choose *Element type* as **Quad** and *Technique* as **Structured** and click Ok.

L. Click on *Element type*. Click and select plate and click *Done*. Element Type dialog appears.

(a) In case of CST/LST elements, choose Standard for Element Library, Linear/Quadratic for Geometric order and choose Tri . **CPS3/CPS6** element gets selected. Click *Done* for select regions.

(b) In case of 4node/8 node quadrilateral elements, choose Standard for Element Library, Linear/Quadratic for Geometric order and choose Quad . **CPS4/CPS8** element gets selected. Click *Done* for select regions.

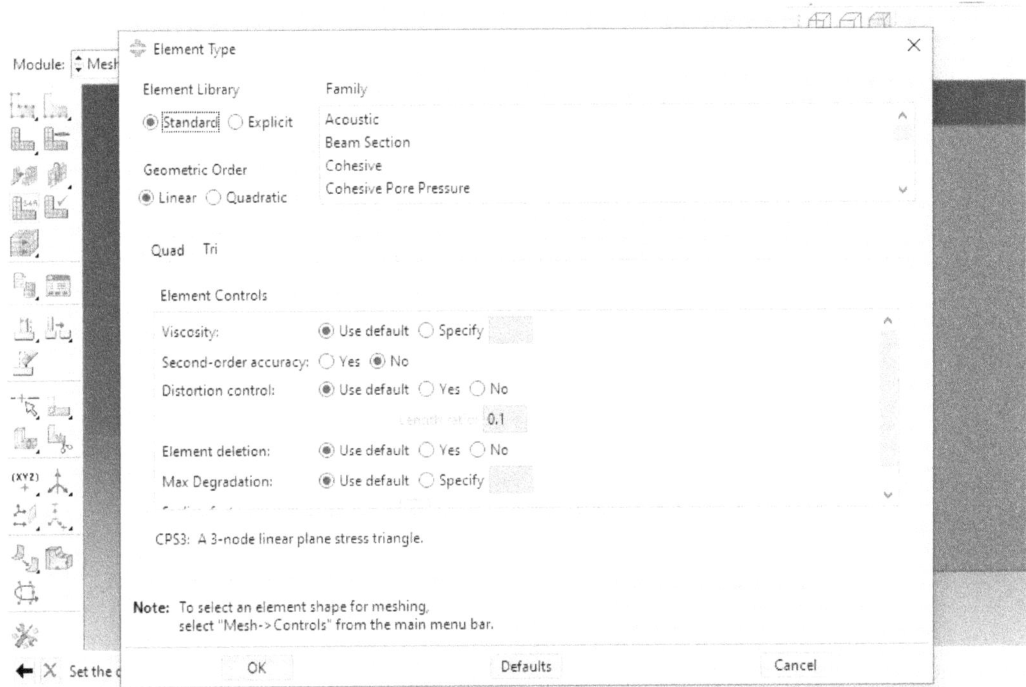

M. Click Create Part Instance Icon and mesh appears.

(a) In case of CST and LST elements the following mesh appears

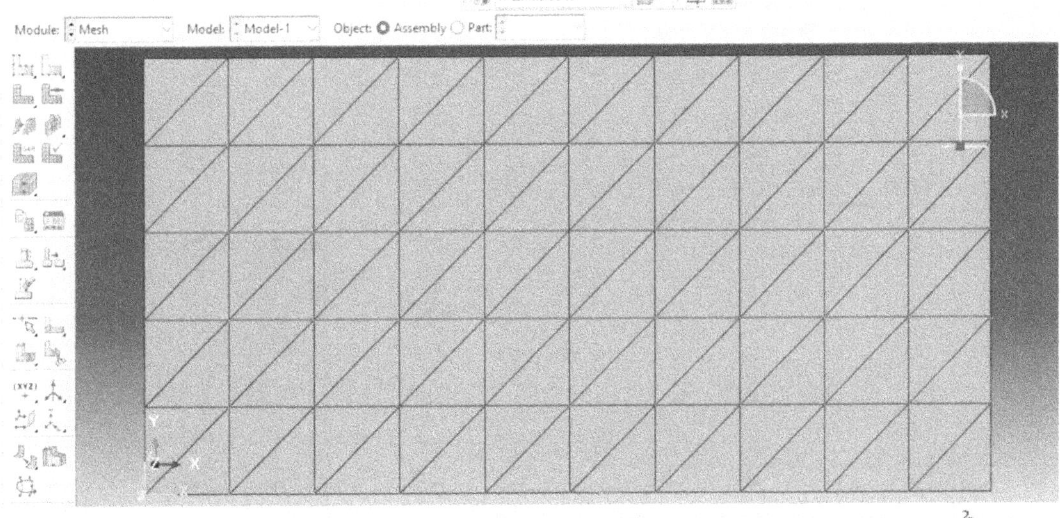

(b) In case of 4-noded/8-noded quadrilateral elements, the following mesh appears:

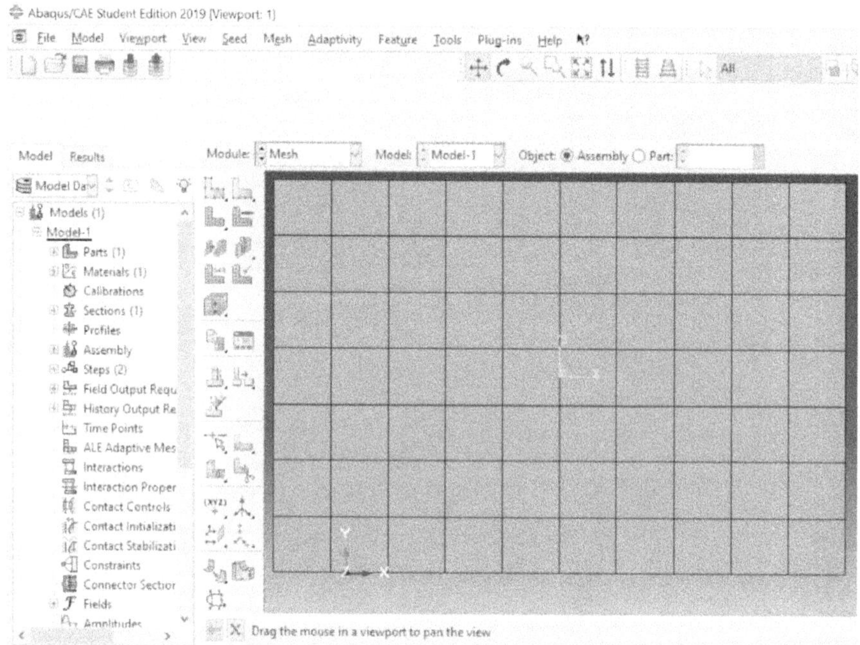

N. Click *Job* in model menu and click on Create Job icon.Type **Pplate** in Name and click *Continue*. Accept everything in Edit Job dialog and click Ok.

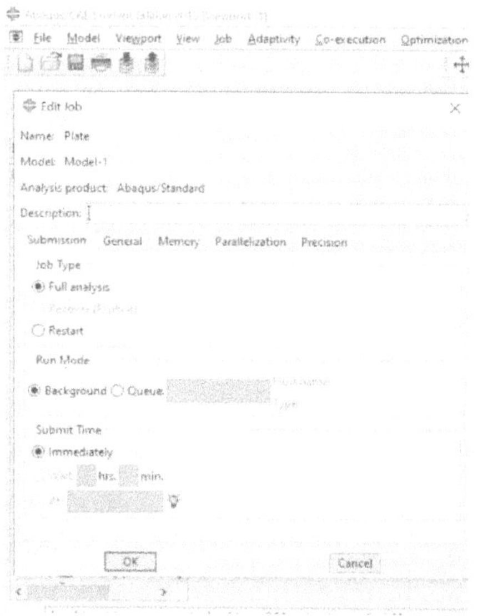

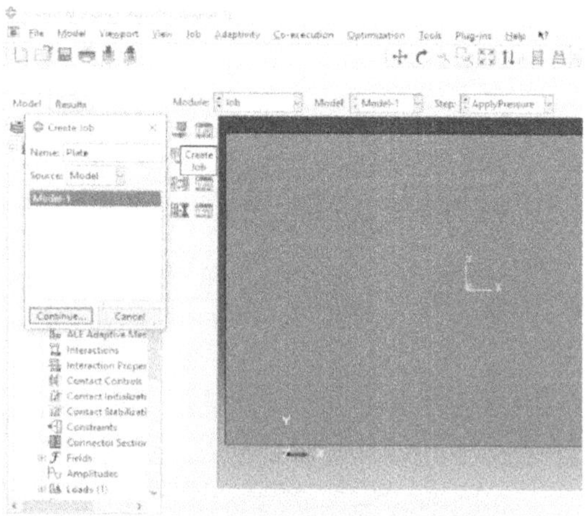

O. Click on Job Manager Icon and Job Manager dialog opens. Click on Submit. After the Job is done, click on *Results*.

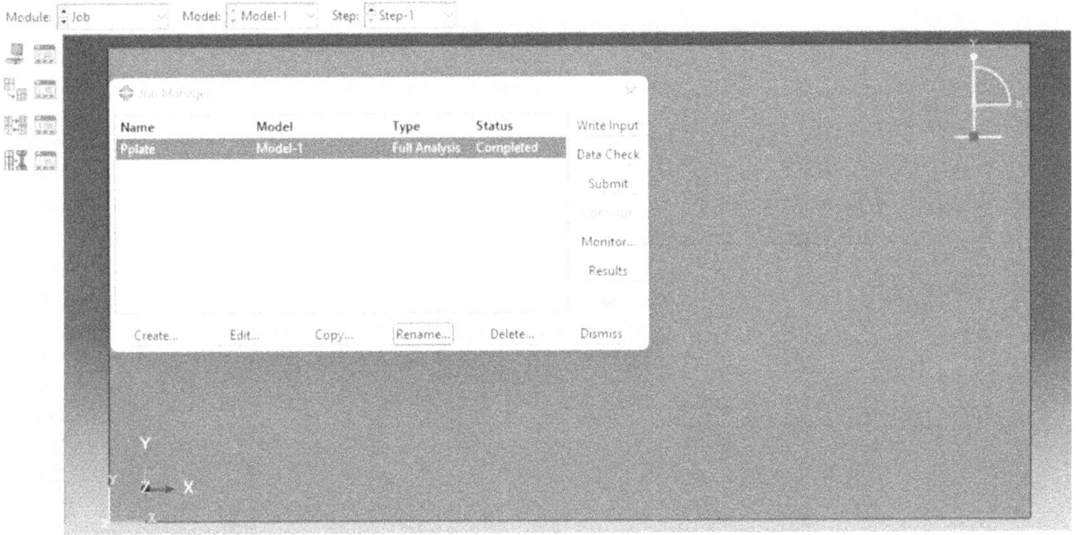

P. Results can be seen as follows:

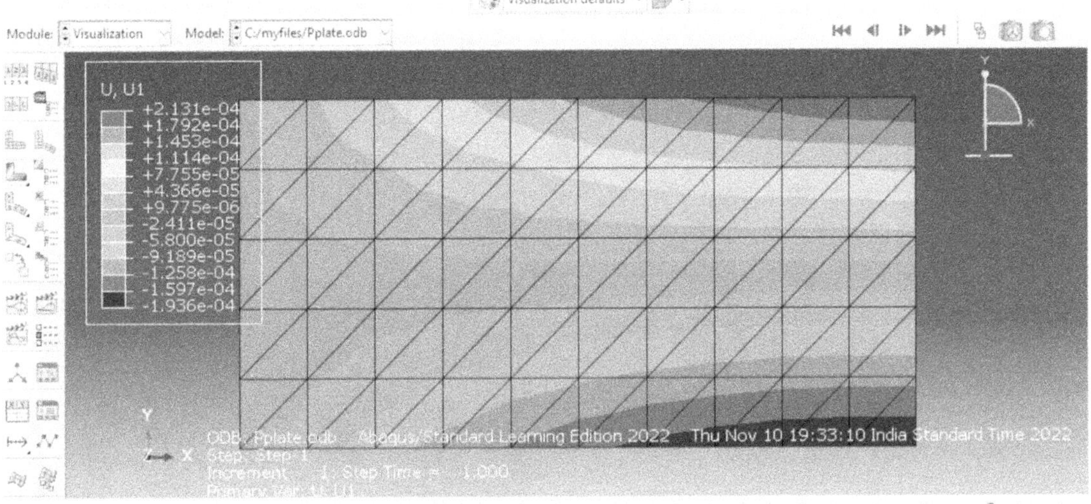

B.5 AXISYMMETRIC PROBLEMS

The following is the common procedure for the analysis of axisymmetric domain using ABAQUS. It shall be noted that that the selection of element type as given in step M will be different for three node triangular element, six node triangular element and eight node quadrilateral element.

Q. Start **ABAQUS**. In **Start Session** dialog, choose **Create Model Database with Standard/Explicit Model**. On the main menu, click on **File** and set **Set Work Directory** to choose your working directory. Click on **Save As** and name the file **axiT.cae**.

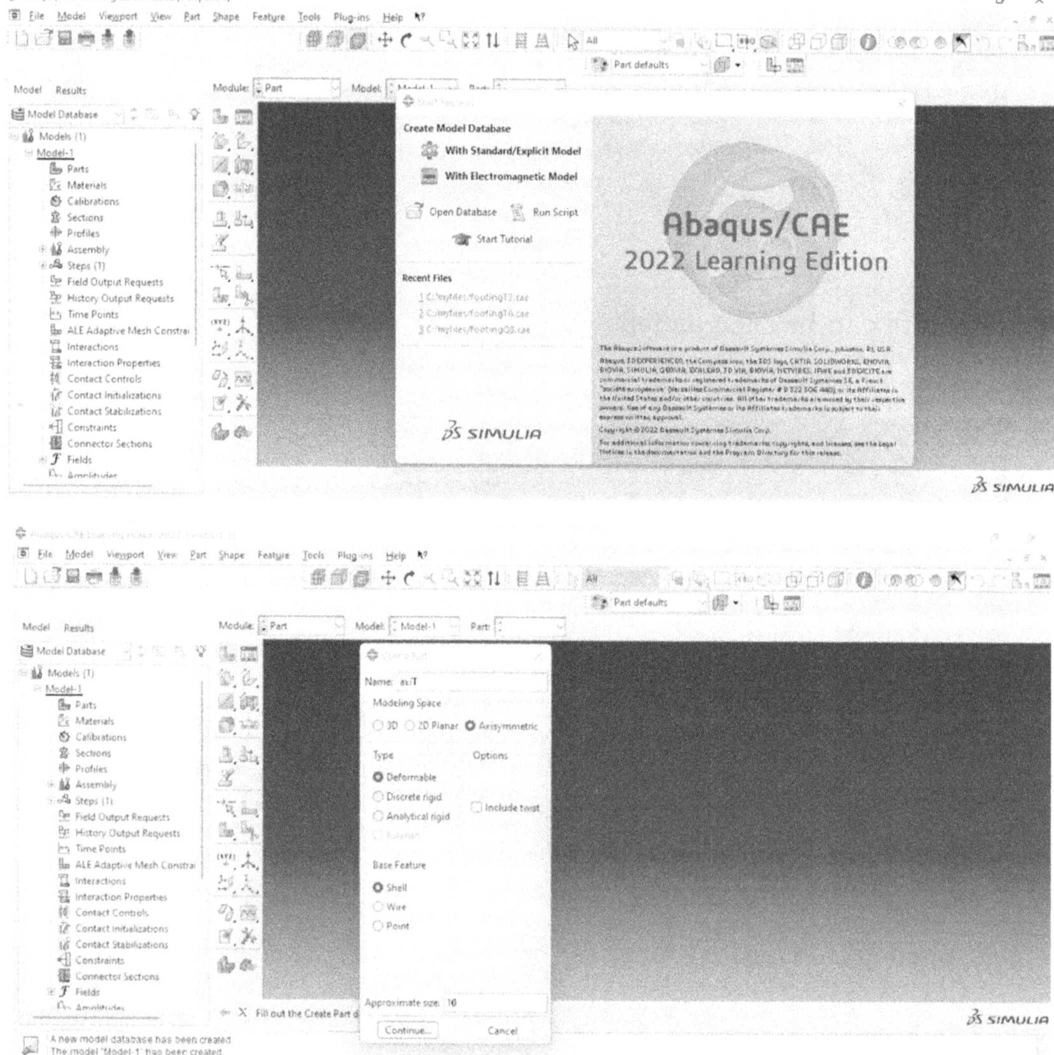

R.   Click on **Create Part** icon in the *Module Icon* menu. In the **Create Part**

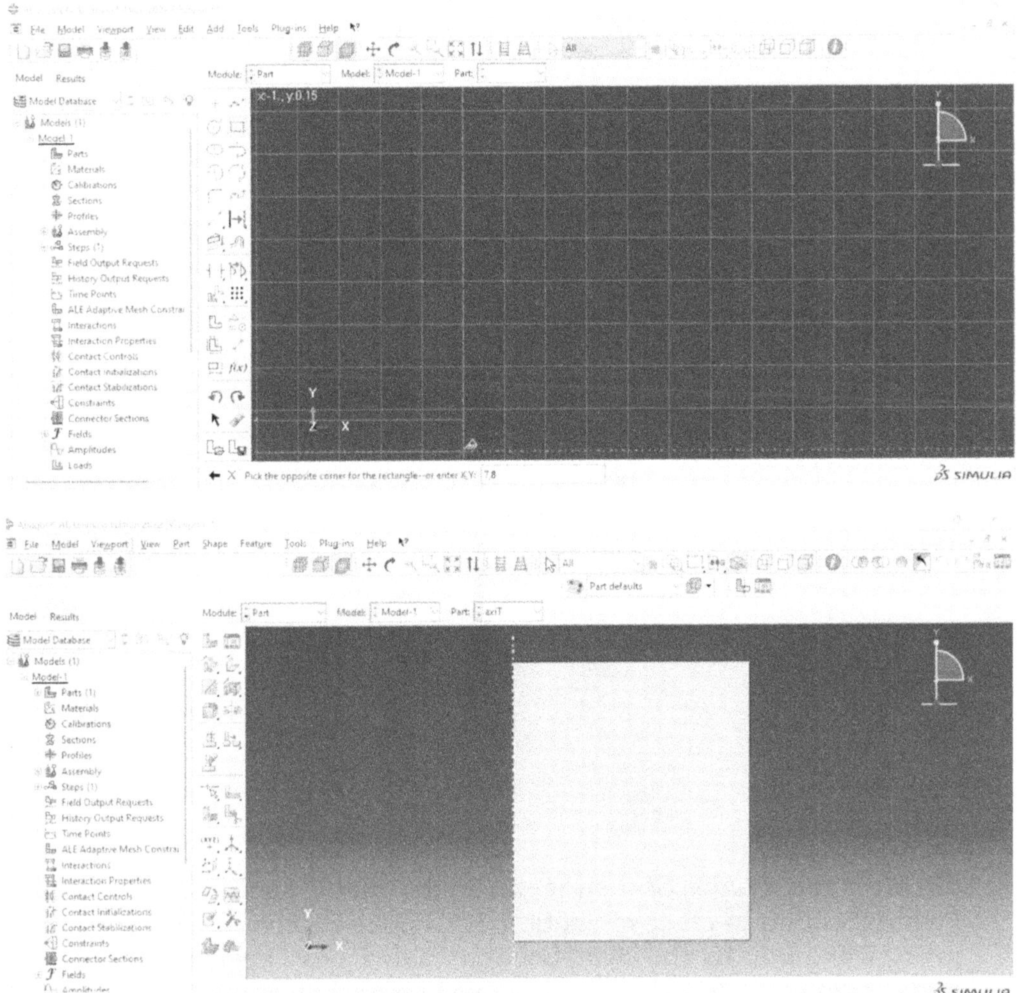

dialog, give name as **axiT**, choose *Modelling Space* as **Axisymmetric** , *Type* as **Deformable** and *Base feature* as **Shell** and *Type* as **Planar**. Indicate *Approximate size* as **10** and press **Continue.**

S.   In the sketcher menu, choose Create Lines: Rectangle(4 lines) and enter **0.0,0.0** in the window *"Pick a starting corner for Rectangle or enter X,Y."* and press **Enter**. In the window *"Pick the opposite corner for Rectangle or enter X,Y."*, enter **7.0,0.5** and press **Enter**. Press × symbol for the next dialog. A prompt *"Sketch the section for the planar shell"* appears. Press **Done** and complete the rectangle. A grey rectangle appears in the sketcher window.

T. Click the *Material Manager* icon, and click *Create* in the Material Manager dialog box that appears. Enter **Soil** in place of Material-1. Edit the description box and enter "***E is in kPa***". Enter Mechanical menu and Elastic tab and enter **1.0e5** for Young's modulus and **0.35** for Poisson's ratio. Press ok to confirm the entered values. Press Dismiss on the Material Manager Dialog box.

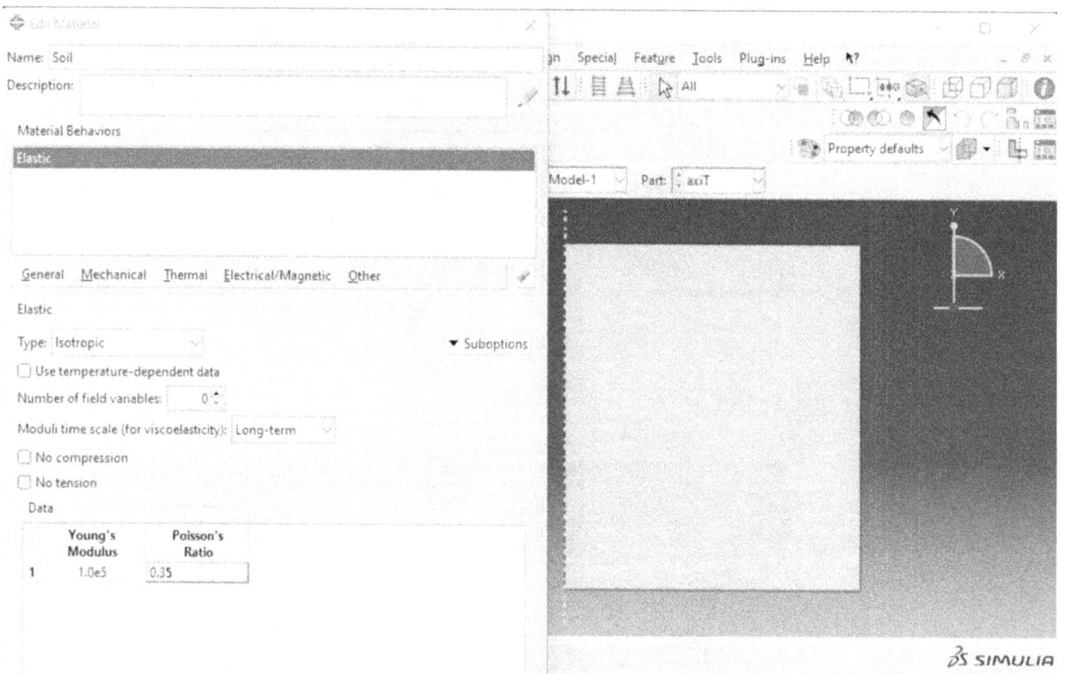

U. Click on the *Create Section* icon. In the dialog box that appears, enter FootingSection in place of Section-1. Choose *Soild* and *Homogeneous* and press *Continue*. In the Edit Section dialog box, keek the checkbox for Plane stress/strain thickness unchecked and click Ok

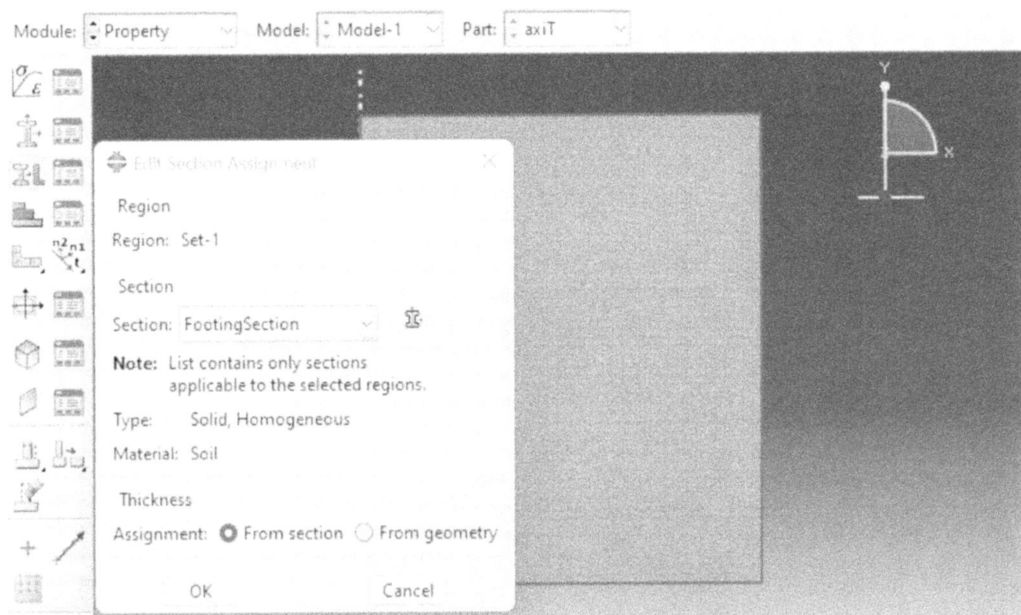

V.   Click on *Assign Section* icon. Use mouse click to select the plate and press *Done*. In the Edit Section Dialog, Plate Section appears. Accept and press Ok. For the prompt: *Select the regions to be assigned a section*, select the plate using the mouse click and press *Done*.

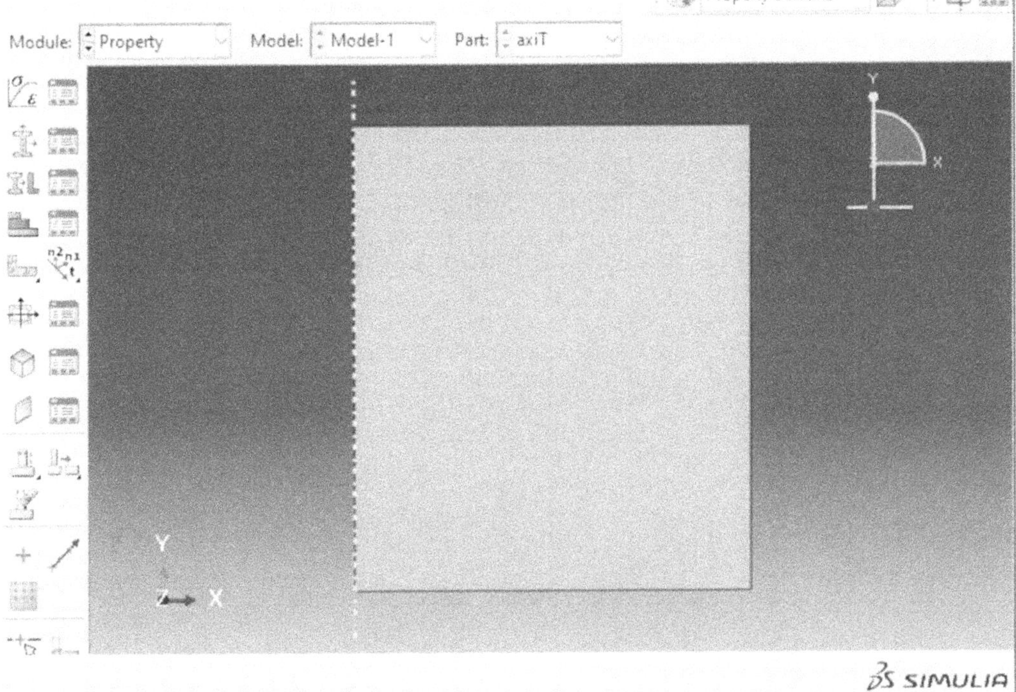

W. Click assembly in module menu and click on *Create Instance* icon. Choose Independent (Mesh on instance) and press Ok. Click *Step* in module menu and click on *Create Step* icon. Rename Step-1 as ApplyPressure, accept *Static, General* in and click *Procedure Type* and click *Continue.* In the description of Edit step dialog, type Apply uniform normal pressure on plate. Click Ok.

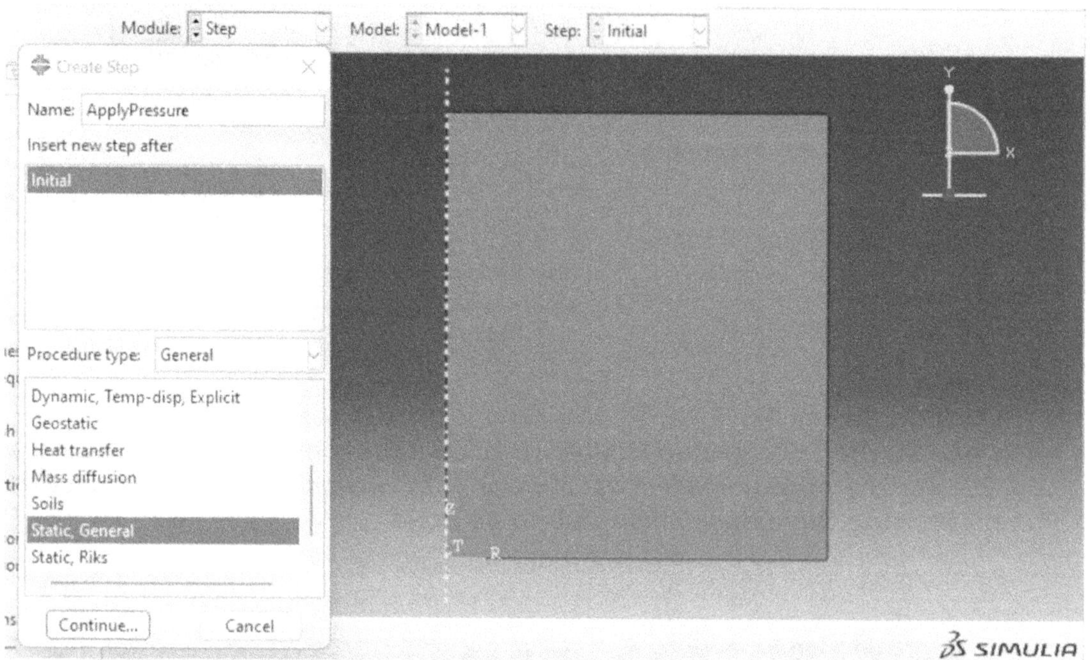

X. Click *Tools* in the main menu and click on *Partition*. In the *Create Partition* menu, choose type as Edge and choose *Enter parameter* as method. Using mouse, select the top edge to partition and press *Done*. Enter a value of $\frac{1}{7} = 0.14285714$ and click *Create Partition*. Press X to end the dialog.

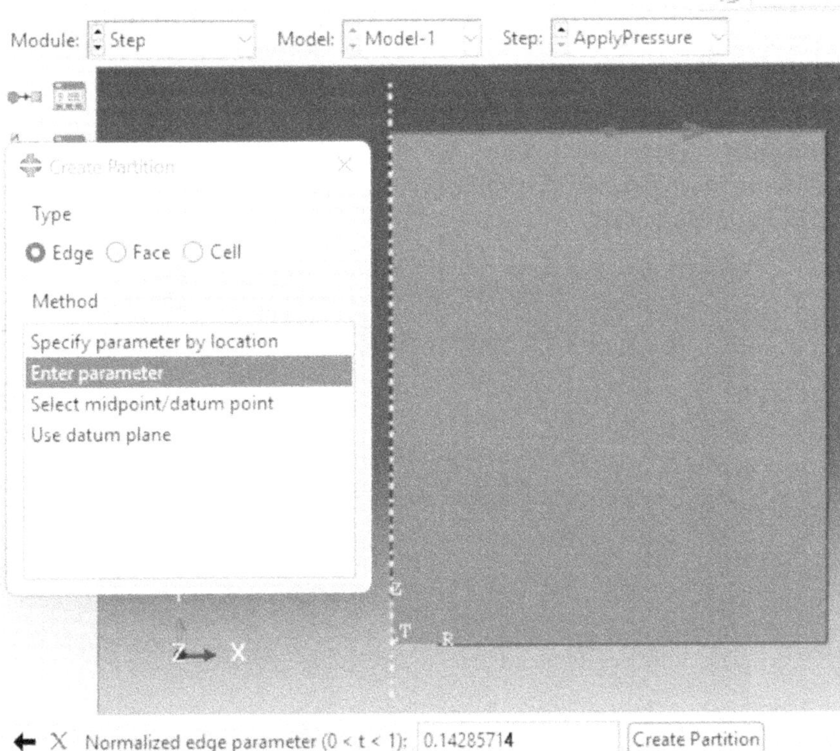

Y.   Click *Load* in module menu and click on *Create Load* icon. Replace Load-1 by typing **Pressure**. Accept Category as *Mechanical* and choose *Pressure* in the *Type of Step* dialog. Press *Continue*. Use mouse click to Select surfaces for the load, click on shorter part of the top edge and click *Done*. The Edit Load dialog appears. Enter pressure as 63.662 and clcik *OK*. Pressure is now displayed.

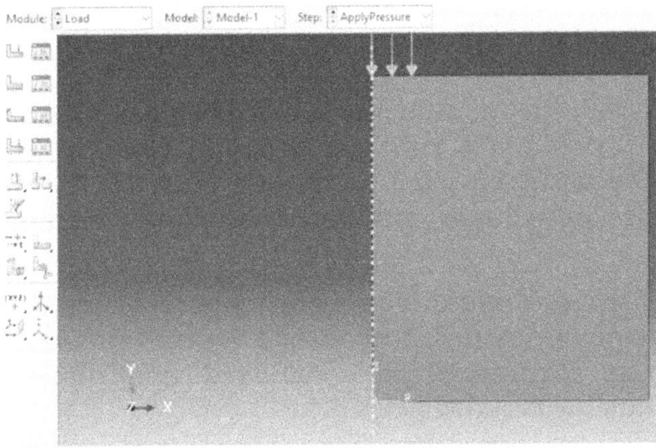

Z. Click *BC* on main menu, and click *Create*. Type **centreline** for name and choose Category Mechanical and Type Displacement/Rotation and click *Continue*. With the mouse, choose the centreline and click *Done*. In the Edit Boundary condition dialog box, click checkbox for U1 and click *OK*. Once again repeat the procedure for right edge by clicking checkbox for U1. Now repeat procedure for bottom edge by clicking checkboxes for U1 and U2 and click OK.

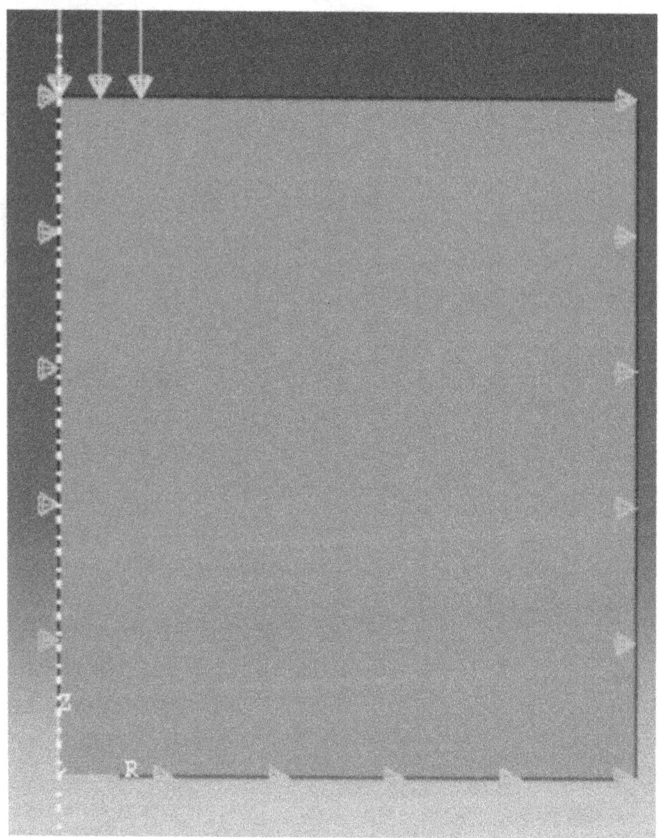

AA. Click Mesh from the Module menu and click on Seed on the main menu and select edges. Click both parts of top edge for the prompt Select Regions to be assigned local seeds and click *Done*. In the Local Seeds menu, select method as by size and approximate element size as 0.5.and click OK.

BB. Select Mesh from Module menu and choose Mesh Controls and select Element shape as Tri (for 3-node and 6-node triangular elements) and Technique as Structured. For 8-noded quadrilateral choose select Element shape as Quad. Click OK.

CC. Select Mech from the main menu and select Element type,select entire domain with mouse and click *Done*. Select Standard for element library, geometric order as Linear and choose Tri (for 3-node triangle). Select Standard for element library, geometric order as Qudradic and choose Tri (for 6-node triangle) and uncheck *Modified Formulation*. Select Standard for element library, geometric order as Linear and choose Quad (for 8-node quadrilateral) and uncheck *Reduced Integration*. Click Mesh the Instance.

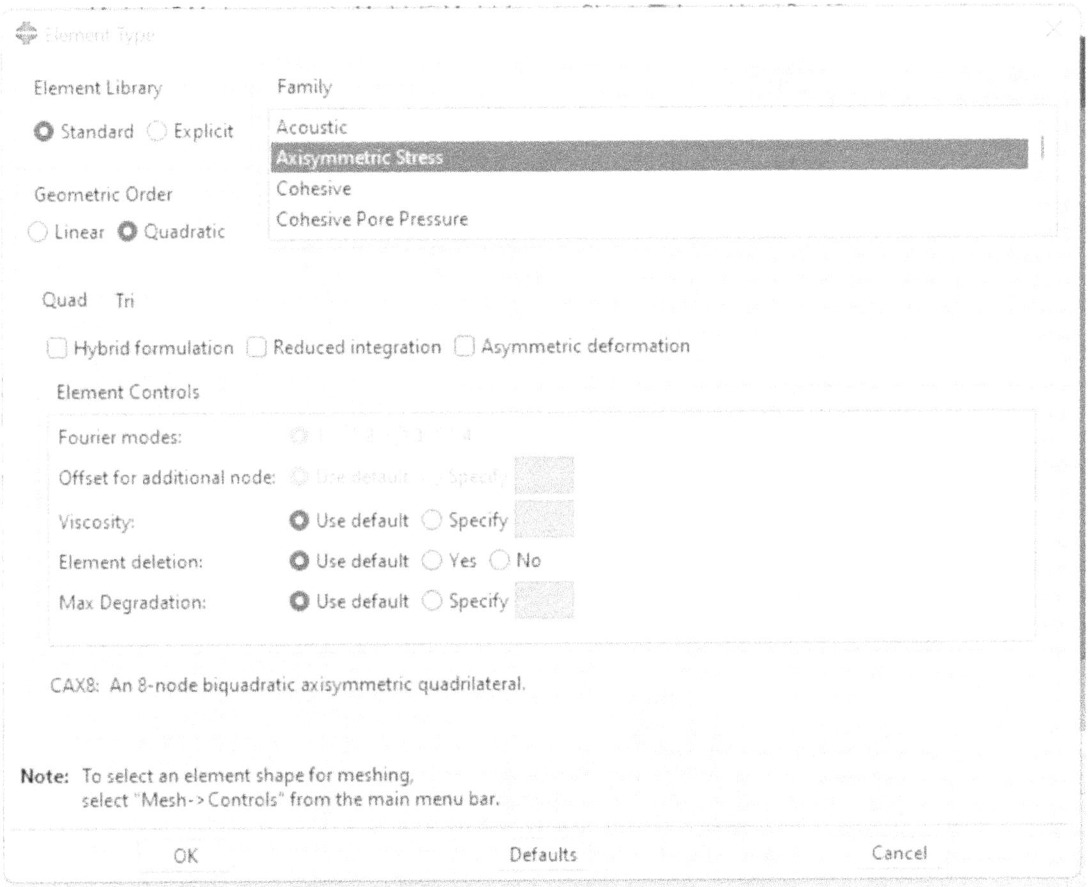

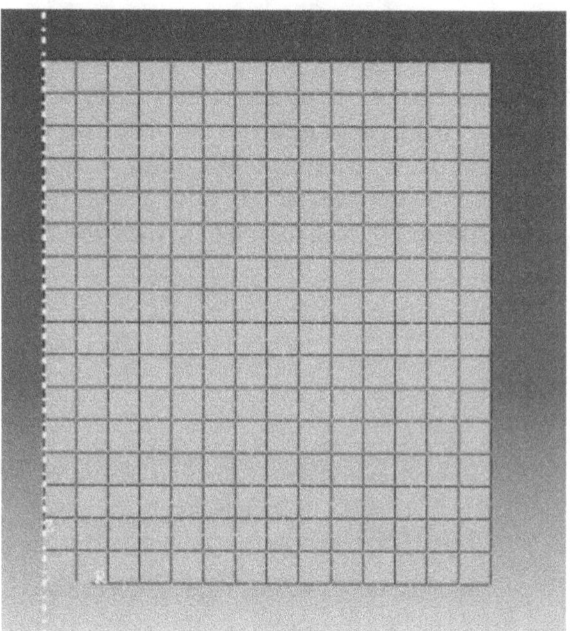

DD. Go to the Job menu under Module and create a new job as ApplyPressure and click Continue. Accept the default options and click OK. Dismiss the Job Manager Dialog. Click *Submit*. After Job is completed, click on *Results* and choose Visualization from Module menu. Choose Plot from the main menu and choose Plot on the undeformed shape. All results can now be seen as per choice.

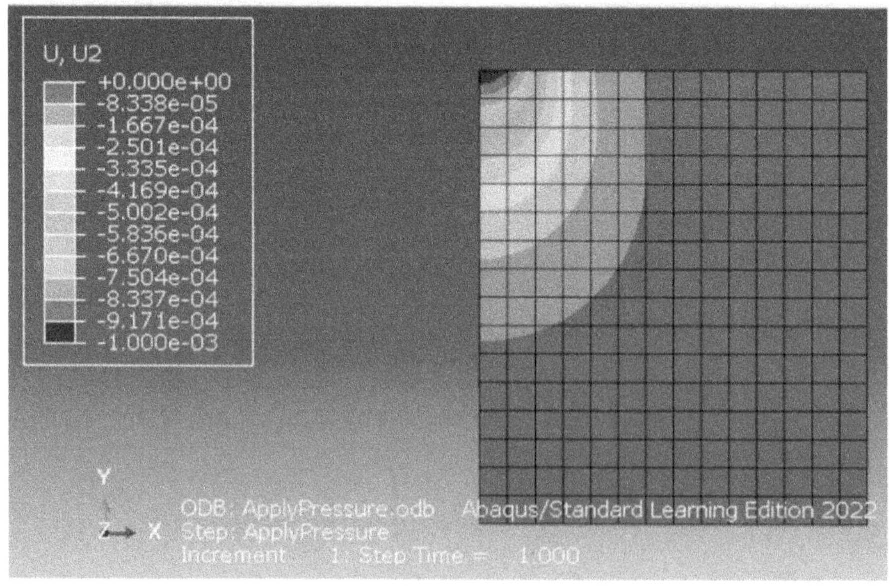

B.6   PLATES

The following is the common procedure for the analysis of plates using ABAQUS. It shall be noted that that the selection of element type as given in step M will be different for four node and eight node elements.

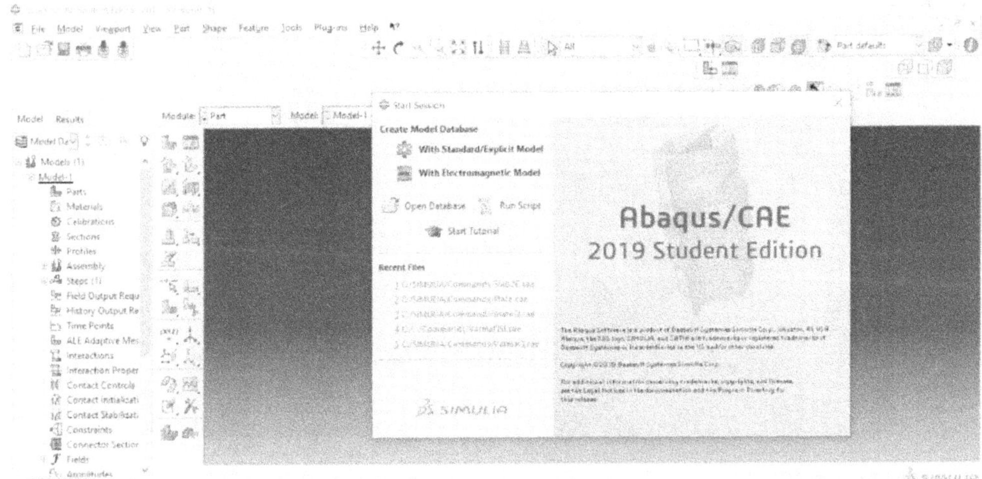

EE. Start **ABAQUS**. In **Start Session** dialog, choose **Create Model Database with Standard/Explicit Model**. On the main menu, click on **File** and set **Set Work Directory** to choose your working directory. Click on **Save As** and name the file **S4Ex1.cae**.

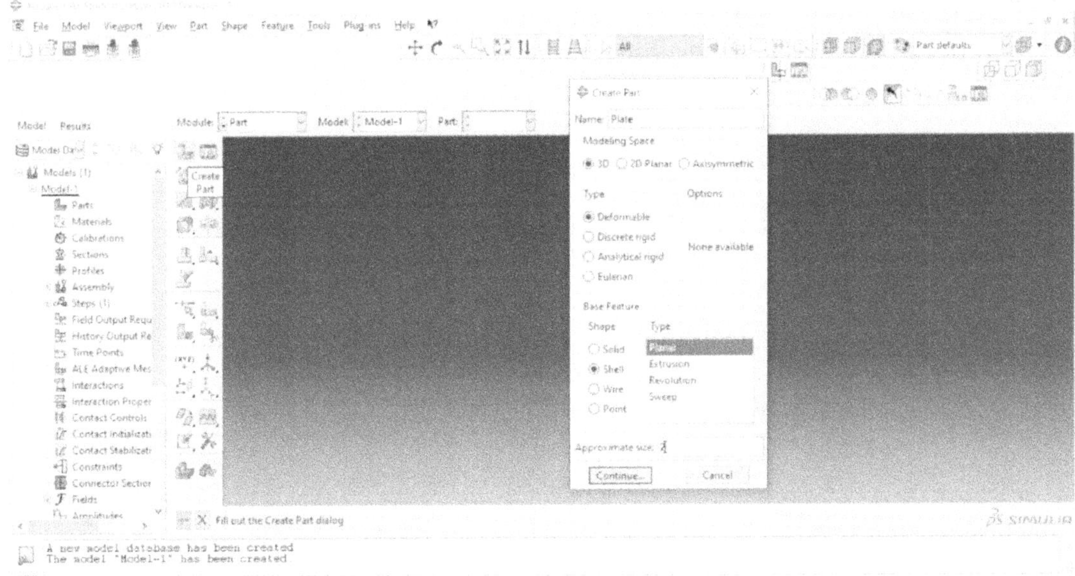

FF.   Click on **Create Part** icon in the *Module Icon* menu. In the **Create Part** dialog, give name as **Plate**, choose *Modelling Space* as **3D** , *Type* as

Deformable and *Base feature* as **Shell** and *Type* as **Planar**. Indicate *Approximate size* as **2** and press **Continue**.

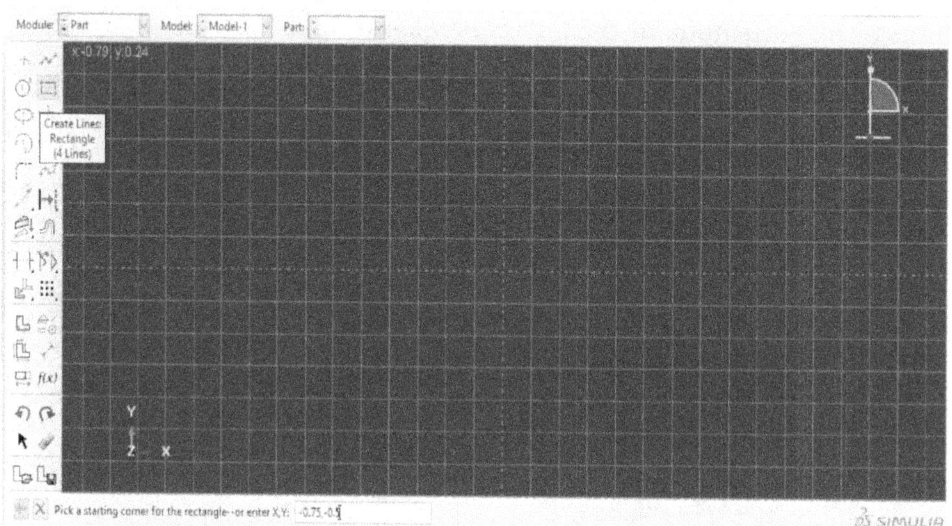

GG. In the sketcher menu, choose Create Lines: Rectangle(4 lines) and enter **-0.75.-0.5** in the window *"Pick a starting corner for Rectangle or Penter X,Y."* and press **Enter**. In the window *"Pick the opposite corner for Rectangle or enter X,Y:"*, enter **0.75,0.5** and press **Enter**. Press × symbol for the next dialog. A prompt *"Sketch the section for the planar shell"* appears. Press **Done** and complete the rectangle. A grey rectangle appears in the sketcher window.

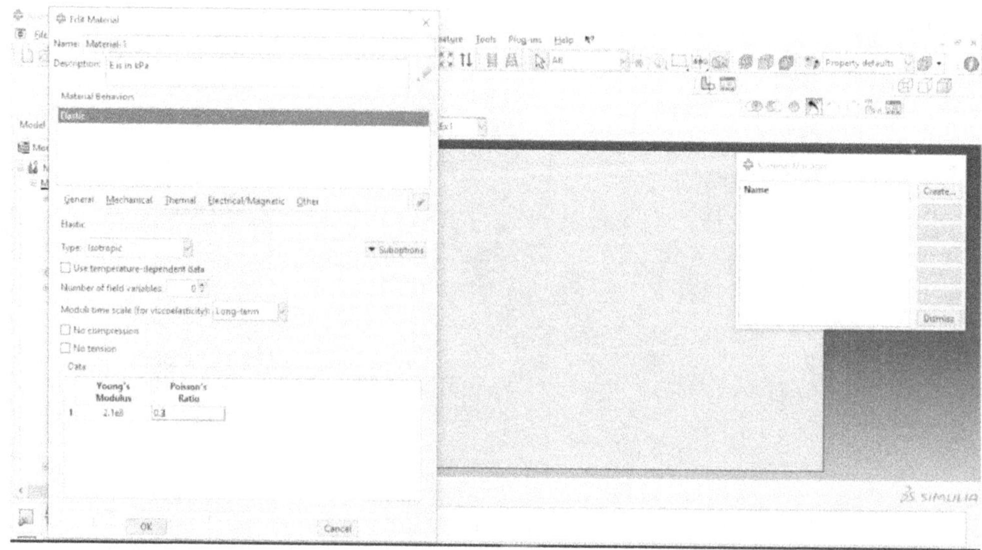

HH.   Click the *Material Manager* icon, and click *Create* in the Material Manager dialog box that appears.   Enter **Steel** in place of Material-1. Edit the description box and enter "***E is in kPa***". Enter Mechanical menu and Elastic tab and enter **2.1e8** for Young's modulus and **0.3** for Poisson's ratio. Press ok to confirm the entered values. Press Dismiss on the Material Manager Dialog box.

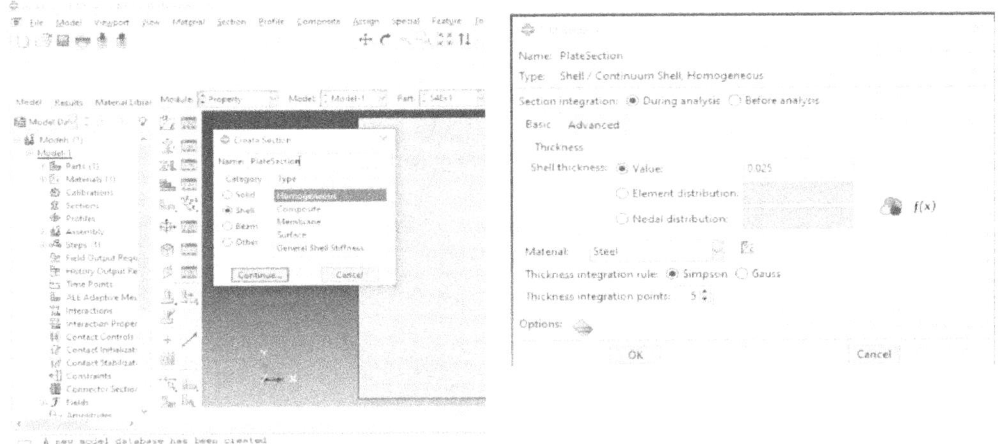

II.   Click on the *Create Section* icon. In the dialog box that appears, enter PlateSection in place of Section-1. Choose *Shell* and *Homogeneous* and press *Continue.* In the Edit Section dialog box, enter thickness of section as 0.025 and click Ok

JJ.   Click on *Assign Section* icon. Use mouse click to select the plate and press *Done*. In the Edit Section Dialog, Plate Section appears. Accept and press Ok.

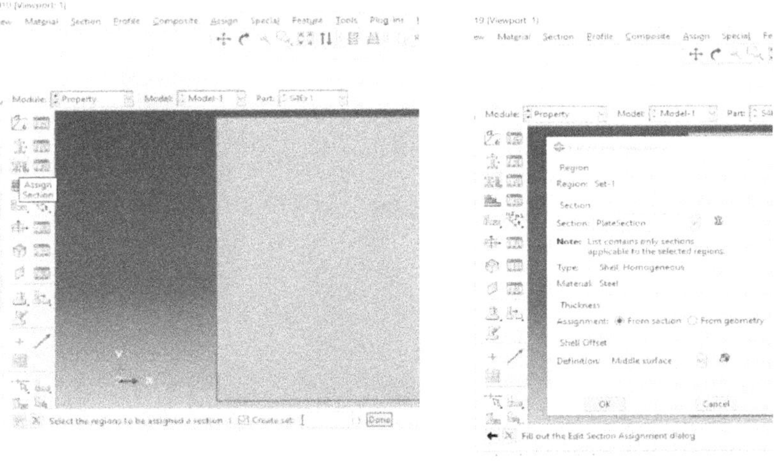

KK. For the prompt: *Select the regions to be assigned a section,* select the plate using the mouse click and press *Done.* Click assembly in module menu and click on *Create Instance* icon. Choose Independent (Mesh on instance) and press Ok. Click *Step* in module menu and click on *Create Step* icon. Rename Step-1 as ApplyPressure, accept *Static, General* in and click *Procedure Type* and click *Continue.* In the description of Edit step dialog, type Apply uniform normal pressure on plate. Click Ok.

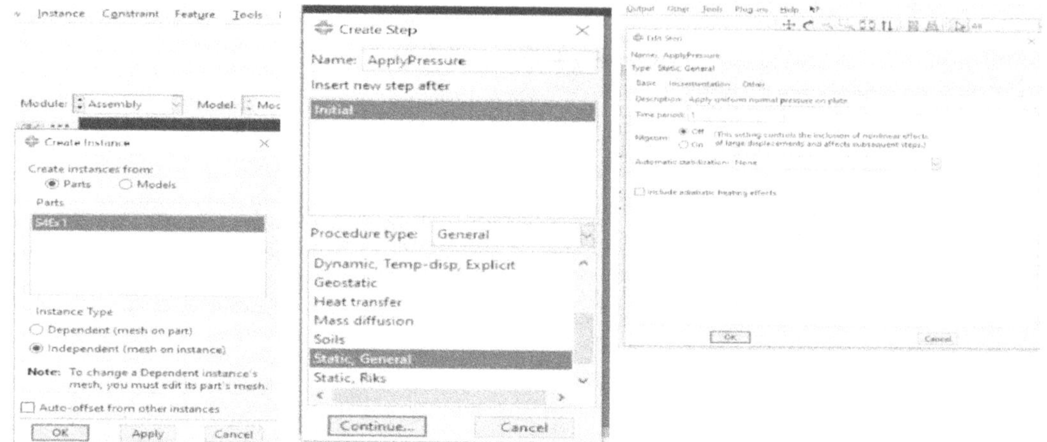

LL. Click *Load* in module menu and click on *Create Load* icon. Replace Load-1 by typing **Pressure**. Accept Category as *Mechanical* and choose *Pressure* in the *Type of Step* dialog. Press *Continue.* Use mouse click to Select surfaces for the load and click *Done.* Choose brown when prompted to choose Brown or Purple. The Edit Load dialog appears.

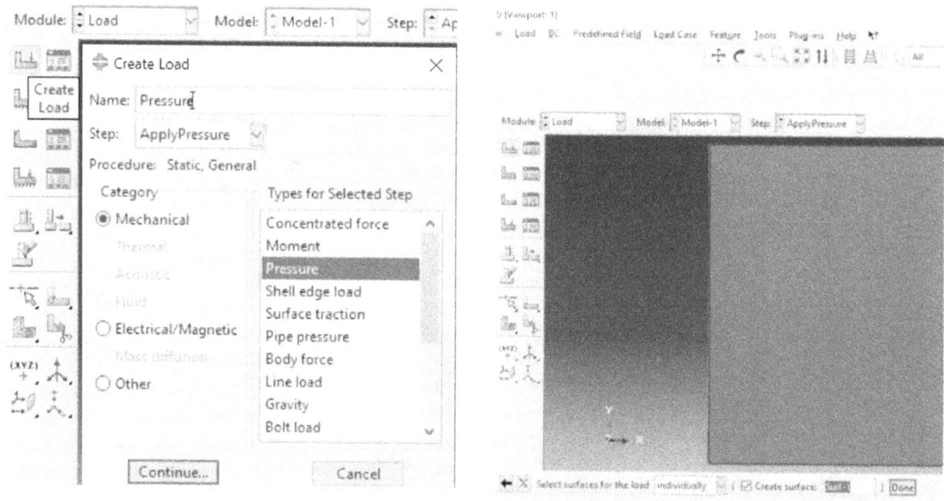

MM. In the Edit Load dialog, enter **140.0** for magnitude and click Ok. Choose *BC* from the main menu and choose submenu *Create*.

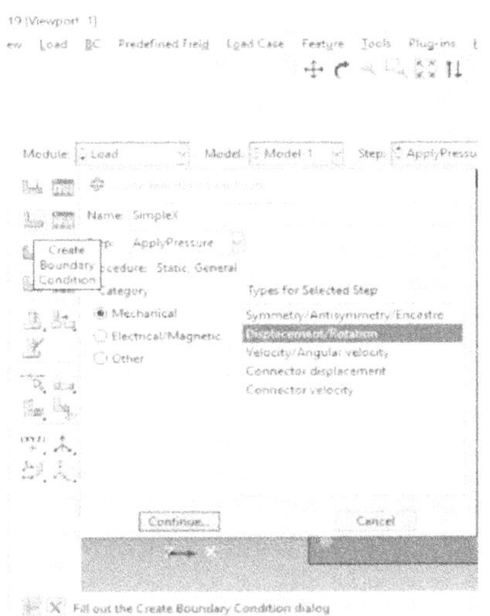 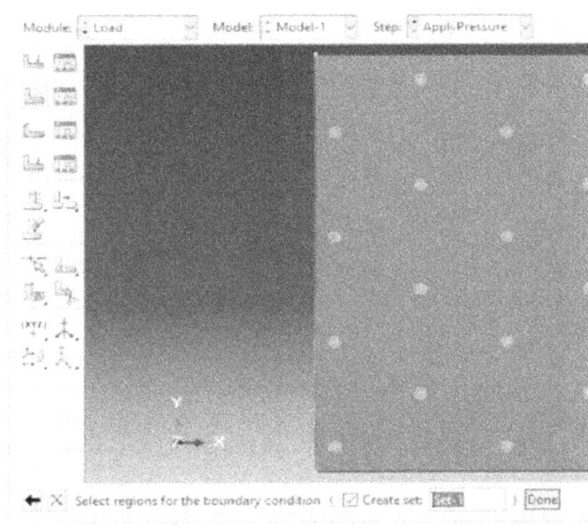

NN. In the *Create Boundary Condition* dialog, type SimpleX and choose *Displacment/Rotation* from *Types for selected step*,and click *Continue*. Using shift+click, choose bottom edge and click *Done*. In the Edit Boundary condition dialog, click the checkboxes corresponding to UR2,U3 and UR3 and click Ok. create boundary condition SimpleY by choosing left edge and clicking checkboxes corresponding to UR1,U3 and UR3 and click Ok.

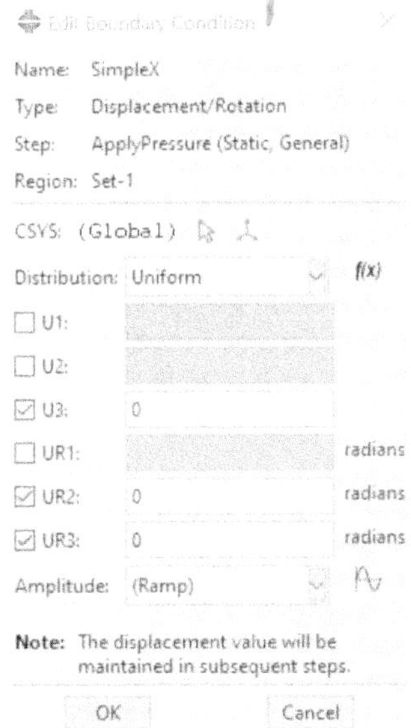 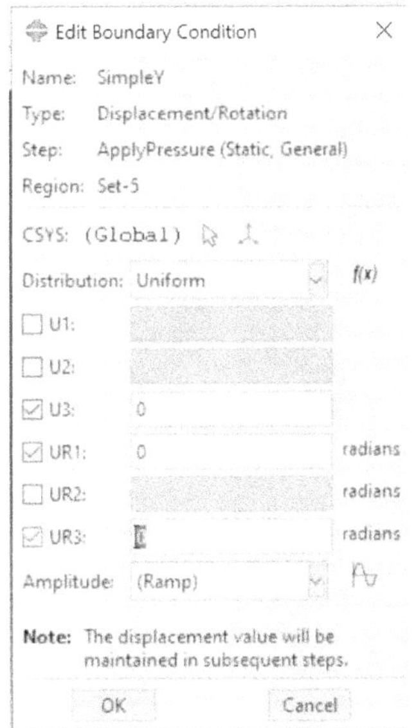

OO. Create boundary condition named Xsym and click on right edge. In the Edit Boundary Condition dialog, choose *Symmetry /Antisymmetry/ Encastre* choose **XSYMM** and click *Ok*. Similarly create boundary condition named Ysym by clicking on top edge and choosing YSYMM.

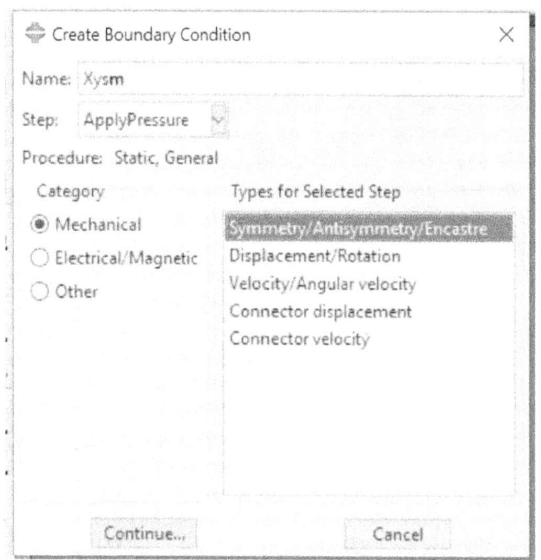
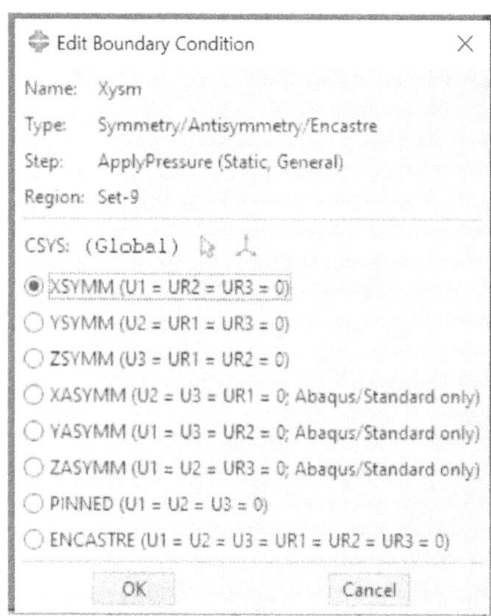

PP. Boundary condtions appear as shown below.

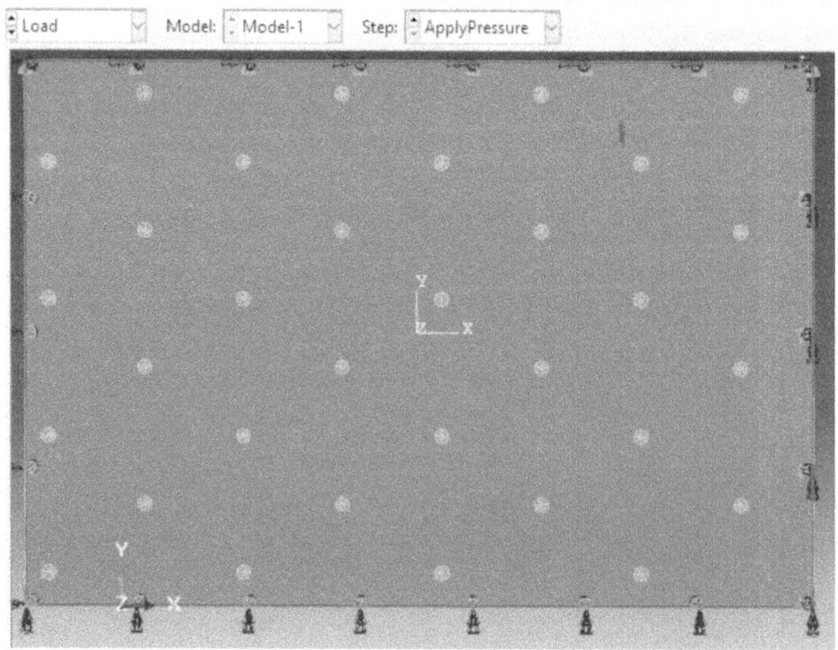

QQ. Click mesh in the model menu and click on *Controls*. In the Mesh Controls dialog box, choose *Element type* as **Quad** and *Technique* as **Structured** and click Ok.Click on *Element type*. Click and select plate and click *Done*. Element Type dialog appears. Choose Standard for Element Library, Linear for Geometric order and choose checkbox Reduced Integration and *small* for *Membrane strains*. **S4R5** element gets selected. Click *Done* for select regions.

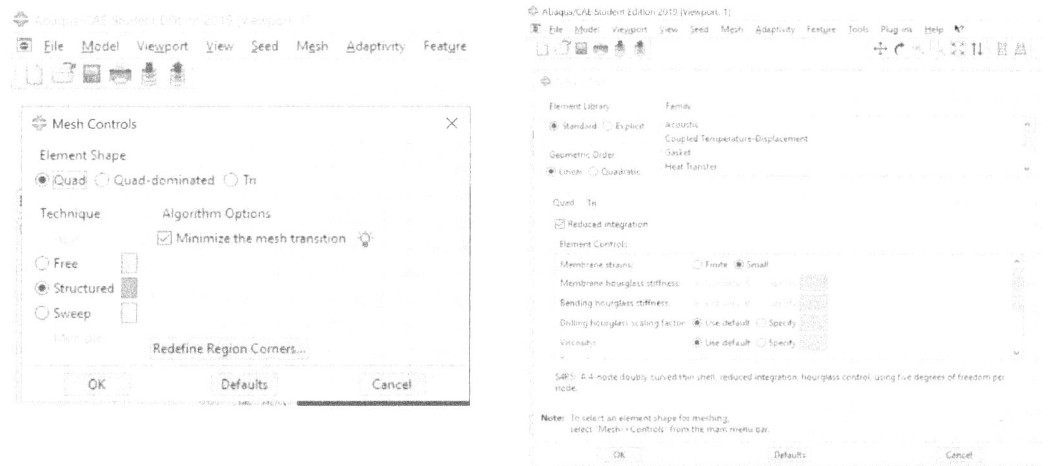

In case of eight noded elements, choose **Quadratic** as Geometric Order and reduced integration.**S8R5** element gets selected.

RR. Click Create Part Instance Icon and mesh appears.

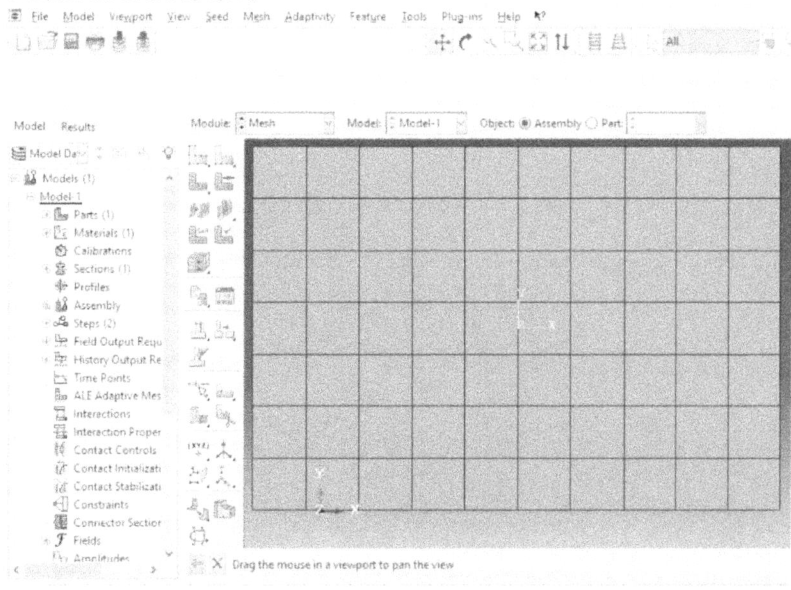

SS. Click *Job* in model menu and click on Create Job icon.Type **Plate** in Name and click *Continue*. Accept everything in Edit Job dialog and click Ok.

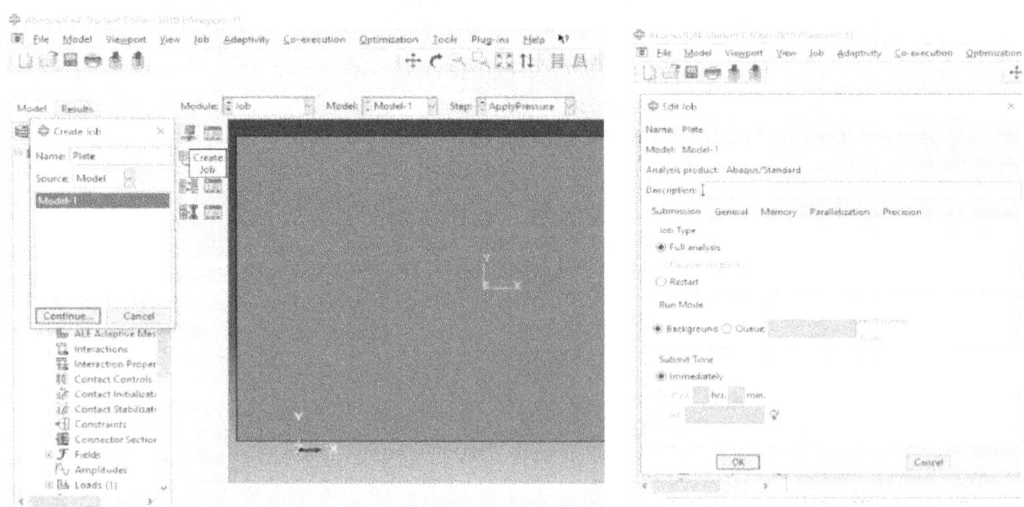

TT. Go to model tree on the left hand side and expand Field Output options. Double-click on F-output-1 and *Edit Field Output Reques*t dialog. Uncheck all items. Select only *U, Translations and Rotations* in *Displacement/Velocity/Acceleration* and *SF,Section forces and Moments* in *Forces and Moments*. Click Ok.

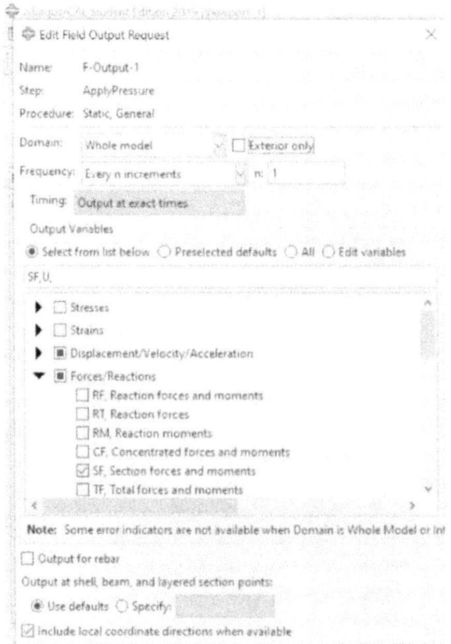

UU. Click on Job Manager Icon and Job Manager dialog opens. Click on Submit. After the Job is done, click on *Results*.

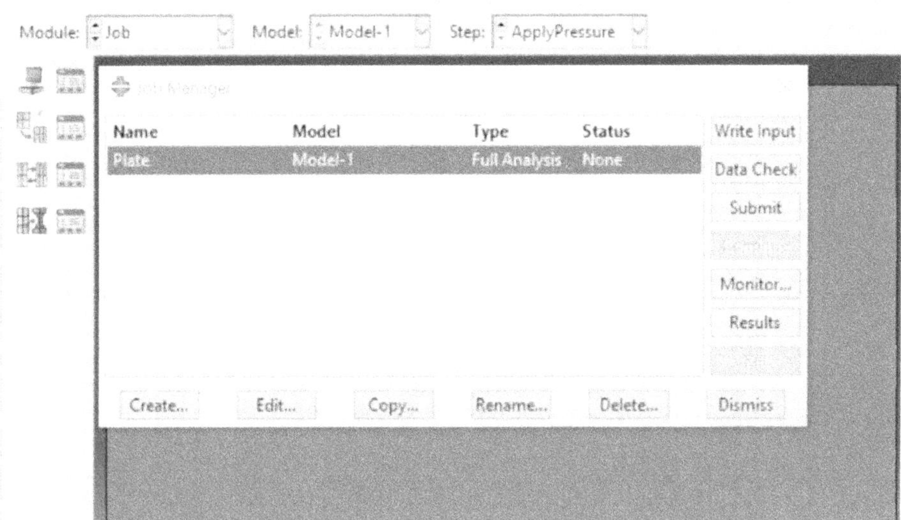

VV. Results can be seen as follows:

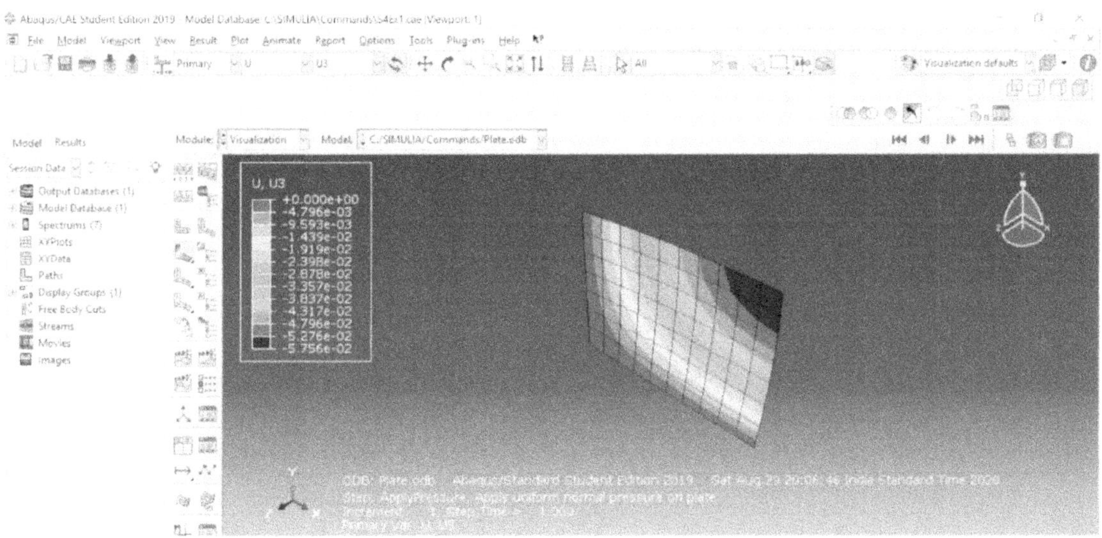

Bibliography

1. Argyris, J. H., and Kelsey, S., *Energy Theorems and Structural Analysis*, Butterworths, London, 1960.

2. Bathe, K.J., Wilson, E.L., *Numerical Methods in Finite Element Analysis*. Prentice Hall, Englewood Cliffs ,1978.

3. Belytschko, T., Chiapetta, R.L. and Bartel, H.D., "Efficient large scale non-linear transient analysis by finite elements," *Int. J. Numer. Meth. Engng.*, Vol. 10, pp. 579-596, 1976.

4. Clough, R. W., "The finite element method in plane stress analysis," *Proceedings of the Second ASCE Conference on Electronic Computation*, Pittsburgh, PA, 1960.

5. Courant, R., "Variational Methods for the Solution of Boundary Value Problems," *In: Symposium on the Rayleigh–Ritz Method and Its Applications*, May 2-3. American Mathematical Society, Washington D.C,1941.

6. Gallagher, R. H., and Padlog, J., "Discrete Element Approach to Structural Stability Analysis," *Journal of the American Institute of Aeronautics and Astronautics*, Vol. 1, No. 6, pp. 1437–1439, 1963.

7. Grafton, P. E., and D. R. Stome. "Analysis of Axi-symmetrical Shells by the Direct Stiffness Method." AIAA Journal Vol. 1, pp 2342-2347, 1963.

8. Hrennikoff, A., "Solution of Problems of Elasticity by the Frame-Work Method," *ASME Journal of Applied Mechanics* 8, A619–A715,1941.

9. Kirchhoff, G. *Uber das Gleichgewicht und die Bewegung einer elastischen Scheibe*. Journal für die reine und angewandte Mathematik, ger, 40, 1850,51-88.

10. Levy, S., "Computation of Influence Coefficients for Aircraft Structures with Discontinuities and Sweepback," *Journal of Aeronautical Sciences*, Vol. 14, No. 10, pp. 547–560, Oct. 1947.

11. Levy, S., "Structural Analysis and Influence Coefficients for Delta Wings," *Journal of Aeronautical Sciences*, Vol. 20, No. 7, pp. 449–454, July 1953.

12. McHenry, D., "A lattice analogy for the solution of stress problems,' *Journal of the Institution of Civil Engineers*, Vol. 21, No. 2, pp. 59-82, 1943.

13. Melosh, R. J., "A Stiffness Matrix for the Analysis of Thin Plates in Bending," *Journal of the Aerospace Sciences*, Vol. 28, No. 1, pp. 34–42, Jan. 1961.

14. R. D. Mindlin, "Influence of rotatory inertia and shear on flexural motions of isotropic elastic plates," *Journal of Applied Mechanics—Transactions of the ASME*, vol. 18, no. 1, pp. 31–38, 1951.

15. Oden, J. T., and Ripperger, E. A., *Mechanics of Elastic Structures*, 2nd ed., McGraw-Hill, New York, 1981.

16. Przemieniecki,K. "Finite element structural analysis of local instability," *AIAA 1972-354. 13th Structures, Structural Dynamics, and Materials Conference.* April 1972.

17. Reddy, J.N. *Theory and Analysis of Elastic Plates and Shells* (2nd ed.). CRC Press,2006. https://doi.org/10.1201/9780849384165

18. E. Reissner, "The effect of transverse shear deformation on the bending of elastic plates," *Journal of Applied Mechanics-Transactions of the ASME*, vol. 12, no. 2, pp. A69–A77, 1945.

19. Szilard.R., *Theories and Applications of Plate Analysis: Classical, Numerical and Engineering Methods.* John Wiley and Sons, Inc, New York. 2004.

20. Strang, G., Fix, G.J., *An Analysis of the Finite Element Method.* Prentice Hall, Englewood Cliffs,1973.

21. Szabo, B. A., and Lee, G. C., "Derivation of Stiffness Matrices for Problems in Plane Elasticity by Galerkin's Method," *International Journal of Numerical Methods in Engineering*, Vol. 1, pp. 301–310, 1969.

22. Timoshenko, S. and Woinowsky-Krieger, S., *Theory of Plates and Shells.* New York: McGraw-Hill,1959.

23. Turner MJ, Clough RW, Martin HC, Topp L.," Stiffness and deflection analysis of complex structures," *Journal of Aeronautical Sciences* 1956;**23.**

24. Zienkiewicz, O.C., Cheung, Y.K, *The Finite Element in Structural and Continuum Mechanics*, McGraw-Hill publishing Co. Ltd, London, 1967.

25. Zienkiewicz, O.C. *The Finite Element Method* (3rd edn). McGraw-Hill: New York, 1977.

Index